D0708746

ANNUAL REVIEW OF GENETICS

ANNUAL REVIEW OF GENETICS

HERSCHEL L. ROMAN, *Editor*
The University of Washington, Seattle

LAURENCE M. SANDLER, *Associate Editor*
The University of Washington, Seattle

ALLAN CAMPBELL, *Associate Editor*
Stanford University, Stanford

VOLUME 7

1973

ANNUAL REVIEWS INC. 4139 EL CAMINO WAY PALO ALTO, CALIFORNIA 94306

ANNUAL REVIEWS INC.
Palo Alto, California, USA

International Standard Book Number: *0-8243-1207-4*
Library of Congress Catalog Number: 67-29891

Assistant Editors	Toni Haskell Virginia Hoyle
Indexers	Mary. Glass Susan Tinker
Subject Indexer	Leonie K. Piternick

PRINTED AND BOUND IN THE UNITED STATES OF AMERICA

PREFACE

The 18 reviews that comprise this volume represent one of the broadest coverages of contemporary genetics that we have published. Three articles deal with diverse aspects of population genetics: the theoretical implications of protein polymorphisms (R. C. Lewontin), the parameters that affect interpretations of gene evolution based on studies of amino acid sequences in proteins (W. M. Fitch), and the diversity among natural populations of *Drosophila* in their response to environmental stresses (P. A. Parsons).

Three papers review different approaches to studying genetic recombination: the genetic approach in fungi, using principally tetrad analysis to sort out the products of single meiotic recombinational events (D. R. Stadler), the use of mutants in bacteria to dissect the recombinational process (A. J. Clark), and the biochemical approach that seeks to establish a molecular basis for recombination (C. M. Radding). A fourth paper considers the problem of replication in relation to recombination in phage T4 (A. H. Doermann).

New insights into eukaryotic chromosome structure and organization are the consequences of technical and interpretative advances that are the subjects of three reviews: the use of autoradiography to localize gene products (RNA) to specific positions along the chromosome (D. E. Wimber and D. M. Steffensen), the interpretation of DNA hybridization experiments in relation to the significance for chromosomal organization of reiterated sequences (C. D. Laird), and the application of new staining techniques to characterize metaphase chromosomes, an advance of particular importance in mammalian cytogenetics (T. C. Hsu).

A substantial portion of this volume is devoted to reviews of regulatory and developmental genetics. The most intensively studied network of gene regulation is represented by bacteriophage λ (I. Herskowitz). The possibilities for the isolation of molecular factors involved in regulation are analyzed in a discussion of in vitro protein synthesis (G. Zubay). Studies of regulatory systems in higher organisms are described for the immune response (W. H. Hildemann), for sex-linked genes in Drosophila (J. C. Lucchesi), and for a directed heritable change in gene expression in maize (R. A. Brink). Progress in research on the control of development is reported in biochemical studies of meiosis (H. Stern and Y. Hotta) and in the exploitation of *Drosophila* as exceptional material for combined genetic and embryological investigations (J. H. Postlethwait and H. A. Schneiderman).

This volume also features a review of human genetics (V. A. McKusick and G. A. Chase) that had to be shortened because of space limitations. Even in its abridged form the article is successful in documenting the continuing high rate of progress in this dynamic area of genetics.

Our authors are our primary resource; we deeply appreciate their efforts and their cooperation. We are indebted to Virginia Hoyle, who provided expert editorial assistance in the preparation of manuscripts for publication, to Toni Haskell, our new Assistant Editor, and to Dr. Leonie Piternick, who prepared the Subject Index. We regret the premature retirement of Drs. R. C. Lewontin and R. Dulbecco from the Editorial Committee and are pleased that Drs. J. F. Crow and R. I. DeMars have accepted appointment as their replacements. Dr. R. D. Owen, a former member of the Committee, responded generously to our invitation to participate in the planning of the volume.

The Editor and the Editorial Committee

CONTENTS

POPULATION GENETICS

3044

R.C. Lewontin

Museum of Comparative Zoology, Harvard University, Cambridge, Massachusetts

Since the last general reviews of population genetics by Lewontin (1) and Spiess (2), which were meant to give an overview of the field up until 1967-68, there has been a pronounced concentration of activity in both experimental and theoretical work. This concentration is a consequence of the discovery, beginning in 1966, of extraordinary amounts of allelic variation of structural genes coding for enzymes and other proteins, segregating in natural populations. The documentation of this variation in a large variety of species has been a major effort of experimental work in the last seven years. Theoretical work has, in turn, been largely preoccupied with explaining how so much variation is to be explained, and determining what its importance in evolution is likely to be. From a broad and diverse science, attempting to establish a theory and phenomenology of all aspects of evolutionary change, population genetics has, in the last few years, turned into one virtually possessed by a single problem, the problem of protein polymorphism and protein evolution.

THE PROBLEM The theory of population genetics is essentially an extension of the laws of Mendel to include the fact that different genotypes leave different numbers of offspring, either because of "natural selection," or differences that are inconsistent over time and space, which we put down to "random genetic drift." The theory of population genetics is then framed in terms of the frequencies of genotypes in populations, usually considering one locus in isolation from the rest of the genome. On the other hand, the observations of population genetics have not, in general, been of the frequencies of individual genotypes except in special cases of striking single-locus polymorphisms or chromosomal variations like inversions and translocations. It is the assumption of the neo-Darwinian theory of evolution, however, that most evolutionary change occurs by small inherited changes in size, shape, behavior, and physiology, rather than by the substitution of alleles with large and obvious effects on morphogenesis and physiology. Precisely because these changes are assumed to be individually small in their effect on phenotypes, it has not been possible to enumerate genotypes unambiguously. This, in turn, means that the variables in population genetic theory, genotype frequencies, have not been susceptible to observation for just those characters of most direct interest to the evolutionary geneticist. In order to make the theory and the observations of population genetics come into direct contact with each other, it is necessary to find methods for enumerating alternative genotypes in populations, even when the genotypic differences result in arbitrarily small changes in morphogenesis and physiology. It was the development of such a method that has led to a major concentration of effort in population genetics in the last seven years.

In 1966, Hubby and Lewontin (3, 4) and Harris (5) showed how the techniques of gel electrophoresis of proteins could be used to characterize genetic variation in natural populations. The technique will distinguish between heterozygotes and homozygotes for allelic substitutions at arbitrarily chosen loci provided (*a*) that the locus codes for an enzyme (or other protein) that can be visualized by a specific dye reaction in vitro and (*b*) that the allelic substitution causes an amino acid substitution changing the net electrostatic change on the protein. In principle any enzymatic protein will satisfy criterion (*a*). An unknown proportion of allelic substitutions satisfy criterion (*b*), but from considerations of the genetic code and the empirical composition of proteins, it is at least 25%. Lewontin & Hubby (4) used the method to study 18 loci in samples from 5 natural populations of *Drosophila pseudoobscura.* They estimated that 30% of the structural gene loci within any population are segregating for electrophoretically detectable alternative alleles (*allozymes*), not including occasional very rare variants, and that 12% of loci are heterozygous in a random individual. If we assume that amino acid substitutions causing detectable changes in electrophoretic mobility are only about one-third of all substitutions, and that the different sorts of substitutions are at random over loci, then the direct estimates of Lewontin & Hubby project to 65% of all structural gene loci polymorphic and 32% of all loci heterozygous per individual. Even if we do not make this projection and use only the observed electrophoretic variation, the amount of genetic diversity in a typical population of *D. pseudoobscura* is immense. Harris's (5) estimates based on 10 loci in a single human population were 30% of loci polymorphic and 10% heterozygosity per individual for man. Not only did the method allow an estimate of total genetic variation, but each locus could be studied separately and in combination with other loci so that a direct enumeration of genotypes could be made, despite the fact that the physiological and morphogenetic effect of the gene substitutions was unknown and might be arbitrarily small. Thus, for the first time, a general method existed for observing the variables that appeared directly in population genetic theory, for genes of small effect.

SURVEYS OF ALLOZYME VARIATION The appearance of these first rough estimates of genic variation in small samples led immediately to a search for similar variation in a wide variety of other organisms and for improved estimates in *Drosophila pseudoobscura* and man. Surveys differ in the kind of information that can be derived from them. One of the earliest, by Johnson, Kanapi, Richardson, Wheeler & Stone (6) showed that 9 out of 20 loci representing 6 different enzyme functions in *Drosophila ananassae* were polymorphic, but no allele frequencies could be calculated from the data. Baker (7) found that 4 out of 10 egg white proteins were polymorphic in 37 domestic chicken breeds and that 24 of the breeds were polymorphic for at least one of the proteins. The number of birds examined in each flock was small (sometimes only 2 or 3) and the variety of loci studied is low. Another difficulty is illustrated by the studies of inbred and feral mouse populations by Ruddle et al (8) and Roderick et al (9). They found that all 16 loci studied in 35 inbred strains showed some genetic variation among the strains, while in natural populations 5 out of 17 loci were polymorphic with an average heterozygosity per population between 7.6% and 14.7%. However, the loci studied were not chosen at random, but were picked because they had already been shown to be variable in some strain in previously published reports.

In order to establish the average genetic variation in a population, a study must have adequate sample sizes, genomes sampled directly from natural populations, or maintained in culture only for a few generations, a large sample of loci, and a diverse sample of loci coding for proteins of a variety of different functions. Table 1 shows the outcome of surveys of variation that best meet these criteria and that are known to me. I have arbitrarily excluded any survey with fewer than 18 loci. The impression from Table 1 is overwhelming that polymorphism for structural genes is the rule in a variety of animals including man, mouse, Drosophila, and the horse-shoe crab. The best studied organism is man, with 71 loci showing 28% polymorphism and 6.7% heterozygosity per individual. The lowest variation is in *Mus musculus domesticus* with 20% of loci polymorphic and 5.6% heterozygosity per individual, while the highest is in *Drosophila willistoni*, virtually all of whose loci are reported as polymorphic and with an average heterozygosity per individual of 18.4%. The median proportion of loci polymorphic in Table 1 is 30% and the median heterozygosity per individual is 10.6%. If we again assume that electrophoresis detects only one-third of all the allelic variation, the projected values become 66% of loci polymorphic and 29% heterozygosity per individual.

While it is tempting to try to rationalize the very large apparent difference in genetic variation between *Mus musculus domesticus* and *Drosophila willistoni*, by reference to their life histories, the standard errors of heterozygosity are very large because so few loci have been studied, so that the differences may be more apparent than real. For example, previous comparisons of *Drosophila melanogaster* and *D. simulans* by O'Brien & McIntyre (23) and Berger (24) based on 10 loci and 6 loci, respectively, gave 23% heterozygosity for *D. melanogaster* in both studies but only 7% and 0% for *D. simulans*. The more complete study of these species shown in Table 1 gives a very different picture.

In addition to the more extensively studied organisms in Table 1, there are many less extensive surveys of a great variety of organisms, whose outcome, at least qualitatively, supports the general picture. Among the animals studied are: *Drosophila athabasca* and *D. affinis* (21), *D. ananassae* and *D. nasuta* (6,25), *D.*

Table 1 Surveys of genic heterozygosity in a number of organisms

Species	Number of populations	Number of loci	Proportion of loci polymorphic per population	Heterozygosity per locus	Standard error of heterozygosity	Ref.
Homo sapiens	1	71	.28	.067	.018	(10)
Mus musculus musculus	4	41	.29	.091	.023	(11)
M. m. brevirostris	1	40	.30	.110		(12)
M. m. domesticus	2	41	.20	.056	.022	(11)
Peromyscus polionotus	7(regions)	32	.23	.057	.014	(13)
Drosophila pseudoobscura	10	24	43	.128	.041	(14) (15)
D. persimilis	1	24	.25	.106	.040	(16)
D. obscura	3(regions)	30	.53	.108	.030	(17)
D. subobscura	6	31	.47	.076	.024	(18)
D. willistoni	{2-21	28	.86	.184	.032	(19)
	{10	20	.81	.175	.039	(20)
D. melanogaster	1	19	.42	.119	.037}	(21)
D. simulans	1	18	.61	.160	.052}	
Limulus polyphemus	4	25	.25	.061	.024	(22)

mimica (26), the harvester ant, *Pogonomyrmex barbatus* (27), the snails *Cepaea nemoralis* and *C. hortensis* (28), the chicken (7), the pheasant, *Phasianus calchicus* (28), the quail, *Coturnix coturnix* (29). Plants, too, appear to be rich in allozyme diversity, but the peculiarity of the breeding systems in those species studied so far make it impossible to compare the results to the animals in Table 1. Thus, there is polymorphism, but very little heterozygosity, in the nearly completely selfed wild oats, *Avena barbata* and *A. fatua* (30, 31); the dandelion, *Taraxacum officinale*, has a number of allozyme combinations in natural populations (32), but these cannot be analyzed genetically because the species is completely apomictic; *Oenothera biennis* is a fixed heterozygote at 5 out of 19 loci, presumably because of the permanent multiple translocation heterozygosity in the species (33).

Despite the overwhelming evidence of high heterozygosity for most species, there are some reports of low genetic variation. In a sample of 138 individuals of *Rattus rattus* from a wild population in Novosibirsk, not a single variant was found at 21 loci (34). The lack of experimental detail, and the contrast with the finding of high variation in *Mus*, make this report doubtful. Seven loci in three Southeast Asian populations of Macaques, *Macacus fascicularis* showed heterozygosities of 1.9%, 3.4%, and 7.7% (35). The Pacific salmon, *Oncorhyncus keta*, was found by Altukov et al (36) to be only about 2% heterozygous, but the number of loci in the study is uncertain since there was no genetic analysis and there were a large number of lens and muscle protein bands observed. There are three convincing published cases of very low variation based upon a sufficient number of loci and individuals. The cricket frog, *Acris crepitans*, is polytypic in southeastern United States with a number of local geographic races differing from each other in enzyme alleles, but with very little heterozygosity within local populations (37). A species of kangaroo rat, *Dipodomys merriami*, has 12% –29% polymorphism in various populations with heterozygosities varying from 2% –7%, while in *D. ordii* the heterozygosity per individual is only 0.4% –1.7% in different populations (38). Finally, Nevo & Shaw (39) have shown that the average heterozygosity in the fossorial rodent, *Spalax ehrenbergi*, is only 3.7%. We must emphasize again that, in view of the large standard errors of average heterozygosity, the meaning of these apparently low levels is uncertain.

The Genome as a Whole

If any general meaning is to be ascribed to the very large amount of allozyme variation that has been observed, it is necessary to inquire into the representativeness of the sample of loci. Loci so far studied are those that code for the primary amino acid sequence of enzymes and a few nonenzymatic proteins like egg albumins and serum proteins. No "structural" proteins like muscle or membrane proteins have been studied, nor have any controlling genes. In view of the fact that the amount of DNA in Drosophila is sufficient to code half a million average sized polypeptides, there is a reasonable chance that as much as 90% of the genome might be taken up with controlling genes. Even among the enzymes there has been a strong bias toward those that are soluble. A few particle bound enzymes have been examined and no great difference has been found. In man (10) two out of three mitochondrial enzymes are polymorphic (average heterozygosity 11%), in *Peromyscus polionatus* (13) none are, out of three tested, while in *Limulus* (22) both mitochondrial enzymes studied were polymorphic (heterozygosity 14%).

Among soluble enzymes there is evidence for heterogeneity between functional classes. Gillespie, Kojima & Tobari (21, 25) have shown that if enzymes are divided between those directly concerned with glycolysis, and all others, that the former are less heterozygous than the latter. This suggestion is borne out in all those species for which enough enzymes have been examined to make a comparison possible. In *Drosophila simulans* (25) the ratio of heterozygosities in Group I and Group II enzymes are at an extreme: .030/.364, while in *Mus musculus* (12) they are nearly equal: .089/.106, but the difference is in the same direction. Since Group I enzymes are proportionally over-represented in all genetic surveys thus far, the average heterozygosity for the structural genes as a whole has been underestimated if anything.

Explanations of Genic Diversity

The incontrovertible demonstration of a tremendous genic polymorphism for enzymes leads immediately to the question of explaining the heterogeneity. There are essentially two schools of explanation that have roots in two older general trends in population genetics. One of these trends, called by Dobzhansky (40) the "classical" school, held that nearly all individuals in a population were genetically identical at nearly all their loci, being usually homozygous for a wild type allele at each locus, but occasionally being heterozygous for a deleterious recessive or partly recessive mutation. This view held that most of the action of natural selection was a cleansing or purifying one, which swept out of the population the recurrent deleterious mutations. This view is best expressed by Muller in his famous article, "Our Load of Mutations" (41). A consequence of this classical view is that adaptive evolution is limited by the occurrence, very rarely, of favorable mutations and that most of the time the species cannot adapt rapidly to the vicissitudes of a changing environment, because adaptive genetic variation is a rarity. The opposite point of view, called by Dobzhansky the "balance" school, supposes that there is a large amount of standing variation at all times in populations, variation that is potentially adaptive. On this view adaptive evolution is not limited by the rate of occurrence of rare favorable mutations, and populations can track a changing environment as a rule, unless a really drastic and remarkable environmental variation has occurred. The observations on allozyme variation would seem to have resolved this conflict clearly in favor of the "balance" school, but not so. The "classical" hypothesis is conserved by the addition of the not unreasonable hypothesis that the variation revealed by electrophoretic studies is, in fact, irrelevant to the physiology and morphology of the organisms and that therefore all of the allelic variants in natural populations are effectively wild type. Such variation is selectively neutral and this newer form of the classical theory is sometimes called the "neutral" theory of genic polymorphism, most clearly expounded by Kimura & Ohta (42, 43). The theory that the genic variation revealed by electrophoresis is selectively neutral is coupled and consistent with a theory of neutral substitution in protein evolution which supposes that most of the differences in amino acid sequence between proteins in different species has occurred without natural selection. This "non-Darwinian evolution" (King & Jukes 44) is proposed as the chief process that underlies the differentiation in evolution of cytochrome-c, haemoglobin, insulin, fibrinopeptide A, and other proteins whose sequence has been determined in a variety of organisms (45 –47). The balance school, on the contrary, regards the standing genetic diversity within populations as being a consequence of balancing forms of natural selection, especially heterosis, and supposes that the substitution of amino acids in proteins in the course of phyletic evolution is also an adaptive process. The attempt to resolve the issue between "balance" and "neutral" (or, better, "neo-classical") schools has been a major preoccupation of much experimental and theoretical work in recent years.

Patterns of Geographic Variation

One kind of evidence brought to bear on the issue of selection is the pattern of gene frequency variation among populations within a species. In general, there is remarkably little spatial variation in gene frequencies. In *Drosophila pseudoobscura*, Prakash, Lewontin & Crumpacker (15) found essentially identical allele frequencies in all populations in North and Central America. Only the isolated population around Bogotá, Colombia, was different, being much more homozygous than the rest of the species, usually for the alleles that were most common elsewhere. Loci associated with the third chromosome did vary from locality to locality and this was shown by Prakash & Lewontin (48, 49) to be the result of the association of particular alleles with particular inversions for which the species is polymorphic and which are differentiated in frequency between populations. A similar situation exists in *D. willistoni* (Ayala, Powell & Dobzhansky 20). Populations on the mainland of South America and in the islands of the Caribbean have similar allele frequencies at 18 out of 20, the remaining two being consistently different between the mainland and the islands. *D. willistoni* has a rich inversion polymorphism on all its chromosomes, but the islands tend to be chromosomally monomorphic (50). At present it is not known what the relationship is between chromosomal differentiation and genic variation in this species. The extensive survey of *D. obscura* in Finland by Lakovaara & Saura (17) showed 14 loci identically monomorphic in all localities, 15 polymorphic loci with no local differentiation at all, and 1 polymorphic locus with marked differences between central Finland and Lappland. The same authors' study of *D. subobscura* (18) gave similar results: all monomorphic loci identically monomorphic in all populations, 9 polymorphic loci with closely similar allele frequencies over all populations, 2 polymorphisms with strong differentiation and 3 polymorphisms which were monomorphic on islands. In *Mus musculus*, there is some geographical variation across North America, more than is found in Drosophila, while the population of the island of Jamaica was virtually monomorphic (11). For these species the pattern emerges of weak differentiation among populations that are not totally isolated from each other, except for loci associated with chromosomal polymorphisms, and a marked increase in homozygosity on islands.

Two cases of marked allele frequency differentiation give a further insight into the situation. There is a clear difference in heterozygosity at 32 loci in *Peromyscus polionotus* between isolated and semi-isolated beach islands and peninsulas of the Florida Gulf Coast panhandle where heterozygosity is around 2%, and the mainland range in Georgia and peninsular Florida where heterozygosity is 8%. On the other hand, insularity and the beach habitat are not alone sufficient to account for these differences since two of the most polymorphic populations are beach islands on the Atlantic coast of Florida. The second case of strong differentiation of gene frequencies is in the Pacific islands in *Drosophila ananassae* and *D. nasuta* (51, 52). Islands of the same group, as for example several islands of Western Samoa or of Fiji are much more similar in allele frequency than widely separated island groups. There is no relation however, of the distance between major island groups and gene frequency differences. nor is there any relation between island size and average heterozygosity.

In man there is an intermediate degree of genic differentiation among geographical populations. By analyzing genic diversity for 17 polymorphic genes in man by the usual information measure of diversity, Lewontin (53) showed that

85% of human genic diversity is within local populations, 8% is between local populations within races, and 7% is between races.

The patterns of similarity and difference can be interpreted as supporting either the balanced or classical view. So, for example, the large differences between island groups and the similarities within groups are taken by Johnson (52) as evidence that selective forces, differing in different localities, are controlling the variation. On the other hand, differentiation between widely separated, isolated, island groups with some similarity between neighboring islands within a group is precisely what is to be expected if purely random differentiation is occurring, with no migration over long distances and a small amount of gene exchange locally. Obviously any attempt to interpret the meaning of geographic variation must take into account the role of migration. For that reason, a great deal of interest has been shown in the theory of differentiation under migration.

Migration and Random Drift

The argument of the neoclassical school is that average levels of heterozygosity and degree of differentiation between populations can be explained as the consequence of new, neutral mutations, a certain proportion of which rise to intermediate gene frequencies in finite populations by chance, and may eventually become fixed. It was shown by Kimura & Crow (54) that the average heterozygosity for neutral alleles in a population of size N is, at steady state

$$H = 1 - \frac{1}{4N\mu + 1} \qquad 1.$$

where μ is the mutation rate of new alleles. There is a single pseudoparameter $N\mu$ which is the product of a large but unknown number, N, and small but unknown number, μ, so that any observed value of H can be satisfactorily explained by a "reasonable" choice of parameter values. So, for example, the observed heterozygosity of 10% in Table 1 is equivalent to $N\mu = .03$, corresponding, say to $\mu = 3 \times 10^{-5}$ and $N = 10^3$. This formulation is strictly accurate only when each mutation is unique. However, Kimura (55) and Crow & Kimura (56) have shown that the formula is very close to correct provided that the number of distinguishable alleles is much larger than $4N\mu$.

While the average heterozygosity in a population can easily be made compatible with random genetic drift, each population should have a different array of particular alleles, since it is a matter of pure chance which equivalent neutral allele is in high frequency.

However, if there is migration among the populations, variation between populations is reduced. A standard measure of genic variation among populations is the ratio of the variance of allele frequencies among populations to the theoretical variance if all populations were fixed for one allele or another:

$$f = \frac{\sigma_p^2}{\bar{p}(1 - \bar{p})} \qquad 2.$$

where $\bar{p}$ is the average allele frequency over the entire collection of populations. It was long ago shown by Wright (57) that if a large number of populations each of size N received migrants at a rate m from a general pool with the average gene frequency, then at equilibrium

$$f = \frac{1}{4Nm + 1} \qquad 3.$$

If Nm is of order 10 or greater, that is 10 migrant individuals per generation, irrespective of population size, then f will be 2% or less and the populations will be essentially identical. This model shows that very little migration will keep populations genetically similar in the absence of selection. Models of continuously distributed populations in one and two dimensions with migration falling off as a continuous function of distance have been examined by Wright (58) and Malécot (59, 60). If the standard deviation of the migration distribution is σ, the density of individuals per unit area is d and the long distance migration from other populations is u, then the correlation in gene frequency between two sample areas x units of distance apart will be proportional to

$$p(x) = \frac{e^{-x\sqrt{2u}/\sigma}}{\sqrt{x}} \qquad 4.$$

and the variation in gene frequency over the entire range is given by

$$f = \frac{1}{(1 + 8\pi d\, \sigma^2 / -\ln 2u)} \qquad 5.$$

So, for example, if *Drosophila pseudoobscura* had a local density of 1 per square meter and its standard deviation of migration were only 10 meters per generation with a long distance migration rate of only 10^{-4}, f is only .004, and the correlation between areas 1 kilometer apart is only 1%.

A yet more realistic model of migration, the stepping stone model, was analyzed by Kimura & Weiss (61) with essentially the results given by expressions 4 and 5 above. In this model individual populations make up a rectangular lattice with exchanges between adjacent populations at a rate $\sigma/4$ and with some long distance migration, u, as well. Further refinements of the model to include unequal migration rates in different directions and to take account of the fact that there are a finite number of populations were made by Maruyama (62–64); but with little effect on the quantitative conclusions. It appears that any degree of similarity or difference between local geographic populations can be explained by appropriate choices of population size and migration rate, without recourse to selection.

General tests of selective and random theories There are a number of consequences of selective and nonselective theories that, in principle at least, could be used to distinguish the causes of polymorphism, although in practice they have turned out to be in conflict or ambiguous. Following a formulation by Kimura & Crow (54), Lewontin & Hubby (4) calculated that *Drosophila pseudoobscura* could not possibly support the estimated 3000 polymorphic loci by heterosis, since that would imply that a multiply heterozygous individual was 10^{43} times as fit as the average individual in the population. This calculation, which assumes selection to be operating independently on each locus, was criticized by King (65), Milkman (66) and Sved, Reed & Bodmer (67) who all pointed out the estimated polymorphism could be heterotic without such excessive fitnesses, if reasonable models of epistatic interaction between genes was supposed. In particular, King (65) showed that truncation selection, in which a fixed proportion of the population with the greatest heterozygosity survived, was consistent with large amounts of polymorphism.

Neutral theory predicts a relationship between proportion of loci polymorphic, P, and average heterozygosity per individual, H, of the form

$$P = 1 - q^{H(1-H)} \qquad 6.$$

where q is the frequency of "variant" alleles at a locus required to classify it as polymorphic. Kimura & Ohta (43) have shown that the observed data fit this curve well. On the other hand, Johnson & Feldman (68) have shown that neutral theory predicts that the relationship between the number of alleles at a locus, k, and the total homozygosity at a locus, h, should be such that hk is an increasing function of k, while the data on several species of Drosophila show it to be a decreasing function. This shows that allelic frequencies at multiple allelic loci are too uniform for the neutral theory.

A third consequence of neutral theory is that the variation of gene frequencies among populations at any locus is entirely a result of the breeding structure of the species and therefore all loci should have the same variation among populations, as expressed by f (see formula 2). Lewontin & Krakauer (69), following a suggestion originally made by Cavalli-Sforza (70), have shown that the variation in f values among loci, over the world population of the human species, is much too large to be explained by chance, so that some of the loci must be under differential selection. The data for *Drosophila pseudoobscura* are also significanty heterogeneous, but the heterogeneity is contributed entirely by two loci.

In a series of papers on substitution of amino acids in evolution, Kimura & Ohta (42, 47, 71, 72), especially in their book (43), have argued that the rate of substitution of amino acids in typical proteins is too great to be the result of natural selection because the genetic load from these substitutions would be more than any population could support. On the other hand, they show that under neutral theory, the rate of amino acid substitution is equal to the mutation rate to neutral alleles and that mutation rates of the order of 10^{-7} per cistron would be sufficient to account for gene substitution in evolution. The difficulty with their argument about substitution rates being too high for genetic load is that the calculation depends critically on the supposition that higher organisms have 10 million structural genes like those coding for haemoglobin, cytochrome-c, etc. If the number of structural genes were, say, 100,000, their argument would lose all force.

Measurements of Selection in Nature

There are formidable problems in directly testing natural selection. Christiansen & Frydenberg (73) have analyzed the statistical problems of fitness estimation in organisms with a synchronized life cycle and have applied their analysis to an esterase polymorphism in the fish *Zoarces viviparus*, but no evidence of selection at any stage was found. For populations with continuous breeding, the problem of defining fitness is complex. In a series of fundamental papers, Charlesworth & Geisel (74 –76) have established the relationship between the age schedules of mortality and fecundity and the selection coefficients and rates of gene frequency change.

Direct evidence from nature of selection of allozyme variation exists for some cases. Berger (77) found clear evidence of an increase in an allele of alpha glycerolphosphate dehydrogenase during the summer and fall in two populations of *Drosophila melanogaster* in two years. Schopf & Gooch (78) found a cline in an allele of leucine amino peptidase in a bryozoan, associated with a 6°C change in water temperature over a short distance and this was repeated over a long north-south geographical cline. There is a strong correlation of temperature and oxygen concentration with lactate dehydrogenase allele frequencies in the Blenny, *Anoplorchus* (79) and of a number of environmental features and several loci in the harvester ant (27). A geographical cline that is very likely the result of temperature change is that of an esterase allele in the fish *Catastomus clarki*, which changes from a frequency of .18 to 1.00 over 7.6 degrees of latitude in the Colorado River basin (80). An unexplained altitudinal cline in an alcohol dehydrogenase locus has been shown in *Drosophila melanogaster* in the Soviet Union (81).

The problem with such observations is that we do not know how typical they may be. Negative results are not usually reported, so that the mere *existence* of cases of selection proves nothing about the generality.

Selection in the Laboratory

While it is easier to demonstrate selection in the controlled condition of the laboratory where individual components of the life cycle can be studied, it is by no means a trivial problem. The fundamental papers on fitness estimation are those of Prout (82 –84) in which it is shown that even with optimal experimental designs, very large samples are needed.

Another problem is that it is difficult to distinguish the effect of the locus being followed from surrounding loci with which it is linked, since laboratory experiments often are started with a very small sample of genomes from nature. Thus, the rapid changes in malic dehydrogenase, alcohol dehydrogenase, alpha glycerolphosphate dehydrogenase, and esterase alleles observed in laboratory populations of *Drosophila melanogaster* (77, 85) were probably the result of linkage of these loci to mutations of low fitness at other loci. On the other hand, when considerable care was taken to include a large number of founding genomes in laboratory populations, no changes in allele frequency were found at two loci in *D. pseudoobscura* (Yamazaki 86). A convincing case for selection can be made if changes in allele frequencies are shown to depend upon the presence of the substrate for the enzyme in question. Thus, Gibson (87) and Wills & Nichols (88, 89) found selection for alcohol dehydrogenase and octanol dehydrogenase in *Drosophila melanogaster* only when the substrates were added in excess to the food medium. Unfortunately, this method is only applicable to loci coding for enzymes that have exogenous substrates, but if an array of such enzymes could all be shown to be selectable, the neutral hypothesis would be in serious difficulty.

One form of selection that successfully meets the objection of excess genetic load is frequency dependent selection in which rare genotypes are more fit. A stable equilibrium of allele frequencies will result, yet at equilibrium there may be little or no difference in fitness between genotypes, and thus no genetic load at equilibrium. In a finite population, however, the fluctuation in allele frequency around the equilibrium may result in a substantial load (90). Experimental demonstrations of frequency dependent selection with rare genotypes most fit, have been carried out for alcohol dehydrogenase and esterase loci in *Drosophila melanogaster* by Kojima and others (91 –94), but in at least one case (91) the apparent frequency dependent selection was probably an artifact of the estimation method. A specific test of frequency dependent selection in *Drosophila pseudoobscura* by Yamazaki (86) gave negative results. In general, such frequency dependent selection cannot be the rule because it would require that the environment have as many dimensions as there are separate loci, each being maintained by the mechanism. In a population segregating at several thousand loci, every genotype is rare!

Multiple Locus Theory

Many of the difficulties of understanding how selection can maintain so many polymorphisms arise because the models of population genetics deal chiefly with single loci, so that the genome is regarded as a collection of individual independent loci each undergoing its separate evolution. But genes are organized on chromosomes and these, rather than individual loci, are the units of genetical segregation, subject to some recombination. In recent years an increasing amount of attention has been given to the dynamics of multiple locus systems when the fact of their chromosomal organization is taken into account. The first exact treatments of two locus problems in which natural selection was taken into account were those of Kimura (95), Lewontin & Kojima (96) and Bodmer & Parsons (97), all of whom considered various forms of symmetric selection models with various genotypes given equal fitnesses in order to make the mathematical analysis tractable. These models have all been subsumed under a general symmetric model which has been completely analyzed by Karlin & Feldman (98). The results of these analysis show that there exist a number of possible stable equilibria depending upon the intensity of linkage and the degree of interaction between the loci in determining fitness. Defining the ratio of the epistatic interaction e to the recombination fraction r, as a *coupling coefficient* $C = e/r$ the nature of the equilibria depends on the magnitude of the coupling. If C is small, because recombination is frequent, or epistatic interaction is weak, the two loci evolve essentially independently and each reaches an independent equilibrium. If C is very large, however, because the genes are tightly linked and strongly interacting, the loci are not independent at equilibrium, but are correlated so that there is an excess of coupling or repulsion gametes. In the most extreme case of $r = 0$ only AB and ab gametes may exist in the population, or only Ab and aB. As recombination is loosened all four gametic types appear at equilibrium but with an excess of coupling or repulsion and $AB = ab$, $aB = Ab$. With still further loosening of linkage the correlation between loci may disappear but for some very loose linkages may reappear again, now with $AB \neq ab$ (the so-called asymmetrical equilibria). As shown by Ewens (99), even more complex behavior is possible with some fitnesses, such that all stable equilibria with both loci segregating may disappear at intermediate levels of recombination, but may be present for both loose and tight linkage. The results of symmetrical viability models have been extended to three loci by Feldman, Franklin & Thomson (100) who have shown analytically that for intermediate values of recombination, there may be alternative equilibria in which loci may be correlated or uncorrelated. Thus, genes may be found correlated in one population and uncorrelated in another, although the selection is the same in both cases.

The only other general equilibrium case treated analytically is the two-locus multiplicative model, for which Bodmer & Felsenstein (102) have shown that if recombination is smaller than the product of the genetic loads at the separate loci, there will be a stable correlation between the loci at equilibrium.

The work on two loci suggests that unless the coupling between pairs of genes is large, there will be no lasting effect of linkage on the equilibrium condition of the population. However, a numerical study by Franklin & Lewontin (101) of systems with large numbers of loci has shown that synergistic effects occur which are totally unpredictable from two-locus theory. If a pair of loci with a weak coupling coefficient is embedded in a chromosome with other segregating genes with equally weak coupling coefficients, there is nevertheless a much more powerful effect of linkage on the equilibrium of gamete frequencies than would be predicted from the separate coupling coefficients. This synergistic effect, which arises from the interactions of each gene of the pair with all the other genes, becomes greater and greater as the number of segregating genes increases. Franklin & Lewontin showed that if the number of loci is allowed to grow larger while the coupling coefficient between pairs of adjacent loci grows smaller and smaller, a limiting condition is reached with only a few dozen loci, such that the number of loci and their individual coupling coefficients become irrelevant to the equilibrium of the population. All that needs to be known is the total map length of the chromosome segment and the inbreeding depression when the entire chromosome segment is made homozygous. On the basis of these two parameters, which are measurable in practice, it can be predicted what the structure of the gametic pool will be at equilibrium. Moreover, for moderate inbreeding depressions and moderate map lengths, a highly correlated structure will evolve in which only a few gametic types will be present in the population although all the loci are polymorphic. The quantitative features of the model have been reproduced analytically by Slatkin (103) using a continuous model of the genome.

In finite populations, the effect of sampling error is such as to produce correlations between loci, even in the absence of selection, and of sufficient magnitude to make a difference to the outcome of selection when it does occur. Thus, Ohta & Kimura (104) have shown that an unselected gene which is correlated in its distribution with a selected one by chance, will appear to be under selection and will be carried along with the selected locus, so that its evolution is strongly affected by selection, even though the locus itself is neutral. This "hitchhiking effect" is of great importance in the interpretation of experiments and natural historical observations on allozyme frequency changes. Hill & Robertson (105) have shown that the random effects of finite population size will be greater for a neutral locus that is linked to a selected one than in the absence of linkage. One consequence is that a new slightly favorable mutant will have a *smaller* probability of eventual fixation if it is linked to a selected locus! The general stochastic theory of linked systems under selection has not been well studied, but for completely unselected loci, the rates of fixation have been worked out under a variety of models (106 –109).

The various studies of linkage and selection have generally shown an increase in mean fitness of the population with tighter linkage, and Lewontin (110) has proved that complete linkage always results in the highest possible fitness for any model of selection. It does not follow, however, that for intermediate values of recombination a small decrease in recombination will raise fitness, nor that modifiers of recombination will always be selected for tightening linkage. Specific analytic investigation of the small variety of models that can be treated exactly have all shown that a gene that reduces the linkage distance between two loci will be favored if the two loci have fitness interactions between them (111 –113). It must be pointed out, however, that the range of models so far investigated is too small to make a definitive statement.

The predictions of multiple locus theory have been found to hold in nature. Three polymorphic allozyme loci on the third chromosome of *Drosophila pseudoobscura* are strongly correlated with each other and with particular inversions in natural populations (114, 115), and these associations are consistent in all geographical populations. There are several reports of correlation between inversions and particular alleles at allozyme loci, although not of allozyme loci with each other within the inversions, in other species of Drosophila (116 –118). The only systematic search for linkage correlation among polymorphic allozyme loci not associated with inversions has been the study of 5 loci on the third chromosome of *Drosophila melanogaster* by Charlesworth & Charlesworth (119), who found consistent, but weak, association among the loci spanning 20 map units.

Prospects

The discovery of vast amounts of genetical protein polymorphism in a wide diversity of organisms has created very difficult problems that will occupy population genetics for some time to come. Methods must be found, despite the enormous statistical difficulties, of determining whether natural selection is operating to mold this variation, or whether it is physiologically meaningless and, in that sense, variation that is irrelevant to adaptive evolution. At present there seem to be only two approaches open. One is an exhaustive study of a group of randomly chosen polymorphic loci in some species, to show that all, or nearly all, do in fact have a reflection of their genetical variation in the physiology and morphogenesis of the organism under conditions that are part of the natural life history of the species. Positive results would be conclusive, but negative findings would still leave the question in the air. Such an exhaustive study requires that a great deal more be known about the physiology and ecology of the species then is currently known about the population geneticist's favorite organism, Drosophila. In one sense man is the obvious species for this study and there is already some evidence from in vitro studies that most allozyme variants are biochemically different (120). The only unambiguous alternative, at least in present theory, is to find consistent linkage correlations among allozyme polymorphisms in natural populations. If loci are consistently correlated in a wide range of geographical localities, then they must be under natural selection. The chance of finding such correlations will remain small, however, until the genetic map of some organism is far better saturated with enzyme loci than is now the case. What is required for this study is a group of five or so polymorphic loci within 5 recombination units.

In theory, a great deal more emphasis must and will be placed on the theory of multiple locus systems in temporally fluctuating environments. It is unlikely that any of the problems and contradictions of evolutionary genetics can be resolved by even more arcane models of single loci, or of constant enviroment. The current and future genetic composition of populations is certainly the result of the evolution of coupled systems of genes in a fluctuating environment, and these interactions and temporal instabilities are first order phenomena, not simply minor modifications of an otherwise simple and constant process.

Literature Cited

1. Lewontin, R.C. 1967. *Ann. Rev. Genet.*, 1:37–70
2. Spiess, E.B. 1968. *Ann. Rev. Genet.*, 2:165–208
3. Hubby, J.L., Lewontin, R.C. 1966. *Genetics*, 54:577–94
4. Lewontin, R.C., Hubby, J.L. 1966. *Genetics*, 54:595–609
5. Harris, H. 1966. *Proc. Roy. Soc.* (*London*) *Ser. B.*, 164:298–310
6. Johnson, F.M., Kanapi, C.G., Richardson, R.H., Wheeler, M.R.,

Stone, W.S. 1966. *Proc. Nat. Acad. Sci. U.S.*, 56:119–125
7 Baker, C.M.A. 1968. *Genetics*, 58:211–26
8. Ruddle, F.H., Roderick, T.H., Shows, T.B., Weigl, P.G., Chapman, R.K., Anderson, P.K. 1969. *J. Hered.*, 60:321–22
9. Roderick, T.H., Ruddle, F.H., Chapman, V.M., Shows, T.B. 1970. *Biochem. Genet.*, 5:457–66
10. Harris, H., Hopkinson, D.A. 1972. *J. Human Genet.*, 36:9–20
11. Selander, R.K., Hunt, W. G., Yang, S.Y. 1969. *Evolution*, 23:379–90
12. Selander, R.K., Yang, S.Y. 1969. *Genetics*, 63:653–67
13. Selander, R.K., Smith, M.H., Yang, S.Y., Johnson, W.E., Gentry, J.B. 1971. *Univ. Texas Publ.*, 7103: 49–90
14. Prakash, S., Lewontin, R.C., Hubby, J.L. 1969 *Genetics*, 61:841–48
15. Prakash, S., Lewontin, R.C., Crumpacker, D.W. 1973. *Genetics* In press
16. Prakash, S. 1969. *Proc. Nat Acad. Sci. U.S.*, 62:778–84
17. Lakovaara, S., Saura, A. 1971. *Genetics*, 69:377–84
18. Lakovaara, S., Saura, A. 1971. *Hereditas*, 69:77–82
19. Ayala, F.J., Powell, J.R., Tracey, M.L., Mourão, C.A., Perez-Salas, S. 1972. *Genetics* 71:113–139
20. Ayala, F.J., Powell, J.R., Dobzhansky, Th. 1971. *Proc. Nat. Acad. Sci. U.S.*, 68:2480–83
21. Kojima, K., Gillespie, J., Tobari, Y.N. 1970. *Biochem. Genet.* 4:627–37
22. Selander, R.K., Yang, S.Y., Lewontin, R.C., Johnson, W.E. 1970. *Evolution*, 24:402–14
23. O'Brien, S.J., McIntyre, R.J. 1969. *Am. Nat.*, 103:97–113
24. Berger, E., *Genetics*, 1970. 66:672–83
25. Gillespie, J., Kojima, K. 1968. *Proc. Nat. Acad. Sci. U.S.*, 61:582–85
26. Rockwood, E.S. 1969. *Univ. Texas Publ.*, 6918:111–25
27. Johnson, F.M., Schaffer, H.E., Gillaspy, J.E., Rockwood, E.S. 1969. *Biochem. Genet.*, 3:429–50
28. Manwell, C., Baker, C.M.A. 1968. *Comp. Biochem. Physiol.*, 26:195–209
29. Baker, C.M.A., Manwell, C. 1967. *Comp. Biochem. Physiol.*, 23:21–42
30. Marshall, D.R., Allard, R.W. 1970. *Heredity*, 29:373–82
31. Clegg, M.T., Allard, R.W. 1972. *Proc. Nat. Acad. Sci. U.S.*, 69:1820–24
32. Solbrig, O. 1971. *Am. Sci.*, 59:686–96
33. Levin, D.A., Howland, G.P., Steiner, E. 1972. *Proc. Nat. Acad. Sci. U.S.*, 69:1475–77
34. Serov, O.L. 1972. *Isozyme Bull.* 5:38
35. Weiss, M.L., Goodman, M. 1972. *J. Human Evol.*, 1:41–48
36. Altukhov, Yu. P., Salmenkova, E.A., Omelchenko, V.T., Satchko, G D., Slynko, V.I. 1972. *Genetika*, 8:67–75
37. Dessauer, G.C., Nevo, E. 1969. *Biochem. Genet.*, 3:171–88
38. Johnson, W.E., Selander, R.K. 1971. *Syst. Zool.*, 20:377–405
39. Nevo, E., Shaw, D.R. 1972. *Biochem. Genet.*, 7:235–41
40. Dobzhansky, Th. 1955. *Cold Spring Harbor Symp. Quant. Biol.*, 20:1–15
41. Muller, H.J. 1950. *Am. J. Human Genet.*, 2:111–76
42. Kimura, M., Ohta, T. 1971. *Nature*, 229:467–69
43. Kimura, M., Ohta, T. 1971. *Theoretical Aspects of Population Genetics.* Princeton, NJ: Princeton Univ. Press
44. King, J.L., Jukes, T. 1969. *Science*, 164:788–98
45. Dickerson, R.E. 1971. *J. Mol. Evol.*, 1:26–45
46. Kimura, M. 1968. *Nature*, 217:624–26
47. Kimura, M., Ohta, T. 1971. *J. Mol. Evol.*, 1:1–17
48. Prakash, S., Lewontin, R.C. 1968. *Proc. Nat. Acad. Sci. U.S.*, 59:398–405
49. Prakash, S., Lewontin, R.C. 1971. *Genetics*, 69:405–08
50. Da Cunha, A.B., Burla, H., Dobzhansky, Th. 1950. *Evolution*, 4:212–35
51. Stone, W.S., Wheeler, M., Johnson, F.M., Kojima, K. 1968. *Proc. Nat. Acad. Sci. U.S.*, 59:102–09
52. Johnson, F.M. 1971. *Genetics*, 68:77–95
53. Lewontin, R.C. 1973. *Evol. Biol.*, 6:381–98
54. Kimura, M., Crow, J.F. 1964. *Genetics*, 49:725–38
55. Kimura, M. 1968. *Genet. Res.*, 11:247–69
56. Crow, J.F., Kimura, M. 1970. *An Introduction to Population Genetics Theory.* New York: Harper & Row
57. Wright, S. 1940. *Am. Nat.*, 74:232–48
58. Wright, S. 1951. *Ann. Eugenics*, 15:323–54
59. Malécot, G. 1955. *Cold Spring Harb. Symp. Quant. Biol.*, 20:52–53
60. Malécot, G. *Proc. Fifth Berkeley Symp. Math. Stat. Prob.*, 4:317–32
61. Kimura, M., Weiss, G.H. 1964. *Genetics*, 49:561–76
62. Maruyama, T. 1969. *J. Appl. Prob.*, 6:463–77
63. Maruyama, T. 1970. *Theoret. Pop. Biol.*, 1:101–19
64. Maruyama, T. 1971. *Ann. Human Genet.*, 34:201–19

65. King, J.L. 1969. *Genetics*, 55:483–92
66. Milkman, R.D. 1967. *Genetics*, 55:493–95
67. Sved, J., Reed, T.E., Bodmer, W. 1967. *Genetics*, 55:469–81
68. Johnson, G., Feldman, M. 1973. *Theoret. Pop. Biol.*, 4. In press
69. Lewontin, R.C., Krakauer, J. 1973. *Genetics*, 74. In press
70. Cavalli-Sforza, L. 1966. *Proc. Roy. Soc., B*, 164:362–79
71. Kimura, M., Ohta, T. 1973. *Sixth Berkeley Symp. Prob. Stat.* In press
72. Ohta, T., Kimura, M. 1971. *J. Mol. Evol.*, 1:18–25
73. Christiansen, F.B., Frydenberg, O. 1973. *Theoret. Pop. Biol.*, In press
74. Charlesworth, B. 1970. *Theoret. Pop. Biol.*, 1:352–70
75. Charlesworth, B., Geisel, J.T. 1972. *Am. Nat.*, 106:388–401
76. Charlesworth, B. 1972. *Theoret. Pop. Biol.*, 3:377–95
77. Berger, E. 1971. *Genetics*, 67:121–36
78. Schopf, T.J ., Gooch, J.L. 1971. *Evolution*, 25:286–89
79. Johnson, M.S. 1971. *Heredity*, 27:205–26
80. Koehn, R.K., Rasmussen, D.I. 1967. *Biochem. Genet.*, 1:131–44
81. Grossman, A.I., Koreneva, L.G., Ulitskaya, L.E. 1970. *Genetika*, 6:91–96
82. Prout, T. 1965. *Evolution*, 19:546–51
83. Prout, T. 1969. *Genetics*, 63:949–67
84. Prout, T. 1971. *Genetics*, 68:127–49
85. Yarbrough, K.M., Kojima, K. 1967. *Genetics*, 57:677–89
86. Yamazaki, T. 1971. *Genetics*, 67:579–603
87. Gibson, J. 1970. *Nature*, 227:959–60
88. Wills, C., Nichols, L. 1971. *Nature*, 233:123–25
89. Wills, C., Nichols, L. 1972. *Proc. Nat. Acad. Sci. U.S.*, 69:323–25
90. Kimura, M., Ohta, T. 1970. *Genet. Res.*, 16:145–50
91. Kojima, K., Yarbrough, K.M. 1967. *Proc. Nat. Acad. Sci. U.S.*, 57:645–49
92. Kojima, K., Tobari, Y.N. 1969. *Genetics, 61:201–09*
93. Huang, S.L., Singh, M., Kojima, K. 1971. *Genetics*, 68:97–104
94. Kojima, K. 1971. *Evolutior*, 25:281–85
95. Kimura, M. 1956. *Evolution*, 10:278–87
96. Lewontin, R.C., Kojima, K. 1960. *Evolution*, 14:458–72
97. Bodmer, W.F., Parsons, P.A. 1962. *Advan. Genet.*, 11:1–100
98. Karlin, S., Feldman, M.W. 1970. *Theoret. Pop. Biol.*, 1:39–71
99. Ewens, W. 1968. *Theoret. Appl. Genet.*, 38:140–44
100. Feldman, M., Franklin, I., Thomson, G. 1973. *Theoret. Pop. Biol.*, 4. In press
101. Franklin, I., Lewontin, R.C. 1970. *Genetics*, 65:707–34
102. Bodmer, W.F., Felsenstein, J. 1967. *Genetics*, 57:237–65
103. Slatkin, M. 1972. *Genetics*, 72:157–68
104. Ohta, T., Kimura, M. 1971. *Genetics*, 69:247–60
105. Hill, W.G., Robertson, A. 1966. *Genet. Res.*, 8:269–94
106. Karlin, S., McGregor, J. 1968. *J. Appl. Prob.*, 5:487–566
107. Ohta, T., Kimura, M. 1969. *Genet. Res. Camb.*, 13:47–55
108. Hill, W.G., Robertson, A. 1968. *Theoret. Appl. Genet.*, 38:226–31
109. Watterson, G.A. 1970. *Theoret. Pop. Biol.*, 1:72–87
110. Lewontin, R.C. 1971. *Proc. Nat. Acad. Sci. U.S.*, 68:984–86
111. Nei, M. 1967. *Genetics*, 57:625–26
112. Feldman, M.W. 1972. *Theoret. Pop. Biol.*, 3:324–46
113. Feldman, M.W., Balkau, B. 1972. In *Population Dynamics*, ed. T.N.E. Greville. 357–83. New York: Academic
114. Prakash, S., Lewontin, R.C. 1968. *Proc. Nat. Acad. Sci. U.S.*, 59:398–405
115. Prakash, S., Lewontin, R.C. 1971. *Genetics*, 69:405–08
116. Loukas, M., Krimbas, C.B. pers. comm.
117. Nair, P.S., Brnčic, D. 1971. *Am. Nat.*, 105:291–94
118. Mukai, T., Mettler, L.E., Chigusa, S. 1971. *Proc. Nat. Acad. Sci. U.S.*, 68:1065–69
119. Charlesworth, B., Charlesworth, D. 1973. *Genetics*, 73:351–59
120. Harris, H. 1971. *J. Med. Genet.*, 8:444–52

GENETICS OF IMMUNE RESPONSIVENESS

3045

W. H. Hildemann

School of Medicine and Dental Research Institute, University of California, Los Angeles

This is a selective review of five major topics intensively investigated during the last several years. They are the phylogeny of immune responsiveness, immune responses to infectious agents, genetic regulation of immune responses in man, specific immune response genes in mice and guinea pigs, and mechanisms of *Ir* (immune response) gene action. Some 70 key references are cited to enable the reader to pursue any of these topics in greater depth. To save space, more recent or definitive articles are listed. Hopefully, none of the now numerous workers in this field will feel slighted by the selective emphasis inherent in this overview.

Phylogeny of Immunological Characteristics

Realization that specific immune responses to infectious agents are decisively influenced by host genetic constitution has evolved mainly as a consequence of selective animal breeding. Diverse strains of domestic animals from honey bees to horses exhibit substantial differences in genetic resistance to particular pathogens (1). Such resistance commonly has a multigenic basis, but often depends primarily upon the specific immune response at least in vertebrates. Specific competence with immunological memory has appeared until recently to be characteristic only of vertebrates and associated with lymphoid cells, a thymus gland, and the ability to produce serum antibodies (2). Hagfish and lampreys, though most primitive among vertebrates, appear to be as highly evolved as other fishes below the teleost level in essential immunologic capacities. They reject skin allografts with concomitant development of immunologic memory and produce IgM-type antibodies to various xenogeneic antigens (Table 1). Specificity and memory are found in all vertebrate immune systems. At progressively higher levels of phylogeny ranging from bony fishes to mammals, the cytoarchitecture of the lymphoid system and the molecular classes of antibodies inducible both increase in complexity (3-5). Whether the genetic library or repertoire of responsiveness to potential antigens increases in phylogeny is unknown. The functional spectrum of cellular or humoral responses might

TABLE 1 Phylogeny of immunological characteristics

Class or group	Reaction to tissue allografts (cell-mediated immunity)		Thymus	Blood lympho-cytes	Intestinal lymphoid tissue	Lymph glands or nodes	Immunoglobulin Antibodies		
	Chronic	Acute					IgM-type	IgN, IgG &/or IgA type	IgE type
Inverte-brates	0 or +[a]	0	0	0 or +	0	0	0	0	0
Hagfish & Lampreys	+	0	?	+	+	0	+	0	0
Sharks & Rays	+	0	+	+	+	0	+	0	0
Bony fishes	+	+	+	+	+	0	+	0 or +	0
Amphibians	+	+	+	+	+	+	+	+	0
Reptiles	+	0	+	+	+	+	+	+	0
Birds	0	+	+	+	+	+	+	+	0 or + (?)
Mammals	0[b]	+	+	+	+	+	+	+	+

[a] Capacity present in many metazoan invertebrate phyla.
[b] Chronic or slow reactivity is characteristic only of selected strains or colonies of mice, rats and Syrian hamsters.

be expected to expand progressively in higher animals, but this assumption is unproved, mainly for lack of investigative effort.

The evolution of vertebrate immunoglobulins as such is gradually becoming better understood, although meaningful structure-function relationships remain conjectural. The ancestral immunoglobulin gene possibly coded for a polypeptide of about 110 amino acids. By gene duplication, diversification, and adaptive selection of the resulting antibodies, the IgM-type found in all vertebrates had already evolved in hagfish and lampreys several hundred million years ago (Fig. 1). Why this most elaborate of antibodies should be primitive is still puzzling, although its polymeric structure associated with 5–10 antigen-binding sites surely increases the probability of encountering antigen. Other molecular classes of antibodies, first evident in the transitional group of advanced bony fishes, may reflect adaptive homeostatic regulation of immune responsiveness (Table 1 and Fig. 1). IgG-type antibodies are known to predominate in prompt secondary responses and to be responsible for specific feedback immunoregulation (7).

A current overview of mammalian immune responses is illustrated in Figure 2. Two major pathways determining cellular (T-lymphocyte) and humoral (B-lymphocyte) immunity may be distinguished. Cell-mediated immunity depends on thymus-derived small lymphocytes which may produce a variety of non-antibody proteins as effector molecules. Humoral immunity depends on bone marrow-derived lymphocytes and plasma cells which secrete immunoglobulin antibodies of multiple molecular classes. The principal antibody classes of mammals (i.e. IgM, IgG, and IgA) exhibit striking differences in carbohydrate content, of unknown function, associated with the heavy (H-) chains. Functional antibody units consist of two H- and light (L-) polypeptide chains joined by disulfide bands, plus carbohydrate. About half of the amino acids of L chains

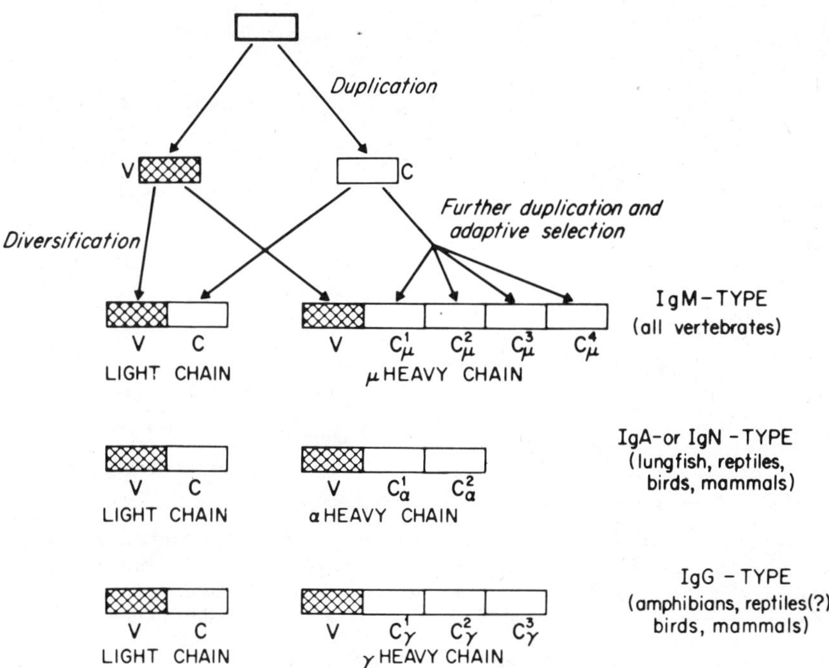

Figure 1 Genetic scheme for origin of vertebrate immunoglobulins and molecular representation of heavy and light polypeptide chains. The lengths of the chains are proportional to the carbohydrate-free molecular weight. The basic unit is a homology of approximately 110 amino acids for which the ancestral gene presumably coded. The variable region (V) of each chain is placed at the amino terminal end of the molecule. C = constant regions. The particular class of C region homology units is given by the subscript. The superscript numeral gives the position. Adapted from Marchalonis & Cone (5). See also Clem & Leslie (6) for further discussion.

and a larger fraction of H chains are constant or invariant. The remainder of both H and L chains, which include antibody combining sites, are highly variable in their amino acid sequences (Fig. 1) (8). At least several amino acids comprise an antibody combining site, but the critical number responsible for complementariness toward an antigen is unknown. At least five structural genes and perhaps many more are required for synthesis of one antibody molecule.

Other naturally occurring molecules such as complement, lysozyme, and histamine can serve as amplifying mechanisms once antigen-antibody reactions have occurred. Immunologic memory resides in separate, long-lived populations of both T-cells and B-cells, which may respond promptly by differentiation and proliferation to initiate a heightened secondary response on renewed contact

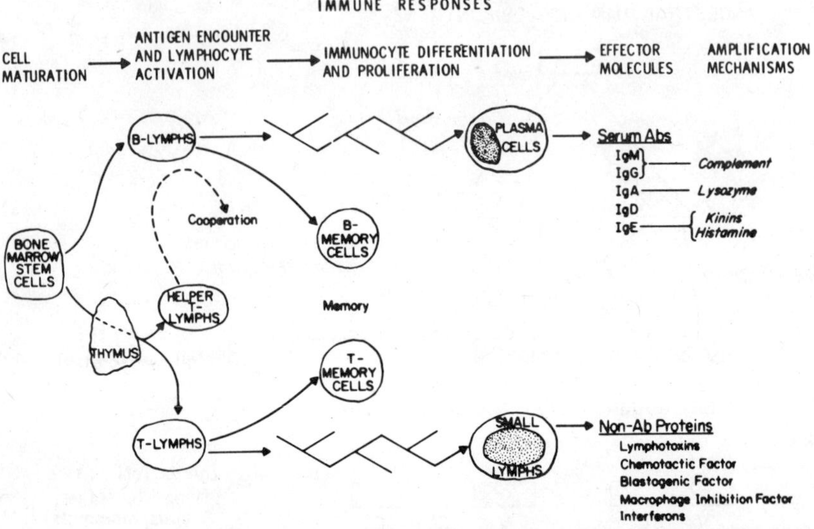

Figure 2 Pathways of mammalian immune responses. Cell-mediated immunity is associated with thymus-derived small lymphocytes which produce nonantibody proteins as effector molecules. Humoral immunity is associated with bone marrow-derived lymphocytes and plasma cells that produce immunoglobulin antibodies of multiple molecular classes.

with a given antigen. Many specific antibody responses by B-lymphocytes require the cooperation of helper T-lymphocytes or of thymosin, a thymic hormone (9). Neonatally thymectomized or congenitally thymusless mice are immunologically crippled (10). The thymus then is the master organ of immunologic responsiveness. Many details of the various immunologic events just summarized remain poorly understood. The most primitive fishes (hagfish and lampreys) have neither a thymus nor bone-marrow as such, but probably have equivalent cell sources because both T- and B-cell functions are present.

Given the well-developed immune capacities of all vertebrates, the early origins of immunocompetence must be sought among the invertebrates. Despite many information gaps, a new post-Metchnikoff conception of invertebrate potentialities is emerging as follows: Essential cellular (T-cell) immunocompetence evolved among metazoan invertebrates before the additional (B-cell) vertebrate capacity to produce immunoglobulin antibodies (2, 3, 11). Capacity for recognition of and reaction to foreign tissue evolved still earlier and is demonstrable even toward allografts in coelenterates (12). Specific cell surface recognition units and perhaps even primordial immunoglobulins may yet be found in invertebrates (e.g. echinoderms and tunicates) ancestrally close to the vertebrates. Diverse types of leukocytes, including small lymphocyte-like cells, are certainly present in advanced invertebrates (2, 11, 12). Research in immunologic phylogeny now appears quite promising and should lead to important new insights.

Immune Responses to Infectious Agents

Genetically determined host resistance to particular pathogens may often depend on single, autosomal, dominant or recessive genes. However, given the complex pathways of immune responses (Fig. 2), multiple genes are more commonly implicated in disease resistance, as might be expected. Resistance of the all-or-none type to a virulent pathogen is uncommon. In other words, a strain of animals highly resistant to a pathogen at low dosage will usually show increasing susceptibility at much higher dosages. To test inherited susceptibility in terms of host genotypes requires a uniform pathogen genotype and a uniform environment. Although immunogenetic studies of resistance to infectious agents necessarily involve complex antigens with multiple determinants, experimental use of inbred host strains and selectively cloned stocks of pathogens controls these major variables most desirably. In the now classic studies of Gowen (13), selectively inbred strains of mice were shown to possess a characteristic degree of resistance to a cloned strain of *Salmonella typhimurium*. The range was from nearly complete resistance to total susceptibility measured as percent of animals surviving infection. Moreover, the same host genotypes were capable in similar degrees of developing specific immunity induced by a killed vaccine of this *Salmonella* strain. Progeny tests revealed dominance of most of the multiple genes facilitating immunity. The immunologic superiority of females over males was also observed in this work. In many more recent studies, females have been found to mobilize both stronger antibody responses and stronger cell-mediated immunity than males (14-16).

The main genetic basis of host resistance toward various viruses and bacteria is summarized in Table 2. The decisive role of phagocytosis by the reticuloendothelial system in curtailing infections by Arbo B viruses and hepatitis virus is well established. The dominant resistance of mice to the antigenically similar, arthropod-borne (Arbo B) viruses responsible for encephalitides is enforced by the virucidal action of histiocytes and macrophages, whereas the recessive allele in phagocytic cells of susceptible mice allows virus multiplication and continuing invasion (17). Similarly, genetic susceptibility to hepatitis virus depends on the selective destruction of macrophages that normally prevent viremia. Crosses between resistant C3H mice and susceptible Pr mice gave F_1 progeny that were quite susceptible like the Pr strain. Susceptible progeny in the F_2 and backcross to C3H generations were sufficiently close to 75% and 50% proportions, respectively, to support the assumption of a single dominant gene for susceptibility. In these genetically "simple" situations, a single lysosomal enzyme is probably involved at an early stage of the virus-host relationship. Predictably, multiple genes are generally evident under other circumstances.

(C3H X C57BL/6)F_1 hybrid mice, like their C57BL/6 (H-2^b) parents, are resistant to leukomogenesis by the Gross Passage A virus, whereas C3H (H-2^k) animals are uniformly susceptible. Segregation of resistant and susceptible mice in backcross and F_2 progenies indicated that at least two independent loci are determinative (31). The observed proportions of resistant animals were 28% in

TABLE 2 GENETIC BASIS OF HOST RESISTANCE DETERMINED FOR PARTICULAR INFECTIOUS AGENTS

Pathogen	Host Species	Genetic Resistance (+)			References
		Single Dominant Gene	Recessive Gene	Polygenic	
Arbo B viruses	Mouse	+			17
Ectromelia virus	Mouse	+			18
Erythroblastosis virus	Chicken	+			19
Hepatitis virus	Mouse		+		20
Rous sarcoma virus	Chicken		+		21,22
Friend leukemia virus	Mouse			+	23,24
Gross leukemia virus	Mouse			+	25,26
Polyoma virus	Mouse			+	27,28
Mammary tumor virus	Mouse			+	29
Salmonella	Chicken			+	13
	Mouse			+	
Bacterial/ Viral Pneumonitis	Mouse			+	30

he backcross $(1/2)^n$ and 55% in the F_2 $(3/4)^n$, consistent with an $n = 2$ locus ıssumption. The incidence of leukemia among $H\text{-}2^k/H\text{-}2^k$ homozygotes was > 90% in both backcross and F_2 mice, while mice carrying the $H\text{-}2^b$ allele were nore resistant. Thus one locus (*Rgv-1*) determining virus susceptibility is closely ınked to *H-2*, the major gene complex governing histocompatibility antigens in mice (32). Failure of some 10% of $H\text{-}2^k$ $H\text{-}2^k$ segregants to develop leukemia within a ten-month observation period could mean that susceptibility is influenced by part of the *H-2* locus itself or that *Rgv-1* is located outside H-2 allowing about 10% recombination to occur.

At least three independently segregating gene loci in addition to the *H-2* locus must be invoked to account for the overall evidence of susceptibility to polyoma virus in neonatal F_1 hybrids and reciprocal backcross progeny of highly susceptible AKR/J and relatively resistant C57BL/6J strains of mice (27, 28). Multiple criteria were used to measure susceptibility to plaque-purified virus after neonatal injection: latent period before appearance of tumors, spectrum and multiplicity of tumors, incidence of runting/tumor development, and capacity of virus to multiply in cell cultures of various host genotypes with cytopathic effect.

High susceptibility of albino AKR mice to the runting syndrome was recessively inherited. Numerous runted mice died before tumors were detectable, but long-surviving runts always developed tumors. High but submaximal susceptibility of (AKR X C57BL/6) F_1 hybrids to tumorigenesis was found at both low and high virus dosages. Moreover, the recessive resistance of the C57BL/6 strain was substantially overcome by infection with polyoma virus in high dosage.

Much increased susceptibility to runting of albino compared with black progeny of the F_1 X AKR backcross was found at both virus dosages tested. Analysis of the segregation data showed that susceptibility to runting involved interaction of the albino gene, or a gene closely linked to it, and an independent recessive gene also of AKR origin. Moreover, the incomplete dominance ascribed to the two genes regulating susceptibility to tumorigenesis and resistance to runting may well mean that other mouse genes operate in the development of polyoma virus-induced disease.

Complete susceptibility of cell cultures irrespective of host genotype to the cytopathic multiplication of polyoma virus indicated that host resistance did not hinge upon absence of vulnerable cells in vivo. Early maturation of immunologic capacity now appears to be the decisive component in the genetic resistance to malignant infection associated with the C57BL genotype (8). Resistance of C57BL mice to tumorigenesis is associated with restriction of tumors to the salivary glands, whereas multiple tumors developed with much greater frequency in AKR mice and their susceptible albino backcross progeny. Increased vulnerability to malignancy reflected either in runting or tumors was also marked by substantial reduction in the average latent period for tumor development. Among the diverse criteria of susceptibility evaluated, far more significant differences were revealed at low virus dosage. At very high virus dosage, genetic resistance to tumorigenesis could be almost entirely overcome.

The major "take-home lesson" from this type of study is the complexity of the immunogenetic basis of infectious disease susceptibility. Mutants of the infecting species are under continual selection pressure to evolve new surface antigen configurations which can circumvent acquired immunity in the host population to previously prevalent strains. Polygenic control of sequential host reactions to a pathogen is surely operative in nearly all instances. When single dominant or recessive genes appear to be decisive, as in macrophage capacity to kill invading Arbo B or hepatitis viruses, this is only because one critical step among many happens to distinguish the responses of the host genotypes studied. Indeed, use of congenic lines differing only by one allele at a known locus has the obvious advantage of avoiding uncontrolled polygenic effects. Murine genes now known to influence susceptibility to leukemia alone include Steel (1), *W*-dominant spotting, dilution (*d*), flexed-tail (*f*), hairless (*hr*), *Fv-1*, *Fv-2*, *AKv-1* as well as various histocompatibility (*H*) loci (24, 33-36). The list will doubtless grow in the future, but the task of determining where and how these genes act remains as formidable as ever. For example, mutant host genes can determine whether genes coding for RNA oncogene viral functions become expressed as cancer (33). Quite possibly, many host genes directly influence the immune response and thereby act also as immune response genes (37, 38).

Genetic Regulation of Immune Responses in Man

Primary or inborn immunodeficiency disorders in infants have recently revealed the genetic control of the mammalian immune pathways diagrammed in Figure 2. Severe combined immunodeficiencies preclude activation of either the T- or B-cell pathways (39). This type of defect, which can be either X-linked or

TABLE 3 SOME HEREDITARY DEFECTS ASSOCIATED WITH IMPAIRMENT OF IMMUNE PATHWAYS IN MAN[a]

Immunologic Disease	Cellular Immunity (T-Cell) Pathway	Serum Antibody (B-Cell) Pathway	Lymphoid Follicle and Germinal Center Development
Hereditary Thymic Aplasia (Autosomal recessive, Swiss-type agammaglobulinemia)	0	0	0
Sex-linked Hypogammaglobulinemia (Bruton-type)	+	0	0
Di George Syndrome (Thymic hypoplasia)	0	+	+

[a] (+) means normal or near normal function;
(0) means capacity absent or greatly impaired.

autosomal recessive, constitutes an early block preventing stem cell differentiation. The result is hereditary thymic aplasia associated with atrophy of all lymphoid tissues and a general failure of immunoglobulin production. In sex-linked hypogammaglobulinemia (Bruton-type), serum antibody synthesis is curtailed because B-cells are absent. Yet cell-mediated immunity dependent on thymus-derived lymphocytes is fully functional. Such patients develop delayed hypersensitivity, resist most viral infections, and reject foreign tissue transplants in a normal manner. By contrast, infants with the Di George Syndrome lack cellular immune competence, and are quite vulnerable to viral infections, but are capable of mounting serum antibody responses dependent upon an intact B-cell pathway. These three distinctive categories of hereditary immunodeficiencies are summarized in Table 3. Many other immunodeficiency disorders with diverse epistatic effects have come to light in recent years (39). Despite uncertainties in disease classification, separate T- and B-cell pathways clearly exist in man as well

TABLE 4 ASSOCIATION BETWEEN HL-A TYPE AND DISEASE SUSCEPTIBILITIES IN MAN

HL-A Antigens or Haplotype	Increased Frequency in Particular Disease	Reference
HL-A 2, 12	Acute Lymphoblastic Leukemia	(43,44)
HL-A 1,8		
HL-A 5	Hodgkin's Disease	(44-46)
	Infectious Mononucleosis	
HL-A 1,8	Childhood Asthma	(44)
HL-A 2	Chronic Glomerulonephritis	(47)
W15 (LND)	Systemic Lupus Erythematosus	(48,49)
HL-A3	Chronic Myeloid Leukemia	(50)
	Multiple Sclerosis	(51)

as other vertebrates. The genes involved control the differentiation of whole populations of lymphoid cells, but provide no insight concerning genes that regulate *specific* immune responses.

The HL-A system of nucleated cell surface antigens is highly polymorphic with some 20,000 genotypes and at least half as many phenotypes possible in human populations (40). Although this was long thought to be the major histocompatibility system in man because organ allografts between HL-A identical siblings are most compatible, recent evidence indicates that a closely linked *MLC* locus detectable by mixed lymphocyte culture reactivity (41) controls or actually reflects the intensity of allogeneic incompatibility. A quite similar association exists between an *MLC* locus and the *H-2* locus in mice (42). *MLC* then behaves as an *Ir* locus in the cell-mediated immunity pathway. However, diverse associations between HL-A type and increased frequency of particular diseases (Table 4) involving specific immune responses suggest that the *HL-A* chromosome region includes immune response genes. The general validity of the disease correlations indicated is still unsettled for two reasons: (*a*) genetic and environmental differences at the population levels studied, and (*b*) lumping together of etiologically different but clinically similar diseases (52). Various unidentified genes besides *HL-A* no doubt influence the occurrence of given diseases.

In Hodgkin's disease or infectious mononucleosis for example, the *HL-A5* allele may allow a defective immune response to a particular virus. However, in systemic lupus erythematosis and multiple sclerosis where both viruses and autoimmune reactions are implicated, the level of influence of *HL-A* genes is obscure. Perhaps the extreme polymorphism of the HL-A alloantigen system exists to defeat molecular mimicry by microorganisms or cancerous cells which might otherwise circumvent host surveillance against "non-self" constituents. Beyond such "big picture" generalization, we know little about genetic regulation of specific immune responses in man. An *Ir* gene specifically controlling HL-A associated IgE antibody responsiveness to an antigen E responsible for ragweed hay fever was recently reported (53). Again, whether an HL-A structural gene coding for cell surface antigens is also functioning as an *Ir* gene remains unknown. Knowledge of specific *Ir* genes is now rapidly accumulating from experimental animal studies.

Specific Immune Response Genes in Mice and Guinea Pigs

The genetic control of immune responses to chemically-defined antigens in diverse inbred strains has recently been extensively investigated by several groups. I shall not attempt to cover all this literature (54-56), but will focus on pathways and possible mechanisms. A thorough review of *H-2* associated immune responses to various antigens in mice has just been completed (57).

Given the cellular and molecular complexity of immune responses, the reader may well ponder the levels of experimental definition to be desired in *Ir* gene studies. First consider the genotypes of the experimental animals. An outbreeding population in Hardy-Weinberg equilibrium would of course reveal a broad range of responses to most antigens. Pairs of ordinary inbred strains with essentially homozygous backgrounds would still express multiple *Ir* gene differences. Selected coisogenic or congenic strains are most desirable because significant differences in any measured immune response can be associated with one known gene locus (55). Use of newly developed recombinant-inbred (RI) strains can also offer decisive advantages in identifying *Ir* gene associations (58). Even clones of given microorganisms or purified preparations of proteins or polysaccharides constitute complex antigens with many determinant groups, each evoking a separate immune response. Progressively greater antigenic definition is achievable with amino acid copolymers or with simple haptens conjugated to the responding animal's own macromolecules. Theoretically, single allelic mutational variants of known alloantigenic macromolecules in inbred strains would be ideal. Maximal definition of immune responses elicited may be achieved by (*a*) measuring purified or fractionated antibodies of given molecular classes by the most sensitive techniques available and (*b*) constant antigen dosage, route, schedule, and environment, each predetermined to provide maximal sensitivity with the tactics of immunization and testing employed. More detailed consideration of these major variables is given elsewhere (8). Ideal experimental definition as just described has yet to be realized in any published study.

Four *Ir* genes have been identified in guinea pigs, PLL, GA, GT, and BSA-1 regulating, respectively, immune responsiveness to hapten-poly-L-lysine conjugates (PLL), to random copolymers of L-glutamic acid and L-alanine (GA), to L-glutamic acid and L-tyrosine (GT), and to limiting doses of bovine serum albumin (BSA) (56, 59). Although the synthetic polypeptides employed have limited structural diversity, they are hardly simple, chemically-defined antigens. Indeed, GAT, which contains both GA and GT antigens determinants, elicits both cellular and humoral immune responses of similar intensities in all guinea pigs. Guinea pigs capable of responding to one hapten-polylysine conjugate can respond fully to polylysine conjugates of the other haptens. The haptens included dimethylaminonapthalenesulfonyl, dinitrophenyl, p-toluenesulfonyl, and benzylpenicilloyl molecules. Strong hypersensitivity of both the delayed and immediate Arthus types, and substantial production of specific hapten antibodies occurred in inbred strain 2 responders whereas nonresponder strain 13 animals showed no specific responsiveness when given immunizing injections in Freund's adjuvant plus repeated booster injections.

Reciprocal test-crosses among progeny of responder X nonresponder guinea pigs indicated that the ability to acquire immunity to dinitrophenyl-L-lysine conjugates is attributable to an autosomal dominant gene carried by strain 2. At least 7–8 lysyl residues in hapten conjugates are required for evocation of both delayed hypersensitivity reactions and for antibody production in responsive animals. However, nonresponder strain 13 pigs carrying the recessive allele of *PLL* can be induced to form antibodies (B-cell function) but not delayed hypersensitivity (T-cell function) against DNP-PLL if immunized with DNP-PLL complexed with immunogenic bovine serum albumin. Thus, DNP-PLL may

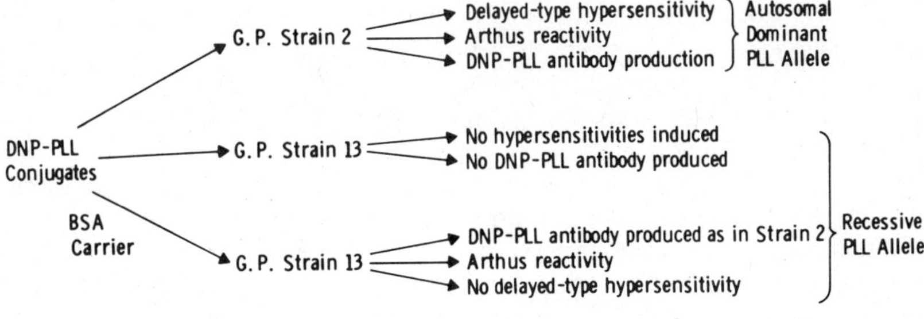

Figure 3 Responses of responder (strain 2) and nonresponder (strain 13) guinea pigs to DNP-PLL. Note that addition of DNP-PLL to immunogenic BSA carrier allows antibody responses but not development of delayed hypersensitivity in strain 13.

behave as a hapten in nonresponders to yield a normal antibody response provided B-cells are activated by an appropriate carrier (Fig. 3). The genetic block in nonresponders is clearly not at the level of specific immunoglobulin production. The recessive *PLL* allele cannot activate the entire T-cell pathway even though helper T-cell cooperation is apparently achieved by the macromolecular carrier (Fig. 4).

The genes controlling immune responsiveness to PLL, GA, and GT appear to be closely linked to each other and to the major histocompatibility locus of guinea pigs. Strain 2 carries the responder alleles of *PLL*, *GA*, and *BSA-1*, whereas strain 13 carries the *GT* responsive allele (56). These genes all appear to

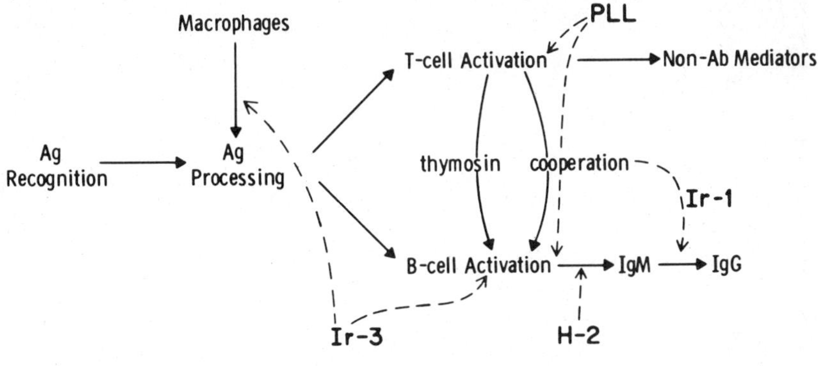

Figure 4 Major sequential steps and components of immune responsiveness. Probable levels of known *Ir* gene regulatory action are indicated by dashed arrows.

have similar immunoregulatory roles, which are connected in some unknown way with major histocompatibility gene function. De Weck (60) postulates that PLL-type genes are alleles of structural genes for cell membrane surface receptors involved in carrier recognition; in this view, the *PLL* gene should determine "some histocompatibility antigen particularly attractive for PLL."

At least 5 *Ir*-loci (*Ir-1*, *Ir-2*, *Ir-3*, *H-2*, and an X-linked gene) have been identified in mice based on experiments with inbred strains and immunogens presumed to possess a limited range of antigenic determinants. In one well-defined investigation, mice from each of 5 strains representing congenic pairs differing at the *H-2* or *H-3* loci and their hybrid progenies were immunized with various haptens coupled to serum albumin obtained from the same strain background (55). Thus the protein carrier itself was not foreign to the responding mice and these animals in turn were identical except for single known differences in histocompatibility alleles. Antibody responses to the 2, 4, 6-trinitrophenyl hapten (TNP) were found to be decisively regulated by products of alleles of the complex *H-2* locus. Low antibody responsiveness was determined by the $H\text{-}2^a$ or a closely linked *Ir* allele as a dominant trait, whereas homozygosity for the $H\text{-}2^b$ allele consistently led to high responsiveness even on diverse strain backgrounds. This study suggests *H-2* control of the *quantity* of anti-TNP produced and therefore a regulatory role in the B-cell pathway (Fig. 4). Possible thymus dependence and differential effects on IgM versus IgG classes invite further study.

The immune responses to at least 6 synthetic polypeptide polymers (GAT^{10}, GAT^4, GL, GLA^5, GAL^{10}, GLT) are regulated by genes near or within the *H-2* region (58). Two separate *H-2* linked Ir genes designated *Ir-IgG* and *Ir-IgA* are reported to control the alloimmune responses to IgG(γ 2a) and IgA myeloma proteins, respectively (61). Although the now extensively studied *Ir-1* region maps between the H-2K and H-2D regions of the *H-2* complex on chromosome 17 (62), the *Ir-2* and *Ir-3* loci have quite separate chromosomal locations. *Ir-2*, which controls the antibody response to an erythrocyte alloantigen, is near the *agouti* and *H-3* loci on chromosome 2 (63). *Ir-3* governing the response to one synthetic polypeptide (Pro-L) is not linked to H-2 (54). The IgM antibody response to Type III pneumococcal polysaccharide is determined in an almost all-or-none manner by an X-linked gene (64). Many *Ir*-genes then may bear no essential relationship to major *H*-loci.

The effect of the *H-2* associated *Ir-1* locus in mice on antibody responses to synthetic amino acid polymers has recently been well explored and provides several new insights. C57BL ($H\text{-}2^b$) mice give high antibody responses to [(T, G)-A--L] and low antibody responses to [(H, G)-A--L]. CBA ($H\text{-}2^k$) mice respond in the opposite directions, poorly to (T, G)-A--L and well to (H, G)-A--L. The difference in antigenic specificity resides only in the substitution of histidine for tyrosine in the polymer. F_1 hybrids ($H\text{-}2^b/H\text{-}2^k$) respond well to both antigens while Fl X CBA backcross progeny show high and low bimodal distributions of antibody responses. The congenic strain pair C3H/Di Sn ($H\text{-}2^k$) and C3H.SW($H\text{-}2^b$) respond in the same manner as a function of the *H-2* allele involved, thereby confirming that the relevant *Ir-1* genes are indeed *H-2* associated (56, 65).

When (T, G)-A--L was initially administered in complete Freund's adjuvant, both H-2^b and H-2^k strains slowly produced (T, G)-A--L antibodies with H-2^b strain antibody levels only slightly greater than those of H-2^k strains. However, antibody titers of H-2^b mice rose rapidly while those of H-2^k mice remained constant after a booster injection of (T, G)-A--L in saline. Fractionation of the antibodies evoked reveals that both H-2 (Ir-1) types yielded equally good IgM (T, G)-A--L responses, but only the high responder strains were able to switchover to IgG antibody production after repeated immunization (65). However, (T, G)-A--L complexed to a protein carrier in the form of methylated bovine serum albumin (MBSA) in the presence of Freund's adjuvant will elicit IgG anti-(T, G)-A--L in otherwise low responders equal to that obtainable in high responders.

Neonatal thymectomy effectively prevents the IgG response to (T, G)-A--L in high responder mice. Adult thymectomy augmented by whole body radiation and bone marrow infusion did not affect the primary IgM response of either high or low responder strains, but did preclude secondary IgG responsiveness (66). Adult thymectomy also blocked the high response to (MBSA)-(T, G)-A--L in low responder mice. The *Ir-1* gene effect then is thymus dependent, apparently at the level of T- and B-cell cooperation which is required for secondary IgG antibody production (Fig. 4).

Mechanism of Ir Gene Action

The immune response genes of mice and guinea pigs just described might better be regarded as *immunoregulatory* genes because none of them determines immunospecificity as such. In other words, no structural genes for antibody combining sites or equivalent T-cell receptors have yet been identified. Subtle effects of *Ir* genes on the specificity of antibody populations have been reported (56), but the available evidence is equivocal. Unconditional all-or-none production of a specific antibody controlled beyond the step of antigen processing could identify structural *Ir* genes of surpassing interest. Such genes might specify amino acid sequence in the variable portions of immunoglobulin H- or L-chains, but effects on constant regions or the carbohydrate portion of an immunoglobulin subclass could also be detectable. The antibody response to dextran is reported to be regulated by a locus linked to the H-chain locus in mice (67). The T-cell pathway reactions to PLL, GA, and GT in guinea pigs appear to be totally under the control of the corresponding genes. Yet the structural antibody genes in these systems are clearly separate, as the B-cell pathway can be turned on regardless of the *Ir* allele present. The characteristics of the *Ir-1/H-2* complex in relation to the synthetic polypeptide (T, G)-A--L may be summarized as follows:

(*a*) *Ir-1* is an autosomal locus within the *H-2* complex.

(*b*) Both high and low responder genotypes give equal IgM responses.

(*c*) Only high responder H-2^b strains produce specific IgG antibodies but low responder H-2^k strains will do so if the antigen is bound to carrier protein.

(*d*) Ir-1 affects specific immunologic memory.

(*e*) T-helper cells are required for high anti-(T, G)-A--L responses.

The blocks in antibody synthesis regulated by the recessive *Ir-1* allele can clearly be circumvented. *Ir-3* gene regulated low responsiveness to Pro--L in (T, G)-Pro-L and (Phe,G)-Pro-L polypeptides is even less decisive than Ir-1 effects, as administration of nonspecific poly A-poly U or isogenic peritoneal macrophages alone is able to convert low responder DBA/1 mice into high responders (68). Use of MBSA carrier protein similarly augmented antibody production. This evidence points toward an effect on antigen processing at the macrophage level. Moreover, limiting dilutions of thymus and bone marrow cells in the presence of excess cells of the complementary type indicated much lower numbers of (T, G)-Pro--L-specific precursors in DBA/1 marrow than in SJL (high responder) marrow. No differences were referable to the strain source of the thymocytes. *Ir-3* then affects the level of specific B-cell activation (Fig. 4).

At least five levels of *Ir*-gene action are now identifiable. First, there are the genes that control early immunocyte differentiation as shown by hereditary impairment of major immune pathways in man (Table 3). Second, there is the class of genes that controls lysosome development or enyzme functions in macrophages. The effects of these genes may range from decisive as in Arbo B, ectromelia or hepatitis virus susceptibility to slight as in Ir-3 functions. Diminished phagocytic or antigen processing capacity is also known to result in impaired immune responsiveness in familial neutropenia and in the Chediak-Higashi anomaly in mammals including man (8). A specific action of the mutant beige (*bg*) and satin (*sa*) genes in mice increases susceptibility to progressive pneumonitis (30). Beige mice show giant lysosomal granules in their leukocytes and this is probably the basis of the immunogenetic lesion. The third and fourth levels include genes such as *PLL* or *Ir-3* which may separately control the main T- and B-cell pathways. Finally, there is *Ir-1* operating at the level of T- and B-cell cooperation by affecting both T-helper cell function and B-cell memory (Figures 2 and 4). These few *Ir* genes proffer only a beginning understanding; many information gaps remain. One may reserve judgment about the evolutionary significance of genes that are activated by synthetic polymers not found in Nature. The conclusion that genes at many loci independently regulate different antibody responses could have been drawn more than a decade ago as a result of Sobey's pioneering studies (69) with complex but naturally occurring antigens in random-bred stocks of rabbits and mice.

The identity of the various *Ir* gene products remains an open question. Association of many different *Ir* functions with *H-2* in mice and with *HL-A* haplotypes in man (Table 4) could mean that cell surface histocompatibility antigens also represent *Ir* gene products. Although such cell surface specificities could well influence susceptibility to infections and especially viral agents as already noted, other *Ir* gene effects appear far removed from this level of antigen recognition. Given abundant evidence of *Ir* operation at numerous biosynthetic and regulatory levels, one can at least be sure that diverse mechanisms are involved. The serum complement system of nonspecific immune amplification characteristic of vertebrates is apparently less important than generally believed, because few genetic deficiencies in complement components have any deleterious effect on overall immunocompetence (70). Personally, I retain a considerable skepticism about the now elaborate lore of complementology. Complement deficiencies deduced from reactions involving antibodies obtained from a different species may represent artificial rather than naturally-occurring entities. Since many xenogeneic combinations of antibodies and complement are functionally incompatible, the skeptical immunogeneticist may well demand that particular complements or components thereof be characterized in relation to antibodies derived from the same species or even from similar genotypes.

Literature Cited

1. Hutt, F.B. 1958. *Genetic resistance to disease in domestic animals*, Comstock: Ithaca, N. Y. 198 pp.
2. Hildemann, W.H. 1972. Phylogeny of transplantation reactivity. Eds. B.D. Kahan, R. A. Reisfeld, 3-73. *Transplantation Antigens*, Academic: N.Y. 538 pp.
3. Hildemann, W.H., Cooper, E. L., eds. 1970. Proc. Symp. phylogeny of transplantation reactions. *Transplant. Proc.* 2:179-341
4. Kampmeier, P.F. 1969. *Evolution and comparative morphology of the lymphatic system.* Charles C. Thomas: Springfield, Ill. 620 pp.
5. Marchalonis, J.J., Cone, R. E. 1973. The phylogenetic emergence of vertebrate immunity. *Aust. J. Exp. Biol. Med. Sci.* In press
6. Clem, L.W., Leslie, G. A. 1969. Phylogeny of immunoglobulin structure and function, 62-88. *Immunology & Development*, ed. M. Adinolfi. The Lavenham Press: England, 187 pp.
7. Hildemann, W.H., Mullen, Y. 1973. The weaker the histoincompatibility, the greater the effectiveness of specific immunoblocking antibodies: A new immunogenetic rule of transplantation. *Transplant. Proc.* 5:617-20
8. Hildemann, W.H. 1970. *Immunogenetics*. Holden-Day: San Francisco, 262 pp.
9. Bach, J.F., Dardenne, M. 1972. Thymus dependency of rosette-forming cells: Evidence for a circulating thymic hormone. *Transplant Proc.* 4:345-50
10. Aden, P.D., Reed, N.D., Jutila, J.W. 1972. Reconstitution of the *in vitro* immune response of congenitally thymusless (nude) mice. *Proc. Soc. Exp. Biol. Med.* 140:548-52
11. Hildemann, W.H., Dix, T.G. 1972. Transplantation reactions of tropical Australian echinoderms. *Transplantation* 15:624-33
12. Hildemann, W.H., Dix, T.G., Collins, J.D. 1973. Tissue transplantation in diverse marine invertebrates. *Current Topics in Immunobiology* vol. 4, Plenum Press: N.Y. In press
13. Gowen, J.W. 1963. Genetics of infectious diseases, 383-404. *Methodology in Mammalian Genetics*, ed. W.J. Burdette, Holden-Day: San Francisco, 646 pp.
14. Hildemann, W.H. 1970. Components and concepts of antigenic strength. *Transplant. Rev.* 3:5-21
15. Ash, L.R. 1971. Preferential susceptibility of male Jirds (*Meriones unguiculatus*) to infection with *Brugia pahangi. J. Parasitol.* 57:777-80
16. Weimer, H.E., Roberts, D.M. 1972. Sex differences in the immunologic and phlogistic responses to *Salmonella typhosa* H antigen in the albino rat. *Comp. Biochem. Physiol.* 41 B:713-22
17. Goodman, G.T., Koprowski, H. 1962. Study of the mechanism of innate resistance to virus infection. *J. Cell. Comp. Physiol.* 59:333-73
18. Roberts, R.A. 1964. Growth of virulent and attenuated ectromelia virus in cultured macrophages from normal

and ectromelia-immune mice. *J. Immunol.* 92:837-42

19. Waters, N.F., Burmester, B.R, Walter, W.G. 1958. Genetics of experimentally induced erythroblastosis in chickens. *J. Nat. Cancer Inst.* 20:1245-56
20. Gallily, R., Warwick, A., Bang, F.B. 1967. Ontogeny of macrophage resistance to mouse hepatitis in vitro and in vivo. *J. Exp. Med.* 125:537-48
21. Nelson, F.W., Burmester, B.R. 1961. Mode of inheritance of resistance to Rous sarcoma virus in chickens. *J. Nat. Cancer Inst.* 27:655-61
22. Bower, R.K., Gyles, N.R., Brown, C.J. 1964. The number of genes controlling the response of chick embryo chorioallantoic membranes to tumor induction by Rous sarcoma virus. *Genetics* 51:739-46
23. Lilly, F. 1970. Fv-2: Identification and location of a second gene governing the spleen response to Friend leukemia virus in mice. *J. Nat. Cancer Inst.* 45:163-69
24. Rowe, W.P. 1972. Studies of genetic transmission of murine leukemia virus by AKR mice. I. Crosses with *Fv -1*^{n} strains of mice. *J. Exp. Med.* 136:1272-85
25. Lilly, F. 1971. The influence of H-2 type on Gross virus leukemogenesis in mice. *Transplant. Proc.* 3:1239-50
26. Gottlieb, C.F., Perkins, E.H., Makinodan, T. 1972. Genetic regulation of the thymus dependent humoral immune response in leukemia prone AKR (*H-2*^{k}) and nonleukemic C3H (*H-2*^{k}) mice. Description of genetic control of the immune response at the level of proliferation. *J. Immunol.* 5:974-81
27. Chang, S.S., Hildemann, W.H. 1964. Inheritance of susceptibility to polyoma virus in mice. *J. Nat. Cancer Inst.* 33:303-13
28. Chang, S.S., Hildemann, W.H., Rasmussen, A.F., Jr. 1968. Immunogenetic aspects of polyoma virus susceptibility in mice. *J. Nat. Cancer Inst.* 40:363-75
29. Nandi, S., Haslam, S., Helmich, C. 1971. Inheritance of susceptibility to erythrocyte-borne Bittner virus in mice. *Transplant. Proc.* 3:1251-57
30. Lane, P.W., Murphy, E.D. 1972. Susceptibility to spontaneous pneumonitis in an inbred strain of beige and satin mice. *Genetics* 72:451-60
31. Lilly, F. 1966. The inheritance of susceptibility to the Gross leukemia virus in mice. *Genetics* 53:529-39
32. Klein, J., Shreffler, D.C. 1971. The H-2 model for the major histocompatibility systems. *Transplant. Rev.* 6:3-29
33. Meier, H., Huebner, R.J. 1971. Host-gene control of C-type tumor virus-expression and tumorigenesis: relevance of studies in inbred mice to cancer in man and other species. *Proc. Nat. Acad. Sci.* 68:2664-68
34. Rowe, W.P., Hartley, J.W., Bremner, T. 1972. Genetic mapping of a murine leukemia virus-inducing locus of AKR mice. *Science* 178:860–62
35. Tennant, J.R., Snell, G.D. 1968. The H-2 locus and viral leukemogenesis as studied in congenic strains of mice. *J. Nat. Cancer Inst.* 41:597-604
36. Taylor, B.A., Meier, H., Myers, D.D. 1971. Host-gene control of C-type RNA tumor virus: Inheritance of the group-specific antigen of murine leukemia virus. *Proc. Nat. Acad. Sci.* 68:3190-94
37. Hildemann, W.H. 1970. The immunogenetics triangle revisited. In *Cellular Interactions in the Immune Response.* 2nd Int. Convoc. Immunol., Buffalo, N. Y., 109-13
38. Lieberman, R., et al. 1972. Genetic factors controlling anti-sheep erythrocyte antibody response and immunoglobulin synthesis in backcross and F_2 progeny of mice genetically selected for "high" and "low" antibody synthesis. *J. Exp. Med.* 136:790-98
39. Fudenberg, R., et al. 1971. Primary immunodeficiencies. Rep. WHO Comm. *Pediatrics* 47:927-46
40. Dausset, J. 1971. The polymorphism of the HL-A system. *Transplant. Proc.* 3:1139-46
41. Ceppellini, R. 1971. Old and new facts and speculations about transplantation antigens in man. *Progress in Immunology*, Academic: N. Y. 973-1025
42. Bach, F.H., Widmer, M.B., Segall, M., Klein, J. 1972. Genetic and immunological complexity of major histocompatibility regions. *Science* 176:1024-27
43. Walford, R., Finkelstein, S., Neerhout, R., Konrad, P. Shanbrom, E. 1970. Acute childhood leukemia in relation to HL-A human transplantation genes. *Nature* 225:461–62
44. Thorsby, E., Engerset, A. and Lie, S.O. 1971. HL-A antigens and susceptibility to diseases. A study of patients with acute lymphoblastic leukemia, Hodgkin's disease, and childhood asthma. *Tissue Antigens* 1:147-52

45. Morris, P.J., Forbes, J.F. 1971. HL-A in follicular lymphoma, reticulum cell sarcoma, lymphosarcoma, and infectious mononucleosis. *Transplant. Proc.* 3:1315-16
46. McDevitt, H.O., Bodmer, WF. 1972. Histocompatibility antigens, immune responsiveness and susceptibility to disease. *Am. J. Med.* 52:1-8
47. Mickey, M.R., Kreisler, M., Terasaki, P.I. 1970. Leukocyte antigens and disease. II. Alterations in frequencies of haplotypes associated with chronic glomerulonephritis. *Histocompatibility Testing 1970*, 237-42
48. Grumet, F.C., Coukell, A., Bodmer, J.G., Bodmer, W.F., McDevitt, H.O. 1971. Histocompatibility antigens associated with systemic lupus erythematosus. *New Engl. J. Med.* 285:193-96
49. Waters, H., Konrad, P., Walford, R.L. 1971. The distribution of HL-A histocompatibility factors and genes in patients with systemic lupus erythematosus. *Tissue Antigens* 1:68-73
50. Degos, L., Drolet, Y., Dausset, J. 1971. HL-A antigens in chronic myeloid leukemia (CML) and chronic lymphoid leukemia (CLL). *Transplant. Proc.* 3:1309-12
51. Naito, S., Namerov, N., Mickey, M.R., Terasaki, P. I. 1972. Multiple sclerosis: Association with HL-A3. *Tissue Antigens* 2:1-4
52. Walford, R.L., Smith, G. S., Waters, H. 1971. Histocompatibility systems and disease states with particular reference to cancer. *Transplant. Rev.* 7:78-111
53. Levine, B.B., Stember, R.H., Fotino, M. 1972. Ragweed hay fever: Genetic control and linkage to HL-A haplotypes. *Science* 178:1201-03
54. McDevitt, H.O, Benacerraf, B. 1969. Genetic control of specific immune responses, 31-74. *Advan. Immunol.*, Vol. 11, eds. F. J. Dixon, H. G. Kunkel, Academic: N. Y. 371 pp.
55. Rathbun, W.E., Hildemann, W.H. 1970. Genetic control of the antibody response to simple haptens in congenic strains of mice. *J. Immunol.* 105:98-107
56. Benacerraf, B., McDevitt, H.O. 1972. Histocompatibility-linked immune response genes. *Science* 175:273-79
57. Klein, J. 1973. Immunology and genetics of the *H-2* system. Springer-Verlag: N. Y. In preparation
58. Merryman, C. F., Maurer, P. H., Bailey, D. W. 1972. Genetic control of immune response in mice to a glutamic acid, lysine, phenylalanine copolymer. III. Use of recombinant inbred strains of mice to establish association of immune response genes with *H-2* genotype. *J. Immunol.* 108:937-40
59. Bluestein, H. G., Green, I., Maurer, P. H., Benacerraf, B. 1972. Specific immune response genes of the guinea pig. V. Influence of the GA and GT immune response genes on the specificity of cellular and humoral immune responses to a terpolymer of L-glutamic acid, L-alanine, and L-tyrosine. *J. Exp. Med.* 135:98-109
60. De Weck, A. L. 1972. Molecular models for induction of the immune response and their relationship to the genetic control of histocompatibility antigens. *Transplant. Rev.* 10:3-35
61. Lieberman, R., Paul, W. E., Humphrey, W., Jr., Stimpfling, J. H. 1972. H-2-linked immune response (Ir) genes. Independent loci for *Ir-IgG* and *Ir-IgA* genes. *J. Exp. Med.* 136:1231-40
62. McDevitt, H. O., et al. 1972. Genetic control of the immune response. Mapping of the *Ir-1* locus. *J. Exp. Med.* 135:1259-78
63. Gasser, D. L. 1969. Genetic control of the immune response in mice. I. Segregation data and localization of the fifth linkage group of a gene affecting antibody production. *J. Immunol.* 103:66-70
64. Amsbaugh, D. F., et al. 1972. Genetic control of the antibody response to type III pneumococcal polysaccharide in mice. I. Evidence that an X-linked gene plays a decisive role in determining responsiveness. *J. Exp. Med.* 136:931-49
65. Grumet, F. C. 1972. Genetic control of the immune response. A selective defect in immunologic (IgG) memory in nonresponder mice. *J. Exp. Med.* 135:110-25
66. Mitchell, G. F., Grumet, F. C., McDevitt, H. O. 1972. Genetic control of the immune response. The effect of thymectomy on the primary and secondary antibody response of mice to Poly-L(Tyr,Glu)-Poly-D,L-Ala--Poly-L-Lys. *J. Exp. Med.* 135:126-35
67. Blomberg, B., Geckeler, W.R., Weigert, M. 1972. Genetics of the antibody response to dextran in mice. *Science* 177:178-80
68. Shearer, G.M., Mozes, E., Sela, M. 1972. Contribution of different cell types to the genetic control of immune

responses as a function of the chemical nature of the polymeric side chains (Poly-L-Prolyl and Poly-DL-Alanyl) of synthetic immunogens. *J. Exp. Med.* 135:1009-27

69. Adams, K.M., Sobey, W.R. 1961. Inheritance of antibody response. V. Correlated antibody responses to various related and unrelated antigens. *Aust. J. Biol. Sci.*, 14:594-97

70. Alper, C.A., Rosen, F.S. 1971. Genetic aspects of the complement system, 252-90. *Advan. Immunol.*, vol. *14*, eds. F. J. Dixon, H. D. Kunkel, Academic: N. Y. 377 pp.

BIOCHEMICAL CONTROLS OF MEIOSIS 3046

Herbert Stern and Yasuo Hotta

Department of Biology, University of California, San Diego, California

INTRODUCTION

This review is addressed to the biochemical events that underlie the early phases of the meiotic cycle. Late activities such as chromosome disjunction and special processes involved in gametogenesis are excluded from this article. Of primary concern are the prophase stages of meiosis during which homologous chromosomes pair and presumably undergo crossing-over. These two events are fundamental to genetic recombination and constitute a major and universal feature of meiosis. In reviewing the field from the standpoint of biochemical mechanisms we have one general concern which is best to state at the outset. The sources of evidence on biochemical activities during meiosis are few. Biochemical analyses of meiosis in yeast are just beginning and will not be reviewed here (20). The bulk of our information comes from analyses of liliaceous plants, but their study has been pursued in very few laboratories. The credibility of the conclusions drawn from biochemical studies depends to a large extent upon the degree to which they are consistent with the information provided by genetics and cytology. Genetic approaches to intragenic recombination in yeast were surveyed two years ago (20). The cytology and fine structure of meiotic cells from various sources have been fully reviewed in the previous volume (68) and genetic aspects of meiosis will be discussed in the succeeding one.

Formulation of biochemical questions concerning the process of crossing-over is guided by genetic data on transmission of traits and by cytological data on pairing and chiasma formation. These classical disciplines continue to define the course of meiotic analyses. Additional pointers to possible mechanisms are provided by studies on the fine structure of the synaptonemal complex. The simplest picture of meiosis that may be drawn from these disciplines and that can serve as a basis for biochemical investigation is that chromosomes synapse during the zygotene stage and undergo crossing-over during the pachytene stage.

This picture, though highly attractive because of its apparent simplicity, has been challenged by various investigators (22, 23, 45). Indeed, the simple and direct relationship between synapsis and crossing-over is unproven even though it is generally favored (27, 43, 56, 57).

No incontrovertible evidence yet exists for the occurrence of crossing-over during the pachytene stage and no demonstration is yet available that zygotene synapsis is an essential condition for crossing-over. The controversies surrounding the relationship of synapsis to crossing-over and/or disjunction are rather lively and have been competently presented elsewhere (22, 27). No attempt will be made in this review directly to challenge the substance of these controversies. Instead, the reader will be put in company with appreciable circumstantial evidence which supports the simple tie between synapsis and crossing-over. In writing this article, we have favored the view that crossing-over between homologous chromosomes follows and is dependent upon chromosome synapsis. The data presented may indeed have considerable bearing on the correctness of the underlying proposition.

SYNAPSIS

The Cytological Framework

The first functionally significant and microscopically observable event of meiosis is the synapsis of chromosomes. In general, pairing between homologs occurs with a high degree of specificity. In extensively studied species such as corn, it is a matter of record that inversions within a chromosome or translocations between chromosomes affect the pattern of pairing in such a way as to maximize pairing between homologous segments and minimize association between nonhomologous ones (57). There are conditions under which nonhomologous pairing may occur, but these conditions will not be discussed in this review (47). In our opinion, the precise nature of such pairing is not yet sufficiently understood to help in our analysis of the normal situation. There are excellent discussions which place considerable emphasis on nonhomologous pairing (22, 59). We wish to emphasize that normally, synapsis is highly specific even with respect to relatively small stretches of chromosome. The fact that homologs behave in this way adds to the burden of accounting biochemically for the deceptively simple presentation of cytological data.

The morphology of paired chromosomes, whether examined by light or electron microscopy, excludes the possibility that synapsis represents a total gene-gene alignment between pairs of homologous chromosomes. When chromosomes begin to pair at the zygotene stage they are already considerably condensed and although no precise figure is available, it may be estimated that less than 1% of the DNA could be located at the periphery of the chromosome and thus available for interaction with its homologous counterpart (68). It is virtually impossible to account for the longitudinal specificity of homolog pairing by requiring total gene for gene matching. If prior chromosome condensation excludes such matching, then specificity of pairing must reside in special regions of the chromosomes. These hypothetical regions would remain exposed in the partially condensed chromosomes and so function as matching sites for synaptic alignment. Prezygotene condensation would be a highly ordered process in which each chromosome consistently folds in such a way that corresponding DNA regions (or derivatives of these) are exposed in each member of a homologous set. If prezygotene condensation were loosely organized so that its pattern in any particular chromosome were variable, it would be impossible to account for synapsis by an apposition of specialized DNA regions. If presynaptic contraction of chromosomes is a real event, as it appears to be, both homologs must undergo identical patterns of condensation if subsequent synapsis is in fact gene specific.

The presynaptic events in question begin no later than the leptotene stage. The lateral elements that appear at leptonema represent a major modification of chromosomal structure. Since the leptotene chromosome is also partly condensed, it is in this chromosome that regions must already be positioned so as to function later in aligning homologous segments. Regardless of the particulars of the models used to account for synapsis, if the prima facie evidence for chromosome condensation prior to pairing is accepted, the relatively small portion of a chromosome that remains exposed for homologous matching must not only correspond to its counterpart in the homolog but must in some way reflect the overall linear order of genes in that chromosome. The hypothetical regions, if they do in fact serve as specialized synaptic stretches, must be distributed more or less uniformly along the entire length of each chromosome, and each of the stretches must display a matching specificity.

Homologous sites need to be juxtaposed in order to facilitate chromosome interaction for synapsis. Just how juxtaposition is achieved is a question of major interest, which has been a prime target for biochemical analysis even though we are ignorant of the nature of synaptic sites. Furthermore, the process of alignment as such should be distinguished from events concerned with the stabilization of chromosome pairs once synapsis has been effected. Presumably, the synaptonemal complex functions as the stabilizing structure. The biochemical composition of the complex still awaits clarification (51, 68). Such clarification must account not only for stabilization but also for the rapid destabilization that occurs at the end of pachytene. From a biochemical standpoint the synaptonemal complex, although functioning as a stabilizer of synapsis, is itself a metastable structure, which appears to be shielded from degradation during zygotene and most of pachytene, after which it disperses except in those cases where it acquires novel post-meiotic functions (18).

Biochemical Descriptions

PREZYGOTENE STAGES Once cells have entered leptotene they do not revert to a mitotic division even if they are subjected to external conditions that have strong effects on chromosome behavior during subsequent stages of meiosis (66). The apparently irreversible commitment of the leptotene chromosome to meiosis may well be related to the process of presynaptic folding discussed earlier. Whether this stage or the preceding S-phase should be designated as the beginning of events associated with synapsis is difficult—and perhaps unnecessary—to determine (67). There is ample experimental evidence for affecting the degree of synapsis and/or the frequency of crossing-over by perturbing cells during premeiosis or perhaps even leptonema (66). Up to the present, however, very few biochemical experiments have been reported that throw any light on the relationships between these perturbations and their subsequent effects on chiasma formation. Moreover, little is known of those biochemical events that occur during leptonema and are associated with the presynaptic modification of chromosomes.

The first distinguishable biochemical difference between the meiotic and mitotic cycles is in the duration of S-phase. In spermatocytes of *Triturus* the premeiotic S-phase is about three times as long as the premitotic one (6). This is also the case for mouse spermatocytes (40) and for lily meiocytes (our unpublished results). A relatively long S-phase for meiotic cells may well be general and, if so, such prolongation must have special significance to meiosis. Analyses of DNA fiber autoradiographs from *Triturus* spermatocytes indicate that the respective rates of DNA synthesis in meiotic and somatic cells are very similar. The extended premeiotic S-phase would appear to be due to a much reduced number of initiation points in the meiotic nucleus. The relationship of a reduction in the number of sites at which replication is initiated to the requirements for synapsis and/or crossing-over is not yet evident from these studies. The view that sites of crossing-over in *Chlamydomonas* are determined by incompletely replicated chromosomal regions remaining from premeiotic S-phase would fit well with the observation of extended premeiotic chromosome replication (8). One may also speculate that the special pattern of presynaptic chromosome folding depends upon localized modifications possibly involving the presence of special proteins for which a prolonged S-phase would be required. These proposals, though speculative, serve to illustrate how synapsis and/or crossing-over can be influenced by perturbing cells in premeiosis.

Various studies have in fact demonstrated that perturbations of cells during or close to premeiotic S-phase may influence the frequency of crossing-over or chiasma formation. X-irradiation during this stage has been found to cause increases in crossing-over frequency in *Chlamydomonas* and chiasma formation in grasshopper (11, 69). Inhibitors of DNA synthesis, if applied temporarily during premeiotic S-phase, may increase the frequency of crossing-over (13, 61). There are conflicting claims about the effectiveness of temperature shock in altering chiasma or crossing-over frequency but the conflicts are more apparent than real inasmuch as different species respond differently to temperature shock (22, 27). Indeed, different loci may respond in opposite direction to abrupt shifts in temperature (1). There is no doubt, however, that cells of some species can be subjected to temperature shock during premeiosis or early meiosis and usually manifest the effects of such shock in a reduced frequency of chiasmata or crossing-over.

It is understandable that the direction and intensity of effects produced in this way are not uniform. The immediate target of a particular perturbation, such as heat or a DNA inhibitor, is either unknown or imprecisely identified. It is inappropriate to conclude that crossing-over occurs during the premeiotic S-phase because treatment of cells during that interval affects its frequency. On the other hand, it is quite appropriate to conclude that such premeiotic sensitivity is due to the occurrence of either crossing-over itself or processes that are critical to the subsequent events of crossing-over. These processes may not be directly related to the extended S-phase but regardless of the relationship, we conclude that premeiotic S-phase metabolism includes activities that are immediately or subsequently involved in crossing-over.

The most specific change in premeiotic S-phase metabolism thus far reported, and which appears to have a direct bearing on synapsis is the occurrence in *Lilium* of a small fraction of DNA which replicates during meiotic prophase (29, 32). This interruption of S-phase replication differs, at least conceptually, from the one postulated in *Chlamydomonas* (8). In the alga, the interruptions are presumed to be more or less randomly distributed and hence variable, whereas in lily, the interruptions are confined to specific and constant regions of the chromosomes. The generality of the lily pattern has not been tested and its occurrence in other organisms has been questioned (19, 41). There are formidable technical difficulties, however, in demonstrating the phenomenon in other systems. It may therefore be useful to set aside the question of generality and, instead, to identify specific consequences of the delayed replication for meiosis in lily. Such information could provide experimental handles for exploration of the phenomenon in other species.

In lily, a population of DNA molecules that constitutes about 0.3% of the nuclear genome and has a GC composition of approximately 50% as against 40% for total DNA, does not replicate until the beginning of zygotene (29, 32). Chromosome replication during the premeiotic S-phase is therefore only about 99.7% complete. Two kinds of evidence have led to this conclusion, one based on DNA-DNA hybridization techniques, the other on BUdR incorporation. DNA synthesized during zygotene can be exclusively labeled by an appropriate tag since zygotene cells can be harvested free from S-phase cells. The DNA so labeled ("Z-DNA") can be hybridized with nuclear DNA from different stages of the meiotic cycle and the relative proportions of Z-DNA per genome determined. Thus measured, nuclei in prezygotene contain half as many Z-DNA sequences as those in pachytene or later stages. Z-DNA sequences must therefore be replicated during zygotene. Moreover, since the level of Z-DNA in postzygotene nuclei is the same as that in somatic cells, zygotene DNA synthesis must represent a delayed replication rather than an *extra* round of replication.

This conclusion is supported by buoyant density analysis of DNA prepared from cells which had been exposed to BUdR (32). BUdR-DNA obtained from cells exposed during premeiotic S-phase shows the expected heavy shift except for the zygotene component whose position on the gradient remains unchanged. The Z-DNA can be identified either by introducing a radioactive label after S-

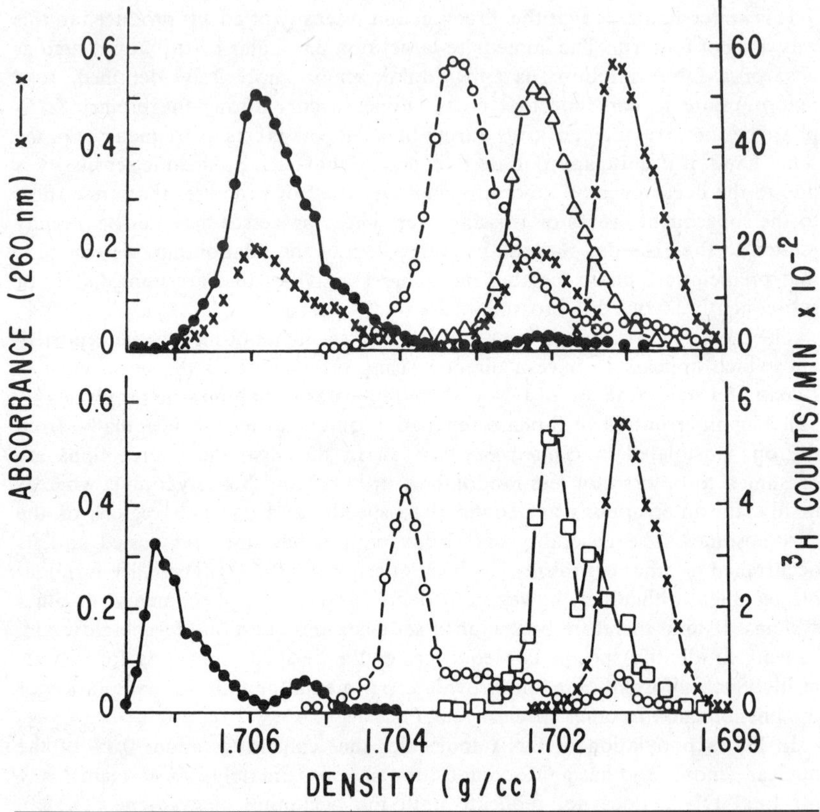

Figure 1.

Patterns of DNA synthesis during premeiotic S-phase (upper curves) and zygonema (lower curves) as displayed in CsCl gradients. Absorbancy of native, *x* — — — *x* and of denatured DNA, △ — — — △; Radioactivity of DNA (no BUdR), □ — — □; Radioactivity of native, *o*– – –*o* and of denatured BUdR-DNA ● — — ●; Absorbancy of denatured DNA from cells grown in presence of BUdR, *xxxxxx*.

The heavy shifts due to BUdR incorporation are shown for both total and Z-DNA. In the absence of BUdR, radioactivity (from ^{3}H-deoxyadenosine) of S-phase labeled DNA tracks the optical density and a separate curve for such labeling is not shown. Labeled Z-DNA, on the other hand, is typically heavier than the bulk of nuclear DNA. Z-DNA does not incorporate either BUdR or radioactive label during premeiotic S-phase. That BUdR incorporation and radioactivity occur, as expected for semiconservative replication, in the same DNA strands is shown by the radioactivity profiles of denatured BUdR-DNA (unpublished analyses).

phase or by hybridizing fractions from the gradient with a preparation of prelabeled Z-DNA. A heavy shift in Z-DNA occurs only if BUdR is supplied to meiocytes during the zygotene stage. The appearance of the heavy shift is shown in Figure 1. It should be noted that although Z-DNA has a high GC composition it is different from r-DNA, which can be identified on the gradient by hybridization with labeled r-RNA (Figure 2). Unlike oocytes, there is no synthesis of r-DNA during the prophase stages of meiosis in microsporocytes.

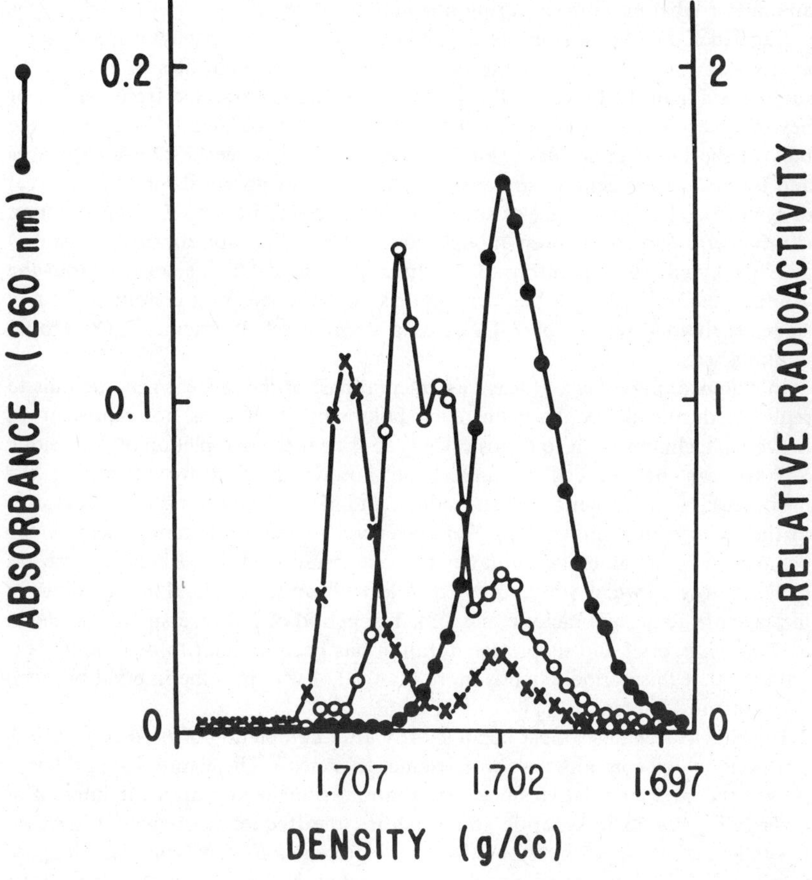

Figure 2.

Buoyant densities of bulk nuclear DNA, ● — — — ●; zygotene-labeled DNA, *o* — — — *o*, and r-DNA, *x* — — — *x* prepared from lily meiocytes. The position of r-DNA was determined by hybridizing each of the fractions with ribosomal RNA according to established procedures (unpublished analyses).

Z-DNA is nuclear in origin. Banding of the zygotene label in a rather narrow region of the CsCl gradient might suggest, by analogy with mouse satellite DNA, that it is concentrated in a particular region of the chromosome. It is almost certain, however, on the basis of recently published studies, that Z-DNA is more or less generally distributed among the chromosomes. Radioautographs show label over various parts of metaphase-I chromosomes in cells that have been exposed to ^{3}H-thymidine during zygotene (36). This finding is in line with earlier reports by Wimber & Prensky (71) demonstrating the incorporation of a small amount of label into nuclei during meiotic prophase.

Labeled Z-DNA prepared as an alkaline extract from pachytene cells sediments as a more or less discrete population of DNA molecules in an alkaline sucrose gradient. If, however, the Z-DNA is similarly extracted from cells at or beyond the metaphase-I stage, it behaves as though covalently bonded to the bulk of the DNA molecules. Thus, covalent bonds between Z-DNA sequences and the rest of the genome are either highly labile or absent during the interval of synapsis. The significance of such bonding characteristics to Z-DNA function is unknown. Moreover, even though covalent bonding appears to be restored after pachytene, thus pointing to the integration of Z-DNA sequences into the genome, the linkage must be fairly labile since mild shearing conditions at any stage result in a release of Z-DNA as a discrete set of fragments (Y. Hotta, unpublished).

In *Lilium* and *Trillium*, at least, an integral part of the nuclear genome fails to replicate during the S-phase but such failure by itself does not represent an irreversible commitment to meiosis. Cells at or near the completion of premeiotic S-phase can be induced to undergo a mitotic division by explanting the meiocytes into a synthetic culture medium and, if so, Z-DNA is synthesized prior to their entry into mitosis (32). This reversion to mitosis is reminiscent of the behavior of the ameiotic mutant in maize described by Rhoades and his collaborators, in which presumptive meiotic cells undergo a final mitotic division in place of the normal meiotic one (52). The period of full reversibility in *Lilium* or *Trillium* is brief and its precise duration has been difficult to determine. It is unlikely that this period extends more than a 1/4 way into the interval between S-phase and leptotene.

Intranuclear changes must begin shortly after completion of premeiotic DNA synthesis, which commits the chromosomes to meiosis. The changes are probably progressive and cumulative since perturbation of meiocytes after the interval of reversibility results in synaptic abnormalities, the frequency of such abnormalities falling off sharply when the cells enter leptotene (66). Unfortunately, it is not possible to describe precisely the relationship between severity of meiotic abnormality and the time in premeiosis at which cells are perturbed. The total interval of susceptibility is too short and the degree of synchrony not sufficient to permit a precise correlation. With information currently available, it can only be stressed that the interval between S-phase and Z-DNA replication is characterized by changes in chromosome organization, which can be experimentally disrupted to a greater or lesser degree, thereby affecting the process of pairing and chiasma formation.

The *timing* of Z-DNA synthesis in *Lilium* is a significant factor in chromosome behavior during zygotene. If a DNA inhibitor is applied to leptotene cells, these cells will not enter the zygotene stage and the chromosomes will not pair (29). It is difficult to determine whether the inhibition of DNA synthesis interferes directly or indirectly with the pairing process. However, in a separate study, Roth & Ito were able to show that addition of a DNA inhibitor during zygotene interrupted formation of the synaptonemal complex without destroying those stretches of the complex which had already formed (60). In lilies, at least, a close relationship appears to exist between Z-DNA synthesis and the process of synapsis. An attractive speculation is that Z-DNA is present in the hypothetical synaptic regions of the chromosomes that were discussed earlier. At the moment, we have no direct way to test the speculation in *Lilium*, nor is there any evidence establishing the generality of Z-DNA behavior. It has been possible, on the other hand, to examine in some detail the association between Z-DNA synthesis and other biochemical processes which accompany chromosome synapsis.

THE PROCESS OF ALIGNMENT In attempting to analyze synapsis, it is essential to distinguish between chromosomal matching sites and the process by which matching is effected, at least until we have a much better understanding of the actual mechanism involved. The final product of the matching process, the synaptonemal complex, is elaborate enough in structure to make it highly probable that an appreciable number of reactions is involved. Moreover, it is unlikely that base-matched DNA sites remain associated in view of the 1000 Å distance separating the chromosomes and the absence of demonstrable DNA within the inter-chromosome region (9, 51). Direct DNA-DNA matching must therefore be a transient event which can lead to synapsis only by activating processes that stabilize pairing in forming a synaptonemal complex.

Several temporal correlations have been established between macromolecular biosyntheses and synapsis (30, 31, 33, 53). Although these do not necessarily reveal causal connections, they do suggest likely relationships and thus contribute to our understanding of synapsis. The continued pairing of chromosomes during zygotene may depend upon continued synthesis of protein (37, 54). The occurrence of protein synthesis during zygotene has been demonstrated in various organisms either by autoradiography or biochemically (50). Moreover, inhibition of protein synthesis inhibits synapsis (54). However, since inhibition by agents such as cycloheximide totally arrests meiosis, inhibitor studies provide no information on specific associations between protein synthesis and synapsis. Radioautographic analyses of protein synthesis also provide no specific pointers, inasmuch as labeling occurs in all regions of the cell (55). Chromatographic analyses of the labeled proteins indicate that numerous proteins are synthesized (30). How many of the proteins synthesized during zygotene are essential to synapsis cannot be determined from the biochemical data. Quite possibly, the large variety of proteins synthesized during synapsis reflects the fact that other activities besides synapsis are being supported at the same time.

A striking and possibly significant characteristic of DNA and protein biosyntheses during zygotene is their association with the nuclear membrane. Although artifacts of fractionation cannot be entirely excluded, they appear to be unlikely and, as will be seen, membrane-associated events can be more directly tied to synapsis than can zygotene protein synthesis in general. The DNA synthesized during zygotene has been found to be transiently associated with newly synthesized protein in a lipoprotein complex. If zygotene cells are exposed to radioactive thymidine, approximately half of the labeled, but no more than a few percent of the total, DNA is released from the chromatin mass by a gentle homogenization of the nuclei. Z-DNA is selectively released in association wth protein and lipid. The association is resistant to disruption by 2 M NaCl, 8 M urea, or nonionic detergents. The complex can be disrupted with sodium dodecyl sulfate, and if so, Z-DNA may be recovered in the native state (25, 30). The complex so isolated has a characteristic and constant buoyant density in cesium sulfate gradients thus indicating a more or less fixed composition. It can be readily distinguished from associations present in either S-phase or pachytene nuclei. Pachytene-labeled DNA extracted in this way is associated with very little if any protein. An S-phase labeled DNA complex has been identified but it bands consistently at a different density in a $CsSO_4$ gradient (25).

Double labeling and chase experiments indicate that the association between Z-DNA and lipoprotein is transient (25). If cells labeled in zygotene are harvested at the pachytene stage, the label is no longer selectively extracted as in zygotene but remains associated with the bulk of the DNA. These and other observations on the transient association of Z-DNA with a more or less constant amount of lipoprotein, suggest a transient role of the nuclear membrane in synapsis. Moreover, since the proteins present in the complex are also labeled during the zygotene interval, the data could be interpreted as indicating the biosynthesis of a lipoprotein-DNA complex in support of the synaptic process. However, although the data are suggestive, additional evidence is required to establish that the complex is actually a part of the nuclear membrane and that the lipoprotein complex is an essential factor in synapsis. The temporal correlation between complex formation and synapsis could be fortuitous, or even if not, it could be entirely related to the replication of Z-DNA rather than to any role it might have in the pairing process.

Other data favoring membrane participation in synapsis come from analyses of fractionated nuclear membranes prepared by the method of Kashnig & Kasper (38). Using this method, membranes are disrupted into vesicles by sonication and are then centrifuged to equilibrium in a discontinuous sucrose gradient. In this way, inner and outer membranes may be separated, as they band at different buoyant densities. A distinguishing and probably significant difference between meiotic and somatic nuclei is in the respective banding patterns of their membranes (33, 34). In addition to the outer and inner membrane fractions present in both types of nuclei, a third band of denser material is found in preparations from meiotic nuclei. This heavier band forms at a density of 1.22–1.20 g/cc as against 1.20–1.18 and 1.18–1.16 for inner and outer membranes respectively. A heavy fraction has been found in nuclei from a variety of meiotic cells including those from man, bull, and rat (34, 46). It has not been observed to any significant extent in preparations from somatic cells regardless of species. A change that may well be universal must occur in the composition or association of nuclear membranes when cells enter meiosis. The details of the compositional change are still unknown, but most probably some parts of the nuclear membrane become associated with comparatively high concentrations of protein. The high protein/lipid ratio would account for the membrane vesicles with high buoyant density.

The significance of the heavy band to synapsis is partly revealed by the presence of two specific proteins whose properties are consistent with a meiotic function. The first of these is a DNA-binding protein which, in a number of respects, is similar to the gene 32-protein identified by Alberts and coworkers in T4 infected bacterial cells (3). The similarity is of meiotic interest inasmuch as gene 32-protein is essential to recombination. The meiotic binding protein that is localized in the heavy membrane fraction has been partially purified by solubilization with deoxycholate or digitonin and by passing the solution through DEAE- or DNA-cellulose columns (33, 34). The protein obtained from lily meiocytes can, like the gene 32-protein, facilitate the renaturation of single-stranded DNA. The protein activity is not specific with respect to source of DNA and it acts equally on T7 and lily DNA. Although several types of DNA-binding proteins are present in meiotic and somatic cells, this particular one appears to be unique to meiocytes and consequently may be presumed to have a special relevance to meiosis. The occurrence of this protein both in lily microsporocytes and mammalian spermatocytes reinforces that conclusion. Moreover, like many other factors operating in meiotic prophase, the binding protein is cyclical in behavior. It begins to rise during leptotene and reaches a plateau at zygotene. By mid-pachytene the concentration begins to decrease and eventually falls below the level of detectability at the termination of the pachytene stage. The temporal association of this protein with the interval of chromosome synapsis and its capacity to facilitate strand matching makes it reasonable to suppose that it functions in synapsis by facilitating homolog alignment at selected regions in the nuclear membrane. Implicit in this interpretation is the assumption that alignment is first effected by a matching of specific DNA sequences and that these sequences find their way to the nuclear membrane at appropriate times during zygotene.

The view that the nuclear membrane functions in synapsis has been reinforced by the identification of a colchicine-binding protein in the membrane of meiotic nuclei (31). This protein is of particular interest to meiosis because of the known effects of colchicine on chiasma frequency. A basis for these effects may become apparent from the biochemical studies to be discussed here.

Nuclear membranes and also chromatin from a variety of somatic tissues have the capacity to bind colchicine (65). In meiocytes of *Lilium* this capacity is largely localized in the nuclear membrane and, if expressed relative to total protein, the concentration in the membrane is about 100 × that in the nucleus as a whole. In contrast with the DNA-binding protein, the colchicine-binding one is present in all three subfractions of the nuclear membrane. However, since the heavy membrane fraction appears only in meiotic cells, there is a considerable increase in total amount of colchicine-binding protein during meiotic prophase (31). The protein has been dissociated from the membrane by means of nonionic detergents and some of its properties have been characterized. No differences are evident between preparations made from meiotic and somatic nuclei. Generally, the protein extracted from the nuclear membrane behaves like tubulin except for two of its properties. It has a lower molecular weight than its cytoplasmic counterpart, 100,000 vs 120,000 daltons, and also unlike its counterpart, it is not precipitated by vinblastine. The differential precipitability is retained even if the nuclear colchicine-binding protein is mixed with several times its concentration of cytoplasmic protein. Thus, the nuclear protein can be retained in solution while its cytoplasmic counterpart is precipitated by vinblastine. The nature of these differences in solubility and in apparent molecular weight is unknown. The difference is, nevertheless, significant inasmuch as it indicates that the nuclear protein is not a cytoplasmic contaminant.

The presence of a colchicine-binding protein in the nuclear membrane, particularly during meiotic prophase, could provide a basis for the capacity of colchicine to interfere with the formation of chiasmata. The effect of colchicine on chiasma formation was first described by Levan, who attributed its action to an interference with chromosome pairing (42). Studies of wheat have confirmed Levan's findings but have led to the general conclusion that colchicine exerts its effect by acting on premeiotic association of homologs rather than on the pairing process itself (12, 15).

More than one view has been advanced about how premeiotic association relates to synapsis. One prominent interpretation is that somatic association prealigns chromosome pairs in a way that facilitates synapsis (17). Since colchicine disrupts somatic association it is presumed that its effect on meiosis derives from that disruption. There is, however, an alternative interpretation, which rejects the reality of somatic association on experimental grounds and identifies premeiotic alignment as a special event that is linked directly to the meiotic cycle. The most elegant evidence in favor of this second interpretation is the behavior in wheat of an isochromosome constructed from the two long arms of chromosome 5D (14). The two homologous arms of the chromosome form chiasmata under conditions in which colchicine virtually eliminates chiasmata in the other chromosomes. More recent analyses of the timing of colchicine action in wheat place the sensitive interval before the premeiotic S-phase (12). The distinctive behavior of the isochromosome is explained by the fact that the centromere necessarily prealigns the two arms under conditions where prealignment of normal homologs is abolished by colchicine treatment.

Whether prealignment is a universal feature of meiosis remains an open question, although various bits of evidence have been adduced in support of premeiotic association (5, 17, 49, 52). The timing of premeiotic sensitivity to colchicine in wheat is puzzling from a mechanistic standpoint and may relate to the special requirements in this group of plants for assuring homologous as against homoeologous pairing of the meiotic chromosomes. In general, the effect of 5B in wheat and of accessory chromosomes in other species such as those of *Lolium* (16) on homologous pairing, has not yet been satisfactorily explained despite many efforts to rationalize the action of these chromosomes on synapsis between other members of the karyotype. Nevertheless, the existence of synaptic effects between nonhomologous chromosomes lends support to the still sketchy biochemical evidence on the elaborateness of the synaptic process. Presumably, B chromosomes could affect synapsis by metabolic activities, which yield products that modify the pairing properties of other chromosomes.

Detailed temporal analysis of meiosis in *Lilium* supports Levan's conclusion that colchicine interfaces with the *process* of pairing (31, 62). Colchicine applied at leptotene disrupts the pairing process and such disruption can be effected even up to the early part of the zygotene stage. The evidence that colchicine reduces chiasma frequency by disrupting pairing is based on studies of zygotene and pachytene stages by light and electron microscopy. Chromosomes in colchicine-treated cells that suffer a reduction in chiasma frequency or become achiasmatic are either partially or entirely unpaired when examined at pachytene. There is a consistent relationship between the level of chiasma frequency and the degree of prior synapsis. When colchicine-treated meiocytes are examined by the EM, lateral elements are invariably present but synaptonemal complexes are only partially formed or, if formed, may appear abnormal. The extent of the upset is roughly proportional to the decrease in chiasma frequency. Colchicine has no effect on the synaptonemal complex once formed, which is consistent with the observation that chiasma frequency is unaffected in cells treated during late zygotene and pachytene. Whether colchicine acts by interfering with the initiation of pairing in individual chromosome segments or whether it disrupts the process as a whole is not known. Indeed, we do not know whether an analogy can be drawn between replicons and pairing regions with respect to initiation. It would be difficult, if not impossible, to identify cytologically initiating points of synapsis. If these do exist, the distinction between colchicine acting on the initiation of pairing as against the process of pairing may have little operational value.

Biochemical analyses of meiocytes exposed to colchicine indicate that the achiasmatic effect is not accompanied by a general disruption of one or more major biosynthetic pathways (31). Rates of RNA, DNA, and protein synthesis are essentially unaffected by exposure of cells to colchicine and if affected, the effects are slight and clearly not the cause of chiasma failure. In view of the localization of the colchicine-binding protein in the nuclear membrane, it is attractive to suppose that colchicine acts by disturbing the normal course of events at the nuclear membrane. This supposition is given substance by the observation that colchicine-treated cells that become achiasmatic lack the normal concentration of DNA-binding protein in the heavy membrane fraction during meiotic prophase. The effect is restricted to membrane association because the total DNA-binding protein within the meiocytes is equal to that in the controls. It is, of course, possible that the DNA-binding protein is actually associated with the nuclear membrane in vivo, but that the association is weakened by colchicine and is destroyed in the course of fractionation. Whether or not this is so may be less important for the present than the fact that the association between nuclear membrane and DNA-binding protein is altered by colchicine treatment. The full extent of the alteration needs to be determined by a more complete analysis of protein and other components in the different subfractions of nuclear membranes from treated and untreated cells. It is important to note that the decrease in membrane-associated DNA-binding protein occurs only if cells are treated either before or no later than early zygotene. Once the binding protein is present in the nuclear membrane it does not appear to be dislodged by the introduction of colchicine. In this respect there is a near-perfect parallel between the respective actions of colchicine on the formation of the synaptonemal complex and on the localization of DNA-binding protein.

RESUME The biochemical data on metabolic events during chromosome synapsis are clearly insufficient to provide telling insights into specific synaptic mechanisms. An involvement of the nuclear membrane in the process of synapsis is probable but far from proven. That involvement appears to be transient, and should not be confused with EM evidence from various animal systems that telomeres remain attached to the nuclear membrane during the prophase of meiosis (64). The nature of membrane function as deduced from the biochemical analyses can be described in superficial terms only but in so doing a framework can be provided for further probes of the synaptic process. Several investigators have already assigned the nuclear membrane a special role in synapsis (10, 64).

We visualize a more or less uniform distribution of specific DNA stretches along the entire length of each chromosome. These stretches are presumed to serve as sites for matching of partially condensed homologous chromosomes. Each site would necessarily contain its own unique sequence and thus account for longitudinal specificity in chromosome pairing. Whether Z-DNA might function in this way is conjectural but it is conceptually convenient to assign it such a role. If, as the evidence strongly indicates, synapsis occurs when chromosomes are already appreciably condensed, then the folding process must be a precise one in which the postulated synaptic sites in a chromosome remain exposed for interaction with those in the corresponding homolog. Moreover, if the synaptic process occurs in association with the nuclear membrane, the chemistry of the pairing regions must be such as to favor an interaction between membrane and one or more components of each synaptic site during the interval when pairing occurs. Since the membrane association is transient, a mechanism must exist for altering membrane affinity of individual synaptic sites just prior to and immediately following synapsis in a particular chromosome region. The presence of a DNA-binding protein in the nuclear membrane may be regarded as a mechanism for promoting the matching of homologous DNA stretches. Since formation of a duplex from members of homologous chromatids would be facilitated by the presence of DNA stretches in each of the chromatids, which can readily become single-stranded, it is attractive to speculate that membrane association coincides with the replication of Z-DNA. Once matching has occurred, it must trigger a set of events in the nuclear membrane that tie the lateral elements into the coherent structure of the synaptonemal complex. The internal surface of the nucleus may thus function as a framework within which the processes promoting stabilization of paired chromosomes follow the transient ones facilitating base sequence matching.

CROSSING-OVER

Introduction

Many of the discussions and controversies about crossing-over in meiosis turn on the issue of when it occurs in the meiotic cycle (22, 26, 45, 56). The evidence in support of the view that crossing-over follows synapsis is necessarily limited because no adequate method is available for interrupting meiosis and directly determining whether crossing-over had occurred prior to the interruption. Commonly, cells have been perturbed by one technique or another at different meiotic stages and the frequency of chiasmata or crossing-over in such cells compared with normal frequencies (27). The advantages as well as pitfalls of such physiological approaches were discussed in the section on synapsis. What emerges from these and other types of study is the conclusion that certain events that are decisive in determining the pattern of crossing-over occur well before pachytene. One intriguing and attractive point of view has been presented in connection with analyses of premeiotic DNA synthesis in *Chlamydomonas* (8). The essentials of the viewpoint were described previously, its principal feature being that potential sites of crossing-over are determined during the premeiotic S-phase. The sites are believed to consist of incompletely replicated DNA stretches, which remain as gaps until repaired later during meiotic prophase. Treatment of zygotes with phenylethyl alcohol toward the end of the premeiotic S-phase causes an increase in recombination frequency, the effect being attributed to an increase by the drug in the number of late replicons that fail to complete replication. An alternative but less elegant proposal is that the inhibitors affect pre- or early zygotene events that are essential to synapsis. An increase in chiasma frequency by treatment of lily meiocytes with inhibitors of DNA synthesis during early as well as mid-zygotene has, in fact, been reported (61). The changes that occur in *Chlamydomonas* zygotes during the interval between termination of S-phase and early zygotene might be hard to separate with respect to time of occurrence because of the relatively short duration of meiosis. It should be stressed, nevertheless, that there are several pieces of cytogenetic evidence that point to a prepachytene determination of the sites of crossing-over but not necessarily to crossing-over itself (5, 44).

Regardless of the nature of premeiotic treatments that affect crossing-over frequency, a key consideration is the unambiguous demonstrations in several species that cells that are perturbed during late zygotene and early- to mid-pachytene suffer a reduction in chiasma or crossing-over frequency (26, 27, 43, 55, 56). Such an effect requires that an essential phase of crossing-over follow synapsis. It is possible that crossing-over occurs during both premeiotic S-phase and pachytene. If so, the experimental evidence in favor of the premeiotic period is weaker than the evidence for crossing-over during meiotic prophase.

Background of Biochemical Studies

Biochemical analysis of the pachytene stage supports the classical cytogenetic sequence of homolog synapsis followed by crossing-over. However, before turning to the molecular details of the sequence it should be emphasized that no obvious relationship exists between the morphology of a chromosome bivalent during pachytene and the molecular juxtapositions required for crossing-over between homologous DNA strands. None of the data to be presented here will add materially to our understanding of how DNA heteroduplexes might be formed over the 1000 Å distance that separate the two homologous chromatin masses in the synaptonemal complex.

Biochemical experiments with lilies, designed to elucidate the relationship between synapsis and crossing-over, are in agreement with the physiological data obtained in other organisms. Experiments in which cycloheximide was used to inhibit part or all of protein synthesis during different stages of meiosis have proved to be particularly revealing (54, 55). All intervals of meiotic prophase depend upon continued synthesis of protein, so that substantial inhibition of synthesis during any of these intervals invariably results in meiotic arrest. In lilies, at least, arrests thus effected are reversible but the nature of meiotic development following removal of the inhibitor depends upon the stage and duration of inhibitor treatment. If treatment and removal are done between leptotene and mid-zygotene, meiosis proceeds more or less normally. If, however, the interval of treatment falls during late zygotene to early pachytene, meiosis proceeds at a normal rate following inhibitor removal but chiasma formation is appreciably reduced or even eliminated. There is thus a relatively short interval of susceptibility to inhibitors of protein synthesis which coincides with the time at which pairing has been or is nearly completed and which may lead to a sharp reduction in chiasma frequency. This interval corresponds to the one found by Chandley & Kofman-Alfaro in mammalian spermatocytes during which irradiation causes an unusually high increase in thymidine incorporation (7, 41). No effect on chiasmata is observed if inhibitors are applied to lily meiocytes after early pachytene.

EM studies indicate that the protein inhbitor causes a precocious dissociation of bivalents accompanied by a disruption of the synaptonemal complex (54). Upon removal of the inhibitor during late zygotene to early pachytene, chromosome condensation begins without restoring the original synaptic association. These results can be explained by assuming that crossing-over does not occur until early pachytene and that destabilization of synapsis just prior to the normal time of crossing-over is too late to permit restoration of the synaptonemal complex. The absence of stabilized bivalents eliminates the essential pairing association and thus prevents crossing-over between homologs. A similar relationship between a lack of stabilized bivalents and low chiasma frequency may be deduced from the effects of colchicine on chiasma formation (14, 62). As already described, the defective synapsis caused by colchicine may be assumed to account for the ultimate reduction in chiasma frequency.

In practice, sharp separations of zygotene and pachytene cells for biochemical analyses are difficult to obtain. It is nevertheless clear that transition from zygotene to pachytene is accompanied by a major change in DNA metabolism (32, 35). In general, characteristics of that metabolism are best explained by the occurrence of crossing-over during the pachytene stage. Enzymes have been identified in lily meiocytes, which produce single-stranded nicks in native DNA and can repair the gaps that are subsequently formed (35). Presumably, the significant consequence of repair activity associated with pachytene is the occurrence of crossing-over. An idealized interpretation of the evidence that bears on mechanisms for effecting crossing-over is as follows. Once homologous chromosomes have become associated by synapsis, a new set of metabolic events is initiated among which a nicking of chromosomal DNA is prominent. Simultaneously with the introduction of DNA breaks, or nearly so, appropriate repair enzymes become active and eventually seal the breaks. The outcome of such repair is that a more or less constant number of crossovers per meiotic nucleus is formed. What follows is a discussion of the experimental evidence that points to this interpretation.

The Nicking of Chromosomes

Evidence for a programmed nicking of chromosomal DNA during pachytene comes from analyses of deoxyribonuclease activities during the meiotic cycle (35). Two types of deoxyribonucleases have been found in *Lilium*, one with a pH optimum of 5.6 and the other with an optimum of 5.2. The two activities have been separated by column chromatography and the basis for regarding them as two separate enzymes is at least tentatively established. The distinction, as will be seen, has special relevance to meiosis. Studies have not been made to determine whether one activity arises by a modification of the other. It is known, however, that the "pH 5.2" enzyme is found exclusively in meiotic cells. The "5.6" enzyme, on the other hand, has been found in all vegetative tissues of lily that have been examined, whereas it is present in the meiotic cells of some but not all varieties. The "5.2" enzyme has not been found in somatic cells and it therefore appears to be a distinctive constituent of meiotic cells (35).

The exclusive presence of a particular deoxyribonucleolytic activity in meiotic cells alone is highly suggestive of its special role in meiotic metabolism. That at least 50% of this activity has been localized in the nucleus adds likelihood to the possibility (35). Moreover, the cyclical display of activity by this enzyme further reinforces the view of its special function. The nuclease shows transient activity, which begins to rise in zygotene, reaches a maximum at pachytene, and disappears at the termination of that stage. The temporal profile of activity is consistent with a role of promoting DNA nicking during late zygotene to mid-pachytene. The nature of the nicks introduced by the enzyme has been determined from analyses of partially purified preparations. The enzyme is an endonuclease, which acts only on native DNA and makes single-stranded nicks that expose 3′-phosphate and 5′-hydroxyl ends. Although exonucleases that can act on the free ends are known to be present, they have not yet been studied. Several enzyme activities that could be involved in the repair of the endonucleolytic breaks have been identified and described (35). There is an ATP-dependent polynucleotide ligase present in meiocytes, which specifically seals 5′-phosphate to 3′-OH ends. As the meiotic endonuclease produces 3′- rather than 5′-phosphate ends, direct action by the ligase on the product of nuclease activity is not possible. Enzymes are nevertheless present which, in theory at least, can correct the situation. The 3′-phosphate ends can be, and perhaps are, removed by phosphatase which is abundantly present in meiocytes. The 5′-OH end can be phosphorylated by a polynucleotide kinase, which is highly active during meiotic prophase. However, it is significant that of all the enzymes discussed in connection with pachytene metabolism, only the endonuclease has a strong cyclical pattern of activity. None of the other enzymes appears to undergo major shifts in activity during meiotic prophase although both polynucleotide kinase and ligase fall to low levels at the end of pachytene.

The restriction of the endonuclease to meiotic cells and the peaking of its activity during pachytene nicely support a scheme requiring the deliberate formation of nicks during meiosis. Failure to detect meiotic endonuclease activity during the premeiotic S-phase may or may not be significant to the debate concerning the occurrence of crossing-over during the interval of chromosome replication. Presumably, numerous interruptions in the DNA strands would in any case be present because of ongoing replication. Interestingly, polynucleotide ligase activity, which might serve to seal these interruptions, is relatively high in late premeiotic S-phase (35).

Of major significance to interpretations of meiotic endonuclease function are some recent studies in our laboratory of DNA extracted from alkali-treated meiotic cells and analyzed on alkaline sucrose gradients. The choice of denatured DNA for analysis is determined by the need to minimize shearing of DNA during extraction. Satisfactory preparations of native DNA from the microsporocytes have not yet been obtained. The much lower susceptibility of denatured DNA to shearing action has made it feasible to determine approximately the average in situ lengths of single DNA strands at different stages of meiosis. Although the data obtained in this way provide no information on whether single- or double-stranded breaks are present, the frequency of breaks at any stage of meiosis may be inferred from strand length. A comparison of the respective sedimentation patterns of pachytene and postpachytene denatured DNA is shown in Figure 3. This figure illustrates a typical and characteristic difference between DNA extracted from pachytene cells and that extracted from cells in either earlier or later stages of the meiotic cycle. Assuming that alkaline DNA extracts from cells reflect, in at least a relative way, actual in situ strand sizes, the data reveal a substantially higher number of relatively short DNA strands during the pachytene stage than during earlier or later stages. An obvious and reasonable explanation for the lower molecular weight fragments during pachytene is an endogenous nicking of DNA strands at this stage. Thus interpreted, there is a gratifying correspondence between the activity of the meiotic endonuclease and the frequency of in situ DNA nicks at different stages of the cycle. The correspondence supports the view that an endonuclease is activated at pachytene to induce breaks, which may then serve as initiating sites

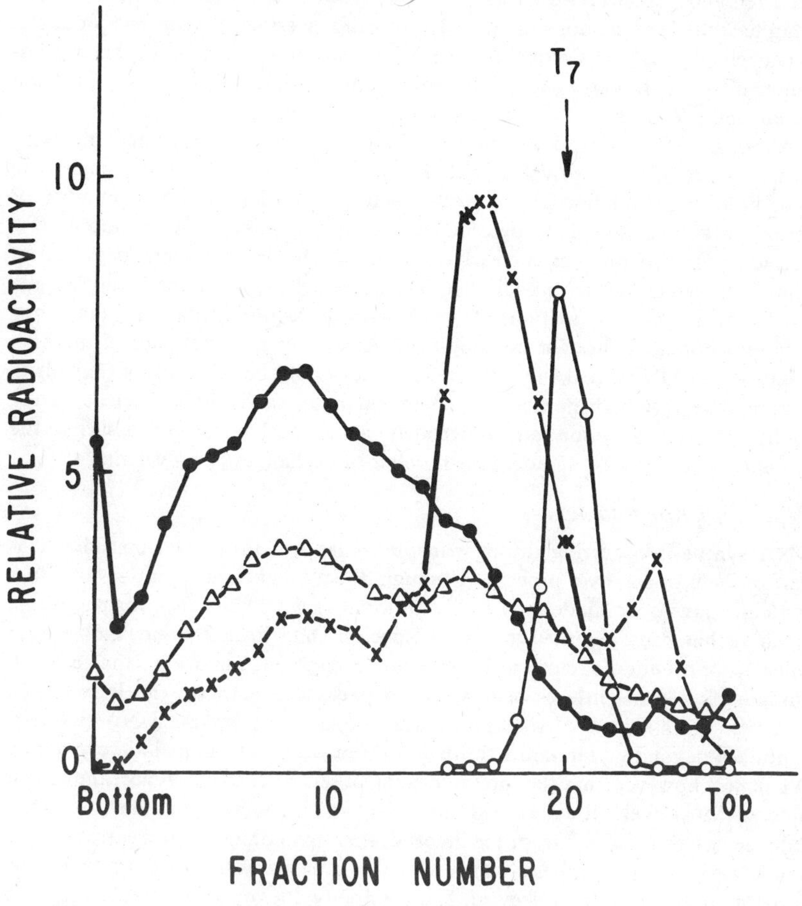

Figure 3.

Sedimentation profiles in an alkaline sucrose gradient of denatured DNA prepared from prezygotene cells, △ — — — △, pachytene cells *x* — — — *x*, and post-diakinesis cells ● — — — ●. The DNA from each of the groups of cells was radioactively labeled during the premeiotic S-phase so as to provide uniformly labeled DNA for analysis. T7 DNA is included as a marker (unpublished analyses).

for repair replication and, thereby, for crossing-over. The disappearance of short DNA fragments after pachytene implies the occurrence of such repair replication. It should be pointed out, however, that the detectability of low molecular weight fragments during pachytene depends upon the occurrence of an appreciable lag between the introduction of a nick and its ultimate repair. If the two occurred in quick succession, fragments would not be seen. The existence of such a lag has not been demonstrated and it remains possible, though not probable, that pachytene DNA is more fragile rather than more fragmented. The significance of such fragility would be much more difficult to assess than the significance of nicks.

Although DNA strand nicking provides an attractive basis for initiating crossing-over, it is conceivable that the breaks that appear during pachytene have little or no relationship to recombination. Indeed, the actual number of breaks, although difficult to determine, have been estimated to be about 10^5/nucleus (35). This number is clearly in excess of the 36 crossovers measured as chiasmata, which occur in each cell. We nevertheless prefer the view that the breaks occur in anticipation of crossing-over requirements and that they represent potential sites for crossing-over. Any other interpretation of the high frequency of DNA breaks in pachytene nuclei would be much more difficult to sustain. The very high number of repair regions in each *Lilium* nucleus during pachytene may function advantageously as a mechanism for assuring the ultimate occurrence of an adequate number of exchanges per bivalent.

Repair of Chromosomes

DNA synthesis occurring during pachytene has generally been considered as part of the crossing-over process. Although direct biochemical analyses of DNA synthesis have been made only in lily, autoradiographic evidence for pachytene synthesis has been obtained in various species. Data from *Triturus*, mouse, and wheat support the conclusion, though not always unequivocally, that a small amount of DNA synthesis occurs during pachytene (41, 58, 71). It has been suggested from studies of wheat autoradiographs that prophase DNA synthesis continues beyond metaphase I through the remainder of the meiotic cycle (58). We doubt, however, that the incorporations observed correctly reflect the in situ process. Most likely, the observed incorporations are either stimulated by DNA damage due to radiation from the incorporated thymidine or they represent very low levels of repair synthesis which may occur in many tissues during some or perhaps all phases of the cell cycle. The possibility does exist that wheat and even other organisms have a modified schedule of meiotic repair synthesis if they have a relatively short meiotic cycle (4). It is nevertheless difficult to visualize how repair synthesis concerned with crossing-over could be effective after bivalent separation. Generally, investigators have had difficulty in demonstrating pachytene DNA synthesis by autoradiography, let alone repair synthesis that was not artificially stimulated. It would therefore seem advisable for the present to set apart the phenomenon of background repair that may be common to all cells from repair synthesis that occurs as a distinctive activity of pachytene cells.

Biochemical analyses of pachytene DNA metabolism support the conclusion that synthesis is of the repair-replication rather than the semi-conservative kind (32). Hybridization of pachytene-labeled DNA (P-DNA) with DNA extracted from cells at different stages of the meiotic cycle shows that the proportion of P-DNA hybridized is the same for all meiotic stages and also for somatic cells. This result contrasts with the behavior of Z-DNA, which shows a doubling of hybridizable sequences between the beginning and end of zygotene. The constant proportion of P-DNA sequences in the genome despite their synthesis during pachytene, is best explained as a display of turnover associated with repair replication (32).

The conclusion that pachytene is the prime interval for repair replication is of major significance in relating crossing-over to the meiotic cycle. To be convincing, the conclusion requires a great deal of additional documentation. Chandley and coworkers have interpreted the strong stimulation by UV or X-rays of ^{3}H-thymidine incorporation in mouse and human spermatocytes at late zygotene-early pachytene as evidence for an ongoing repair-replication during that interval (7, 41). The interpretation is based upon the general observation that thymidine incorporation during S-phase is decreased by irradiation. Since incorporation is increased by irradiation of zygotene-pachytene cells, the normally small synthesis is presumed to be of the repair type. The increased incorporation of ^{3}H-thymidine at this stage is in itself of great interest, even though the conclusion that it indicates normal repair-replication is very much open to question. Indeed, it is by no means certain that repair of radiation-induced damage is identical in mechanism with repair syntheses leading to crossing-over. Other explanations, which would throw no additional light on crossing-over, could be advanced to explain the enhanced response to radiation. Thus, regardless of the validity of the particular interpretation, the observations provide much needed confirmation of the occurrence of DNA synthesis during early pachytene and also of the presence of a suitable complement of repair-replication enzyme activities during that interval.

The buoyant density of BUdR substituted P-DNA provides additional evidence for repair-replication (32). In contrast with the behavior of zygotene DNA (Figure 1), BUdR-containing pachytene DNA does not shift significantly in position when sheared to about 5×10^5 daltons and analyzed on a CsCl gradient. These results suggest that stretches containing BUdR are too small a proportion of the DNA fragments to have an observable effect on their buoyant density. Since labeling of DNA during pachytene is distributed, though not uniformly, over the entire DNA profile, it may be supposed that short lengths of newly synthesized DNA are scattered through many regions of the genome. The actual size of the stretches synthesized has not been determined, and it can only be conjectured that the size is compatible with the length of gaps that might be formed after insertion of nicks in the chromosomes.

The conclusion that pachytene synthesis resembles repair-replication has been given additional support by comparing the effects of several inhibitors on DNA synthesis at different meiotic stages (32). Hydroxyurea, for example, which has been shown to inhibit semi-conservative replication preferentially, is a comparatively poor inhibitor of P-DNA synthesis whereas it is a very effective inhibitor of Z-DNA and premeiotic DNA synthesis. Generally, hydroxyurea and other selective inhibitors, such as 6-(*p*-hydroxyphenylazo)-uracil, affect S-phase and Z-DNA synthesis in one way and radiation-induced and P-DNA synthesis in another. The similarity between radiation-induced and P-DNA synthesis in response to these inhibitors holds irrespective of the meiotic interval during which the cells have been irradiated. Although, as mentioned earlier, we cannot assume an identity between the mechanisms underlying repair-replication with those underlying crossing-over, a significance must be attributed to the fact that P-DNA synthesis behaves more like a repair system than a semi-conservative replication system in its response to various inhibitors.

Taken together, the several lines of evidence, biochemical and autoradiographic, point toward the conclusion that P-DNA synthesis is general in occurrence and that it closely resembles repair replication. Such replication would be expected to occur in the course of crossing-over. The evidence for meiotic crossing-over taking place at the pachytene stage is thus strong even though circumstantial. Nevertheless, such evidence does not preclude the possibility of crossing-over also occurring in premeiotic S-phase. If a repair type of replication did occur during the S-phase it would be masked by the much higher activity of semi-conservative replication and remain undetected. Moreover, the process of semi-conservative replication may itself provide special conditions different from those at pachytene which nevertheless make crossing-over possible. The point we most wish to emphasize is that the characteristics of P-DNA synthesis alone, however closely they reflect repair activity, are inadequate to establish a single time for crossing over during meiosis.

There is, in brief, no direct way now available by which to demonstrate a relationship between P-DNA synthesis and crossing-over. The fact that biochemical data fall nicely into place within the classical cytogenetic framework may be intellectually gratifying but it does not qualify as a definitive demonstration. Indeed, as discussed earlier, a conservative estimate of the number of sites undergoing repair synthesis during pachytene is about 10^4 times the number of chiasmata formed. If, therefore, all of the DNA synthesis occurring at pachytene is associated with the process of crossing-over, it must be concluded that most of the breaks and/or gaps that are present in the pachytene nucleus are repaired without an exchange between homologous DNA strands. If this characterization of P-DNA synthesis is correct, then crossing-over must be regarded as a most unlikely consequence of repair replication at any particular site even though it is a virtually certain, although rare, consequence of repair in 10^4 sites.

Various models could be devised to rationalize the relationship of a very high ratio of repair to crossing-over sites. It might be supposed that the rarity of crossing-over is a function of the very low probability of two homologous nicked strands being properly juxtaposed. A more sophisticated model might be one in which the leptotene folding of the chromosomes prior to synaptonemal complex formation leaves a very small number of chromosomal stretches entrapped in the lateral elements and subsequently in the complex. If the entrapped stretches were the only potential sites for crossing-over at pachytene, and if nicking by endonuclease were a much more widespread activity than strand entrapment, an excess of repair regions over chiasmata would be expected. These and other models, however, cannot be adequately evaluated until a better understanding is available of the nature of P-DNA metabolism.

Regulation of Pachytene DNA Synthesis

Two new features of pachytene metabolism, which have an important bearing on the regulation of P-DNA synthesis, have been described in recent studies. The triggering of pachytene events and the loci of pachytene repair are the two regulatory features concerned. At the time of writing, these features of regulation can only be sketchily described but they are considered here because they have an important bearing on biochemical models of crossing-over. In discussing P-DNA synthesis, we have treated it as an event that occurs more or less uniformly over the entire chromosome and, like the genetic model of crossing-over, is equally probable for any particular segment of the chromosome. We have also treated pachytene as a meiotic interval that is more or less independently programmed for a set of specific metabolic activities. The data now being obtained do not support either of these views but, on the contrary, indicate that there is an appreciable degree of selectivity in sites of P-DNA synthesis and that there is functional dependence of pachytene synthesis on zygotene synapsis.

Profiles of P-DNA on CsCl gradient regularly show a clustering of radioactivities at various positions along the density gradient (32). Even though the radioactivity clusters are distributed over the entire profile of total nuclear DNA and do not form discrete bands, the curve of P-DNA radioactivity, unlike that of radiation-induced repair, is rarely, if ever, smooth. It appears as though synthesis occurs preferentially among certain classes of DNA fragments, each having sufficiently distinctive base compositions to provide for the clustering observed. $C_o t$ curves of P-DNA confirm this conclusion (63). P-DNA label does not reanneal like the bulk of nuclear DNA. Instead, a major portion of the radioactivity reanneals at a comparatively low $C_o t$ value, indicating that P-DNA contains a higher proportion of reiterated sequences than the nuclear DNA as a whole. Moreover, the shape of the $C_o t$ curve reflects a low degree of complexity, suggesting that P-DNA synthesis usually involves regions containing a class of sequences that has a limited number of varieties in the genome. The length of these common sequences is undetermined, but the results obviously suggest that if DNA nicking marks the beginning of the crossing-over process, then that nicking does not occur randomly. It is inviting to speculate that nicking occurs in or close to regions containing one from among the several special varieties of repeated sequences present in the genome. Such an arrangement in which unique sequences are juxtaposed to repetitious ones can be easily visualized as facilitating hydrogen bonding between nicked single strands from pairs of homologous chromosomes. Such matching, unlike that involved in synapsis, would be functional at the molecular level and presumably encompass relatively short stretches of DNA of an order appropriate for crossing-over. Ideally, the segments concerned should contain a multiple of simple repeats so as to relax the conditions for homologous strand association. Analyses of DNA in several organisms do in fact show that unique sequences are joined to relatively short stretches of repetitive ones (24).

A very different aspect of pachytene metabolism has been touched upon by evidence indicating that the initiation of endonucleolytic activity and of repair replication depend upon the prior completion of synapsis. In situations where synapsis is defective or incomplete, as in achiasmatic hybrids or in colchicine-induced achiasmatics, relatively little DNA synthesis has been detected during the pachytene stage (31). In colchicine-treated systems, for example, the proportion of achiasmatics corresponds roughly to the amount of P-DNA·synthesis inhibited. Superficially, at least, it appears as though P-DNA synthesis depends upon the formation of bivalents and that in the absence of synapsed bivalents repair replication does not occur to an appreciable degree. Observations on P-DNA synthesis are reinforced by the behavior of alkaline extracts of DNA. If alkaline extracts are made from achiasmatic cells the DNA thus extracted shows only a small proportion of low molecular weight fragments when analyzed on an alkaline sucrose gradient (unpublished studies). It appears therefore that nicking is not properly triggered if synapsis is lacking or if it is abnormal. The succession of synapsis and crossing-over has been regarded as a meaningful temporal sequence from the standpoint of meiotic functions. Little thought has been given to the possibility that mechanisms exist that can and do prevent the initiation of normal pachytene DNA synthesis in the absence of appropriate synapsis of homologs. To be sure, there are no a priori grounds for supposing that nicking of chromosomes and their repair require the prior occurrence of synapsis. The evidence that such a requirement does in fact exist, although in need of corroboration, points to the general conclusion that pachytene synthesis depends upon prior synapsis not only for juxtaposing homologous DNA strands but also for activating the mechanisms essential to repair replication.

CONCLUDING CONSIDERATIONS

In this review, we have attempted to analyze synapsis and crossing-over in biochemical terms. The data presented are clearly insufficient fully to resolve the major controversies concerning meiotic mechanisms, notably the time of crossing-over and its relation to chromosome synapsis. Moleular genetics has provided insights into mechansims of recombination but, as yet, no solutions to the problem in higher organisms. The task of effecting cross-overs during gametogenesis is more than a process of effecting recombination at the molecular level. Sealing of gaps between DNA strands terminates a relatively long sequence of events that begins at or close to the premeiotic S-period. It would appear as though the action of numerous genes is required to guarantee a consistent frequency of randomized exchanges between homologous chromosomes. The techniques of electron microscopy, radioautography, and conventional biochemistry have now contributed enough to our understanding of meiosis so that models of crossing-over have become increasingly sophisticated in outline (2, 10, 28, 39). Although the biochemical data discussed in this review do not add decisively to one particular model, they merit being set in some conceptual, if speculative, perspective. This is what we will attempt to do now.

We favor the view that zygotene synapsis is an essential condition for crossing-over during meiosis but there are no a priori grounds for excluding other perhaps less effective mechanisms in some meiotic systems. Our choice is based on various pieces of circumstantial evidence among which we include the correlation between experimentally induced asynapsis or desynapsis and reduction in chiasma frequency. The in situ evidence for nicked DNA and its repair during pachytene, and the in vitro evidence for appropriate enzyme activities to sustain these events, make it very difficult to assign crossing-over to the premeiotic S-phase even though it cannot be entirely excluded. A primary and largely unanswered question is how synapsis provides the appropriate structural framework for crossing-over between the separated members of a bivalent. The evidence from some cytogenetic studies that the sites of crossing-over are determined at a stage prior to pachytene (44) may be conveniently explained by supposing that the potential sites of crossing-over are already determined at the time that the chromosomes synapse.

Synapsis itself is the end-product of a sequence of processes. It begins with a modification of chromosomes during leptotene and ends with a stabilization of pairs by formation of the synaptonemal complex. We do not favor the view of chromosome pairing as a progressive process that begins by a coarse alignment of homologs, possibly at the telomeres or centromeres, and ends with very precise matching. The behavior of wheat meiocytes in particular points to some form of premeiotic alignment that can be interpreted as a process of coarse pairing. There is, indeed, ample evidence for the existence of premeiotic associations in various species (5). We find it difficult, however, to reconcile the idea that a coarse prealignment is essential to synapsis with the fact that translocated segments generally pair effectively with their homologous counterparts in different chromosomal settings. Prealignment may thus be an accessory rather than a central event in the process of synapsis. We prefer a system that basically functions by random collisions between synaptic sites but in which effective collisions are facilitated by localizing the process within a suitable framework such as the nuclear membrane.

If the cytological evidence is taken at face value, then chromosomes are unpaired at leptotene. The modifications they undergo during this stage, lateral element formation and partial contraction, are therefore preparatory to, rather than a consequence of, incipient synapsis. We set forth reasons in the body of the review for believing that each chromosome must be equipped with synaptic sites and that these sites consist of relatively short stretches of DNA interspersed more or less regularly along the length of each chromosome. Such sites in particular would remain exposed for homolog interaction after leptotene contraction. We tentatively suggested that Z-DNA might play such a synaptic role and regarded its delayed replication as facilitating strand matching. A much broader experimental basis for Z-DNA behavior would be required, however, before its synaptic role could be treated as anything more than pure speculation.

The morphological changes that leptotene chromosomes undergo make it reasonable to suppose that the potential sites for crossing-over are already determined by the manner in which the DNA strands are folded. If it is assumed that the synaptic sites are a constant feature of the chromosome surface at leptotene, it is not difficult to visualize an occasional stretch of intersynaptic strand being trapped at the pairing surface of the chromosome and thus available for interaction with a strand similarly trapped in the homolog. We are in effect suggesting that the leptotene chromosome has already specified possible sites of crossing-over and that the activities effecting crossing-over during pachytene operate within lengths of DNA prescribed at leptotene.

Studies or speculations by several investigators point to the nuclear membrane as an appropriate framework for facilitating matching between homologous regions of chromosomes (10, 21, 48, 70). We distinguish between matching sites, the process of effecting matching, and the stabilization of matched regions. The first of these can be assigned to complementary base sequences regardless of whether the matching is direct or indirect. The second of these is necessarily complex because the matching process must provide favorable conditions for several related events. Corresponding synaptic regions must find one another, presumably as a consequence of collision, and such regions must then rapidly align with each other once the collision has occurred. We suppose that the DNA-binding protein facilitates the post-collision alignment. We presume that the association of lateral elements into a synaptonemal complex is a stabilizing process that is triggered by alignment. In our opinion, the initially aligned structure that forms on collision of homologous synaptic sites does not function as an integral part of the synaptonemal complex but serves only to trigger the joining of the preexisting but now matched lateral elements. Alignment in this sense regulates rather than contributes to the formation of the stable synaptic pair. Conceivably, factors other than strand matching could trigger synaptonemal complex formation. Whether the complex is structurally normal in the absence of strict homolog alignment remains unclear even though apparently well documented claims have been made for normality in some hybrids (47). From the standpoint of meiotic function, the most important feature of synapsis as a whole is that it can be disturbed or modified in at least three different ways. Synaptic sites may be partly or improperly exposed at zygotene due to abnormal contraction of the leptotene chromosome; the mechanisms for facilitating alignment may be disturbed by structural lesions in the nuclear membrane or in some of the cofactors such as the DNA-binding protein; the regulation of synaptonemal complex formation may be disrupted either by defects in the triggering mechanism or by defective structural components. The appearance of a heavy nuclear membrane fraction at meiotic prophase most probably reflects the many activities that occur there in connection with synapsis of homologs and may thus serve to house many synaptic lesions.

The relatively sharp changes in metabolism that accompany the progression of cytological stages during meiotic prophase almost certainly reflect a programmed sequence of events that are essential to the ultimate occurrence of crossing-over between homologs. Even though the location of cross-overs is randomized, their timing is not, and it is reasonable to suppose that the timing is regulated at least in part by the initiation of DNA nicking. Although synapsis may be a mechanism that provides for efficient crossing-over, it does not assure its initiation. If crossing-over is initiated by DNA nicking, initiation may be regulated by completion of synapsis. Furthermore, we favor the conclusion that the nicks introduced at pachytene are not entirely random but are associated with regions of repeated sequences. Randomness probably applies to the crossing-over region as a whole which, as discussed, may be determined prior to zygotene synapsis. The small stretches of repeated sequences presumed to be present within these regions would relax the conditions for homologous strand association by permitting overlaps in the DNA heteroduplexes. The overlaps would lead to individual variations in length of the repeated sequences associated with crossing-over, but the variations would not affect genetic correspondence between homologous strands. Statistically, repeated sequences would remain constant in length, shortened stretches being balanced by extended ones. Theoretically, complete loss of a repeated sequence stretch would significantly reduce the probability of a cross-over for the associated region and would manifest itself as a mutation affecting exchange in a particular locus. Such speculation on the microstructure of the regions involved in a cross-over already strains the capacities of biochemical analyses. The details of the speculation are best left to genetic studies.

ACKNOWLEDGMENTS

We wish to acknowledge a major indebtedness to the National Science Foundation for having been the primary source of financial support for our meiotic studies over many years. We also thank the N. I. C. H. D. for generous supplementary support of our program. One of us (HS) wishes to express his appreciation to Professors Mogens Westergaard and Diter von Wettstein who furnished him with much intellectual stimulation while on sabbatical leave at the Genetics Institute, University of Copenhagen. Both of us are grateful to our graduate student, Ms. Jennie Mather, who provided many helpful criticisms.

Literature Cited

1. Abel, W. O. 1964. Untersuchungen über den einfluss der temperatur auf die rekombinationshäufigkeit bei *Sphaerocarpus*. *Z. Vererbungslehre* 95:306-17
2. Ahmad, A. F., Bond, D. J., Whitehouse, H. L. K. 1972. The effect of an inverted chromosome segment on intragenic recombination in another chromosome of *Sordaria brevicollis*. *Genet. Res. Camb.* 19:121-27
3. Alberts, B. 1970. Function of gene 32-protein, a new protein essential for the genetic recombination and replication of T4 bacteriophage DNA. *Fed. Proc.* 29:1154-63
4. Bennett, M. D. 1971. The duration of meiosis. *Proc. Roy. Soc. London B.* 178:277-99

5. Buss, M. E., Henderson, S. A. 1971. Induced bivalent interlocking and the course of meiotic chromosome synapsis. *Nature New Biol.* 234:243-46
6. Callan, H. G. 1972. Replication of DNA in the chromosomes of eukaryotes. *Proc. Roy. Soc. London B.* 181:19-41
7. Chandley, A. C., Kofman-Alfaro, S. 1971. "Unscheduled" DNA synthesis in human germ cells following UV irradiation. *Exp. Cell Res.* 69:45-48
8. Chiu, S. M., Hastings, P. J. 1973. Premeiotic DNA synthesis and recombination in *Chlamydomonas reinhardii. Genetics* 73:29-43
9. Comings, D. E., Okada, T. A. 1970. Mechanism of chromosome pairing during meiosis. *Nature* 227:451-56
10. Comings, D. E., Okada, T. A. 1972. Architecture of meiotic cells and mechanisms of chromosomal pairing. *Advan. Cell Mol. Biol.* 2:310-84
11. Davies, D. R. 1968. Radiation studies on meiotic cells of *Chlamydomonas reinhardii*. In *Effects of radiation on meiotic systems*: *Int. Atomic Energy Agency*. 123-33
12. Dover, G. A., Riley, R. 1973. The effect of spindle inhibitors applied before meiosis on meiotic chromosome pairing. *J. Cell Sci.* 12:143–61
13. Davies, D. R., Lawrence, C. W. 1967. The mechanism of recombination in *Chlamydomonas reinhardii*. II. The influence of inhibitors of DNA synthesis on intergenic recombination. *Mutation Res.* 4:147-54
14. Driscoll, C. J., Darvey, N. L. 1970. Chromosome pairing: Effect of colchicine on an isochromosome. *Science* 169:290-91
15. Driscoll, C. J., Darvey, N. L., Barber, H. N. 1967. Effect of colchicine on meiosis of hexaploid wheat. *Nature* 216:687-88
16. Evans, C. M., Macefield, A. J. 1973. The effect of B chromosomes on homoeologous pairing in species hybrids. I. *Lolium temulentum* × *Lolium perenne. Chromosoma* 41:63-73
17. Feldman, M. 1968. Regulation of somatic association and meiotic pairing in common wheat. *Proc. 3rd Intern. Wheat Genet. Symp. Aust. Acad. Sci., Canberra.* 169-78
18. Fiil, A., Moens, P. B. 1973. The development, structure and function of modified synaptonemal complexes in mosquito oocytes. *Chromosoma* 41:37-62
19. Flavell, R. B., Walker, G. W. R. 1973. The occurrence and role of DNA synthesis during meiosis in wheat and rye. *Exp. Cell Res.* 77:15-24
20. Fogel, S., Mortimer, R. K. 1971. Recombination in yeast. *Ann. Rev. Genet.* 5:219-36
21. Gillies, G. B. 1972. Reconstruction of the *Neurospora crassa* pachytene karyotype from serial sections of synaptonemal complexes. *Chromosoma* 36:119-30
22. Grell, R. F. 1969. Meiotic and somatic pairing. In *Genetic Organization, Vol. 1*, ed. E. W. Caspari, A. W. Ravin, 361-492. New York: Academic
23. Grell, R. F., Bank, H., Gassner, G. 1972. Meiotic exchange without the synaptinemal complex. *Nature New Biol.* 240:155-57
24. Hearst, J. E., Botchan, M. 1970. The eukaryotic chromosome. *Ann. Rev. Biochem.* 39:151-81
25. Hecht, N. B., Stern, H. 1971. A late replicating DNA protein complex from cells in meiotic prophase. *Exp. Cell. Res.* 69:1-10
26. Henderson, S. A. 1966. Time of chiasma formation in relation to the time of deoxyribonucleic acid synthesis. *Nature* 211:1043-47
27. Henderson, S. A. 1970. The time and place of meiotic crossing-over. *Ann. Rev. Genet.* 4:295-324
28. Holliday, R. 1971. Biochemical measure of the time and frequency of radiation-induced allelic recombination in *Ustilago. Nature New Biol.* 232:233-36
29. Hotta, Y., Ito, M., Stern, H. 1966. Synthesis of DNA during meiosis. *Proc. Nat. Acad. Sci. USA* 56:1184-91
30. Hotta, Y., Parchman, L. G., Stern, H. 1968. Protein synthesis during meiosis. *Proc. Nat. Acad. Sci. USA* 60:575-82
31. Hotta, Y., Shepard, J. 1973. Biochemical aspects of colchicine action on meiotic cells. *Mol. Gen. Genet.* 122:243-60
32. Hotta, Y., Stern, H. 1971. Analysis of DNA synthesis during meiotic prophase in *Lilium. J. Mol. Biol.* 55:337-55
33. Hotta, Y., Stern, H. 1971. A DNA-binding protein in meiotic cells of *Lilium. Develop. Biol.* 26:87-99
34. Hotta, Y., Stern, H. 1971. Meiotic protein in spermatocytes of mammals. *Nature New Biol.* 234:83-86
35. Howell, S. H., Stern, H. 1971. The appearance of DNA breakage and repair activities in the synchronous

meiotic cycle of *Lilium*. *J. Mol. Biol.* 55:357-78

36. Ito, M., Hotta, Y. 1973. Autoradiographic analysis of thymidine incorporation during meiotic prophase in microsporocytes of *Lilium*. *chromsoma.* In press
37. Ito, M., Hotta, Y., Stern, H. 1967. Studies of meiosis in vitro. II. Effect of inhibiting DNA synthesis during meiotic prophase on chromosome structure and behavior. *Develop. Biol.* 16:54-77
38. Kashnig, D. M., Kasper, C. B. 1969. Isolation, morphology, and composition of the nuclear membrane from rat liver. *J. Biol. Chem.* 244(14):3786-92
39. King, R. C. 1970. The meiotic behavior of the *Drosophila* oocyte. *Int. Rev. Cytol.* 28:125-68
40. Kofman-Alfaro, S., Chandley, A. C. 1970. Meiosis in the male mouse. An autoradiographic investigation. *Chromosoma* 31:404-20
41 Kofman-Alfaro, S., Chandley, A. C. 1971. Radiation-initiated synthesis in spermatogenic cells of the mouse. *Exp. Cell Res.* 69:33-44
42. Levan, A. 1939. The effect of colchicine on meiosis in *Allium*. *Hereditas* 25:9-26
43. Lu, B. C. 1970. Genetic recombination in *Coprinus*. II. Its relation to the synaptinemal complexes. *J. Cell Sci.* 6:669-78
44. Maguire, M. P. 1966. The relationship of crossing over to chromosome synapsis in a short paracentric inversion. *Genetics* 53:1071-77
45. Maguire, M. P. 1968. Evidence on the stage of heat-induced crossover effect in maize. *Genetics* 60:353-62
46. Mather, J., Hotta, Y. 1973. A eukaryotic DNA-binding protein in meiotic cells. Proc. from NATO Advanced Study Inst. on DNA Replication. In press
47. Menzel, M., Price, J. M. 1966. Fine structure of synapsed chromosomes in F_1 *Lycopersicon esculentum-Solanum lycopersicoides* and its parents. *Am. J. Bot.* 53:1079-86
48. Moens, P. B. 1969. The fine structure of meiotic chromosome polarization and pairing in *Locusta migratoria* spermatocytes. *Chromosoma* 28:1-25
49. Moens, P. B. 1970. Premeiotic DNA synthesis and the time of chromosome pairing in *Locusta migratoria*. *Proc. Nat. Acad. Sci. USA* 66:94-98
50. Monesi, V. 1971. Chromosome activities during meiosis and spermiogenesis. *J. Reprod. Fert. Suppl.* 13:1-14
51. Moses, M. J. 1968. Synaptinemal complex. *Ann. Rev. Genet.* 2:363-412
52. Palmer, R. G. 1971. Cytological studies of ameiotic and normal maize with reference to premeiotic pairing. *Chromosoma* 35:233-46
53. Parchman, L. G., Lin, K. 1972. Nucleolar RNA synthesis during meiosis of lily microsporocytes. *Nature New Biol.* 239:235-37
54. Parchman, L. G., Roth, T. F. 1971. Pachytene synaptonemal complexes and meiotic achiasmatic chromosomes. *Chromosoma* 33:129-45
55. Parchman, L. G., Stern, H. 1969. The inhibition of protein synthesis in meiotic cells and its effect on chromosome behavior. *Chromosoma* 26: 298–311
56. Peacock, W. J. 1968. Chiasmata and crossing over. In *Replication and Recombination of Genetic Material*, ed. W. J. Peacock, R. D. Brock, 242-52. Canberra: Australian Acad. Sci.
57. Rhoades, M. M. 1968. Studies on the cytological basis of crossing over. In *Replication and Recombination of Genetic Material*, ed. W. J. Peacock, R. D. Brock, 229-41. Canberra: Australian Acad. Sci.
58. Riley, R., Bennett, M. D. 1971. Meiotic DNA synthesis. *Nature* 230:182-85
59. Riley, R., Chapman, V., Young, R. M., Belfield, A. M. 1966. The control of meiotic chromosome pairing by the chromosomes of homeologous group 5 of *Triticum aestivum*. *Nature* 212: 1475–77
60. Roth, T. F., Ito, M. 1967. DNA dependent formation of the synaptinemal complex at meiotic prophase. *J. Cell Biol.* 35:247-55
61. Sen, S. K. 1969. Regulation of chiasma frequency in *Lilium* microsporocytes in vitro. *Nature* 224:178-79
62. Shepard, J., Boothroyd, E. R., Stern, H. The action of colchicine on synapsis in meiotic cells of *Lilium*. In preparation
63. Smyth, D. R., Stern, H. Reannealing of DNA synthesized during pachytene in microsporocytes of *Lilium henryi*. *Nature*. In press.
64. Solari, A. J. 1971. The behavior of chromosomal axes in Searle's X-autosome translocation. *Chromosoma* 34:99-112
65. Stadler, J., Franke, W. W. 1972. Colchicine-binding proteins in chromatin and membranes. *Nature New Biol.* 237:237-38

66. Stern, H., Hotta, Y. 1967. Chromosome behavior during development of meiotic tissue. In *The Control of Nuclear Activity*, ed. L. Goldstein, 47-76. Englewood Cliffs: Prentice Hall
67. Walters, M. S. 1972. Preleptotene chromosome contraction in *Lilium longiflorum* "Croft" *Chromosoma* 39:311-32
68. Westergaard, M., von Wettstein, D. 1972. The synaptinemal complex. *Ann. Rev. Genet.* 6:71-110
69. Westerman, M. 1972. Dose-response of chiasma frequency to X-radiation in *Chorthippus brunneus*. *Mutation Res.* 15:55-65
70. Wettstein, R., Sotelo, J. R. 1971. The molecular architecture of synaptonemal complexes. *Advan. Cell Mol. Biol.* 1:109-52
71. Wimber, D. E., Prensky, W. 1963. Autoradiography with meiotic chromosomes of the male newt (*Triturus viridescens*) using H^3-thymidine. *Genetics* 48:1731-38

RECOMBINATION DEFICIENT MUTANTS OF *E. COLI* AND OTHER BACTERIA 3047

Alvin J. Clark

Department of Molecular Biology, University of California, Berkeley, California

1. INTRODUCTION

Characterization of mutants defective in genetic recombination can help clarify the molecular nature of recombination and its relationships to replication, repair, and other processes of DNA metabolism. A key assumption of this approach is that recombination deficient mutants can be recognized. The main focus of this article will be an examination of the assumption as it applies to certain mutants of *E. coli*. Although it will be possible to mention results obtained with other bacteria, an emphasis on *E. coli* is required by space limitations. This is particularly regrettable because of the large amount of information now available on recombination deficient mutants of nonenteric bacteria and the power of the comparative approach. An accompanying article by Radding (1) adopts the comparative approach and reviews what is known about the molecular nature of the steps of recombination of viruses and eukaryotes as well as prokaryotes.

2. DETECTION OF RECOMBINATION DEFICIENT MUTANTS AND A SUMMARY LIST OF THE PERTINENT GENES

The most direct method of detecting mutants that might be blocked in genetic recombination is to screen survivors of mutagenic treatment for strains with reduced ability to produce recombinants when exposed to cells of an Hfr donor (2), generalized transducing phage (Yajko & Clark, unpublished results), or purified DNA from an appropriate donor (3). Mutants detected in this fashion must be shown to have only the growth factor requirements of the parent strain and to have no additional mutation affecting the characteristic selected in test crosses. They must also be shown to retain their recipient ability for transferred DNA. By use of radioactive and density labels transferred DNA may be detected directly. Alternatively biological methods may be used. For example transfection (i.e. transformation with phage DNA), formation of infectious

TABLE 1. LIST OF GENES AFFECTING DNA METABOLISM IN *E. coli*[a]

Genes	Approximate map location	Reference for mapping	Function of gene, distinguishing characteristics of mutants or other information	Representative reference
A. *rec* Genes - all mutations reduce recombinant formation in appropriate genetic backgrounds and have little or no effect on plasmid inheritance.				
recA	51	(86)	Complete recombination deficiency and many other phenotypic defects including suppression of *tif-1*.	(4)
recB	54	(15, 87)	Structural gene of exonuclease V.	(16)
recC	54	(15, 87)	Structural gene of exonuclease V.	(16)
recD	?	—	Reserved for the unmapped mutation *rec-34*.	(7)
recE	?	—	By definition the structural gene of the ATP-independent nuclease characteristic of *sbcA* mutants.	(60)
recF	73	(62)	Recombination deficiency of *recB*⁻ *recC*⁻ *sbcB*⁻ strains; blocks UV-induction of lambda prophage.	(62)
recG	72-74	(17)	Recombination deficiency.	(17)
recH	50-53	(17)	Recombination deficiency.	(17)
recJ	54-56	(62)	Uncharacterized mutation: recombination deficiency of *recB*⁻ *recC*⁻ *sbcB*⁻ strain.	(62)
recK	25-40	(62)	Uncharacterized mutation: recombination deficiency of *recB*⁻ *recC*⁻ *sbcB*⁻ strain.	(62)
recL	74-75	(62)	Recombination deficiency of *recB*⁻ *recC*⁻ *sbc*⁻ strain.	(62)
B. *uvr* Genes - all mutations confer sensitivity to UV-irradiation.				
uvrA	81	(88)	Determines one of the steps in pyrimidine dimer excision.	(88)
uvrB	18	(88)	Determines one of the steps in pyrimidine dimer excision.	(88)
uvrC	36	(88)	Determines one of the steps in pyrimidine dimer excision.	(88)
uvrD	75	(89, 90)	Mutation dominant, no effect on mutation rate.	(91)
uvrE	75	(92)	Mutation produces high mutation rate; possibly identical to *uvrD*.	(93, 94)
uvrF	70-72	(95)	Mutation inhibits UV-induction of prophage lambda (may be identical to *recF*).	(95)

C. *dna* Genes - all mutations confer a conditional block in DNA synthesis.				
dnaA	73	(96)	Initiation of new rounds of chromosome replication.	(97)
dnaB	79-81	(96)	Elongation of DNA chains.	(97)
dnaC	89	(96)	Inititation of new rounds of chromosome replication.	(96, 98)
dnaD	89	(96)	Mutation may lie in *dnaC* but confers inability of lambda to replicate.	(96, 98) (99)
dnaE	4	(96)	Structural gene for DNA polymerase III.	(100)
dnaF	42	(96)	Structural gene for ribonucleoside diphosphate reductase: subunit B1.	(101)
dnaG	61-62	(96)	"Regulates initiation of the synthesis of Okazaki pieces"	(102)
D. Other significant genes				
cetC	89	(103)	Colicin E2 refractivity; conjugational but not transductional recombination deficiency, UV-sensitivity.	(104)
lex	81	(105)	Mutations dominant; mimic *recA* mutations except do not cause appreciable recombination deficiency.	(105)
lig	45	(106)	Structural gene for DNA ligase.	(107)
lon	10	(108)	UV-sensitivity and radiation induced filament formation.	(108)
polA	76	(109)	Structural gene for DNA polymerase I.	(110)
polB	2	(111, 111a)	Structural gene for DNA polymerase II.	(112)
rac	29	(50)	Suppressor *recB* and *recC* mutant phenotype in conjugational merozygotes.	(50)
sbcA	?	—	Suppressor of *recB* and *recC* mutations; mutants contain high ATP-independent DNase activity (see *recE*).	(60)
sbcB	38	(51)	The structural gene for exonuclease I; *recB*$^-$ *recC*$^-$ *sbcB*$^-$ strains are Rec^+ Uv^R Mit^R.	(61)
xonA	38	(63)	Probably identical to sbcB; *recB*$^-$ *recC*$^-$ *xonA*$^-$ strains are Rec^- Uv^R Mit^R.	(63)
xthA	31	(113)	Structural gene for exonuclease III.	(114)
tif	51	(83)	One mutation known causes thermal derepression of wild type lambda and other defects mimicking UV-irradiation.	(83)
zab	51	(81)	Suppressor of *tif-1*.	(81)

[a] The location of each gene is given to the nearest whole unit on the map of *Escherichia coli* drawn by Taylor & Trotter (85b); exact positions and additional references are given by Taylor & Trotter (85b). The list is not meant to be all inclusive but includes genes which in the author's opinion are pertinent to work in recombination.

centers by generalized transducing phage, or zygotic induction may be used to indicate the occurrence of transformation, transduction, or conjugation respectively. The inheritance of plasmid DNA, such as that of sex factors or resistance factors, by a process called repliconation (4) is another useful biological method to indicate that DNA transfer occurs successfully to a putative recombination deficient mutant. In this case an additional fact may be learned about the mutant. If inheritance by repliconation is successful then it is clear that zygotes formed from the mutant as a result of transfer are capable of multiplying vegetatively. Thus it seems likely that mutants that show reduced recombinant frequencies but show normal ability to receive DNA and to inherit it by repliconation are defective in one of the steps in genetic recombination. The genes in which mutations produce this phenotype are known as *rec* genes. At least ten such genes have been identified in *E. coli* and they are listed in Table 1A along with their approximate map locations and some distinguishing characteristics.

In addition to their detection by reduced recombination frequency *rec* mutants may also be detected by some of their pleiotropic properties. Sensitivity to ultraviolet light, Xrays, or γ rays are for example properties characteristic of many *rec* mutants (5); consequently some *rec* mutants have been discovered among survivors of mutagenic treatment isolated because of their sensitivity to these treatments (e.g. 6–8). Inability of prophages to be induced by various treatments is conferred by certain *rec* mutations (9, 10) and the apparent inability to be lysogenized by appropriate temperate phage is conferred by other *rec* mutations (11, 12). Hence these properties may also be used to detect mutants among which there may be some *rec*$^-$ strains (13).

Many mutants sensitive to ultraviolet light, which suffer no apparent decrease in recombination frequency, are also found. In *E. coli* these are called *uvr* mutants and at least six responsible genes are recognized (Table 1B). Mutants sensitive to other agents known to attack DNA but suffering no or little decrease in recombination frequencies have also been detected and some of these are included in Table 1D. These genes are listed because there is a possibility their mutations may affect recombination frequencies in the appropriate genetic background. As will be discussed in section 3 (Multiple Pathways for Recombination are Inferred from Indirect Suppression of *recB* and *recC* Mutations) these genes may affect sites in minor pathways of recombination or in one of two or more major alternative pathways. If this is true then only by testing mutations of particular genes in genetic backgrounds in which alternative pathways are blocked will the role of these genes in recombination be detectable. This is particularly true of genes determining known enzymes of DNA metabolism such as polymerase II and exonuclease III which were detected by examining survivors of mutagenic treatment for the absence of the enzymes (Table 1D).

Also included in Table 1C are the six or seven genes in which mutations block the initiation or continuation of chromosome replication. In one case (*dnaB?*) the mutations have been found to decrease recombination frequencies (14). This opens the possibility of the involvement of DNA synthesis in bacterial genetic recombination.

3. THE *recB* AND *recC* GENES OF *E. coli*: IS THEIR ATP-DEPENDENT DNA NUCLEASE A RECOMBINATION ENZYME?

The Widespread Occurrence and the Properties of ATP-dependent DNA Nucleases

Among the mutations causing an appreciable decrease in recombinant frequencies in wild type *E. coli* are a group that cluster between *thyA* and *argA*. Nearly three dozen alleles are known; the first 20 that were tested fell into two distinct complementation groups (15): *recB* and *recC*. There are two temperature sensitive mutations (16) and two amber suppressible mutations (Margossian & Clark, unpublished results) which fall, one each, in *recB* and *recC*. Experiments on another series of mutations in this region (17) have resulted in a much more complex complementation pattern in which six complementation groups can be distinguished (i.e. four in addition to *recB* and *recC*).

All of the mutations in this cluster have one feature in common; they inactivate a nuclease which has been named exonuclease V. A more extensive description of the properties of *E. coli* exonuclease V is presented in the accompanying review by Radding (1) but brief mention of them is necessary here. The enzyme is an ATP-dependent exonuclease degrading double- or single-stranded DNA processively to oligonucleotides. The action on double-stranded DNA appears to occur in the 3′ to 5′ and 5′ to 3′ directions simultaneously and there appears to be no preference for the 3′ or 5′ terminus of single-stranded DNA. The enzyme is also an endonuclease stimulated by ATP in its scission of single-stranded closed circular DNA. In addition the enzyme will degrade ATP in the presence of double-stranded DNA whether or not digestion of the DNA occurs (18). When they have been tested, all four properties of the enzyme are absent in *recB* and *recC* mutant extracts. The enzymes isolated from the temperature sensitive *recB* and *recC* mutants may show differential effects of heat on the four activities (Kushner, personal communication).

The distinctive ATP-dependency of exonuclease V has led to very rapid discovery of similar enzymes in a wide variety of species. In fact the first such activity was described in extracts of *Micrococcus luteus* (19), and the properties of the purified micrococcal enzyme were described at the same time as those of the purified *E. coli* enzyme (20–22). ATP-dependent nuclease activity has also been described in extracts of *Diplococcus pneumonia* (23), *Hemophilus influenzae* (e.g. 24), and *Bacillus subtilis* (25)—all of which undergo recombination following transformation—*Pseudomonas aeruginosa* (Miller & Clark, unpublished results)—which undergoes recombination following transduction and conjugation —and *Bacillus laterosporus* (26) and *Mycobacterium smegmatis* (27). These findings imply that exonuclease V activity will be found to be ubiquitous in the eubacteria and possibly throughout the prokaryotic world. Other enzymes of nucleic acid metabolism known to be this widespread are DNA and RNA polymerases. The necessity for these enzymes is well known. The ATP-dependent nucleases may also be of crucial importance.

Extensively purified ATP-dependent nucleases from different species have similar properties. In addition to those mentioned for the *E. coli* exonuclease V another property has been described for the *Hemophilus* enzyme (28). In a low ionic environment (5–10 mM Tris-HCl, 10–50 mM NaCl) and at an ATP concentration of 1 mM this enzyme will degrade T7 DNA to oligonucleotides through a series of different sized double-stranded fragments with single-stranded termini. The number of nucleotide moieties in the single-stranded portions range around 4200 and permit the annealing of a sizeable fraction of the fragments with one another to produce DNAs of increased molecular weights. Since this is suggestive of the results of genetic recombination, the hypothesis has been offered that ATP hydrolysis by the enzyme catalyzes an unwinding of the double-stranded DNA from its termini followed by a rapid endonucleolytic cleavage of one of the strands (28). When *E. coli* exonuclease V is tested under similar conditions degradation of linear duplex DNA also results in double-stranded fragments with single-strand termini (29). At higher salt and lower ATP concentrations these intermediates in degradation are not found (Linn, personal communication) perhaps because the ratios of the various activities of exonuclease V are sensitive to ATP-concentration (30). Before we can be certain of the activity of the ATP-dependent DNAses in vivo, therefore, we may need to know the intracellular conditions under which the enzymes operate.

A Vital Role for Exonuclease V

Cultures of strains carrying *recB* or *recC* mutations are characterized by a high proportion of inviable cells [i.e., cells incapable of multiplying (31) or of forming a macroscopic colony (15, 32)]. This proportion may reach 80% in the case of *recB* and *recC* mutants (32) so that only 20% of the cells are capable of forming a colony. The reason for this inviability is not known but experiments are underway to determine the physiological abnormalities of the inviable cells by first separating them from the viable cells (Capaldo & Barbour, personal communication).

Other genes must contribute to the remaining viability of the *rec* mutants as seen from the observation that *polA*$^-$ *recB*$^-$ double mutants are far more inviable than *recB*$^-$ single mutants (33). In one case (Emmerson, personal communication) there was enough residual viability to perpetuate the double mutant strain but in other cases the double mutant cannot be isolated (33). When the *polA* mutation (*polA12*) determines a heat sensitive polymerase I the biochemical events accompanying inviability can be measured following a shift from permissive to nonpermissive temperature (33). Recent studies by Monk, Kinross & Town (34) show that no defect in the rate of DNA synthesis or the rate of joining of Okazaki fragments can be detected. There is also no evidence of the accumulation of nicks or gaps in the newly synthesized DNA of the *polA12 recB* double mutant. From the absence of predicted effects on these features or normal metabolism the conclusion is reached that perhaps chromosomal segregation or cell division is affected in the double mutant at the nonpermissive temperature (34).

There is recent evidence on the complex structure of the bacterial chromosome (35) which may enable testable hypotheses of this nature to be formulated. The evidence indicates that the chromosome at the end of replication rounds is divided into about fifty supercoiled loops separated from each other by interactions between local regions of DNA and a core material, presumed to be RNA. Each supercoiled loop is capable of folding into a compact structure or of unfolding when nicked.

It seems worthwhile to speculate on the possible relations between recombination and replication or segregation of such a complex compact chromosomal structure. For example catenanes may result from the replication of the DNA in all or a fraction of the supercoiled loops; in this case the separation of the catenanes preparatory to segregation may depend upon recombination enzymes. Alternatively the interactions between the local chromosome regions and the core may involve recombination enzymes either in their formation or loosening to permit replication. In either case metabolic defects produced by the absence of exonuclease V may have been undetectable by the methods employed by Monk, Kinross & Town (34) because of their infrequent occurrence in the DNA of the chromosome. Chromosome organization at levels higher than the primary and secondary structure of DNA may provide a basis for hypotheses to explain the vital metabolic role of exonuclease V.

One additional cautionary thought, however, is pertinent in interpreting results with a *recB*$^-$ *polA*12 double mutant. At the permissive temperature for polymerase I the majority of these double mutant cells are already inviable because of their *recB* mutation. Hence the biochemical events leading to inviability of the fraction of the population that dies at the nonpermissive temperature may be masked. It would seem desirable, therefore, first to study the biochemical events leading to inviability of conditional single *recB* and *recC* mutants so that the additional effects of a *polA* mutation may be seen in perspective. When the inviability of *recB* and *recC* single mutants is understood, the vital role of exonuclease V in DNA metabolism (broadly interpreted to include events like chromosome segregation) or in cell division will be clearer, as will its relationships to other enzymes of DNA metabolism.

Degradative Activities of Exonuclease V in vivo

Although the vital role of exonuclease V in cell metabolism can not as yet be correlated with its in vitro activities, there are in vivo effects of the enzyme for which a correlation can be made. These are degradation of infecting phage DNAs under a variety of circumstances and degradation of endogenous DNA following UV irradiation or the interruption of DNA synthesis.

Phage DNA not carrying the modifications specific to its host is subject to restriction. Purified restriction enzymes catalyze double-stranded scissions at particular locations on unmodified phage DNA producing high molecular weight fragments, but restriction in vivo results in rapid breakdown of unmodified DNA to mononucleotides and nucleosides that are excreted (36). The rate of this complete degradation is reduced by *recB* mutations (37). This indicates that exonuclease V can degrade restricted DNA rapidly but that other enzymes can perform the degradation more slowly. Also a mutation inactivating exonuclease I lowers the extent without changing the kinetics of degradation (37). Exonuclease I acts specifically on single-stranded DNA degrading it to mononucleotides beginning from a 3′ hydroxylated terminus (37a). Thus a fraction (i.e. about 20%) of the products of exonuclease V digestion of restricted DNA may be single-stranded polynucleotides and contain 3′ hydroxylated termini. The activity described by Friedman & Smith (28) for the *Hemophilus* ATP-dependent nuclease recently demonstrated for exonuclease V (29) may therefore be responsible for at least part of the postrestrictive degradation of lambda DNA.

Unrestricted DNA is also attacked by exonuclease V. Experiments using helper-phage promoted transfection of half-molecules of lambda indicate that exonuclease V can degrade linear molecules (38, 39). Helper-phage can prevent this degradation when they are *gam*$^+$. Unger & Clark (40) have shown that the phage gene *gam*$^+$ is responsible for the disappearance of exonuclease V activity which occurs shortly after infection. Recently the *gam* protein has been purified and its action is now under investigation (Sakaki, Y., A. Karu, S. Linn & H. Echols, personal communication). Experiments by Enquist & Skalka (41) provide a rationale for understanding this interaction between phage and bacterial proteins by indicating that a step in lambda replication, perhaps the transition from Cairns-type to rolling-circle-type replication or the production of linear concatemers, is peculiarly sensitive to exonuclease V especially in *recA* mutants.

Degradative effects of exonuclease V on endogenous DNA have been inferred from differences between *recB*$^+$ and *recB*$^-$ mutants. Following UV-irradiation of *recA* or *lex* single mutants degradation of significant and dose-dependent fractions of the DNA is observed greatly in excess of the degradation in wild type cells (42, 43). This UV-induced degradation is reduced greatly if *recA recB* (44) or *lex recB* (45) double mutants are treated, thus indicating the involvement of exonuclease V. A similar exonuclease V-dependent degradation of DNA occurs without irradiation when *dnaB* thermally sensitive or *polA12 recA* mutants are incubated at the nonpermissive temperature (33, 46). It seems possible in these situations that DNA metabolism is arrested in such a way that an intermediate subject to exonuclease V degradation is produced. Alternatively internal conditions leading to a change in the activity spectrum of exonuclease V might occur in these situations exposing the DNA to degradation without an arrest of synthesis. In these four situations cell death is involved as well as DNA degradation. Death is not dependent on degradation, however, because the *recB*$^-$ derivatives of these strains show very little degradation but are as sensitive to UV or heat as their *recB*$^+$ relatives.

Finally, exonuclease V protects its host from a product of the *old* gene of temperate phage P2. This conclusion stems from the observation that P2 lysogens of *recB* or *recC* mutants cannot be isolated unless the infecting P2 phage is an *old* mutant (12, 47). Apparently lysogeny occurs but the *recB*$^-$ or *recC*$^-$ lysogens are prevented from multiplying by the constitutively produced *old* product. Perhaps the *old* gene determines the production of a polynucleotide inhibitory to cell growth and exonuclease V degrades the polynucleotide.

Effects of recB and recC Mutations on Bacterial Recombination

Following conjugation and transduction of *recB* and *recC* mutants, recombinant colonies are observed at frequencies from 0.3% to 20% of that observed with wild type recipients (48, 49). In certain crosses the high frequencies can be explained by the transfer of *recB*$^+$ or *recC*$^+$ alleles to zygotes. In other crosses the high frequencies can be explained by transfer of *rac*$^+$ (a mnemonic for this gene or set of genes is "*r*ecombination *ac*tivation") to zygotes in which it may be transiently derepressed (50, 51). Hence the impression has grown that in the absence of transfer of *recB*$^+$, *recC*$^+$, or *rac*$^+$ alleles no recombinant colonies would be observed (Howard-Flanders, personal communication).

Recombination between F′ plasmids and the chromosome of their *recB* and *recC* mutant hosts, however, can not be explained in this manner. Chromosome mobilization by F*lac* in *recB*$^-$ and *recC*$^-$ single mutants, for example, results in transfer of chromosome markers with frequencies 10% to 50% of those of a wild type strain (52). In another example formation of recombinant F′ plasmids has been detected following transfer of a suitably marked F′ plasmid to a *recB* or *recC* mutant recipient (53). Frequencies of from 8% to 60% of those observed with wild type F′ strains are observed at different times after formation of the plasmid-carrying cells. A different type of experiment also indicates that quite high levels of recombination can occur in *recB* and *recC* mutants. The amount of β-galactosidase produced in zygotes formed by crossing a *lacZ*$^-$ Hfr (in which *lacZ* is transferred early to zygotes) with a series of *F*$^-$ strains carrying a noncomplementary *lacZ* mutation is measured (Low & Birge, personal communication). Intragenic recombination is presumed necessary for β-galactosidase production in this situation. The amounts of enzyme in *recB*$^-$ and *recC*$^-$ zygotes were nearly as much as in wild type zygotes. By contrast, *recA*$^-$ zygotes formed no detectable enzyme.

In these experiments when the viability of a primary recombinant cell, i.e. its multiplication to produce a visible colony, was not required to detect a recombinant chromosome, the recombination frequencies measured were reasonably high. When the viability of the recombinant cell was required, however, the recombination frequencies were far lower. Hence the hypothesis has grown that the absence of exonuclease V reduces severely the viability of recombinant cells while not reducing the amount of recombination nearly as much. From this point of view exonuclease V appears to be more involved with cell viability than with recombination.

Detection of Joint and Recombinant Molecules

The most direct method for establishing that a mutation blocks recombination is to show that joint or recombinant molecule formation is blocked in the mutant. Joint molecules have fragments of each parental DNA hydrogen-bonded to one another, while in recombinant molecules the parental fragments are covalently joined (53a). If DNA is transferred successfully to a mutant but joint molecule formation fails to occur, then recombination may be blocked in a presynaptic or synaptic step (5). If joint molecules are formed but not recombinant molecules then a heteroduplex lengthening or postsynaptic step (5) may be blocked. If recombinant molecules are formed but recombinant clones are rare then it is possible that a postrecombinational step in the conversion of a zygote to vegetatively multiplying cell is blocked. Alternatively the recombinant molecules detected may be abnormal byproducts of a blocked recombination pathway and may be unable to replicate.

Paul & Riley (54) are now engaged in this type of study. Following conjugation of Hfr cells whose chromosomes are labeled with ^{3}H with F^{-} recipient cells whose chromosomes are labeled with ^{32}P, ^{2}H, and ^{15}N, joint molecules have been detected by CsCl density gradient centrifugation. Such molecules occur in *recB*⁻ recipients reportedly in lower amounts than in *recB*⁺ recipients, indicating that recombination occurs up through the stage of synapsis in the absence of exonuclease V. Whether or not the absence of exonuclease V blocks postsynaptic events leading to recombinant molecules cannot as yet be assessed. It is important to determine the formation of recombinant molecules in *recB* mutants because their occurrence at a frequency approaching that in wild type cells may indicate that exonuclease V is necessary for postrecombinational events connected with viability of recombinant cells and not with recombination itself. In the case of mutant of *Bacillus subtilis* thought to be recombination deficient conversion of joint to recombinant molecules was found but the recombinant DNA failed to replicate (55). This may mean that the mutation affected postrecombinational methabolism of *Bacillus subtilis* zygotes. On the other hand the recombinant DNA might be abnormal enough not to be replicatable. In one *rec* mutant of *Hemophilus influenzae* covalent bonds are formed between some donor and recipient molecules but these must be abnormal recombinant molecules, since the donor marker cannot be recovered as active transforming DNA (56). Should recombinant molecules be formed at wild type frequencies in *recB* or *recC* mutants it will therefore be necessary to establish that these are not abnormal byproducts of a blocked recombination pathway before the conclusion is justified that exonuclease V is not a recombination enzyme.

Effects of recB and recC Mutations on Bacteriophage Lambda Recombination

Under normal circumstances general recombination of phage lambda occurs at the same frequencies in wild type and *recB*⁻ or *recC*⁻ hosts (57); it does not require exonuclease V. There are two reasons for this independence of lambda recombination from the *recB* and *recC* genes. The first is that lambda carries at least two genes that are involved in its own recombination: *redX* and *redB*. *redX* determines an exonuclease with a well-defined function, which can be used to carry out recombination in vitro (1). *redB* determines a protein of unknown function which has been hypothesized to play the role in lambda recombination played by the *recA* gene product in *E. coli* recombination (e.g. 58). The second reason is that lambda also carries a gene whose product inhibits exonuclease V activity, the *gam* gene (40). Thus *gam*⁺ lambda-phage-infected cells are essentially RecB⁻ RecC⁻ phenocopies, and hence cannot recombine using exonuclease V even when *redX*⁻ or *redB*⁻ phage are used. When *gam*⁻ *red*⁻ lambda phage are used, lambda recombination appears as dependent upon exonuclease V as it was dependent upon the *redX* nuclease under normal circumstances (59). Since bacterial cell viability is not necessary to detect recombinants of phage lambda, this result supports the hypothesis that exonuclease V is a recombination enzyme. The support is weak, however, because an alternative explanation of the results may be offered without hypothesizing a role in recombination for exonuclease V (Skalka, personal communication). In *recB*$^+$*recC*$^+$ hosts, exonuclease V interferes with the formation of maturable phage DNA by rolling-circle-type replication if the infecting phage is *red*$^-$*gam*$^-$. Under these conditions, most maturable DNA is produced by recombination that might not directly involve exonuclease V. In *recB*$^-$ hosts, maturable DNA of *red*$^-$*gam*$^-$ phage is also produced by replication, effectively diluting the maturable DNA formed by recombination. Under these conditions recombinant frequencies would be lower than in *recB*$^+$*recC*$^+$ hosts, although exonuclease V might not be directly involved in recombination.

Multiple Pathways of Recombination are Inferred from Indirect Suppression of recB and recC Mutations

If exonuclease V is a recombination enzyme then an explanation of the successful recombination that proceeds in its absence must be provided. The existence of pathways of recombination that do not involve exonuclease V and that carry out the detected residual recombination would be such an explanation. In a previous review article (5) the concept of a pathway of recombination was discussed and the outlines of work in *E. coli* were described that lead to the hypothesis that multiple pathways of recombination occur in that organism. Indirect suppression of *recB* and *recC* mutations by *sbcA* (60) and *sbcB* (61) mutations had been found leading to recombination proficient strains genetically and enzymatically distinct from wild type *E. coli*. In the case of the *sbcB* mutants exonuclease I was defective and in the case of the *sbcA* mutants an ATP-independent nuclease activity was present at higher concentrations than in *sbcA*$^+$ strains. We are still uncertain whether this ATP-independent activity represents a new nuclease or a change in amount or nature of an already described nuclease, e. g. exonuclease III. Fractionation of extracts of an *sbcA*$^-$ strain has resulted in more than 100-fold purification of the enzyme, but comparable studies of *sbcA*$^+$ extracts are still to be concluded (Nagaishi, Kushner & Clark, unpublished results). In any event this ATP-independent nuclease activity may perform the recombination function normally performed by exonuclease V. Consequently it is difficult at this time to argue convincingly from the *sbcA*$^-$ suppression that multiple pathways of recombination exist in *E. coli*. The case of the *sbcB*$^-$ suppression is different, however.

In *sbcB* mutants the absence of ExoI compensates for the absence of ExoV and enables recombinant colonies to be detected at wild type frequencies (61). In itself this does not indicate multiple pathways of recombination. For example an argument might be offered based on a single pathway to explain the recombination proficiency which exonuclease I deficiency (produced by *sbcB*$^-$) confers on exonuclease V deficient (i.e. *recB*$^-$*recC*$^-$) strains. Accordingly exonuclease I is seen as an enzyme degrading an intermediate in the recombination pathway to a side product that can become a recombinational intermediate again through the action of exonuclease V (Fig. 1a). Note that exonuclease V is hypothesized not to be on the direct recombination pathway. Note also that

A. Hypothesis of one recombination pathway

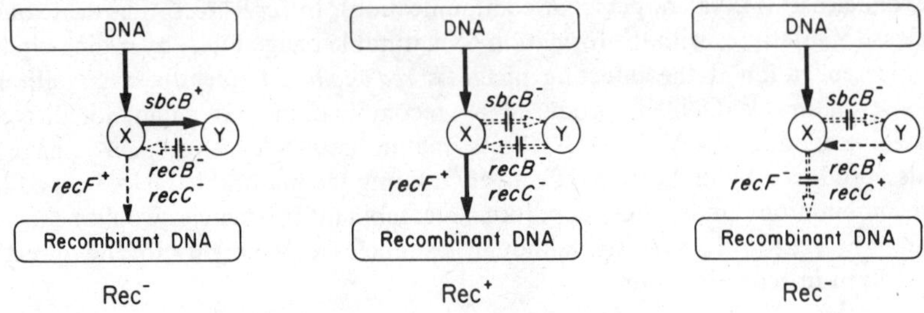

B. First hypothesis of two recombination pathways

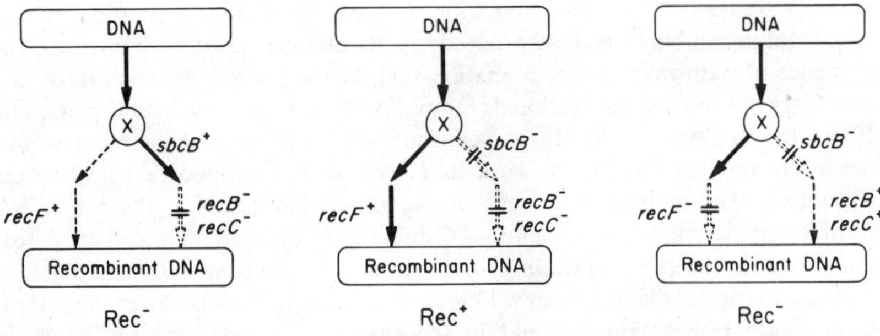

C. Second hypothesis of two recombination pathways

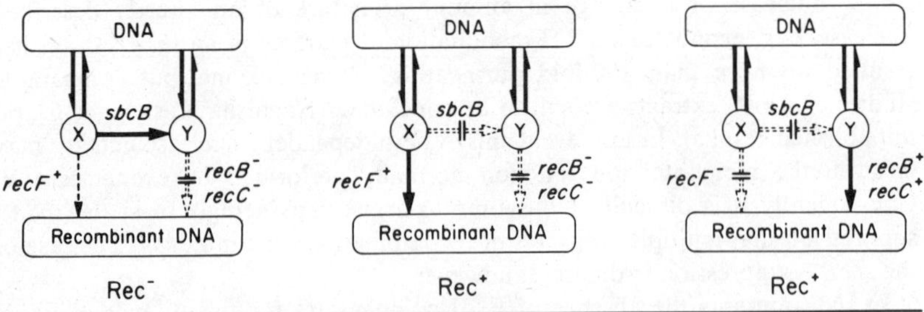

Figure 1 Hypothetical relationships between *recB*, *recC*, *sbcB*, and *recF* genes. Each frame is a diagrammatic representation of the recombination pathways in a strain of a different genotype. The genes involved are indicated next to the arrows. Solid arrows represent an indefinite number of enzymatic steps operating with high effectiveness; a dashed arrow represents an indefinite number of enzymatic steps operating with low effectiveness. A lightly dashed arrow crossed by two parallel lines represents a genetic block at some point in a series of enzymatic steps. Rec^+ stands for "recombination proficient" and Rec^- for "recombination deficient".

exonuclease I, operating in the absence of functional exonuclease V would inhibit recombination and make *recB*$^-$ or *recC*$^-$ mutants recombination deficient. Hence inactivation of exonuclease I would prevent degradation of the recombination intermediate and would restore recombination proficiency. A gene directly involved in recombination, the *recF* gene, has been assigned to the pathway in the figure for purposes that will become clear.

There is, however, a two-pathway hypothesis which provides the same explanation of indirect suppression by *sbcB*$^-$ (Fig. 1b). According to that hypothesis exonuclease V is considered to be on the main line of one of the pathways of recombination which is called the RecBC pathway after the genes determining exonuclease V. The second pathway involves the hypothetical *recF* gene and is consequently called the RecF pathway. One of the intermediates of the RecF pathway is acted upon by exonuclease I in its role as an enzyme of the RecBC pathway. Thus exonuclease I diverts material from the RecF pathway and might lead to recombination deficiency in *recB* and *recC* mutants. Inactivation of exonuclease I would therefore restore recombination proficiency by preventing the diversion of the branch point intermediate from the RecF pathway.

To distinguish which of these hypothetical situations pertains, recombination deficient mutants were selected beginning with the Rec$^+$ strain which is genotypically *recB*$^-$ *recC*$^-$ *sbcB*$^-$ (62). Among the Rec$^-$ mutants were *recA*$^-$ and *sbcB*$^+$ mutants. In addition three mutations were located in one gene (called *recF*) between *pyrE* and *ilv* and another mutation (in a gene called *recL*) was located near *metE*. Other mutations decreasing recombinant frequencies were also found but we have chosen to concentrate our analysis on the *recF* and *recL* mutations for the time being. In the *recB*$^-$ *recC*$^-$ *sbcB*$^-$ genetic background each *recF* and *recL* mutation blocks all but 0.05–0.5% of normal recombination as detected by recombinant colony formation (62). In a *recB*$^+$ *recC*$^+$ *sbcB*$^+$ genetic background neither *recF* nor *recL* mutations have any detectable effect on the formation of recombinant colonies (62). This result is incompatible with the one pathway hypothesis in Figure 1a because according to this hypothesis all mutations blocking recombination in *recB*$^-$ *recC*$^-$ *sbcB*$^-$ strains should also block recombination in *recB*$^+$ *recC*$^+$ *sbcB*$^+$ strains. Hence it seems that there must be a RecF pathway operating alternatively to the RecBC pathway as expressed by the diagram in Figure 1b.

Another datum on the role of exonuclease I requires that we modify the diagram in Figure 1b, however. According to the diagram an *sbcB*$^-$ *recF*$^-$ double mutant would be expected to be Rec$^-$ even when exonuclease V is present because exonuclease I is hypothesized to be directly involved in the RecBC pathway. Such a strain is, however, fully recombination proficient (62). This indicates that *sbcB* cannot be directly involved in the RecBC pathway even though its action antagonizes the RecF pathway. Hence we have arrived at the working hypothesis that exonuclease I converts an intermediate in the RecF pathway into an intermediate in the RecBC pathway in a shunt reaction and have presented the diagram in Figure 1c for publication (62).

It should be emphasized that the role of exonuclease I in recombination is still hypothetical. We have yet to test the shunt hypothesis by constructing the appropriately mutant strains. We also must satisfactorily explain the existence of *xonA* mutations (63). These mutations inactivate exonuclease I and, like *sbcB* mutations, suppress the sensitivity to UV characteristic of *recB* and *recC* mutants. Unlike *sbcB* mutations, *xonA* mutations fail to suppress completely the recombination deficiency of *recB* and *recC* mutants. It is unclear whether *xonA* mutations leave more exonuclease I activity in vivo than *sbcB* mutations, whether they inactivate completely only one of two different activities which might be possessed by exonuclease I, or whether they are nonpolar or structural gene mutations in contrast to *sbcB* mutations which might be polar or regulatory gene mutations.

Although we cannot specify with certainty the role of exonuclease I, the hypothesis of two recombination pathways seems the simplest to explain the relationships among the various genes whose mutations we have examined. This hypothesis leads us to expect that the amount of residual recombination by the RecF pathway in *recB*$^-$ *sbcB*$^+$ strains might vary with the type of recombinant product detected. For example the production of recombinant colonies following conjugation may involve an intermediate more susceptible to exonuclease I than the production of transmissible recombinant F′ plasmids or the type of recombinant product between F′ plasmids and the chromosome that results in chromosome mobilization. Intragenic recombination leading to enzyme production might also involve an intermediate less sensitive to exonuclease I than intragenic recombination leading to the formation of recombinant colonies. In this way we seek to understand the effects of *recB* and *recC* mutations on different types of bacterial recombination discussed in the section on "Effects of *recB* and *recC* Mutations on Bacterial Recombination" without at the same time denying that exonuclease V is a recombination enzyme.

The type of analysis applied to define the RecF pathway may also be applied to establish the existence of multiple sets of recombination pathways in other bacteria. In fact this is what was done to demonstrate the difference between general (i.e. Red pathway) and site-specific (i.e. Int pathway) recombination of phage lambda (57). In this case however the demonstration was made more convincing by the observation that recombination between two markers on the right half or between two markers on the left half of lambda occurred only by the Red pathway while recombination between one marker on the left and another on the right occurred by both the Red and Int pathways. Such support of the idea of two alternative general recombination pathways in *E. coli* has not yet been demonstrated although there is ample opportunity expressed in the hypotheses of the previous paragraph. One method already tried by Crawford & Preiss (64) involved measuring the distribution of *trpA* and *trpC* markers among transductants inheriting *trpB*$^+$ by virtue of recombination between different *trpB* mutations. High negative interference was just as great in *recB*$^-$ *recC*$^-$ *sbcB*$^-$ strains as it was in *recB*$^+$ *recC*$^+$ *sbcB*$^+$ strains. The conclusion reached was that "the mechanism of recombination used by *sbcB*-suppressed *recB*$^-$ *recC*$^-$ strains yields the same distribution of closely linked nonselected markers as the normal mechanism" (64).

A Repair Function for the RecF Pathway

Besides being recombination deficient, *recB* and *recC* mutants are also more sensitive to ultraviolet irradiation than their *rec*$^+$ ancestor (5). The inferred repair deficiency is suppressed by exonuclease I deficiency produced either by *sbcB* or *xonA* mutations (61, 63), and *recF* and *recL* mutations in the *sbcB*$^-$ *recB*$^-$ *recC*$^-$ background produce ultraviolet sensitivity comparable to that produced by a *recA* mutation alone (62). The *recF* and *recL* mutations in an otherwise wild type genetic background confer a small amount of sensitivity to UV so that it is only in combination with *recB* or *recC* mutations that full sensitivity is conferred (62).

An inference from these results is that the RecF and RecBC pathways can function as alternatives in repairing lesions in DNA resulting from UV-irradiation. Gaps in daughter strands produced by replication of DNA containing pyrimidine dimers have been implicated as the lesions repaired by a recombinational mechanism (65, 66). *recA* mutants show no such gap repair (67) while *recB*, *recC*, and *recF* single mutants repair gaps normally (67; Rothman & Clark, unpublished results). The *recB recC sbcB recF* quadruple mutant has been tested recently (Rothman & Clark, unpublished results) and gap-repair is slow and incomplete. This leads to the conclusion that at least part of gap repair is performed either by the RecF and RecBC pathways so that both must be blocked to see repair deficiency.

Transformability of E. coli and Some of its Implications

It has long been known that whole cells of *E. coli* were subject to transfection by phage DNA immediately following helper phage infection (68). Recently it was found that the requirement for helper can be circumvented by treatment of cells with Ca^{++} solution (69). Using this method to produce competence Oishi & Cosloy (70) have demonstrated transformation (i.e. uptake of purified DNA followed by integration of donor markers in the recipient's chromosome) of strain K-12 of *E. coli*. To achieve transformation the genotype of the recipient strain is important. An *sbcB*$^-$ *recB*$^-$ *recC*$^-$ strain (lacking exonucleases I and V respectively) shows the highest frequency of transformation, while *sbcB*$^-$ *recB*$^+$ *recC*$^+$ (i.e. ExoI$^-$ExoV$^+$) and *sbcB*$^+$ *recB*$^+$ *recC*$^+$ (i.e. ExoI$^+$ExoV$^+$) strains produce respectively 20% and 1% the maximum number of transformants (71). The *sbcB*$^+$ *recB*$^-$ *recC*$^-$ strain (i.e. ExoI$^+$ExoV$^-$) shows no ability to be transformed as expected from its *Rec*$^-$ phenotype (71). These phenomena were discovered independently by Wackernagel (72) and we have confirmed them (Kato & Clark, unpublished results). Additionally we have shown that *recF*, *recL*, and *recK* mutations in an *sbcB*$^-$ *recB*$^-$ *recC*$^-$ genetic background reduce transformation frequencies by 80% to 95%. Thus it appears that the RecF pathway is responsible for the transformational recombination observed.

The results of Oishi & Cosloy leave the impression that exonuclease V inhibits transformation of *E. coli* perhaps because it degrades the donor DNA which is taken up by Ca^{++} treated cells. This appears to set *E. coli* apart from *Hemophilus influenzae*, *Bacillus subtilis*, and *Diplococcus pneumoniae*, all of which are transformable (73). In cells of each of these species activity similar to exonuclease V has been found. It seems unlikely that these activities will be found to inhibit transformation. The mutants of *Diplococcus pneumoniae* (8) and *Bacillus subtilis* (25) lacking ATP-dependent nuclease activity will be helpful in testing the effects of this enzyme on transformation frequencies.

4. THE *recA* GENE OF *E. coli*

Mutations in *recA* appear to block recombination in *E. coli* almost completely (5). Unless the *recA*$^+$ allele is transferred to the zygote, the rare progeny inheriting markers from an Hfr donor are primarily produced by repliconation, not recombination (49). In a recent study (Guyer & Clark, unpublished results) the maximum frequency of recombinants was 10^{-4}% the frequency observed with a Rec$^+$ strain but the conditions used do not permit distinction of recombinants from revertants. In another study true recombinant progeny occur when an F prime plasmid is inherited by a *recA*$^-$ Hfr recipient at a level of about 4×10^{-2}% of that expected with a *recA*$^+$ Hfr (74). A very low level of recombination may also occur at either of two duplications present in certain *recA*$^-$ lysogens of a thermally inducible lambda prophage thereby producing cells that survive at the inducing temperature (75). It is unclear what this very low level of recombination signifies. Possibly the *recA* mutations are slightly leaky or are slightly suppressed and some residual functional gene product permits the low level of recombination observed. On the other hand very inefficient *recA*-independent pathways may be operative.

Most *recA* mutations have not been tested for possible suppressibility by informational suppressors, but one amber suppressible *recA* mutation has been described (76). Many attempts to obtain thermally conditional mutations have failed (13, 17, and Nagaishi & Clark, unpublished results) but one heat sensitive *recA* mutant is available (H. Ogawa, personal communication).

recA mutations are intriguing not only because they block recombination so completely but also because they lead to a set of startling pleiotropic effects. These include high sensitivity to UV (2) and X-irradiation (6), enhanced DNA breakdown following UV-irradiation (42), and insensitivity to the mutagenic action of UV (77, 78). In addition lambda lysogens of *recA* mutants can not be induced directly by UV (9, 10) or mitomycin treatment (79), thymine starvation (13), or thermal treatment when carrying the mutation *tif-1* (80, 81), nor can they be induced indirectly by conjugation with a UV-irradiated F$^+$ or F′ donor or by infection with UV-irradiated P1 phage (82). All *recA* mutations thus far tested show these properties. Until recently all *recA* mutations also had been found to block spontaneous production of lambda phage during growth of a lysogenic culture (4). Now two alleles, *recA142* (Margossian, Clark, & Horii, unpublished results) and *recA169* (17) have been found to permit spontaneous production but still to block UV-induction of lambda. The effects of these two alleles on other types of induction have not been tested.

Further complexities connected with *recA* have been revealed by recent experiments of Castellazzi, George & Buttin (81, 83). The thermal conditional mutation *tif-1*, which lies very close to *recA*, confers a number of properties that mimic the action of UV on wild type cells: at the nonpermissive temperature induction of wild type lambda prophage occurs in a *tif-1* lysogen and reactivation and mutation of UV-irradiated lambda phage occur in a *tif-1* nonlysogen. Revertants of a *tif-1* mutant lysogenic for wild type lambda which survive exposure to the nonpermissive temperature fall into four classes: back mutants to tif^+, and forward mutants to $recA^-$, lex^-, or zab^-. Since *recA* mutations block UV-induction of wild type lambda, it is of no surprise that they block the mock UV-induction enabled by *tif-1*. Also *lex* mutations are probably the same as the *exr* mutation in *E. coli* B, which reduces or inhibits UV-induction of wild type lambda (84). *zab* mutants constitute an unexpected fourth class and have some properties that justify their provisional relegation to a new gene. They lie very close to *recA* and *tif-1*, which distinguishes them from *lex* mutations with which they otherwise share many properties. Included in these is the slight reduction of recombination proficiency (down to 30% of wild type levels) caused by *zab* mutations; hence *zab* mutations are also distinguishable from *recA* mutations.

It is unclear at present how to interpret this complex behavior of the *recA* region. It might reflect functional complexity of either the *recA* gene product or the *recA* genetic region. In the latter case mutations may have greater or lesser polar effects, or some mutations may affect regulatory regions (e.g. a promotor of transcription), and others may affect structural regions determining the protein. Another possibility is that the mutations may be more or less suppressible by the amber suppressor(s) present in some of the strains in which they occur (e.g. AB1157 in which many *recA* mutations are studied carries *sup*-37 suspected to be a *supE* allele).

Another facet of the complex phenotype afforded by *recA* mutations is their effects on cell division. Only one or two alleles have been tested in this manner but they are found to permit cell division in strains which otherwise would form snakes following UV-irradiation (84, 85). In these strains it therefore appears that the presence of functional *recA* product leads to a block in cell division. It should also be noted that *tif-1* nonlysogenic strains filament at the nonpermissive temperature and that this is blocked by *lex*, *recA*, and *zab* mutations (81, 83).

It is unclear how the pleiotropic effects of *recA* mutations are caused by the absence of the *recA* gene product. The suggestion has been heard repeatedly that the *recA* gene is a regulatory gene controlling synthesis or activity of a set of proteins. It is also possible that *recA* mutations lead to differential activation or inactivation of various enzymes through a failure to regulate changes in concentrations of intracellular metabolites (like ATP) or ions (like Mg^{++}). Finally it should be remembered that Tomizawa & Ogawa (85a) have suggested that *recA* mutations affect the structure of the bacterial chromosome by making it less rigidly held or folded. Their argument was based on differential efficiencies of ^{32}P decay. The chromosomes of $recA^+$ bacteria are inactivated less efficiently than phage or transducing DNAs. To explain this Tomizawa & Ogawa hypothesized that the $recA^+$ chromosome is rigidly held while phage and transducing DNAs are not. The rigidity of the former would permit repair of simultaneous breaks in both strands of DNA if the breaks were staggered such that the strands were held together by a few hydrogen-bonded base pairs. On the other hand the looseness of the others would prohibit repair due to a tendency for the strands to separate. Then Tomizawa & Ogawa observed ^{32}P decay inactivated *recA*$^-$ chromosomes at the same high efficiency that it inactivated the phage and transducing DNAs. This led naturally to the hypothesis that *recA*$^-$ chromosomes are not rigidly held. Recently Worcel & Burgi (35) have shown that the *recA*$^+$ chromosome is folded and coiled in a way that might give substance to Tomizawa and Ogawa's hypothesis of rigidity. Using Worcel & Burgi's methods it may be easy to test the hypothetical lack of rigidity of *recA*$^-$ chromosomes.

In light of the pleiotropic properties of *recA* mutations it is difficult to know whether this gene and its product exert their effects directly in one of the steps of recombination or indirectly through a regulatory role, a role determining the structure of the chromosome or some other role. What does seem clear, however, is that *recA* mutants show no or virtually no residual recombination. Thus *recA* and its product appear to be involved in both pathways of recombination we have recognized through investigation of indirect suppression of *recB* and *recC* mutations.

5. CONCLUDING REMARKS

Mutants of *E. coli* and other bacterial species that are altered in their ability to form recombinant progeny have been found. Evidence that these mutants are defective in genetic recombination rather than in pre- or post-recombinational metabolism is available. In this light it seems reasonable to categorize these mutants as recombination deficient and to use information obtained from their characterization to formulate working hypotheses on the molecular nature of recombination and its relationships to other forms of DNA metabolism. The controversial nature of the evidence should be recognized, however, and hence the tentative nature of conclusions in this area.

Acknowledgments

The author has enjoyed and benefited from discussions with and suggestions offered by many people and is especially grateful to the following: Ann Emerick, Jane Gillen, Mark Guyer, Takeshi Kato, Linda Margossian, Bob Miller, Haruko Nagaishi, John Pemberton, Bob Rothman, and Ann Templin. The figure was designed and drawn by Nick Story.

This investigation was supported by Public Health Service Research Grant No. AI 05371 from the National Institute of Allergy and Infectious Diseases.

Literature Cited

1. Radding, C. 1973. *Ann. Rev. Genet.* 7:87–111
2. Clark, A. J., Margulies, A. D. 1965. *Proc. Nat. Acad. Sci. USA* 53:451–59
3. Beattie, K. L., Setlow, J. K. 1971 *Nature New Biol.* 231:177–79
4. Clark, A. J. 1967. *J. Cell Physiol.* 70: Suppl. I. 165–80
5. Clark, A. J. 1971. *Ann. Rev. Microbiol.* 25:438–64
6. Howard-Flanders, P., Theriot, L. 1966. *Genetics* 53:1137–50

7. van de Putte, P., Zwenk, H., Rorsch, A. 1966. *Mutation Res.* 3:381–92
8. Vovis, G. F., Buttin, G. 1970. *Biochim. Biophys. Acta* 224:42–54
9. Brooks, K., Clark, A. J. 1967. *J. Virol.* 1:283–93
10. Hertman, I., Luria, S. E. 1967. *J. Mol. Biol.* 23:117–33
11. Holloway, B. W. 1969. *Bacteriol. Rev.* 33:419–43
12. Sironi, G. 1969. *Virology* 37:163–76
13. Devoret, R., Blanco, M. 1970. *Mol. Gen. Genet.* 107:272–80
14. Stallions, D. R., Curtiss III, R. 1971. *J. Bacteriol.* 105:886–95
15. Willetts, N. S., Mount, D. W. 1969. *J. Bacteriol.* 100:923–34
16. Tomizawa, J., Ogawa, H. 1972. *Nature New Biol.* 239:14–16
17. Storm, P. K., Hoekstra, W.P.M., De Haan, P. G., Verhoef, C. 1971. *Mutation Res.* 13:9–17
18. Karu, A. E., Linn, S. 1972. *Proc. Nat. Acad. Sci. USA* 69:2855–59
19. Tsuda, Y., Strauss, B. S. 1964. *Biochemistry* 3:1678–84
20. Anai, M., Hirahashi, T., Takagi, Y. 1970. *J. Biol. Chem.* 245:767–74
21. Anai, M., Hirahashi, T., Yamanaka, M., Takagi, Y. 1970. *J. Biol. Chem.* 245:775–80
22. Hout, A., Oosterbaan, R. A., Pouwels, P. H., De Jonge, A.J.R. 1970. *Biochim. Biophys. Acta* 204:632–35
23. Vovis, G. F., Buttin, G. 1970. *Biochim. Biophys. Acta* 224:29–41
24. Smith, H. O., Friedman, E. A. 1972. *J. Biol. Chem.* 247:2854–58
25. Prozorov, A. A., Kalinina, N. A., Naumov, L. S., Chestukhin, A. V., Shemyakin, M. F. 1972. *Genetika* 8:142
26. Anai, M. 1967. *Seikagaku* 39:167
27. Winder, F. G., Lavin, M. F. 1971. *Biochim. Biophys. Acta* 247:542–61
28. Friedman, E. A., Smith, H. O. 1973. *Nature New Biol.* 241:54–58
29. Karu, A. E., MacKay, V., Goldmark, P. J., Linn, S. 1973. unpublished
30. Goldmark, P. J., Linn, S. 1972. *J. Biol. Chem.* 247:1849–60
31. Haefner, K. 1968. *J. Bacteriol.* 96: 652–59
32. Capaldo-Kimball, F., Barbour, S. D. 1971. *J. Bacteriol.* 106:204–12
33. Monk, M., Kinross, J. 1972. *J. Bacteriol.* 109:971–78
34. Monk, M., Kinross, J., Town, C. D. 1973. *J. Bacteriol.* 114:1014–17
35. Worcel A., Burgi, E. 1972. *J. Mol. Biol.* 71:127–47
36. Boyer, H. W. 1971. *Ann. Rev. Microbiol.* 25:153–76
37. Simmon, V. F., Lederberg, S. 1972. *J. Bacteriol.* 112:161–69
37a. Lehman, I. R. 1960. *J. Biol. Chem.* 235:1479–87
38. Pilarski, L. M., Egan, J. B. 1973. *Virology*. In press
39. Wackernagel, W., Radding, C. M. 1973. *Virology*. In press
40. Unger, R. C., Clark, A. J. 1972. *J. Mol. Biol.* 70:539–48
41. Enquist, L. W., Skalka, A. 1973. *J. Mol. Biol.* 75:185–212
42. Clark, A. J., Chamberlin, M., Boyce, R. P., Howard-Flanders, P. 1966. *J. Mol. Biol.* 19:442–54
43. Howard-Flanders, P., Boyce, R. P. 1966. *Radiat. Res. Suppl.* 6:156–84
44. Willetts, N. S., Clark, A. J. 1969. *J. Bacteriol.* 100:231–39
45. Moody, E.E.M., Low, K. B., Mount, D. W. 1973. *Mol. Gen. Genet.* 121:197–206
46. Buttin, G., Wright, M. R. 1968. *Cold Spring Harbor Symp. Quant. Biol.* 33: 259–69
47. Lindahl, G., Sironi, G., Bialy, H., Calendar, R. 1970. *Proc. Nat. Acad. Sci. USA* 66:587–94
48. Emmerson, P. T., Howard-Flanders, P. 1967. *J. Bacteriol.* 93:1729–31
49. Low, B. 1968. *Proc. Nat. Acad. Sci. USA* 60:160–67
50. Low, B. 1973. *Mol. Gen. Genet.* 122:119–30
51. Templin, A., Kushner, S. R., Clark, A. J. 1972. *Genetics* 72:205–15
52. Wilkins, B. M. 1969. *J. Bacteriol.* 98: 599–604
53. Hall, J. D., Howard-Flanders, P. 1972. *J. Bacteriol.* 110:578–84
53a. Anraku, N., Tomizawa, J-I. 1965. *J. Mol. Biol.* 11:501–08
54. Paul, A. V., Riley. M. 1973. unpublished
55. Davidoff-Abelson, R., Dubnau, D. 1971. *Proc. Nat. Acad. Sci. USA* 68: 1070–74
56. Notani, N. K., Setlow, J. K., Joshi, V. R., Allison, D. P. 1972. *J. Bacteriol.* 110:1171–80
57. Signer, E. 1971. *The Bacteriophage Lambda*, ed. A. D. Hershey, 139–74. Cold Spring Harbor, New York: Cold Spring Harbor Laboratory
58. Sobell, H. M. 1973. *Advan. Genet.* 17. In press
59. Unger, R. C., Echols, H., Clark, A. J. 1972. *J. Mol. Biol.* 70:531–37
60. Barbour, S. D., Nagaishi, H., Templin, A., Clark, A. J. 1970. *Proc. Nat. Acad. Sci. USA* 67:128–35
61. Kushner, S. R., Nagaishi, H., Templin, A., Clark, A. J. 1971. *Proc. Nat. Acad. Sci. USA* 68:824–27

62. Horii, Z., Clark, A. J. 1973. *J. Mol. Biol.* In press
63. Kushner, S. R., Nagaishi, H., Clark, A. J. 1972. *Proc. Nat. Acad. Sci. USA* 69:1366–70
64. Crawford, I. P., Preiss, J. 1972. *J. Mol. Biol.* 71:717–33
65. Rupp, W. D., Howard-Flanders, P. 1968. *J. Mol. Biol.* 31:291–304
66. Rupp, W. D., Wilde III, C. E., Reno, D. L., Howard-Flanders, P. 1971. *J. Mol. Biol.* 61:25–44
67. Smith, K. C., Meun, D. H. C. 1970. *J. Mol. Biol.* 51:459–72
68. Kaiser, A. D., Hogness, D. S. 1960. *J. Mol. Biol.* 2:392–415
69. Mandel, M., Higa, A. 1970. *J. Mol. Biol.* 53:159–62
70. Cosloy, S. D., Oishi, M. 1973. *Proc. Nat. Acad. Sci. USA* 70:84–87
71. Oishi, M., Cosloy, S. D. 1972. *Biochem. Biophys. Res. Comm.* 49:1568–72
72. Wackernagel, W. 1973. *Biochem Biophys. Res. Comm.* 51:306–11
73. Hotchkiss, R. D., Gabor, M. 1970. *Ann. Rev. Genet.* 4:193–224
74. De Vries, J. K., Maas, W. K. 1971. *J. Bacteriol.* 106:150–56
75. Feiss, M., Adhya, S., Court, D. L. 1972. *Genetics* 71:189–206
76. Mount, D. W. 1971. *J. Bacteriol.* 107:388–89
77. Miura, A., Tomizawa, J. 1968. *Mol. Gen. Genet.* 103:1–10
78. Witkin, E. M. 1969. *Mutation Res.* 8:9–14
79. Hertman, I. M. 1969. *Genet. Res. Camb.* 14:291–307
80. Kirby, E. P., Jacob, F., Goldthwait, D. A. 1967. *Proc. Nat. Acad. Sci. USA* 58:1903–10
81. Castellazzi, M., George, J., Buttin, G. 1972. *Mol. Gen. Genet.* 119:153–74
82. Rosner, J. L., Kass, L. R., Yarmolinsky, M. B. 1968. *Cold Spring Harbor Symp. Quant. Biol.* 33:785–89
83. Castellazzi, M., George, J., Buttin, G. 1972. *Mol. Gen. Genet.* 119:139–52
84. Green, M.H.L., Greenberg, J., Donch, J. 1969. *Genet. Res. Camb.* 14:159–62
85. Inouye, M. 1971. *J. Bacteriol.* 106:539–42
85a. Tomizawa, J., Ogawa, H. 1968. *Cold Spring Harbor Symp. Quant. Biol.* 33:243–51
85b. Taylor, A. L., Trotter, C. D. 1972. *Bacteriol. Rev.* 36:504–24
86. Willetts, N. S., Clark, A. J., Low, B. 1969. *J. Bacteriol.* 97:244–49
87. Emmerson, P. T. 1968. *Genetics* 60:19–30
88. Howard-Flanders, P., Boyce, R. P., Theriot, L. 1966. *Genetics* 53:1119–36
89. Ogawa, H. 1970. *Mol. Gen. Genet.* 108:378–81
90. Siegel, E. C. 1970. *J. Bacteriol.* 104:604–05
91. Ogawa, H., Shimada, K., Tomizawa, J. 1968. *Mol. Gen. Genet.* 101:227–44
92. Smirnov, G. B., Skavronskaya, A. G. 1971. *Mol. Gen. Genet.* 113:217–21
93. Mattern, I. E. 1971. *First European Biophysics Congress*, ed. E. Broda, A. Locker, H. Springer-Lederer, 237–40
94. Siegel, E. C. 1973. *J. Bacteriol.* 113:145–60
95. Storm, P. K., Zaunbrecher, W. M. 1972. *Mol. Gen. Genet.* 115:89–92
96. Wechsler, J. A., Gross, J. D. 1971. *Mol. Gen. Genet.* 113:273–84
97. Gross, J. D. 1972. *Curr. Top. Microbiol. Immunol.* 57:39–74
98. Carl, P. L. 1970. *Mol. Gen. Genet.* 109:107–22
99. Wechsler, J. 1973. unpublished
100. Gefter, M. L., Hirota, Y., Kornberg, T., Wechsler, J. A., Barnoux, C. 1971. *Proc. Nat. Acad. Sci. USA* 68:3150–53
101. Fuchs, J. A., Karlström, H. O., Warner, H. R., Reichard, P. 1972. *Nature New Biol.* 238:69–71
102. Lark, K. G. 1972. *Nature New Biol.* 240:237–40
103. Threlfall, E. J., Holland, I. B. 1970. *J. Gen. Microbiol.* 62:383–98
104. Holland, I. B., Threlfall, E. J., Holland, E. M., Darby, V., Samson, A.C.R. 1970. *J. Gen. Microbiol.* 62:371–82
105. Mount, D. W., Low, K. B., Edmiston, S. J. 1972. *J. Bacteriol.* 112:886–93
106. Gottesman, M. M., Hicks, M. L., Gellert, M. 1973. *J. Mol. Biol.* 77:531–47
107. Konrad, E. B., Modrich, P., Lehman, I. R. 1973. *J. Mol. Biol.* 77:519–29
108. Howard-Flanders, P., Simson, E., Theriot, L. 1964 *Genetics* 49:237–46
109. Gross, J., Gross, M. 1969. *Nature* 224:1166–68
110. DeLucia, P., Cairns, J. 1969. *Nature* 224:1164–66
111. Campbell, J. L., Soll, L., Richardson, C. C. 1972. *Proc. Nat. Acad. Sci. USA* 69:2090–94
111a. Hirota, Y., Gefter, M., Mindich, L. 1972. *Proc. Nat. Acad. Sci. USA* 69:3238–42
112. Kornberg, T., Gefter, M. L. 1971. *Proc. Nat. Acad. Sci. USA* 68:761–64
113. Milcarek, C., Weiss, B. 1973. *J. Bacteriol.* 113:1086–88
114. Milcarek, C., Weiss, B. 1972. *J. Mol. Biol.* 68:303–18

MOLECULAR MECHANISMS IN GENETIC RECOMBINATION 3048

Charles M. Radding

Departments of Medicine and Molecular Biophysics and Biochemistry, Yale University School of Medicine, New Haven, Connecticut

Introduction

Genetic recombination is the set of processes that results in new linkage relationships of genes or parts of genes (1). Three subclasses of recombination are currently recognized: (*a*) *General recombination* describes exchanges that occur between homologous genophores, more or less anywhere along their length. (*b*) *Site-specific recombination* describes exchanges that occur at a limited number of specific sites, as seen for example in the integration of some viral genomes into the host genome (2, 3). (*c*) *Illegitimate* or *nonhomologous recombination* are terms used for exchanges between genophores that are *largely* nonhomologous. Examples include the generation of specialized transducing particles (4) and the random chromosomal insertion of the bacteriophage Mu genome (5, 6). Although most of this review will deal with general recombination, the other classes of recombination are no less important (7).

The macromolecular metabolism of DNA includes the partially overlapping areas of replication, repair, and recombination. The recent recognition of the role of transcription in replication and the possible implication of transcription in recombination indicate even greater complexity in these overlaps. Moreover, the degree of overlapping varies from one organism to another (8). As examples, the replication and recombination of phage T4 are extensively overlapping (9–11). The genes required for recombination are those required for replication. By contrast, the genes that govern recombination in bacteriophage λ can be deleted without eliminating replication (12), although there are important relationships between replication and recombination in λ as well (see below).

Figure 1 illustrates the probable role of the heteroduplex joint in producing variations in the genetic output from an exchange. New DNA synthesis in the joint or asymmetry in the formation of joints may be the source of unequal recoveries of genetic information from the respective parents. In organisms like fungi, where the products of meiosis can be isolated, gene conversion is the term used to describe unequal recoveries of genetic markers in the region of the exchange. When observations are made instead on large mating populations, as in bacteriophage crosses, similar events involving heteroduplex joints may appear in the guise of close multiple exchanges (cf Fig. lb). The splicing

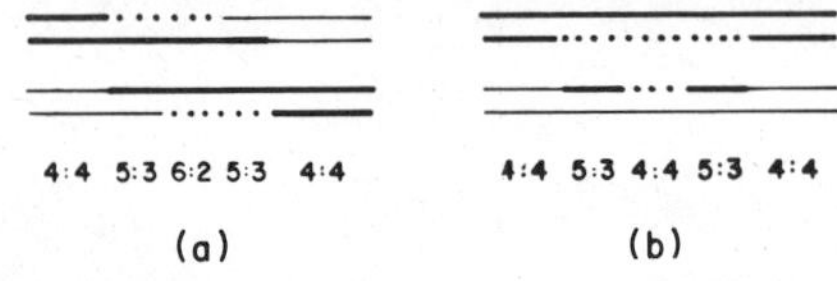

Figure 1 Paradigms of recombination and gene conversion by heteroduplex formation, after Meselson (14). Parental contributions are indicated by solid lines, newly synthesized DNA by broken lines. Two patterns are shown: (*a*) An exchange that produces two heteroduplex joints flanked by the recombinant configuration of distal arms. (*b*) An exchange that produces only one insertion heteroduplex and one recombinant strand (cf Fig. 3d). Resynthesis within the heteroduplex region is illustrated. A related pattern results from an exchange that produces two reciprocal insertion heteroduplexes with the parental configuration of distal arms (cf Fig. 4).

The numbers represent the recoveries of allelic markers that would be produced by the first mitotic division following meiosis. (Two nonrecombinant molecules must be added to each illustration to represent the four chromatids that determine the genetic output.) Resynthesis may result from the mechanism of the strand exchange (cf Figs. 2, 3, 5) or from the excision and repair of mismatched bases. In either case, resynthesis of part of only one of the four participating strands in the heteroduplex region produces post-meiotic segregation (5:3, or 3:5).

These diagrams illustrate potential sources of nonreciprocity. Another potential source of nonreciprocity, in prokaryotes, is simply the loss of one of the partners to the exchange.

paradigm, based on the reunion of homologous sequences (13), also accounts for the association of heterozygosity with recombination (14), and the usual precision of general recombination. On the other hand, models have been proposed, in which the heteroduplex joint plays a much less prominent role in determining the genetic output (15, 16). Nonetheless, for general recombination, the basic problem remains to discover how the heteroduplex joint is made, and what happens to it on the way to becoming physically indistinguishable from the rest of the DNA molecule. The molecular mechanisms that seem relevant to this question form the subject of the following discussion. The order of some parts of this discussion resembles a logical sequence of steps in recombination, but consideration of mechanisms of strand exchange leads to the view that recombination may occur by highly concerted reactions. Constituent processes may occur simultaneously rather than sequentially.

In another article in this volume John Clark reviews recombination deficient mutants of bacteria (8). The continuing challenge of understanding genetic recombination is reflected in numerous excellent reviews written from various viewpoints (1, 13, 14, 17–23).

Joint Molecules

Direct information about intermediates in recombination has come mainly from the study of coliphage T4. Tomizawa and his colleagues first isolated intermediates, called joint molecules, in which the parental contributions were linked by noncovalent bonds, presumably hydrogen bonds in a heteroduplex region (24). Isolation of such intermediates required either chemical or mutational inhibition of DNA synthesis. Anraku et al showed that joint molecules isolated after infection with T4 *ligase*$^-$ *polymerase*$^-$ mutants were about one fourth the size of the T4 genome and contained six gaps of 300–400 nucleotides (25). Approximately half of these gaps could be closed by DNA polymerase and polynucleotide ligase in vitro (25, 26). Subsequently Broker & Lehman observed by electron microscopy that up to 25% of joint molecules isolated after infection by T4 *lig*$^-$ *pol*$^-$ were branched molecules which contained three or four arms; in some instances even more (11). Single stranded as well as double stranded branches were seen. At the junction of double stranded branches, single stranded regions of 200–1400 nucleotides in length were seen in about half of the cases (27). Joint molecules, including the branched variety, were not formed by gene 32 mutants of T4 (11, 24, 28, and see *Initiation of Strand Exchange*). The branched molecules were further related to genetic recombination by the observation that they were less frequent after single infection by nonreplicating T4 *lig*$^-$ *pol*$^-$ than after multiple infection. The interpretation of these branched structures is discussed below (*Strand Exchange*).

Noncovalent union of donor and recipient DNA preceding covalent union has also been demonstrated during conjugation in *E. coli*. Joint molecules with a molecular weight of 40×10^6 contained polynucleotide chains from the F$^-$ recipient and the Hfr donor with molecular weights distributed about 18×10^6 and 9.5×10^6 respectively (29–31).

Complexes of donor and recipient DNA isolated during transformation of *B. subtilis* contained a small proportion of donor DNA that was not covalently linked to the recipient DNA and that may represent joint molecules (32).

Recombinant Molecules: Old vs New DNA

Early work on mechanisms of recombination included efforts to determine whether recombinant molecules are formed by the reunion of parental material (breakage and reunion), or by the copying of information first from one template and then from another (copy choice) (13, 14). Recombination in a number of prokaryotic systems has been shown to involve major contributions of DNA from both parents (14,24,29). Recombination in bacterial transformation produces an insertion heteroduplex in which one strand from the donor replaces a homologous strand in the recipient duplex DNA. The donor strand, which may be several million daltons or more in molecular weight (19, 33) is covalently linked to the recipient in the absence of extensive DNA synthesis (19). In a recent study of transduction by the *Salmonella* phage P22, Ebel-Tsipis et al observed the integration of double stranded segments of donor DNA, the size of which was estimated to be 4 to 8 million daltons. Since two thirds of the donor DNA that was incorporated as sizeable fragments was associated with DNA lighter than the density labeled donor, heteroduplex DNA may also have been formed (34, 35).

Although the copy-choice model in its simplest form has long been discarded, a mechanism involving breakage followed by copying remains a prominent possibility in the production of recombinants (see for example ref. 16). More generally the role of new DNA synthesis in recombination remains an important and unsettled issue. There are a number of ways in which newly synthesized DNA might get incorporated into rejoined molecules: (*a*) in the repair of gaps in or about the heteroduplex joint (Fig. 1, also *Joint Molecules*), (*b*) as an intrinsic part of the mechanism of strand exchange (36, Fig. 5), (*c*) in the generation, by replication of strand interruptions that serve as favorable substrates for recombination enzymes or (*d*) in the generation, by recombination, of replication forks (37, 38).

Subsequent to the demonstration that recombinant λ DNA is produced by contributions of material from both parents, it was recognized that three systems of recombination might be involved: (*a*) the site-specific Int system which can mediate reciprocal recombination at the *att* site (*b*) the general recombination system (Red) determined by phage genes and (*c*) the general recombination system (Rec) determined by genes of the host *E. coli* (21). In addition, and for reasons that probably have more to do with replication than with recombination, phage λ makes a protein called gamma which antagonizes part of the Rec system, making infected cells into Rec^- phenocopies (8). Kellenberger-Gujer & Weisberg therefore reinvestigated the transfer of material from parents to recombinant progeny and demonstrated that transfer occurs in the Red, Int, and Rec systems respectively (39).

The role of new DNA synthesis has been studied by Stahl and his colleagues by using density labeled phage under conditions of limited DNA synthesis (38, 40–43). The most stringent limitation of DNA synthesis was obtained by using mutations in replication genes of both phage λ and *E. coli* (41). A less stringent limitation of DNA synthesis was obtained by performing crosses in an immune lysogen. In this case, active genes to promote recombination and packaging of repressed phage genomes were provided by coinfection with heteroimmune phage that were genetically and physically unlabelled (40, 43). The experiments, which depended upon assays of viable phage progeny, revealed that some maturation step required either replication or recombination. In a cross of heavy parents under conditions of limited DNA synthesis, the production of phage containing fully conserved parental DNA required recombination; the production of phage in which both strands of DNA were newly synthesized did not require recombination; and the production of phage in which only one strand of DNA was newly synthesized was partially dependent on recombination (40). Thus, if recombination was limited, the requirements of maturation selected for progeny in which replication was prominent. Reciprocally, if replication was limited, the requirements of maturation selected for progeny that were recombinant. Further results of these experiments may be summarized as follows: (*a*) The general recombination system (Red) of λ produced some recombinants, the atoms of which were 95% or more of parental origin (43, 44). (*b*) Recombinants produced by the Red system under the selective conditions imposed by limited DNA synthesis varied in the amount of newly synthesized DNA, depending upon the region of the map within which recombination took place. The least new DNA synthesis was associated with recombination in the $C_I - R$ region; appreciable amounts of new DNA synthesis were associated with recombination in the rest of the map (38, 44). (*c*) Severe limitation of DNA synthesis selectively depressed recombination in the region that displayed the most associated DNA synthesis, suggesting that the associated DNA synthesis played an essential role. Other laboratories have studied the influence of recombination on λ replication, which appears to occur in two stages. During the first stage the replicating circular DNA remains circular and its topography resembles the greek letter θ (45). During the second stage concatemers are formed. Simultaneous elimination of the recombination systems, Rec, Red, and Int did not abolish the formation of concatemers (46a), but the combination *red⁻ recA⁻* resulted in a quantitative reduction in late DNA synthesis (37). An additional mutation in the gamma (*gam*) gene of phage λ imposed a severe block in the transition from the early mode to the late mode of replication, as well as a block in the packaging of phage DNA (37, 47). The role of the *gam* gene may be purely negative, through its antagonistic action on the *rec*BC nuclease (see ref. 8). Enquist & Skalka, and Stahl et al proposed that recombination enhances the transition from early to late replication (37, 44). The alternative requirement for either replication or recombination to produce phage (see Experiments of Stahl et al above) can be explained if mature progeny can be formed only from DNA in its concatemeric or late mode of replication. To account for the role of DNA synthesis in the production of recombinants, Stahl et al proposed that Red mediated recombination between circular molecules of λ DNA produces a rolling circle. The extent to which the circle's replication must be driven by new DNA synthesis in order to produce a matureable genome may explain the topological variation in the amount of newly made DNA that is associated with recombinants (44).

Using the conditions that severely limit DNA synthesis (41) Russo studied recombination in the part of the λ map that is associated with the least new synthesis (48). For a marker located 18.5% from the right end of λ DNA, the contribution of DNA from one parent was insufficient to account for recombination by a single exchange involving breakage and reunion of double stranded DNA. Russo proposed instead that the recombinants were formed by transfer of a single strand, as in the strand assimilation model (see *Strand Exchange* below, and Fig. 3b-d). Additional support for this idea was provided by examining the segregation of heterozygotes formed by unselected markers flanking the selected recombination event. The segregation of heterozygotes that involved *both* flanking markers occurred as if heterozygosity had been created by the continuous transfer of a single strand comprising at least 30% of the genome. Similar observations have been made by White & Fox (cited in ref. 48). In Russo's experiments, recombination of the interposed selected markers must be explained by mismatch-repair (see *Gene Conversion*, below). These observations appeared to hold for both the Red and Rec systems (48).

Strand Breakage

Breaks in DNA leading to recombination may result from: (*a*) the action of specific recombination enzymes; (*b*) the processes of replication, repair, or restriction; or (*c*) accidents. Breaks in one or both strands of DNA stimulate recombination. Howard-Flanders and his colleagues showed that recombination restores the integrity of strands that acquire interruptions following replication in ultraviolet irradiated *uvr* mutants, which are defective in the excision of UV lesions (49). The efficiency of repair per thymine dimer suggests that recombination was stimulated by the lesion consisting of a dimer opposite a gap. Recombination was not stimulated without replication of the DNA that contained dimers (50–52). Interstrand cross links may also be repaired by recombination (53). In this case gaps caused by partial excision of cross-links can be repaired prior to replication (50).

The appearance of single strand interruptions in the DNA of bacteriophage T4 was blocked by the addition of chloramphenicol soon after infection (28, 54), and mutations in genes 46 and 47 blocked the apparent enlargement of nicks to gaps (55). Single strand interruptions accumulated in the DNA of T4 *polymerase*$^-$, *ligase*$^-$ mutants prior to the formation of branched intermediates in recombination (11, and see below). Observations first made several years ago, indicated that recombination of phage T4 is enhanced by ligase deficiency, a stimulation attributed to the increased lifetime of nicks (9, 10, 56, 57). Experiments that support this theory (58) were made possible by the observation that a mutation in the T4 rII region restored the viability of T4 ligase mutants (56, 59). Infection by T4 *lig rII* resulted in an increased rate of formation of recombinants for an extended period, a change that did not appear to be attributable to alterations in the timing of DNA synthesis or maturation. Moreover, the frequency of recombination by T4 *lig rII* was related inversely to genetic variations in the levels of host ligase activity (58, 60).

Recombination occurs at a higher frequency near molecular ends of T4 genophores (61, 62). Even when one member of an initial mating pair had an incomplete genome, all of the genetic information of both parents was transmitted to progeny by reiterative recombination that appeared to be spearheaded by the ends of fragments (61).

Recently S. Lederberg observed that unmodified λ DNA survived in a restricting host in which the *rec*BC enzyme (see *Strand Exchange*) had been inactivated by mutation (63). In this circumstance, restriction produced fragments by double strand breaks, but instead of suffering further degradation, the fragments were efficiently rescued by recombination with a consequent decrease in the apparent linkage of genetic markers. Sheared λ DNA, used for transfection, was also rescued by its own or several different host systems of recombination in the absence of the *rec*BC enzyme (64). Similarly, decreased linkage and dependence on recombination has been seen in transfection by certain subtilis phages (65–67). The association of increased recombination with broken DNA shows that double strand breaks play a positive role in recombination and that recombination is essential for survival in some circumstances.

The enzymic mechanisms that produce single stranded or double stranded breaks in duplex DNA have been termed *haplotomic* (single break) and *diplotomic* (double break) respectively by Bernardi (68). The haplotomic mechanism is shown, for example, by pancreatic DNAse, and by endonucleases that are involved in repair by excision. The diplotomic mechanism is characteristic of endonuclease I of *E. coli*, acid DNAse of spleen (DNAse II), and restriction endonucleases (69–73). Endonuclease II of *E. coli* produces single and double strand breaks in alkylated DNA, but only a limited number of single strand breaks in nonalkylated DNA (74). Melgar & Goldthwait showed that pancreatic DNAse can make either single strand or double strand breaks, depending upon which divalent cations are present (75). Endonuclease I of *E. coli* is inhibited by tRNA but the inhibited complex makes a limited number of single strand breaks in the presence of high salt (76). A model for the diplotomic mechanism is provided by the restriction endonucleases, which appear to recognize specific symmetrical sequences in DNA, presumably on the basis of two-fold symmetry of the dimerized enzyme (72, 77, 78). It is likely that the physical configuration of an end plays an important role in genetic recombination: A single stranded end is immediately available for base pairing, for instance. The reunion of broken molecules of λ DNA after transfection is facilitated when the fragments have single stranded ends (64). The diplotomic mechanism need not necessarily break DNA at exactly opposing sites on both strands. The R-1 restriction enzyme has been shown to make staggered cuts that are 4 nucleotides apart, producing fragments with single stranded complementary ends very much like the ends of lambdoid phages (78, 79). Although this kind of *site-specific* cleavage is clearly not the basis of general recombination, it may be related to site-specific recombination and to infrequent events such as inversions, translocations, and the generation of specialized transducing particles.

Initiation of Strand Exchange

The process that is commonly called synapsis remains one of the most obscure aspects of genetic recombination. How are long, densely packed, and folded molecules of DNA (80, 81) brought into register such that homologous sequences can interact? And apart from random breakage and collision, what starts and drives the interaction? We are far from answers to these questions.

Interactions between DNA molecules may be based upon specific sequences of bases as well as homologous sequences of bases. The prototype of sequence recognition is site specific recombination of certain temperate phages (2, 3). The recognition of specific sequences, as opposed to the recognition of homology, could conceivably play a role in general recombination by providing the basis for a preliminary and approximate alignment of homologous sequences. Specific proteins bound to operators, initiation sites, or replication origins might form highly specific configurations of DNA and protein that are capable of mutual recognition via specific interactions of these same bound proteins. Recently Sobell (82) has suggested that symmetrical sequences of nucleotides could result in a tertiary configuration of DNA which might be a specific target for endonucleolytic opening while still keeping the two arms of the DNA linked by hydrogen bonds (72, 83, also *Gene Conversion*, below).

Mutual recognition of sequence homology between two double helixes is possible in principle (84), but because of winding problems, it is difficult to conceive of the formation of four-stranded structures by unbroken double helixes. Lacks proposed a model for synapsis based on sequence homology in which a single strand interacts with double helical DNA by insertion into the large groove (85). A new experimental approach to studying synapsis has come from the discovery, first by Alberts and his colleagues, of proteins that bind preferentially to denatured DNA. By interacting with AT rich regions in native DNA such proteins lower considerably the T_m. On the other hand, the same melting effect on intrastrand base pairing in denatured DNA removes a kinetic barrier to the reformation of double helical molecules and speeds reannealing as much as one thousandfold (86).

The first melting protein studied was the gene 32 protein of T4. The action of gene 32 is necessary for the formation of hydrogen bonded intermediates in T4 recombination (28, 87) including branched molecules (11). The stoichiometric relationship between gene 32 protein and replication is reflected in the synthesis of some 10,000 molecules of protein per cell (86, 88). Similar proteins have been reported in uninfected *E. coli* (89), *E. coli* infected by filamentous phage (90, 91), meiotic cells of the lily (92), and mammalian spermatocytes (93).

The gene 32 protein, which has a molecular weight of 35,000, binds cooperatively to denatured DNA at a ratio of one molecule of protein per 10 nucleotides (86). The complex of protein and single stranded DNA has a flexible rod-like conformation, 60 Å in diameter and 4.6 Å per nucleotide in length (94). In the presence of 32 protein, dAT copolymer melted even at 25°, which is 40° below the T_m. T4 DNA, because of its higher melting point, was not melted by 32 protein at temperatures up to 37° but single stranded fragments of T4 DNA were renatured rapidly at 25° in the presence of 32 protein. When 32 protein was mixed with native λ DNA and fixed with glutaraldehyde, denatured loops with bound protein were seen by electron microscopy (94). The partial denaturation map produced was very similar to that resulting from partial denaturation by alkali (45), showing that 32 protein reacts preferentially with AT rich regions. Although the necessity for fixation by glutaraldehyde obscured the mechanism of the reaction, it is likely that 32 protein reacted with transiently single stranded AT rich regions. No evidence was found that 32 protein preferentially invades ends of DNA molecules or regions around a nick (94).

At high concentrations, 32 protein polymerizes with itself (86, 95). The existence of a stable dimer, as well as polymers, suggests the involvement of two distinct self-interactions in the absence of DNA (96). Gene 32 protein increased, by as much as ten times, the rate of synthesis by T4 polymerase, which uses single stranded DNA as template. The stimulation was clearest under conditions of ionic strength and temperature at which 32 protein might have had the greatest effect on the secondary structure of single strands (95). Gene 32 protein did not stimulate any of the *E. coli* polymerases, nor did the *E. coli* melting protein stimulate synthesis by the T4 polymerase (see below). The specific physical complex which was formed between T4 polymerase and 32 protein (95) is perhaps as intriguing as the complex of 32 protein with DNA. Although 32 protein did not stimulate synthesis by T4 polymerase from a template of native DNA (95), the affinity of 32 protein both for polymerase and single stranded DNA suggests a model for the initiation of strand exchanges (97).

Gene 32 protein also makes single stranded DNA resistant to attack by the $3' \rightarrow 5'$ exonucleolytic activity of T4 polymerase (98), an observation that correlates with the observed absence of single stranded regions in phage DNA extracted from cells infected by gene 32 mutants of T4 (28).

The gene 5 protein, a melting protein purified from *E. coli* infected by phages fd or M13, binds cooperatively with single stranded DNA to form a linear rod-like complex which involves two strands of DNA and has occasional branches (86, 91, 94). This protein did not promote renaturation of DNA, but it doubled the rate of synthesis by polymerase II when the template was replicative intermediate (91). The latter observation may reflect strand displacement, a reaction not normally associated with polymerase II (see Strand Exchange, below).

The melting protein that has been purified from uninfected *E. coli* forms yet another kind of complex with DNA in which the length per nucleotide is only 1.8 Å. DNA synthesis by polymerase II on templates with long single stranded regions is stimulated tenfold by the *E. coli* melting protein whereas purified polymerases I and III are not (89).

From the nuclei of meiotic cells of *Lilium*, Hotta & Stern have isolated a protein with properties like those of T4 gene 32 protein (92). The protein from *Lilium* is synthesized in early prophase stages of meiois, a period during which chiasma formation can be eliminated by inhibition of protein synthesis. The lily protein, which has not been extensively purified, binds specifically to denatured DNA and promotes extensive renaturation of either T7 DNA or lily DNA at 25°. In addition, Hotta & Stern have detected proteins that bind specifically to denatured DNA from nuclei of mammalian spermatocytes. The probable significance of melting proteins for genetic recombination is enhanced by their reported occurrence in meiotic nuclei of higher plants, and mammals. Only the T4 gene 32 protein (86), the protein from *Lilium* (92) and one of the proteins from mammalian spermatocytes (preliminary observation cited in ref. 93) have been reported to promote renaturation of DNA. A good deal of complexity and specificity within this class of proteins is suggested by the distinctly different complexes formed with single stranded DNA, and the varying interactions with polymerases.

The discovery of the role of RNA synthesis in DNA synthesis adds new interest to the possible relationship between transcription and recombination.

Von Hippel & McGhee have suggested that RNA polymerase can be viewed as a melting protein (99). On the basis of electron microscopic observations of DNA during transcription in vitro, Bick et al have suggested that transcription results in temporarily destabilized regions of 300–1200 base pairs in length (100).

Helling observed that recombination within the arabinose operon, mediated by Pl transduction, was stimulated several fold by derepression (101). In experiments on transduction of *gal* by phage Tl Drexler observed large effects of neighboring λ genes which are consistent with the hypothesis that recombination in a given region is stimulated by transcription of that region (102). On the other hand, Shestakov & Barbour saw no effect of induction upon recombination in the *lac* operon following conjugation (103); and Herman observed a negative effect of induction on *lac* recombination in a merozygote (104).

Davies et al have provided indirect evidence that the excision of prophage λ by site specific recombination is lowered by mutations that block transcription of one of the crossover sites (105). While it has seemed appropriate to raise the issue of transcription in the discussion of synapsis, there are clearly other ways in which transcription might be involved in general recombination (105) if indeed it is.

Strand Exchange

Transfer of a portion of a strand from one DNA molecule to another involves breaking a set of hydrogen bonds and making a new set that is identical or nearly so. The disruption of one set of hydrogen bonds can be coupled, by a displacement reaction or by simultaneous degradation and resynthesis, to the formation of an equivalent set of hydrogen bonds (Figs. 2–5).

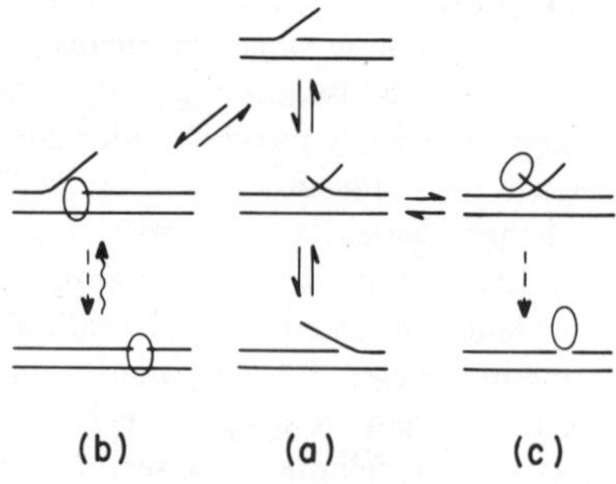

Figure 2 Branch migration (*a*), strand displacement (*b*, read from bottom to top) and strand assimilation (*b* and *c*, read from top to bottom). Reactions involving: new synthesis (⇝), degradation (--→), neither (⇌). Ellipses represent molecules of enzyme, either polymerase (*b*, read from bottom to top) or exonuclease (*b* and *c*, read from top to bottom).

A number of enzymes are capable of exposing single strands by exonucleolytic digestion from the 3′ or 5′ ends of double stranded DNA (106). A new mechanism has been described which appears to produce long single stranded ends by melting hydrogen bonds between staggered cleavages (107). This property is associated with one of the ATP-dependent DNAses which have been implicated in genetic recombination (8, 108). These enzymes, which are found in many bacterial species (8), act exonucleolytically on double stranded and single stranded DNA. The enzyme from *E. coli* also acts endonucleolytically on circular single stranded DNA. In the cases that have been studied, enzymes in this class produce 5′ phosphoryl terminated oligonucleotides, with an average chain length under ten nucleotides. The evidence suggests that they all act processively. The enzyme from *H. influenzae* will not initiate digestion at nicks or gaps; a double stranded substrate is made equally resistant by either 5′ or 3′ single stranded ends. Similarly, nicks are resistant to the action of the enzyme from *E. coli*. Neither the enzyme from *Hemophilus* nor the one from *E. coli* appears to discriminate between the ends of double stranded DNA; labels at the 5′ and 3′ termini of double stranded DNA are released equally rapidly (109–121).

A remarkable property which these enzymes share is a requirement for ATP: 20–40 moles of ATP are cleaved to ADP plus inorganic phosphate per mole of phosphodiester bond cleaved. The expenditure of 200–400 kilocalories in a reaction that is itself exothermic suggests immediately that the ATP plays a role that is only indirectly related to cleavage of phosphodiester bonds. Indeed Karu & Linn observed that chemically cross-linked DNA was resistant to hydrolysis by the *E. coli* enzyme but fully supported the ATPase activity (122). The observations of Friedman & Smith suggest a rationale for the ATPase activity. Working with the enzyme from *H. influenzae* these investigators found conditions under which 16s intermediates accumulated. Examination of the 16s material revealed that it consisted of double stranded DNA fragments with single stranded ends about 2000 nucleotides long. The single stranded ends were not produced by digesting acid soluble nucleotides from one strand at each end of the DNA, since only 3.5% of the DNA was made acid soluble at a time when single stranded regions comprised 23% of the DNA. Friedman & Smith further observed that the kinetics of hydrolysis of the ATP corresponded closely with the kinetics of the production of large fragments with single stranded ends. They suggested that the ATP is involved in melting double stranded DNA between nicks that are made in opposite strands about 2000 nucleotides apart (107).

Lee, Davis & Davidson (123) first described the phenomenon of branch migration, which is illustrated in Fig. 2a. Their observations were made by electron microscopy of denatured and reannealed DNA from a permuted and terminally redundant phage genome. An unforked branch (top and bottom diagrams of Fig. 2a) was 4 times more prevalent than a forked, migrating branch (middle diagram Fig. 2a), from which they estimated a difference in free energy of an unforked branch vs a forked branch of 3.5 kcal/mole. An upper limit to the jump time was estimated at 10^{-3} seconds per nucleotide, the same order of magnitude as the relaxation time per nucleotide during renaturation. Branch

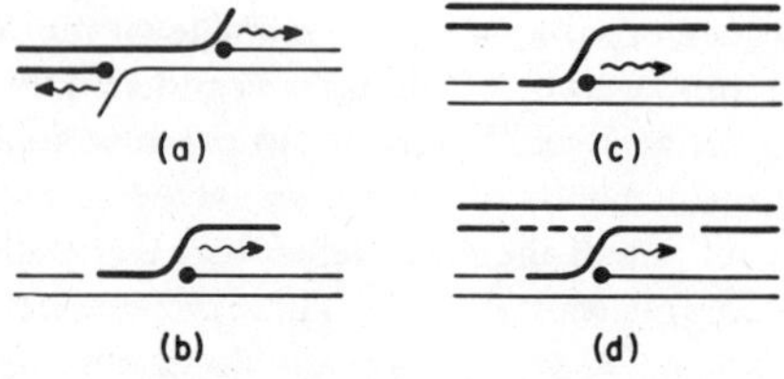

Figure 3 Possibilities for the involvement of strand assimilation and displacement in genetic recombination. Arrows indicate the direction of exonucleolytic degradation at the terminus marked by a dot (•). (*a*) Splicing of homologous fragments. (*b*) Assimilation of a single stranded fragment. (*c*) Strand transfer between helices (143, 144). (*d*) Strand transfer simultaneously driven by displacement (36) from the donor helix and assimilation in the recipient helix (144). The transfers illustrated in *b*, *c*, and *d* produce insertion heteroduplexes. Thus any of these diagrams are possible models for the kind of exchange that occurs in bacterial transformation (19). The strand transfers illustrated in (*c*) and (*d*) can also be reconciled with the production of joint molecules in which the flanking arms are in the recombinant rather than the parental configuration (Fig. 5 and 144). In (*c*) and (*d*) strand transfer is terminated by distal nicks in the donor helix.

migration as illustrated in Fig. 2 involves one helix and one migrating point. A hypothetical form of branch migration has been recognized for a number of years in the form of the Holliday model (Fig. 4, refs. 124, 125) in which a crossed strand exchange between two helices generates a migrating point on each helix. Meselson has recently supplied theoretical arguments in favor of the branch migration aspect of the Holliday model (126). Treating the movement of the crossed strands as a problem in rotary diffusion and considering DNA as a simple cylinder, Meselson calculated that at 37° about 20 seconds would be

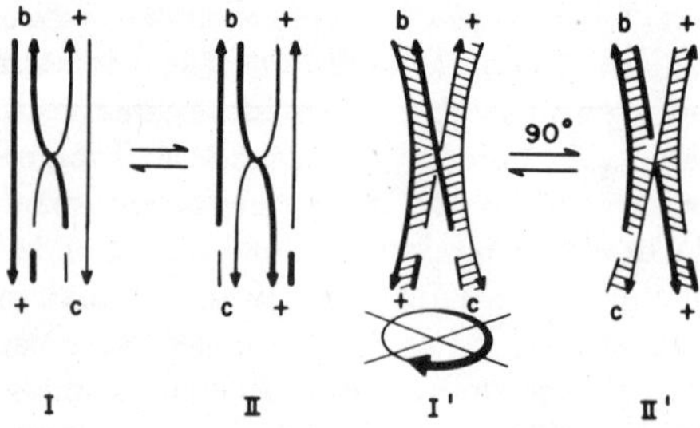

Figure 4 Strand equivalence in a crossed strand exchange, after Sigal & Alberts (36), and Emerson (189). Rotation of the helical arms about each other at the level of the strand exchange makes equivalent: (1) the two pairs of crossed and uncrossed strands, and (2) the parental (I) and recombinant (II) arrangement of distal markers. These effects are graphically illustrated by the optical illusion produced by viewing another representation of a crossed strand exchange (after Emerson, 189) from different angles (I′ and II′). The configuration at the time of cutting the crossed strands would determine the nature of the exchange.

needed for the migration of crossed strands through 1000 base pairs. Further assumptions included, (*a*) the absence of viscous drag other than that of water, and (*b*) the lack of hindrance due to the crossover itself. Another hypothetical form of branch migration has been proposed by Broker & Lehman to explain the origin of multiply branched structures of T4 DNA (11). In its simplest form, this kind of branch migration can be described as two homologous helices melting into each other, end to end, to form a structure with four double stranded arms.

Enzymatically driven forms of branch migration have been described. DNA polymerase I of *E. coli* is capable of introducing new nucleotides into double stranded DNA starting at the site of a nick. As a new chain is synthesized and new hydrogen bonds are formed "behind" the enzyme, old hydrogen bonds are broken "ahead" of the enzyme. The latter can occur in two ways, either by nucleolytic degradation ahead of the enzyme resulting in translation of a nick from one place to another, or by displacement of the strand ahead of the enzyme (127–131). This kind of enzymatic strand displacement is another plausible way in which a strand exchange might be initiated in recombination. Indirect arguments for a role of polymerase I in T4 recombination have been made by Mosig et al (132).

The reverse of strand displacement is strand assimilation (Fig. 2b and c, read from top to bottom). Among the enzymes that may be capable of catalyzing strand assimilation is the exonuclease made by bacteriophage λ. Before discussing this reaction, we will digress briefly on the properties of λ exonuclease and another protein, *β*, which are products of the recombination genes of phage λ (64, 133–135). Exonuclease and *β* protein are each composed of a single polypeptide chain, the respective molecular weights of which are 24,000 and 28,000 (136). Active exonuclease is probably a tetramer (137). A metastable association exists between the exonuclease and *β* protein in vitro, but little is known either about a separate enzymic function of the *β* protein or about any significant modification of the properties of exonuclease by the *β* protein (136, 138, 139). The exonuclease has been highly purified both with and without associated *β* protein (136, 140). The λ exonuclease processively cleaves 5′ mononucleotides from the 5′ phosphoryl end of double stranded DNA (138, 140). The enzyme acts at least a hundredfold more rapidly on native DNA than on denatured DNA, but continues to degrade slowly the single stranded remnants of its initial action (140). Moreover, all known preparations of the enzyme will digest the 5′ single stranded ends of λ DNA which are 12 nucleotides long (140, 141). The λ exonuclease will not initiate digestion at a nick and initiates digestion poorly at a gap (127, 138, 142).

Branched substrates have been constructed to test the ability of λ exonuclease to catalyze strand assimilation. Preparations of λ exonuclease, with or without *β* protein, catalyzed the slow degradation, at a branch of a specifically labeled 5′ terminated strand while the homologous strand was assimilated in place of the degraded strand (Fig. 2b and c, read from top to bottom; also 143–145). Digestion ceased when assimilation was completed, an observation that will be discussed more fully below (see *Closure*). Because of the much more rapid digestion of double stranded than single stranded DNA by λ exonuclease, its action at a single stranded branch has previously been interpreted as illustrated in Fig. 2b, but the data are also compatible with the mechanism illustrated in Fig. 2c. Some observations on λ recombination in vivo that are consistent with a strand assimilation model have already been discussed (see experiments of Russo, under *Recombinant Molecules*).

In spite of the observed bias in favor of an unforked branch (123) the reversible nature of branch migration makes 3′ and 5′ ends equivalent at the site of a single stranded branch (Fig. 2a), at least over short distances. Further support for this idea has been provided by observations on the early product of polymerase I synthesis at a nick. When 100 nucleotides were incorporated at a nick by polymerase I, 10–30 of them were susceptible to exonuclease I which is highly specific for single stranded DNA with a free 3′ terminus. Yet all of the newly synthesized material was also sensitive to exonuclease III, which is specific for double stranded DNA (128). If one of the arms of a migrating branch is anchored in a second molecule of DNA, as illustrated in Fig. 3c and d, then of course the equivalence of 3′ and 5′ ends does not exist. In that case, branch migration may play the role illustrated in Fig. 2c, producing a forked branch with one exposed single stranded terminus. Exonucleolytic degradation of this terminus in a forked branch may be equivalent to exonucleolytic degradation of an unforked branch (146). Thus it is possible that any exonuclease that acts on single stranded DNA is capable of catalyzing strand assimilation by the mechanism shown in Fig. 2c. For example the DNA polymerases I, II, and III of *E. coli* and the polymerase of phage T4 are all associated with a 3′ → 5′

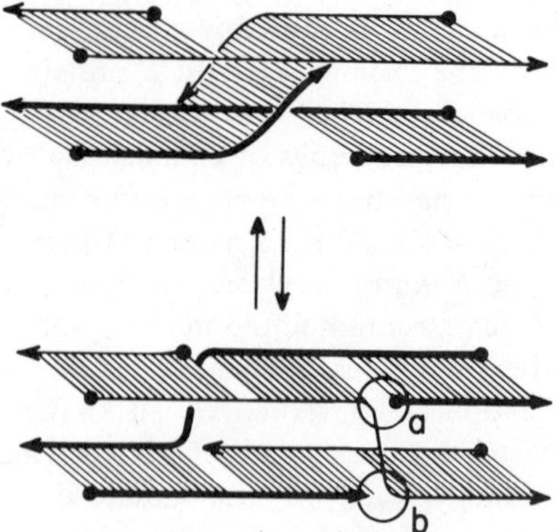

Figure 5 Rotational rearrangement in an exchange initiated by strands of opposite polarity, after Sigal & Alberts (36). The rearranged structure, which is fully paired, has two unitary strand exchanges which can be driven by the concerted action of strand assimilation at the two sites corresponding to (*a*), and strand displacement at the two sites corresponding to (*b*). As in Fig. 3(*c*) and (*d*), distal nicks might be expected to terminate exchange, producing two recombinant structures, one with newly synthesized DNA in the joint, and the other with DNA that is wholly parental (cf 144).

hydrolytic activities with varying degrees of specificity for single stranded DNA (98, 127, 128, 146–151). The properties of several exonucleases suggest another interesting point. Although exonuclease III is specific for native DNA, it will rapidly remove up to three mispaired nucleotides (146). The 3′ → 5′ nucleolytic activity of the T4 polymerase works a hundredfold faster on short single stranded polynucleotide chains than on long ones (98). Branch migration driven by exonucleolytic action, might continuously regenerate an optimal substrate for the driving enzyme.

Branch migration, and its enzymically driven variants, strand displacement, and strand assimilation represent concerted mechanisms for making and breaking hydrogen bonds, and represent plausible mechanisms of strand exchange in genetic recombination, as illustrated in Figs. 3 and 5. A further implication is that constituent steps in recombination, including the uncovering of homologous nucleotide sequences, exchange of strands, elongation of heteroduplex joints, repair of gaps, trimming of surplus strands, and preparation for covalent sealing, may be subsumed in highly concerted mechanisms of exchange resembling those illustrated (Figs. 3, 5, also *Closure*, below).

Gene Conversion[1]

Meiotic gene conversion describes the unequal recoveries of alleles that are observed from crosses in which all of the products of meiosis can be recovered, as in fungi. Also included under the rubric gene conversion are changes from heterozygosity to homozygosity by mechanisms other than segregation, changes that occur in mitotic cells in eukaryotes, as well as in prokaryotes. In principle, unequal recoveries of alleles could result from asymmetry in the mechanism of exchange (15, 16, 82, 144, 152) or from revision of a heteroduplex joint (Fig. 1). The latter is usually conceived as the excision of part of one strand, stimulated by the presence of mismatched base pairs, and repair of the excised region by new DNA synthesis, a theory called mismatch-repair. When the four haploid products of meiosis undergo a post-meiotic division, a further instance of unequal recoveries of alleles is sometimes seen. Among the eight haploid products of the first post-meiotic division, alleles are sometimes recovered in the ratios 5:3 or 3:5. By inference this means that one of the four meiotic chromatids was heterozygous. This phenomenon, which is called half-chromatid conversion, or post-meiotic segregation, means that imperfections in recombinant DNA, such as mismatched base pairs, are not all eliminated prior to replication.

Although the evidence is indirect, recent studies of conversion in *Ascobolus immersus* provide new reason to think that the excision of mismatched bases is at least part of the explanation of meiotic gene conversion. Leblon has studied the pattern of conversion of spore color mutants produced by different mutagens. When both the direction of conversion, wild type to mutant or vice versa,

[1] For complete discussion of gene conversion, see the review in this volume of "The Mechanism of Intragenic Recombination" by David Stadler.

and the time of segregation, meiotic or post-meiotic, were studied, characteristic conversion patterns were seen which correlated with the chemical origin of the mutation. Among 25 acridine induced mutations, which were presumably frameshift mutations, none resulted in post-meiotic segregation, and all but three were converted to the mutant genotype. Reversion studies of two of these mutants suggested that they were opposite in type, with respect to addition or deletion of bases. The direction of conversion was also opposite (153, 154). The mechanism of conversion was explored further by examining the interaction of two mutations in *cis* configuration, in a system in which the phenotypes of the wild type, single mutant, and double mutant were distinguishable. In conversions that involved both sites, a mutation that showed little post-meiotic segregation strongly decreased such segregation of the coupled mutant, but not vice versa; and coupled mutants mutually influenced each other's direction of conversion (155). All of the above observations are most compatible with a mechanism of conversion that involves specific recognition of mismatched bases and excision of an extensive segment of one strand. Hurst, Fogel & Mortimer have recently summarized their observations of gene conversion in yeast (156). They have estimated that conversion may involve segments of DNA as long as a thousand nucleotide pairs. Half of the observed conversions were associated with reciprocal recombination of more distant flanking markers (see *Models*). They concluded that the reciprocal recombination of distant markers that is associated with gene conversion is sufficient to account for all reciprocal meiotic recombination. This supports the view that the asymmetrical events of gene conversion and the symmetrical events of crossing-over are parts of the same overall process of recombination. Unlike the observations on *Ascobolus* (153–155) however, there was no bias in the direction of conversion in yeast; the frequency of tetrads of the kind 3 mutant: 1 wild type equaled that of 1 mutant: 3 wild type (156, 156a). Such parity is at odds with the hypothetical process in which the frequency and direction of conversion are determined by the specific pairs of bases that are mismatched. The isolation in eukaryotes of mutants with various alterations of recombination provides an approach to analyzing the relationships between the mechanisms of conversion and crossingover (125, 157).

The position of a mutation in the region of conversion also influences the frequency of conversion. The effect of position is polarized, and the pattern of this polarity differs, at least in appearance, from one system to another (17). Polarity can be understood as a variation in the efficiency of formation or revision of heteroduplex regions as a function of the distance away from special fixed points at which formation of heteroduplex regions begins (17, 125, 158, 159). Little is known about the actual existence of such special sites, but Sobell (22, 82) has extended the suggestion of Gierer (83) that short inverted duplications of base sequences in DNA might produce intrastrand base pairing and hence tertiary structures similar to those that exist in RNA. Sobell proposed that such structures, distributed throughout DNA, might provide fixed starting points for recombination. Further extensions of the hypothesis yielded a scheme for the generation of two nearby crossed-strand exchanges (Fig. 4) and a scheme for asymmetric formation of heteroduplex joints by exonucleolytic action (144). By itself, asymmetry in the formation of heteroduplex regions must lead exclusively to post-meiotic segregation (Fig. 1). Therefore, this model requires the additional assumption of excision of mismatched bases.

Phenomena resembling conversion have been studied in prokaryotes mostly by examining (*a*) segregation from heterozygotes produced by transformation, or (*b*) segregation from biologically active heteroduplex DNA prepared in vitro. Excluding trivial causes, such as a physiological advantage of one genotype, the unequal yields of progeny have been likened to gene conversion in fungi (19, 160). According to the theory of mismatch-repair, the efficiency of excision varies for different mismatches (19, 160–163). When marker specificity exists, it is essential that numerous markers be examined to get a complete picture of any system, and some apparent discrepancies in the literature probably result from the observation of too few markers. Mismatch repair may be related to variations in the efficiency of bacterial transformation by different markers. Clonal analysis of transformation in *B. subtilis* agrees well with the predictions of mismatch repair. In such experiments by Bresler and his colleagues, specific transformants were isolated nonselectively, although streptomycin treatment was sometimes used to increase the general abundance of transformed cells by 10–50-fold (161, 162). Bresler et al showed that the efficiency of incorporation of markers during transformation of *B. subtilis* correlated, as anticipated, with the frequency of mixed clones and did not correlate with the frequency of pure clones that had been converted to the donor genotype (162). This agrees with the idea that efficiently *transformed* markers are those that are not excised from the insertion heteroduplex formed during transformation. In *Pneumococcus* Roger observed cotransformation of markers arranged in *trans* configuration in heteroduplex donor DNA (164). Conversion in *B. subtilis* has also been studied by using heteroduplex phage DNA in transfection (160, 165). In this case, in which conversion to the genotype of H strand vs L strand can be determined, it was seen that the direction of conversion was more related to map position than to the nature of the mutation itself. An exception was the conversion of heteroduplex DNA for deletion mutants. In most instances, correction was toward the mutant genotype, indicating efficient recognition and elimination of gross distortions in helical structure. Asymmetric replication of the subtilis phage DNA was excluded as an explanation for conversion by the observation of independent and opposite conversion of markers in *trans* position in heteroduplex DNA. In addition, inhibition of replication by FUDR increased the frequency of conversion, possibly by prolonging the lifetime of heteroduplex structures. It has been suggested that conversion may account for marker specific effects in the recombination of very close markers in *E. coli* (163). In recent studies of Pl transduction in *E. coli*, Crawford & Preiss observed an excess of recombinants with a parental arrangement of flanking donor markers, a phenomenon which they likened to gene conversion (166, cf Fig. lb). Earlier observations on heteroduplexes or heterozygotes in *E. coli*, involving a few mutants, showed conversion in one case (167) and none in another (168). Recently, Wildenberg & Meselson have studied progeny resulting from infection with heteroduplex λ DNA under conditions of normal DNA synthesis and no recombination (*rec*⁻, *red*⁻, *int*⁻). They observed conversion at frequencies as high as a few percent, depending on the particular mismatch, over distances of several base pairs (169). White & Fox have also observed unequal recoveries of markers in single bursts after transfection with heteroduplex λ DNA under conditions of limited DNA synthesis (170).

The mismatch-repair theory of conversion is frequently compared to the repair of UV lesions by excision. But some experiments suggest that conversion and excision of UV lesions are only indirectly related. Lacks isolated mutants (*hex*) of *D. pneumoniae* that made all markers equally efficient in transformation. The *hex* mutation had no effect on UV sensitivity and was genetically separable from a mutation (*uvr*) that affected UV sensitivity (171). Conversion of heteroduplex phage DNA in *B. subtilis* was unchanged by host mutations affecting recombination, host-cell reactivation, or UV sensitivity (160). In *Pneumococcus*, UV irradiation or mitomycin-C treatment of the recipient increased the apparent conversion of a marker that otherwise was not converted at all (172). On the other hand, in *B. subtilis* UV irradiation of the donor DNA inhibited the conversion of markers that were otherwise converted efficiently (162). A similar instance was observed in UV irradiated *E. coli*, with a phage marker that was efficiently converted in the absence of UV irradiation (167).

No enzyme has been directly implicated in conversion, but enzymes involved in the excision of UV lesions provide a model for the excision of mismatched bases (51, 73, 173, 174). Two kinds of process have been distinguished: (*a*) *excise and patch*, and (*b*) *patch and excise* (73, 129, 130). *Excise and patch* is the pattern of repair shown by the enzyme system from *M. luteus*. An endonuclease makes an incision on the 5′ side of a thymine dimer, and an exonuclease excises some ten bases including the lesion generated by radiation. The resulting gap must then be closed by a DNA polymerase and a ligase (73). *Patch and excise* describes the reaction catalyzed by DNA polymerase I of *E. coli* in which new nucleotides are polymerized behind the enzyme while thymine dimers are excised ahead of the enzyme by the 5′ → 3′ exonucleolytic activity (129,130). An instance of specific recognition of mismatched base pairs is provided by the experiments of Brutlag & Kornberg. Mismatched bases at a 3′ terminus were specifically removed by the 3′ → 5′ hydrolytic activity of *E. coli* polymerase I (146, also see *Strand Exchange* above). A 3′ → 5′ hydrolytic activity that acts preferentially on denatured DNA is a frequent property of polymerases (see *Strand Exchange*). This activity may have a similar significance for all polymerases, namely the ability to remove mismatched, hence single stranded, bases adjacent to a 3′ terminus. Closely related is the ability of the DNA polymerase of phage T4 to remove and replace the nucleotide at the growing 3′ terminus.

Mutational increases or decreases in this 3′ → 5′ activity are correlated respectively with antimutator and mutator phenotypes (175). Thus the 3′ → 5′ hydrolytic activity, which is a property of many polymerases, has been related to mutagenesis, and may in addition be related to strand exchange and gene conversion.

Although much indirect evidence favors the mismatch repair theory of conversion, direct evidence is still lacking. Even if one grants that mismatches are corrected, it remains to be seen whether correction is specifically triggered by the configuration of the mismatched bases. A proper test of the theory requires the isolation of the putative enzyme systems.

Closure

The final step in the formation of a recombinant molecule from a joint molecule is covalent closure by a polynucleotide ligase. A consideration of strand exchange reveals that covalent closure may be accomplished without new DNA synthesis; the exchange process itself may leave interrupted strands with neither an excess nor a deficiency of nucleotides (see Figs. 2, 3). The way in which this is done is particularly interesting with respect to the processive mode of action of some nucleases, such as the λ exonuclease. A nuclease that acts processively degrades one molecule of polynucleotide extensively before it attacks another molecule (176).[2] What, if anything, stops a processive enzyme from completely degrading its substrate? Experiments with two types of branched substrate have suggested that λ exonuclease stops when strand assimilation has been completed, and that the remaining interruption can be sealed directly by polynucleotide ligase (143, 144, Figs. 2, 3). It is not clear yet how the processive action of λ exonuclease is arrested, but there are several ways in which strand assimilation might be brought to a halt precisely when the interrupted strand has neither too many nor too few nucleotides to be closed by polynucleotide ligase. As one possibility, the enzyme might specifically recognize a nick as a stopping place (Fig. 2b). More general possibilities include high specificity for single stranded DNA, or inability to initiate digestion at a nick (Fig. 2c). In this regard, the action of T4 DNA polymerase is interesting. This enzyme strongly prefers denatured DNA but will also hydrolyze native DNA in the 3′ to 5′ direction (98, 128, 151). In the presence of deoxynucleoside triphosphates however, the 3′ → 5′ hydrolytic activity on native DNA is replaced by synthesis in the opposite direction, 5′ → 3′ (146, 151). Lacking either the ability to degrade in the 5′ → 3′ direction, or to displace the strand ahead of it, the enzyme stops after it has removed a single stranded branch and leaves a nick that can be closed by polynucleotide ligase (127, 128). The same arrest mechanism might operate equally well for polymerases II and III of *E. coli* which hydrolyze single stranded DNA in the 3′ to 5′ direction but will not add nucleotides at a nick (147, 149, 177).

[2] Processiveness is not necessarily all-or-none. An expression derived by Bailey & French (192) can be used to evaluate the number of nucleotides removed per encounter of enzyme and polynucleotide (193).

Random cleavage of nucleotides from different molecules in a population, which is the alternative to the processive mode, is exhibited by exonuclease III of *E. coli* (137) and by the exonucleolytic activity of T4 polymerase (178). Masmune & Richardson showed that the combined action of exonuclease III and polynucleotide ligase on a branched molecule produced covalently sealed molecules (128). (A similar reaction was demonstrated for the exonucleolytic activity of T4 polymerase in the presence of ligase even when deoxynucleoside triphosphates were omitted.) Thus in spite of the preference of exonuclease III for native DNA, and its ability to initiate digestion at a nick, it was able to produce a substrate that could be sealed by ligase. In the absence of processive action, the relative affinities of ligase and exonuclease III for a nick apparently suffice to produce sealed polynucleotide chains. Except for purely degradative exonucleases, it would appear that processiveness plus the ability to act at a nick should be a forbidden combination of properties.

The essential role of polynucleotide ligase for the viability of *E. coli* and of bacteriophage T4 has been demonstrated (177–182). The role of the enzyme in the recombination cycle of phage T4 has already been discussed (*Strand Breakage* above). Recent research reveals a number of novel features of ligases that may also bear on recombination. Modrich, Lehman & Wang (183) have demonstrated the reversibility of the ligase reaction, thus relating ligase, in principle, to the ω protein from *E. coli* (184) and a similar activity from secondary mouse embryo cells (185). Possibly related is the observation that T4 polynucleotide ligase can catalyze the joining of DNA molecules end to end without the intermediate formation of a heteroduplex joint (186, 187). The existence of enzymes that reversibly make and break phosphodiester bonds, and of a mechanism for joining double stranded DNA end to end, speak for the possibility of polynucleotide transferase reactions in some kinds of genetic recombination (1) such as site-specific recombination and illegitimate recombination where homology and heteroduplex joints may play no role. Finally, the recent discovery and purification of an RNA ligase from *E. coli* infected by phage T4, suggests the possibility of recombination between RNA molecules (188).

Models

The complex interactions among recombination, replication, and repair (*Introduction* and ref. 8), make it unlikely that any single model of recombination will ever suffice. Various models have been cited above, and Signer has recently surveyed the numerous formulations of Models (21).

Sigal & Alberts (36) have constructed the physical model that corresponds to the Holliday formulation (124, 125, Fig. 4) namely a crossed strand exchange between two helices. This structure is sometimes called a half chromatid chiasma (189). The physical model revealed that the crossed strand exchange can occur without the loss of base pairs, and confirmed the supposition that a cross connection can readily diffuse along the joined helices (*Strand Exchange*, above).

But most important and interesting was the finding that a simple rotation of the joined helices relative to each other makes the two pairs of crossed and uncrossed strands interchangeable and at the same time makes the parental recombinant configurations of the flanking arms interchangeable (Fig. 4). This observation was anticipated by Emerson who suggested that the crossed strand exchange of helical molecules might behave like the optical illusion also pictured in Fig. 4 (189). As a consequence of this plastic property of DNA, a nuclease that cleaves a single target, the crossed strands, may produce the parental or recombinant configuration of flanking genetic markers with equal frequency (124, 125, 156). Alternatively, distal nicks in the double stranded arms may terminate the crossed strand exchange in one or two steps (of Figs. 3, 5). Another consequence of rotational movement of joined helices is illustrated in Fig. 5. In this example, exchange is initiated in a way that resembles the proposals of Whitehouse (190). Subsequent rotation of the arms on one side of the interconnection produces (*a*) helices joined by two unitary strand exchanges (*b*) full base pairing and (*c*) the recombinant configuration of flanking markers (36). The dynamic properties of DNA, which are revealed by branch migration (*Strand Exchange*) and strand equivalence (Fig. 4), should make one wary of artificial distinctions that can be created by two dimensional representations of recombining molecules.

In crosses of phage fl that selected for wild type recombinants in the recipient host, Boon & Zinder observed the frequent occurrence of bursts from single cells that included only one parental type in addition to the selected recombinant (16). [Similar single bursts have been observed in crosses of phage λ by Melechen & Hudnik-Plevnik (191)]. The segregation of an unselected marker suggested that the parental type was not a passive freeloader but had participated in the recombination event. These observations resemble gene conversion (cf 166). The most general, although not unique, explanation for the data of Boon & Zinder was provided in a model with the following features: (*a*) Exchange, initiated as in the Whitehouse model (190) leads to a single replication fork (*b*) The replication fork, which generates two copies of one allele, is resolved by a second exchange which discards the bit of DNA containing the unduplicated allele. Like an earlier model of Stahl's (15) this one emphasizes asymmetric exchange as a basis for conversion, rather than long heteroduplex regions and mismatch-repair.

Acknowledgments

I am indebted to my colleagues K. S. Sriprakash and W. Wackernagel for numerous helpful criticisms and discussions.

Work in the author's laboratory was supported by grants from the National Institutes of Health (USPHS AI 08160), The Jane Coffin Childs Memorial Fund for Medical Research (268), and the American Cancer Society (ACS NP 90).

Literature Cited

1. Clark, A.J. 1971. *Ann. Rev. Microbiol.* 25:437–64
2. Gottesman, M.E., Weisberg, R.A. 1971. *The Bacteriophage* λ, ed. A.D. Hershey, 113–38. Cold Spring Harbor. 792 pp.
3. Yarmolinsky, M.B. 1971. *Advances in the Biosciences 8*, ed. Gerhard Raspe, 31–67. Pergamon Press, Vieweg
4. Franklin, N.C. 1971. *The Bacteriophage* λ, ed. A.D. Hershey 175–194, Cold Spring Harbor. 792 pp.
5. Bukhari, A.I., Zipser, D. 1972. *Nature New Biol.* 236: 240–43
6. Daniell, E., Roberts, R., Abelson, J. 1972. *J. Mol. Biol.* 69: 1–8
7. Dove, W.F. 1971. *The Bacteriophage* λ, ed. A.D. Hershey, 297–312 Cold Spring Harbor. 792 pp.
8. Clark, A.J. 1973. *Ann. Rev. Genet.* 7: 67–86
9. Bernstein, H. 1968. *Cold Spring Harbor Symp. Quant. Biol.* 33: 325–31
10. Berger, H., Warren, A.J., Fry, K.E. 1969. *J. Virol.* 3: 171–75
11. Broker, T.R., Lehman, I.R. 1971. *J. Mol. Biol.* 60: 131–49
12. Manly, K.F., Signer, E.R., Radding, C.M. 1969. *Virology* 37: 177–88
13. Davern, C.I. 1971. *Progress in Nucleic Acid Research and Molecular Biology*, eds. J.N. Davidson, W.E. Cohn, 229–58 New York: Academic
14. Meselson, M. 1967. *Heritage from Mandel*, ed. A. Brink, 81–104 Univ. Wisconsin Press, Madison
15. Stahl, F.W. 1969. *Genetics* Supp. 61: 1–13
16. Boon, T., Zinder, N.D. 1971. *J. Mol. Biol.* 58:133–51
17. Fincham, J.R.S. 1970. *Ann. Rev. Genet.* 4: 347–72
18. Fogel, S., Mortimer, R.K. 1971. *Ann. Rev. Genet.* 5: 219–36
19. Hotchkiss, R.D., Gabor, M. 1970. *Ann. Rev. Genet.* 4: 193–224
20. Mosig, G. 1970. *Advan. Genet.* 15:1
21. Signer, E. 1971. *The Bacteriophage* λ, ed. A.D. Hershey, 139–74 Cold Spring Harbor, 792 pp.
22. Sobell, H.M. 1973. *Advan. Genet.*, ed. E. Caspari. In Press
23. Whitehouse, H.L.K. 1970. *Biol. Rev.* 45: 265–315
24. Tomizawa, J. 1967. *J. Cell. Physiol.* 70: Supp. 1, 201–13
25. Anraku, N., Anraku, Y., Lehman, I.R. 1969. *J. Mol. Biol.* 46: 481–92
26. Anraku, N., Lehman, I.R. 1969. *J. Mol. Biol.* 46: 467–79
27. Broker, T.R. 1971. *An Electron Microscopic Analysis of Bacteriophage T4 DNA Recombination* Ph.D. Thesis, Stanford University. Stanford, California
28. Kozinski, A., Felgenhauer, Z.Z. 1967. *J. Virol.* 1: 1193–1202
29. Oppenheim, A.B., Riley, M. 1967. *J. Mol. Biol.* 28: 503–11
30. Cooper, A.D., Burgan, M.W., White, C.W., Herrmann, R.L. 1971. *J. Bacteriol.* 107: 433–41
31. Paul, A.V., Riley, M. Personal Communication
32. Dubnau, D., Davidoff-Abelson, R. 1971. *J. Mol. Biol.* 56: 209–21
33. Dubnau, D., Cirigliano, C. 1972. *J. Bacteriol.* 111: 488–94
34. Ebel-Tsipis, J., Botstein, D., Fox, M.S. 1972. *J. Mol. Biol.* 71: 433–48
35. Ebel-Tsipis, J., Fox, M.S., Botstein, D. 1972. *J. Mol. Biol.* 71: 449–69
36. Sigal, N., Alberts, B. 1972. *J. Mol. Biol.* 71: 789–93
37. Enquist, L.W., Skalka, A. 1973. *J. Mol. Biol.* 75:185–212
38. Stahl, F.W., McMilin, K.D., Stahl, M.M., Nozu, Y. 1972. *Proc. Nat. Acad. Sci., USA.* 69: 3598–3601
39. Kellenberger-Gujer, G., Weisberg, R.A. 1971. *The Bacteriophage* λ. ed. A.D. Hershey, 407–15, Cold Spring Harbor 792 pp.
40. Stahl et al 1972. *J. Mol. Biol.* 68: 57–67
41. McMilin, K.D., Russo, V.E.A. 1972. *J. Mol. Biol.* 68: 49–55
42. Stahl, M.M., Stahl, F.W. 1971. *The Bacteriophage* λ, ed. A.D. Hershey 431–42, Cold Spring Harbor 792 pp.
43. Stahl, F. W., Stahl, M.M. 1971. *The Bacteriophage* λ, ed. A.D. Hershey 443–53, Cold Spring Harbor 792 pp.
44. Stahl, F.W. et al 1973. *Proc. 1973 ICN-UCLA Symp. Mol. Biol. In Press*
45. Inman, R.B., Schnös, M. 1970. *J. Mol. Biol.* 49: 93–98
46. Skalka, A. 1971. *The Bacteriophage* λ, ed. A. D. Hershey, 535–47, Cold Spring Harbor, 792 pp.

46a. Segawa, T., Tomizawa, J. 1971. *Mol. Gen. Genet.* 111: 197–201

47. Hobom, B., Hobom, G. 1972. *Mol. Gen. Genet.* 117: 229–38
48. Russo, V.E.A. 1973. *Mol. Gen. Genet.* In Press
49. Rupp, W.D., Wilde, C.E., III, Reno, D.L., Howard-Flanders, P. 1971. *J. Mol. Biol.* 61: 25–44
50. Howard-Flanders, P., Lin, P.F., 1973. *Genetics* In Press

51. Howard-Flanders, P., 1973. *Brit. Med. Bull.* In Press
52. Radman, M., Cordone, L., Krsmanovic-Simic, D., Errera, M. 1970. *J. Mol. Biol.* 49: 203–12
53. Cole, R. 1973. *Proc. Nat. Acad. Sci. USA* 70:1064–68
54. Kozinski, A. W. 1968. *Cold Spring Harbor Symp. Quant. Biol.* 33: 375–91
55. Prashad, N., Hosoda, J. 1972. *J. Mol. Biol.* 70: 617–35
56. Berger, H., Kozinski, A.W. 1969. *Proc. Nat. Acad. Sci., USA*. 64: 897–904
57. Ebisuzaki, K., Campbell, L. 1969. *Virology* 38: 701–03
58. Krisch, H.M., Hamlett, N.V., Berger, H. 1972. *Genetics* 72: 187–203
59. Karam, J.D. 1969. *Biochem. Biophys. Res. Commun.* 37: 416–22
60. Gellert, M., Bullock, M.L. 1970. *Proc. Nat. Acad. Sci., USA*. 67: 1580–87
61. Mosig, G., Ehring, R., Schliewen, W., Bock, S. 1971. *Mol. Gen. Genet.* 113: 51–91
62. Doermann, A.H., Parma, D.H. 1967. *J. Cell Physiol.* 70: supp. 1 147–64
63. Lederberg, S. Personal Communication
64. Wackernagel, W., Radding, C.M., 1973. *Virology* 52:425–32
65. Spatz, H.C., Trautner, T.A. 1971. *Mol. Gen. Genet.* 113: 174–90
66. Green, D.M. 1968. *Genetics* 60: 673–80
67. Okubo, S., Strauss, B., Stodolsky, M. 1964. *Virology* 24: 552–62
68. Bernardi, G. 1968. *Advan. Enzymol.* 31: 1–49
69. Studier, F.W. 1965. *J. Mol. Biol.* 11: 373–90
70. Bernardi, G., Cordonnier, C. 1965. *J. Mol. Biol.* 11: 141–43
71. Young, E.T., II, Sinsheimer, R. 1965. *J. Biol. Chem.* 240: 1274–80
72. Meselson, M., Yuan, R., Heywood, J. 1972. *Ann. Rev. Biochem.* 41: 447–66
73. Kaplan, J.C., Kushner, S.R., Grossman, L. 1971. *Biochemistry* 10: 3315–24
74. Friedberg, E.C., Hadi, S.M., Goldthwait, D.A. 1969. *J. Biol. Chem.* 244: 5879–89
75. Melgar, E., Goldthwait, D.A. 1968. *J. Biol. Chem.* 243: 4409–16
76. Goebel, W., Helinski, D.R. 1970. *Biochemistry* 9: 4793–4801
77. Kelly, T.J., Jr., Smith, H.O. 1970. *J. Mol. Biol.* 51: 393–409
78. Hedgpeth, J., Goodman, H.M., Boyer, H. W. 1972. *Proc. Nat. Acad. Sci., USA*. 69: 3448–52
79. Mertz, J.E., Davis, R.W. 1972. *Proc. Nat. Acad. Sci., USA*. 69: 3370–74
80. Worcel, A., Burgi, E. 1972. *J. Mol. Biol.* 71: 127–47
81. Stonington, O.G., Pettijohn, D.E. 1971. *Proc. Nat. Acad. Sci., USA*. 68: 6–9
82. Sobell, H.M. 1972. *Proc. Nat. Acad. Sci., USA*. 69: 2483–87
83. Gierer, A. 1966. *Nature* 212: 1480–81
84. McGavin, S. 1971. *J. Mol. Biol.* 55: 293–98
85. Lacks, S. 1966. *Genetics* 53: 207–35
86. Alberts, B.M., Frey, L. 1970. *Nature* 227: 1313–18
87. Tomizawa, J., Anraku, N., Iwama, Y. 1966. *J. Mol. Biol.* 21: 247–53
88. Sinha, N.K., Snustad, D.P. 1971. *J. Mol. Biol.* 62: 267–71
89. Sigal, N., Delius, H., Kornberg, T., Gefter, M.L., Alberts, B. 1972. *Proc. Nat. Acad. Sci., USA*. 69: 3537–41
90. Alberts, B., Frey, L., Delius, H. 1972. *J. Mol. Biol.* 68: 139–52
91. Oey, J.L., Knippers, R. 1972. *J. Mol. Biol.* 68: 125–38
92. Hotta, Y., Stern, H. 1971. *Dev. Biol.* 26: 87–99
93. Hotta, Y., Stern, H. 1971. *Nature New Biol.* 234: 83–86
94. Delius, H., Mantel, N.J., Alberts, B. 1972. *J. Mol. Biol.* 67: 341–50
95. Huberman, J.A., Kornberg, A., Alberts, B.M. 1971. *J. Mol. Biol.* 62: 39–52
96. Carroll, R.B., Neet, K.E., Goldthwait, D.A. 1972. *Proc. Nat. Acad. Sci., USA*. 69: 2741–44
97. Hotchkiss, R.D. 1971. *Advan. Genet.* 16: 325–48
98. Huang, W. M. Lehman, I.R. 1972. *J. Biol. Chem.* 247: 3139–46
99. Von Hippel, P.H., McGhee, J.D. 1972. *Ann. Rev. Biochem.* 41: 231–300
100. Bick, M.D., Lee, C.S., Thomas, C.A., Jr. 1972. *J. Mol. Biol.* 71: 1–9
101. Helling, R.B. 1967. *Genetics* 57: 665–75
102. Drexler, H. 1972. *J. Virol.* 9: 280–85
103. Shestakov, S., Barbour, S.D. 1967. *Genetics* 57: 283–89
104. Herman, R.K. 1968. *Genetics* 58: 55–67
105. Davies, R.W., Dove, W.F., Inokuchi, H., Lehman, J.F., Roehrdanz, R.L. 1972. *Nature New Biol.* 238: 43–45
106. Richardson, C.C. 1969. *Ann. Rev. Biochem.* 38: 795–840
107. Friedman, E. A., Smith, H.O. 1973. *Nature New Biol.* 241: 54–58
108. Tomizawa, J., Ogawa, H. 1972. *Nature New Biol.* 239: 14–16

109. Friedman, E.A., Smith, H.O. 1972. *J. Biol. Chem.* 247: 2846–53
110. Smith, H.O., Friedman, E.A. 1972. *J. Biol. Chem.* 247: 2854–58
111. Friedman, E.A., Smith, H.O. 1972. *J. Biol. Chem.* 247: 2859–65
112. Wright, M., Buttin, G., Hurwitz, J. 1971. *J. Biol. Chem.* 246: 6543–55
113. Goldmark, P.J., Linn, S. 1970. *Proc. Nat. Acad. Sci., USA.* 67: 434–41
114. Goldmark, P.J., Linn, S. 1972. *J. Biol. Chem.* 247: 1849–60
115. Tanner, D., Nobrega, F.G., Oishi, M. 1972. *J. Mol. Biol.* 67: 513–16
116. Anai, M., Hirahashi, T., Takagi, Y. 1970. *J. Biol. Chem.* 245: 767–74
117. Anai, M., Takagi, Y. 1971. *J. Biol. Chem.* 246: 6389–92
118. Winder, F.G., Lavin, M.F. 1971. *Biochim. Biophys. Acta* 247: 542–61
119. Hout, A., Ooosterbaan, R.A., Pouwels, P.H., De Jonge, A.J.R. 1970. *Biochim. Biophys. Acta* 204: 632–35
120. Oishi, M. 1969. *Proc. Nat. Acad. Sci., USA.* 64: 1292–99
121. Nobrega, F.G., Rola, F.H., Pasetto-Nobrega, M., Oishi, M. 1972. *Proc. Nat. Acad. Sci., USA.* 69: 15–19
122. Karu, A.E., Linn, S. 1972. *Proc. Nat. Acad. Sci., USA.* 69: 2855–59
123. Lee, C.S., Davis, R.W., Davidson, N. 1970. *J. Mol. Biol.* 48: 1–22
124. Holliday, R. 1964. *Genet. Res.* (*Camb.*) 5: 282–304
125. Holliday, R. 1968. *Replication and Recombination of Genetic Material.* eds. W.J. Peacock, R.D. Brock, Canberra: Australian Acad. Sci.
126. Meselson, M. 1972. *J. Mol. Biol.* 71: 795–98
127. Masmune, Y., Fleischman, R.A., Richardson, C.C. 1971. *J. Biol. Chem.* 246: 2680–91
128. Masmune, Y., Richardson, C.C. 1971. *J. Biol. Chem.* 246: 2692–2701
129. Kelly, R.B., Cozzarelli, N.R., Deutscher, M.P., Lehman, I.R., Kornberg, A. 1970. *J. Biol. Chem.* 245: 39–45
130. Kelly, R.B., Atkinson, M.R., Huberman, J.A., Kornberg, A. 1969. *Nature* 224: 495–501
131. Dumas, L.B., Darby, G., Sinsheimer, R.L. 1971. *Biochim. Biophys. Acta* 228: 407–22
132. Mosig, G., Bowden, D.W., Bock, S. 1972. *Nature New Biol.* 240: 12–16
133. Signer et al 1968. *Cold Spring Harbor Symp. Quant. Biol.* 33: 711–14
134. Shulman, M.J., Hallick, L.M., Echols, H., Signer, E.R. 1970. *J. Mol. Biol.* 52: 501–20
135. Radding, C.M. 1970. *J. Mol. Biol.* 52: 491–99
136. Radding, C.M., Rosenzweig, J., Richards, F., Cassuto, E. 1971. *J. Biol. Chem.* 246: 2510–12
137. Little, J. W. 1967. *J. Biol. Chem.* 242: 679–86
138. Carter, D.M., Radding, C.M. 1971. *J. Biol. Chem.* 246: 2502–10
139. Radding, C.M., Carter, D.M. 1971. *J. Biol. Chem.* 246: 2513–18
140. Little, J.W., Lehman, I.R., Kaiser, A.D. 1967. *J. Biol. Chem.* 242: 672–78
141. Radding, C.M., Unpublished Observations
142. Sriprakash, K.S., Her, M.O., Radding, C.M., In Preparation
143. Cassuto, E., Radding, C.M. 1971. *Nature New Biol.* 229: 13–16 23O;128
144. Cassuto, E., Lash, T., Sriprakash, K.S., Radding, C.M. 1971. *Proc. Nat. Acad. Sci., USA.* 68: 1639–43
145. Radding, C.M., Cassuto, E. 1971. *Advances in the Biosciences 8* ed. Gerhard Raspe 13–29 Pergamon Press, Vieweg
146. Brutlag, D., Kornberg, A. 1972. *J. Biol. Chem.* 247: 241–48
147. Wickner, R.B., Ginsberg, B., Berkower, I., Hurwitz, J. 1972 *J. Biol. Chem.* 247: 489–97
148. Gefter, M.L., Molineux, I.J., Kornberg, T., Khorana, H.G. 1972. *J. Biol. Chem.* 247: 3321–26
149. Kornberg, T., Gefter, M.L. 1972. *J. Biol. Chem.* 247: 5369–75
150. Kornberg, T., Gefter, M.L. 1971. *Proc. Nat. Acad. Sci., USA.* 68: 761–64
151. Hershfield, M.S., Nossal, N.G. 1972. *J. Biol. Chem.* 247: 3393–3404
152. Paszewski, A. 1970. *Genet. Res.* (*Camb.*) 15: 55–64
153. Leblon, G. 1972. *Mol. Gen. Genet.* 115: 36–48
154. Leblon, G. 1972. *Mol. Gen. Genet.* 116: 322–35
155. Leblon, G., Rossignol, J.L. 1973. *Mol. Gen. Genet.* 122:165–82
156. Hurst, D.D., Fogel, S., Mortimer, R.K. 1972. *Proc. Nat. Acad. Sci., USA.* 69: 101–05
156a. Fogel, S., Mortimer, R. K. 1969. *Proc. Nat. Acad. Sci., USA* 62: 96–103
157. Roth, R., Fogel, S. 1971. *Mol. Gen. Genet.* 112: 295–305
158. Whitehouse, H.L.K. 1966. *Nature* 211: 708
159. Touré, Bakary 1972. *Mol. Gen. Genet.* 117: 267–80
160. Spatz, H.C., Trautner, T.A. 1970. *Mol. Gen. Genet.* 109: 84–106
161. Bresler, S.E., Kreneva, R.A., Kushev. V.V. 1968. *Mol. Gen. Genet.* 102: 257–68

162. Bresler, S.E., Kreneva, R.A., Kushev, V.V. 1971. *Mol. Gen. Genet.* 113: 204–13
163. Norkin, L.C. 1970. *J. Mol. Biol.* 52: 633–55
164. Roger, M. 1972. *Proc. Nat. Acad. Sci., USA.* 69: 466–70
165. Trautner, T.A., Spatz, H.C., Behrens, B., Pawkel, B., Behncke, M. 1971. *Advances in the Biosciences 8*, ed. Gerhard Raspe 79–87 Pergamon
166. Crawford, I.P., Preiss, J. 1972. *J. Mol. Biol.* 71: 717–33
167. Doerfler, W., Hogness, D.S. 1968. *J. Mol. Biol.* 33: 661–78
168. Russo, V.E.A., Stahl, M.M., Stahl, F.W. 1970. *Proc. Nat. Acad. Sci., USA.* 65: 363–67
169. Wildenberg, J., Meselson, M. Personal Communication
170. White, R., Fox, M. Personal Communication
171. Lacks, S. 1970. *J. Bacteriol.* 101: 373–83
172. Guerrini, F., Fox, M.S. 1968. *Proc. Nat. Acad. Sci., USA.* 69: 1116–23
173. Mahler, I., Kushner, S.R., Grossman, L. 1971. *Nature New Biol.* 234: 47–50
174. Okubo, S., Nakayama, H., Takagi, Y. 1971. *Biochim. Biophys. Acta* 228: 83–94
175. Muzyczka, N., Poland, R.L., Bessman, M.J. 1972. *J. Biol. Chem.* 247: 7116–22
176. Nossal, N.G., Singer, M.F. 1968. *J. Biol. Chem.* 243: 913–22
177. Knippers, R. 1970. *Nature* 228: 1050–53
178. Nossal, N.G., Hershfield, M.S. 1971. *J. Biol. Chem.* 246: 5414–26
179. Modrich, P., Lehman, I.R. 1971. *Proc. Nat. Acad. Sci., USA.* 68: 1002–05
180. Krisch, H.M., Shah, D.B., Berger, H. 1971. *J. Virol.* 7: 491–98
181. Konrad, E.B., Modrich, P., Lehman, I.R. 1973. *J. Mol. Biol.* 77:519–29
182. Gottesman, M.M., Hicks, M.L., Gellert, M. 1973. *J. Mol. Biol.* 77:531–47
183. Modrich, P., Lehman, I.R., Wang, J.C. 1972. *J. Biol. Chem.* 247: 6370–72
184. Wang, J.C. 1971. *J. Mol. Biol.* 55: 523–33
185. Champoux, J.J., Dulbecco, R. 1972. *Proc. Nat. Acad. Sci., USA.* 69: 143–46
186. Sgaramella, V. 1972. *Proc. Nat. Acad. Sci., USA.* 69: 3389–93
187. Sgaramella, V., Khorana, H.G. 1972. *J. Mol. Biol.* 72: 493–502
188. Silber, R., Malathi, V.G., Hurwitz, J. 1972. *Proc. Nat. Acad. Sci., USA.* 69: 3009–13
189. Emerson, S. 1969. *Genetic Organization*, eds. E.W. Caspari, A.W. Ravin, 267–360, New York: Academic
190. Whitehouse, H.L.K. 1963. *Nature* 199: 1034–40
191. Melechen, N.E., Hudnik-Plevnik, T.A. 1972. *Proc. Nat. Acad. Sci., USA.* 69: 3195–98
192. Bailey, J.M., French, D. 1957. *J. Biol. Chem.* 226: 1–14
193. Lindahl, T. 1971. *Eur. J. Biochem.* 18: 415–21

3049

THE MECHANISM OF INTRAGENIC RECOMBINATION

David R. Stadler

Department of Genetics, University of Washington, Seattle, Washington

MARKER EFFECTS

Historical Background

The prudent scientist keeps his hypothesis simple. He endows it with only enough complexity to account for the observations. He stands stubbornly by his simple hypothesis, resisting the complicating results of other peoples' experiments, until there can be no further doubt of their validity. Then he retreats a very short way, taking up a new position with only enough added complexity to accommodate the unwanted findings. There he digs in and prepares for the next attack of the anarchists.

The siege on the mechanism of genetic recombination has been particularly long and telling. For many years it appeared that the event could be described in the simplest terms: a single, reciprocal exchange between two homologous chromosomes which could occur (with similar likelihood) at the junction between any two adjacent genes. From this it followed that measurements of recombination frequency generated genetic maps that revealed the order of successive sites and even an approximation of the distances between them.

When it was discovered that recombination could take place at a site *within* a functional gene (1, 2), some modification of the model was required. The discoverers of intragenic recombination found that this event could not be explained by a single, reciprocal exchange. It required two or three exchanges in a very short distance to account for some of the products. Furthermore, tetrad analysis showed that these exchanges were not exactly reciprocal. Segregating sites in the immediate vicinity of a recombination event showed frequent 3:1 (or 1:3) segregations. This was called "gene conversion". (Indeed, one way to interpret the production of a wild-type recombinant in a cross between allelic mutants was to say it did not involve exchange at all but was the result of gene conversion of one of the mutant alleles to wild type.)

The model was adroitly reworded: for "single exchange": read "multiple exchanges in a very short region (switching)"; before the word "reciprocal" insert the words "not exactly" (3). But one basic assumption remained intact: frequency of recombination depends only on the distance separating the segregating sites; the genetic markers *reveal* the event, but they do not *influence* it in any way.

Soon there were results from meiotic recombination in Ascobolus (4) and Neurospora (5) which required further tinkering with the model. Switching, whether reciprocal or nonreciprocal, was not sufficient. There appeared to be some kind of polarity. Recombination between two segregating sites in the same gene depended, not only on the distance separating them, but also on their relative positions. The polarity involved the site of nonreciprocal segregation (gene conversion). When a number of crosses involving different pairs of mutant sites were analyzed, it was found that the right-hand member of any pair consistently provided most of the gene conversions.

At this point it might have appeared that the pure and detached role of the genetic markers was no longer credible. But Stahl (6) and Murray (7) managed to preserve (even if briefly) the sanctity of the markers by devising the Fixed Pairing Region Model. This involved only a slight modification of the Switching model: switching occurs in regions of special pairing, and these are not distributed at random but have fixed ends, which correspond to some physical structures along the chromosome. Thus a site near the beginning of a pairing region will experience less frequent nonreciprocal segregations than a site farther along in the same region.

According to the Fixed Pairing Region model, the recombination behavior of a particular segregating mutant depended only upon its position, but that meant position with respect to the pairing region as well as position with respect to other mutant sites and the rest of the chromosome.

Marker Effects in Meiotic Recombination

Then came observations that could not be explained by Fixed Pairing regions. Tetrad analyses in fission yeast (8) and in Ascobolus (9) revealed cases in which two alleles were at virtually the same site, but one was a high conversion (and high recombination) mutant while the other was low. Here it appeared that specific marker effects could not be denied. (An important limitation was that the conclusion that the mutants were at the same site was based on recombination—the very process we wish to understand.)

Rossignol (10) studied gene conversion frequencies of a set of allelic spore-color mutants of Ascobolus. When conversion frequencies were graphed according to map position, they did not display the smooth curve expected by the Fixed Pairing Region model: there were repeated ups and downs. However, Rossignol pointed out that the mutants could be separated into several classes according to their relative frequencies of gene conversion in the two directions. Ascobolus has one mitotic division following meiosis before ascospore formation. Thus each ascus contains eight ascospores. Normal segregation in a cross of mutant × wild type yields a 4:4 ascus. Rossignol classified his mutants by their "dissymmetry coefficients": the ratio of conversions to wild type (6 wild-type spores:2 mutant spores) to conversions to mutant (2:6). He showed that within each class there *was* polarity of conversion frequencies, and all classes showed increasing frequency in the same direction on the map.

Therefore Rossignol proposed that conversion frequency was controlled by two factors: the position of the segregating site in a region of polarized recombination, and the specific molecular configuration at the segregating site (the basis of the dissymmetry coefficient). That is, he concluded that within a Fixed Pairing Region there were specific marker effects.

Rossignol's work suffered from the same shortcoming as earlier studies: the starting assumptions were derived from the process (recombination) that he was trying to explain in the conclusions. Specifically, he assumed that the molecular configuration of a given mutant determined the frequency ratio for its conversion in the two directions. Leblon (11) checked this assumption in the same material by studying the gene conversion spectrum for mutants induced by specific mutagens. He observed some striking correlations.

Twenty-five mutants had been isolated from material treated with ICR-170. This mutagen is believed to cause base additions and/or base deletions. Twenty-two of the mutants showed a significantly higher frequency of conversion to mutant (2:6) than conversion to wild type (6:2). The other three had similar conversion frequencies in both directions. None showed an excess of conversion to wild type.

The reverse pattern of dissymmetry was shown by the mutants induced by N-methyl-N′-nitro-N-nitrosoguanidine (NG), a mutagen believed to cause base substitutions. Twelve of eighteen mutants showed a significant excess of conversion to wild type, while the other six showed similar frequencies in both directions.

An even stronger correlation to the mutagen history was shown when the mutants were scored for presence or absence of post-meiotic segregation (5:3 and 3:5 patterns). *All* of the NG mutants and *none* of the ICR mutants gave high frequencies of these patterns.

Mutants induced by ethyl methanesulfonate (EMS) included five in the NG class (excess conversion to wild type, frequent postmeiotic segregation), two in the ICR class (excess conversion to mutant, rare postmeiotic segregation) and four in a new class (excess conversion to wild type, rare postmeiotic segregation).

Leblon (12) has employed reversion studies of these mutants to make specific proposals about the molecular nature of the mutation events. He concludes that the NG class represents base substitutions, the ICR class is base additions, and the EMS-unique class is base deletions. (This assignment of addition and deletion classes is the reverse of what might have been predicted from the recent work of Benz & Berger (13) on progeny ratios from mixed infections of phage T4. They found that mixtures of deletion mutant and wild type gave a majority of mutant progeny, while mixtures of addition mutant and wild type gave a majority of wild type.)

The work of Leblon provides partial confirmation of the hypothesis of Rossignol, and provides tentative molecular classification for the mutants of particular recombination classes. A critical test of these ideas must be made with mutants for which there is reliable molecular information. Unfortunately, there is not sufficient molecular information about any gene in a eukaryote which is also suitable for intragenic recombination analysis, with one exception: the CY_1 (cytochrome c) gene of yeast (14). Even in this case the genetic and biochemical information are still too limited for any conclusions about marker effects with respect to base sequences. Only in prokaryotes do we have studies on recombination frequencies in crosses involving mutants for which there is hard evidence about physical position and base sequence.

Marker Effects in Prokaryotes

Tessman (15) developed an "ultrahigh sensitivity" method to measure recombination between very closely-linked *rII* mutants of phage T4 and found gross departures from additivity in the frequencies. He suggested that specific marker effects might cause large distortions in maps of genetic fine structure.

Ronen & Salts (16) used the Tessman method to measure recombination between adjacent nucleotides. They made crosses between pairs of phage stocks that carried nonsense mutations in the same codon, one being amber (UAG) and the other opal (UGA). They constructed such mutant pairs for twelve different sites in *rII* and found a 1000-fold range of recombination frequencies for the different sites. It was concluded that recombination frequency must be influenced by the base sequences adjacent to the mutant codons.

Ephrussi-Taylor (17) reported marker-specific effects on the efficiency of transformation of drug-resistant mutants of Pneumococcus. She suggested that the low-efficiency mutants were those that were recognized and usually removed by an excision-repair system.

Norkin (18) studied intragenic recombination at the *lacZ* locus of *E. coli* in *Hfr* × *F−* conjugation. The order of 17 point-mutants had been determined by crosses to a series of overlapping deletions. He found that the recombination frequencies in two-point crosses among the point-mutants could not be used to order the sites, and he concluded that distance played little or no part in determining recombination frequencies for markers separated by less than several thousand nucleotide pairs.

A mutant that was involved in one high-recombination cross would be likely to show high recombination when crossed to other mutants. Norkin concluded that specific nucleotide sequences have dramatic effects on recombination at this level.

Norkin compared mutants containing different nonsense triplets in the same codon for recombination frequency when both were crossed to the same third mutant. The members of such pairs of crosses gave significant differences of frequency, but no one of the three nonsense triplets was consistently high in recombination in all the crosses. A disquieting observation was that two separate occurrences of UAG (amber) mutants in the same codon gave significantly different recombination frequencies when both were crossed to the same mutant; Norkin concluded that this indicated that the mutagen had also caused changes at other sites in the gene.

Stadler & Kariya (19) attempted to analyze marker effects at the *trpA* locus of *E. coli*. This study employed the mutants for which Yanofsky and his colleagues had amassed a large amount of molecular information by comparing the wild-type enzyme (tryptophan synthetase) to the altered proteins that are produced by mutants and revertants. Frequencies of intragenic recombination were measured by *P1* transduction. There was clear evidence of marker-specific effects. Three mutants gave high-frequency recombination (HR) in crosses to any other mutants; each of the three was in the same codon as a known mutant that gave the low-frequency recombination (LR) that was characteristic of most of the mutants.

There appeared to be no simple molecular basis for the HR behavior. The three mutants included one missense (AGA), one ochre (UAA) and one amber (UAG). The LR mutants included all the same classes as well as frame-shift mutants. One of the HR mutants resulted from a transition mutation, a second from a transversion, and the origin of the third is not yet known.

One experiment employed a Trp+ strain with an altered base sequence in *trpA*, and it revealed that HR behavior depended not only on the sequence of the mutant but on its combination with the corresponding sequence from the Trp+ parent.

We conclude that the studies in prokaryotes fail to confirm the proposal of Rossignol and Leblon that mutants that arise from the same kind of event should behave alike in recombination. However, the materials of study are very different, so the comparison may not be appropriate.

We can conclude that there is now abundant evidence that specific marker effects are real. However, there is also compelling evidence that they cannot be explained by the molecular composition of the mutant codon alone. If there is to be a molecular explanation, it will have to involve a larger sequence.

Repair of Heteroduplexes

Several models for recombination include the formation of a segment of heteroduplex DNA that is later "corrected" to complementarity by repair. Perhaps specific marker effects involve the recognition of mismatched sites by the repair systems. Direct evidence for heteroduplex DNA as an intermediate in recombination has been hard to obtain. There is indirect genetic evidence in the tetrads that show post-meiotic segregation; these are presumed to represent heteroduplex regions that were left unrepaired.

In T4 phage there is genetic evidence for DNA molecules that are recombinants for the ends and have a hybrid region at the junction point. However, it is difficult to tell whether sites of noncomplementary bases are repaired in these molecules. Such an event is hard to distinguish from normal replication; both events produce "homoduplexes". One way to detect repair would be to start with a cell that was infected with a single phage heteroduplex molecule and no homoduplexes. If the first event were repair, the cell would yield only one of the two parent alleles among the progeny phage, while if the first event were replication it would yield both types. Such an approach has been attempted by Spatz & Trautner (20) by building artificial hybrid molecules of phage DNA: the procedure involves extraction of DNA and separation of light from heavy strands after denaturation; this is done with both a mutant and a wild-type strain; hybrid double-strand molecules are then built by reannealing the light mutant strands with the heavy wild-type strands, and vice versa.

Spatz & Trautner (20) performed "transfection" experiments with phage SPPl. This involves infection of host cells (*B. subtilis*), not with whole phage, but with extracted phage DNA. Denaturation of the donor DNA destroys its activity, but reannealing restores it. They infected cells with hybrid DNA and did a single-burst analysis of the progeny to find out whether the heteroduplexes had been repaired. Twenty-eight different hybrid strains were tested in this manner, and in all cases a significant fraction of bursts contained only one of the parent types. The overall range of uniparental bursts was 34–90%. Therefore, they concluded that hybrid DNA was repaired before replication in many of the infected cells.

The pattern of repair was not random in these experiments. The typical result for the single bursts from a given experiment was a significant excess of pure bursts for one parent type over the other. The experiments were performed in pairs. For any given mutant, two species of hybrid DNA were prepared: mutant heavy strand with wild-type light strand and mutant light strand with wild-type heavy strand. Progeny were analyzed from both of the reciprocal preparations. The usual result was that if the heavy strand parent was the "preferred strand" (more numerous pure bursts) in the first member of a pair, then the heavy strand was also preferred in the reciprocal. (This means the preferred *genetic* type was the opposite one in the second test.)

Such a result would be predicted if (*a*) the mutants were *transitional* changes from wild type, so that the mismatched base pairs were purine-pyrimidine pairs, and (*b*) the excision-repair system had a specificity of substrate so that it usually excised mismatched purines (or pyrimidines).

Spatz & Trautner performed an experiment to test the specificity of excision. They mutagenized separated single strands of SPPl DNA. They treated heavy strands with hydroxylamine and assumed that this mutagen was specific for C to T transitions. The treated strands were reannealed to light wild-type strands and then transfected in host cells. Mutant plaques were isolated, and it could now be assumed that these were transitions with the pyrimidine on the heavy strand. Reciprocal pairs of hybrid DNA preparations were made between four of these mutants and wild type, and transfection progeny were analyzed. Three of the four gave heavy strand preference (the fourth result was ambiguous) indicating purine excision. However, four of the five mutants produced by hydroxylamine treatment of *light* strands also gave heavy strand preference, and this would indicate pyrimidine excision.

Although the observations of Spatz & Trautner appear to demonstrate some kind of repair of heteroduplex DNA, there is a bothersome ambiguity in this experiment. Transfection only occurs when there are at least two phage-equivalents of donor DNA per cell, and the number of infective centers produced goes up with the square of the concentration of donor DNA. These kinetics are characteristic of multiplicity reactivation of phage that have been inactivated by irradiation. It may be presumed that some regions or sites in the transfecting DNA are inactivated, and these sites must be excluded by recombination in order to produce viable progeny. Given this complication of the transfection life cycle, pure bursts might result from the pattern of inactivation and recombination, rather than from specific repair of heteroduplexes.

General Marker Effects

The discussion above concerned specific effects of particular markers. If segregating sites per se had some effect on recombination, this could be termed a general marker effect. There have been suggestions that closely-linked segregating sites have a regular and uniform effect on recombination. Hershey (21) proposed that nearness of the two segregating sites might *stimulate* recombination. This would produce a departure from map additivity of the type that has been called negative interference. For the pairwise crosses of three linked markers *a*, *b* and *c*; freq.*ab* + freq.*bc* $>$ freq.*ac*.

Holliday (22) suggested that very closely-linked sites would *impede* recombination, producing the result he named "map expansion": freq.*ab* + freq.*bc* $<$ freq.*ac*. Numerous examples of both negative interference and map expansion have been reported in interallelic mapping in microorganisms.

Stadler & Kariya (19) measured transduction frequencies in crosses between mutants of known location in the *trpA* gene of *E. coli*. They found that the frequency of recombination per physical distance was somewhat higher for very close pairs than for pairs of mutants that were farther separated. This result would generate negative interference in mapping. However, this and other examples of negative interference do not necessarily result from general marker effects. If recombination frequently involved double exchanges closely clustered, the resulting maps would show negative interference in the absence of any marker effects.

Perhaps the most direct test for general marker effects is accomplished by inserting a "silent" site of heterozygosity between the two sites for which recombination is being monitored. This can be done with a conditional mutant in permissive conditions. Katz & Brenner (23) reported that inserted heterozygosity caused a pronounced reduction in the frequency of recombination between very closely-linked mutants in the *rII* gene of T4. In Neurospora a temperature mutant was inserted between pairs of absolute mutants at the *mtr* locus (24). This caused no pronounced change in the frequency of recombination, but the length of the segments involved in gene conversion seemed to increase.

EXCHANGE OF FLANKING MARKERS

The problem of marker exchange was clearly recognized by Mitchell (1) in the first description of intragenic recombination in a system with flanking markers. There was a correlation with marker exchange: the frequency of recombined markers among intragenic recombinants was well above the map distance between the markers. But it was not an absolute correlation. There were, in fact, similar numbers of all four marker combinations among the intragenic recombinants. Formally this would mean that single, double, and triple exchanges were taking place with similar frequencies in a very short region.

Holliday (22) designed a molecular model (see Figure 1) that could account for the similar frequencies of parental and recombinant markers accompanying intragenic events. He proposed that the event began with single-strand breaks in two homologous chromatids, followed by an exchange of strands that set up two corresponding regions of "hybrid DNA". This left the two chromatids in a tangle that had to be resolved by two further single-strand breaks.

Holliday predicted that the recombination event might or might not result in the reciprocal exchange of outside markers, depending upon which two of the DNA strands were involved in the second break (which was required to disentangle the strands). If they were the same strands broken at the beginning of the process (strands 2 and 3 in the figure), there was no exchange for flanking markers, while a break in the other two strands generated reciprocal exchange. Holliday did not speculate about the relative frequencies of these two alternatives.

Whitehouse (25) designed a recombination model that also involves single-strand breaks and hybrid DNA, but differs from the Holliday model in several important respects. The Whitehouse model generates marker recombination with every hybrid DNA event. He explained those intragenic recombinants that had parental marker combinations as the result of hybrid DNA forming simultaneously in two adjacent segments.

Exchange of flanking markers has been scored in many studies of intragenic recombination in Neurospora, Aspergillus, Ascobolus, and Sordaria. The frequencies show a wide range. Among intragenic recombinants in yeast, however, it has been consistently observed that the frequency of marker exchange is about 50%.

Hurst, Fogel & Mortimer (26) analyzed 11,023 unselected asci from twelve diploids of yeast in which gene conversion could be detected at one or more segregating sites. These sites were flanked by markers no more than 20.5 map units apart. They found a total of 907 asci with gene conversion (3:1 or 1:3) of which 445 had a reciprocal recombination between the flanking markers. This close fit to 50% led them to conclude that conversion is causally related to crossing over; it is one of two equally-likely ways of resolving the chromatids involved in the conversion event.

Sigal & Alberts (27) have used molecular models to study the physical properties of the hybrid (heteroduplex) DNA which is a putative intermediate in

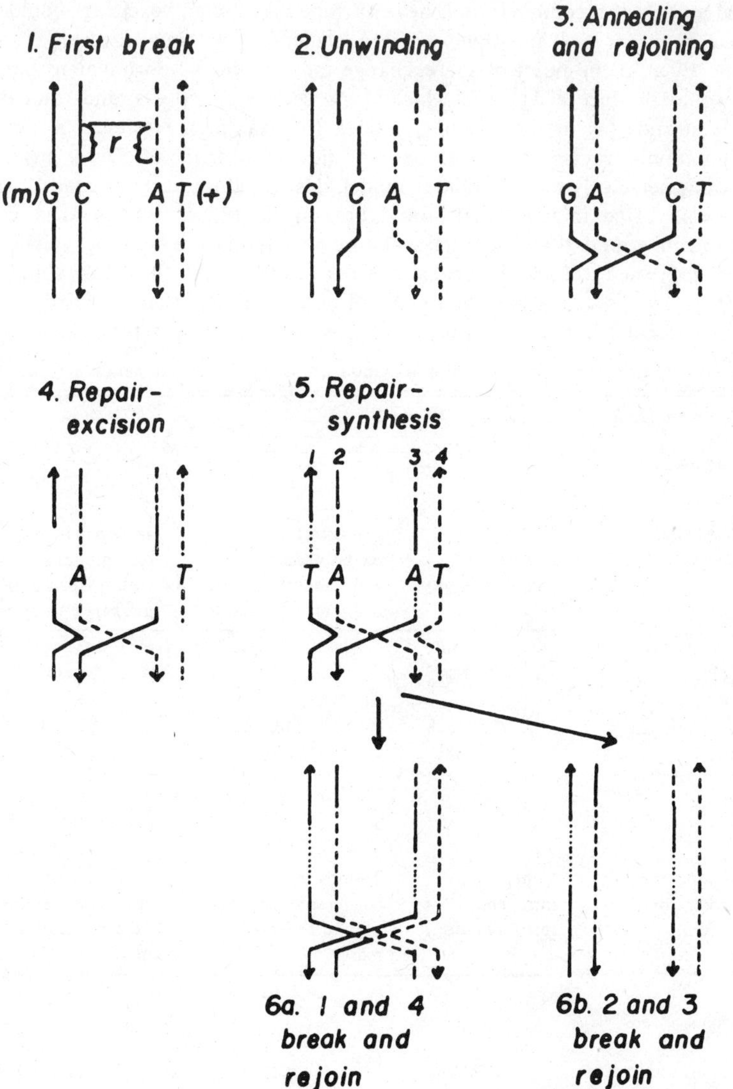

Figure 1 The Holliday model for recombination (38). The region marked *r* in the first drawing has been called the "recombinator" by Holliday, who suggests it may contain a base sequence that is a specific substrate for an enzyme that makes single-strand breaks.

several models of recombination. They conclude that a strand exchange could be formed between two homologous DNA molecules without disturbing base pairing. Even at the point of the exchange there would be no unpaired bases. This would produce a figure in which there were two bridge strands and two outside strands (as in the Holliday model, for example). However, a simple rotation would reverse the positions of these two pairs of DNA strands. Referring specifically to the Holliday model, this would mean that strands 2 and 3 (those involved in the initial strand breaks) would be rapidly exchanging position with strands 1 and 4. If the structure were resolved by cutting whichever strands happened to form the bridge at a particular instant, the result would be a reciprocal crossover in half the cases and no crossover in the other half.

This molecular hypothesis appears to provide an appealing theoretical basis

Table 1 Coincidence of intragenic recombination with separate recombination events.

A m1 B C

a m2 b c

b-c recombination frequencies:	Control (no intragenic recombination)	Intragenic recombinants with parental markers (AB or ab)	Intragenic recombinants with nonparental markers (Ab or aB)
Neurospora random spores (28)	80/724 (11.1%)	48/361 (13.3%)	11/282 (3.9)
Yeast tetrads (tetratype frequencies) (29)	1478/3978 (37.2%)	213/532 (40%)	133/549 (24.2%)
a-b recombination frequencies (for unconverted chromatid):	Control (no intragenic recombination)	Tetrads in which converted chromatid had parental markers	Tetrads in which converted chromatid had nonparental markers
Ascobolus tetrads (30)			
m+ in *m1* × *m2*:	149/1124 (13.2%)	47/275 (17.1%)	12/159 (7.5%)
gene conversion in *m* × *m*+		25/122 (20.5%)	7/85 (8.2%)
Yeast tetrads (29)	1022/8520 (12%)	60/532 (11.3%)	0/549

for the genetic observations from yeast. The trouble is that the genetic results do not really satisfy the predictions.

Coincidence of Separate Recombination Events

The predictions must be adjusted to include the separate crossovers accompanying those intragenic events that produce parental marker combinations. There is considerable genetic evidence that these events do not interfere with crossing over in their proximity. Many years ago (28) it was demonstrated in Neurospora that intragenic recombinants that carried parental combinations of the immediate flanking markers had the normal (no interference) frequency of recombination for a separate but adjacent marked region (see Table 1). The form of the cross was $A\ m1\ B\ C \times a\ m2\ b\ c$, and $m+$ recombinants that carried parental markers at *a* and *b* had normal recombination for the *b-c* interval. The same absence of interference was also demonstrated in yeast (29) in a cross of the same form. In that case whole tetrads were analyzed, so it was possible to conclude that the absence of interference applied not only to the chromatid showing intragenic recombination, but to the other three chromatids as well.

Tetrad analyses of intragenic recombination in Ascobolus (30) have revealed that interference is absent even for the region immediately around the locus under study (the *a-b* region in the cross $A\ ml\ B \times a\ m2\ b$; see Table 1). (This coincident crossing over can only be scored in the chromatids that are not involved in the intragenic event.) The same relationship can be shown for the yeast tetrads of Fogel & Hurst (29).

It seems logical to conclude that intragenic recombination per se exerts no interference. With this assumption, we can predict the frequency of tetratypes (for *a* and *b*) among tetrads with intragenic recombination at *m*, assuming that 50% of such events result in reciprocal recombination for the flanking markers, as predicted by Sigal & Alberts (27). The frequency of tetratypes among the remaining one-half of the tetrads should be the same as in the general population, which is two times the map distance. Thus the overall percent of tetratypes among tetrads with intragenic recombination should be $50 + (1/2)(2)\,ab$, or $50 + ab$, where *ab* is the map distance between *a* and *b*. This calculation ignores double crossovers, but for *ab* distances up to 20 map units they should be infrequent and should permit us to make fairly accurate predictions of the Sigal & Alberts proposal.

Hurst, Fogel & Mortimer (26) reported extensive results with three sets of flanking markers that spanned distances of 14 to 21 map units. These results (Table 2) are compared to the corrected predictions. It appears that marker exchange that is directly related to intragenic recombination occurs with a frequency considerably below 50% in this system.

Recombination Involving Only One Chromatid

One feature of intragenic recombination that has caused difficulties for the hybrid DNA models of Whitehouse (25) and Holliday (22) is the high frequency

Table 2 Analysis of yeast tetrads to test the hypothesis that intragenic recombination events produce marker exchange in 50% of cases. Data from (26)

map distance between markers (*ab*)	number of intragenic recombination tetrads	expected marker exchange (tetratypes) percent	expected marker exchange (tetratypes) number	observed tetratypes
14.2	256	64.2	164	116
17.1	148	67.1	99	74
20.5	81	70.5	57	37

of nonreciprocal recombination. Several studies of tetrads have involved the signaling of intragenic recombination by wild-type ascospores in a cross between allelic mutants. In this situation the simple prediction of the models is that these tetrads will include similar numbers of reciprocal and nonreciprocal recombinants if the two segregating sites are very close together, and rather more reciprocal recombinants if the sites are farther apart. However, the consistent finding of such studies in Neurospora (31), Ascobolus (4, 9), and yeast (29) has been a great preponderance of nonreciprocal recombinants.

A modification of the models has been suggested (32–34) that could account for the high frequency of nonreciprocal events: recombination is sometimes generated from a tetrad containing hybrid DNA in only one chromatid, a situation that could only produce nonreciprocal products. (It is assumed in this case that a donor chromatid has replaced its lost DNA strand by new synthesis of an identical strand.) Stadler & Towe (35) have described a specific modification of the Holliday model involving one-chromatid events. They proposed that these events never produce marker exchange, while the two-chromatid events always do so. They found that such a scheme made accurate predictions about marker distribution in tetrads with intragenic recombination in Ascobolus, if one-chromatid events were about three times as frequent as two-chromatid events. This type of scheme also accounts for a previously puzzling aspect of the tetrad analyses that was especially striking in the yeast results of Fogel & Hurst (29): nonreciprocal recombinants may carry any of the four possible combinations of flanking markers, while reciprocal recombinants nearly all carry the same nonparental combination.

The recombination model that includes both one-chromatid and two-chromatid events can predict the frequency of marker exchange among intragenic recombinants. It depends on the relative frequencies of the two kinds of events and on the map distance between the flanking markers. However, the model has limited predictive value in this regard, because there is no reliable basis for determining the relative frequencies of one-chromatid and two-chromatid events.

The results of tetrad analyses in Sordaria (36, 37) show that intragenic recombination in that fungus sometimes involves two chromatids without producing marker exchange. Aberrant 4:4 asci have two spore pairs showing postmeiotic segregation, so they are assumed to represent two chromatids with "unrepaired" hybrid DNA. About half of the asci of this type are parental ditypes (nonrecombinant) for the flanking markers.

The relationship of marker exchange to intragenic recombination remains troublesome. None of the proposed hypotheses accounts for the results of tetrad analysis from the various ascomycetes. Indeed, it seems that no simple model with any predictive capability could fit all the observations.

Prospects

What are the prospects for significant progress in our understanding of recombination? From which approaches will the important findings arise? There have been dramatic advances in several areas that *may* be important to recombination.

Electron microscopists can now see individual molecules of DNA. They make physical maps of genetic fine structure by a method analogous to that used 35 years ago at the chromosomal level in Drosophila. They see branched molecules that may be intermediates in recombination.

Enzymologists are solving the riddle of the replication of double-stranded DNA by finding out exactly what the different nucleases and polymerases can do. Using UV-sensitive mutants, they are learning the individual roles of these same enzymes in the repair of damaged DNA.

Phage geneticists are tagging one or both parents of a cross with a density label or a radioactive label to construct a detailed description of the parts played by replication and recombination in the life cycle.

The last decade has seen an attack on recombination at the molecular level. But the attackers have been frustrated and dissembled by a basic dichotomy. The battle lines and strategies were designed by Whitehouse (25) and Holliday (22) on the basis of genetic studies in eukaryotes. Specifically, the analysis of meiotic tetrads had given them the crucial intelligence. That was because the most complete and reliable genetic information about the details of a recombination event could only be learned from tetrad analysis. There is good evidence that a tetrad represents *all* the products of a *single* mating between two *complete* haploid genomes. (The italics indicate regularities of the system, which are very hard to obtain in prokaryotes.)

The weapons for the attack were the new techniques mentioned above. They were developed in prokaryotes and have been used there with encouraging success. But there has been very little progress in adapting these methods to eukaryotes. Thus we have models for meiotic recombination, but we have only been able to test them in bacteria and viruses. Success would be possible in this project if recombination had a common basis in these various forms. This is the hope on which our efforts have been based, but, in candor, we must admit that the parallels have not yet become obvious.

Literature Cited

1. Mitchell, M. B. 1955. Aberrant recombination of pyridoxine mutants of Neurospora. *Proc. Nat. Acad. Sci.* 41:215–20
2. Roman, H. 1956. Studies of gene mutation in Saccharomyces. *Cold Spring Harbor Symp. Quant. Biol.* 21:175–85
3. Freese, E. 1957. The correlation effect for a histidine locus of *Neurospora crassa. Genetics* 42:671–84
4. Lissouba, P., Mousseau, J., Rizet, G., Rossignol, J. L. 1962. Fine structure of genes in the ascomycete *Ascobolus immersus. Advan. Genet.* 11:343–80
5. Murray, N. E. 1961. Polarized recombination with the *me-2* gene of Neurospora. *Genetics* 46:886
6. Stahl, F. 1961. A chain model for chromosomes. *J. Chim. Phys.* 58:1072–77
7. Murray, N. E. 1963. Polarized recombination and fine structure within the *me-2* gene of *Neurospora crassa. Genetics* 48:1163–83
8. Gutz, H. 1971. Site specific induction of gene conversion in *Schizosaccharomyces pombe. Genetics* 69:317–37
9. Kruszewska, A., Gajewski, W. 1967. Recombination within the *Y* locus in *Ascobolus immersus. Genet. Res.* 9:159–77
10. Rossignol, J. L. 1969. Existence of homogeneous categories of mutants exhibiting various conversion patterns in gene 75 of *Ascobolus immersus. Genetics* 63:795–805
11. Leblon, G. 1972. Mechanism of gene conversion in *Ascobolus immersus*. I. Existence of a correlation between the origin of mutants induced by different mutagens and their conversion spectrum. *Mol. Gen. Genet.* 115:36–48
12. Leblon, G. 1972. Mechanism of gene conversion in *Ascobolus immersus*. II. The relationships between the genetic alterations in b_1 or b_2 mutants and their conversion spectrum. *Mol. Gen. Genet.* 116:322–35
13. Benz, W. C., Berger, H. 1973. Selective allele loss in mixed infections with T4 bacteriophage. *Genetics* 73:1–11
14. Parker, J. H., Sherman, F. 1969. Fine-structure mapping and mutational studies of gene controlling yeast cytochrome *c. Genetics* 62:9–22
15. Tessman, I. 1965. Genetic ultrafine structure in the T4*rII* region. *Genetics* 51:63–75
16. Ronen, A., Salts, Y. 1971. Genetic distances separating adjacent base pairs in bacteriophage T4. *Virology* 45:496–502
17. Ephrussi-Taylor, H. 1966. Genetic recombination in DNA-induced transformation in pneumococcus. IV. The pattern of transmission and phenotypic expression of high and low-efficiency donor sites in the *amiA* locus. *Genetics* 54:211–22
18. Norkin, L. C. 1970. Marker-specific effects in genetic recombination. *J. Mol. Biol.* 51:633–55
19. Stadler, D. R., Kariya, B. 1973. Marker effects in the genetic transduction of tryptophan mutants of *E. coli*. Unpublished
20. Spatz, H. C., Trautner, T. A. 1970. One way to do experiments on gene conversion? Transfection with heteroduplex *SPP1* DNA. *Mol. Gen. Genet.* 109:84–106
21. Hershey, A. D. 1958. The production of recombinants in phage crosses. *Cold Spring Harbor Symp. Quant. Biol.* 23:19–46
22. Holliday, R. 1964. A mechanism for gene conversion in fungi. *Genet. Res.* 5:282–304
23. Katz, E. R., Brenner, S. 1969. Intracodon recombination in the T4 *rII* gene. *Genetics* 61:s30
24. Stadler, D. R., Kariya, B. 1969. Intragenic recombination at the *mtr* locus of Neurospora with segregation at an unselected site. *Genetics* 63:291–316
25. Whitehouse, H. L. K. 1963. A theory of crossing-over by means of hybrid deoxyribonucleic acid. *Nature* 199: 1034–40
26. Hurst, D. D., Fogel, S., Mortimer, R. K. 1972. Conversion-associated recombination in yeast. *Proc. Nat. Acad. Sci. USA*. 69:101–05
27. Sigal, N., Alberts, B. 1972. Genetic recombination: the nature of a crossed strand-exchange between two homologous DNA molecules. *J. Mol. Biol.* 71:789–93
28. Stadler, D. R. 1959. The relationship of gene conversion to crossing over in Neurospora. *Proc. Nat. Acad. Sci. USA* 45:1625–29
29. Fogel, S., Hurst, D. D. 1967. Meiotic gene conversion in yeast tetrads and the theory of recombination. *Genetics* 57:455–81

30. Stadler, D. R., Towe, A. M., Rossignol, J. L. 1970. Intragenic recombination of ascospore color mutants in Ascobolus and its relationship to the segregation of outside markers. *Genetics* 66:429–47
31. Stadler, D. R., Towe, A. M. 1963. Recombination of allelic cysteine mutants in Neurospora. *Genetics* 48: 1323–44
32. Whitehouse, H. L. K. 1967. Secondary crossing over. *Nature* 215:1352–59
33. Paszewski, A. 1970. Gene conversion: observations on the DNA hybrid models. *Genet. Res.* 15:55–64
34. Roman, H. 1971. Induced recombination in mitotic diploid cells of *Saccharomyces. Genetics Lectures* (Oregon State Univ.) Vol. 2, pp 43–59
35. Stadler, D. R., Towe, A. M. 1971. Evidence for meiotic recombination in Ascobolus involving only one member of a tetrad. *Genetics* 68:401–13
36. Kitani, Y., Olive, L. S. 1967. Genetics of *Sordaria fimicola*. VI. Gene conversion at the *g* locus in mutant × wild-type crosses. *Genetics* 57:767–82
37. Kitani, Y., Olive, L. S. 1969. Genetics of *Sordaria fimicola*. VII. Gene conversion at the *g* locus in interallelic crosses. *Genetics* 62:23–66
38. Holliday, R., 1968. In *Replication and Recombination of Genetic Material*, ed. W. J. Peacock, R. D.Brock,Canberra, Australia: Australian Academy of Science

PARAMUTATION 3050

R. Alexander Brink

Laboratory of Genetics, University of Wisconsin, Madison, Wisconsin

Introduction

Paramutation is an interaction between alleles that leads to directed, heritable change at the locus with high frequency, and sometimes invariably, within the time span of a generation. It is a seeming exception to Mendel's First Law, namely, that contrasting genetic factors emerge from a heterozygote without having influenced each other. Doubtless, however, this particular manifestation of paramutation is only a semblance. The phenomenon may reflect the action in atypical fashion of elements normally involved in some still obscure chromosome process. These elements may be labile and subject to reversible changes that arise, and are transmissible, in somatic cells, in contrast to the accompanying structural genes that characteristically are invariant. Or the chromosome components underlying paramutation may be parasitic or commensal episomes initially of exogenous origin.

Special interest attaches to the possibility that paramutation is an abnormal manifestation of chromosome components that control, but do not specify, developmental and metabolic reactions. The nature, magnitude, spatial relations, and mode of action of constituents that regulate the replication and expression of the structural genes in the complex chromosomes of eukaryotes remain open to wide speculation. A unifying principle that encompasses significantly more than the DNA double helix is needed to advance research in this field (1). The goal is a conceptual framework within which the chromosome can be meaningfully portrayed as a vehicle of both gametic and somatic cell heredity. The place that paramutation would occupy within such a framework is not now apparent. The phenomenon is mainly dealt with in this review, therefore, on an ad hoc basis.

Paramutation has also been referred to in the literature as "somatic conversion" (2), "conversion", or "conversion-type inheritance". Following Mitchell's (3) usage, however, the term conversion is now generally applied to a quite different process involving the nonreciprocal recombination of elements within a locus at meiosis. Hagemann (4) expressed the view, now prevailing, that misunderstanding would be avoided if somatic conversion and conversion-type inheritance were abandoned in the present context in favor of paramutation.

The term paramutation is to be thought of in a literal sense as implying a phenomenon distinct from, but not entirely unlike, mutation. Mutations, although varying widely in kind, always are sporadic and are undirected. Paramutation regularly occurs under specified sets of conditions, and the resulting shift in phenotype in a given case is unidirectional.

Early Observations of Paramutation

Paramutation has been observed infrequently, considering the diversity of animals and plants that has come under genetic surveillance, and the few examples studied have been characterized to widely varying extents.

THE RABBIT EAR ROGUE IN THE GARDEN PEA The first reported instance of the irregular allelic interaction here termed paramutation was the rabbit ear rogue in the garden pea, *Pisum sativum* (5–8). Rogue plants arise sporadically in many pea varieties. They are distinguished by pointed leaflets, upward curving pods, and reduced stipule size. Self-pollinated rogues breed true. F_1 hybrids between normal plants and rogues from the same variety often are intermediate in form at the base of the plant, and become progressively more rogue-like as development proceeds, so that they are often typically rogue in appearance at the first flowering node. Such plants, after selfing, give only rogue offspring. In certain varieties the change in F_1 hybrids occurs more gradually, and the transformation into rogue form may be completed only at the apex of the plant, if at all. Genetic behavior in these instances corresponds to the amount of alteration in morphology of the plant. A progressive increase usually occurs in the proportion of rogue offspring from seed taken at successively higher nodes. It was also observed, following reciprocal crosses between rogues and normals, that the frequency of rogues among the offspring of F_1 individuals of intermediate form increases more rapidly through the pollen than through the female gametophyte as the plant axis is ascended.

These observations were confirmed by Brotherton (9, 10) who offered an explanation of the phenomenon in terms of "mass somatic mutation" in rogue × control heterozygotes. Primary rogues were assumed to arise by infrequent mutation of a single gene, x, to X. In Xx heterozygotes, x became highly unstable, and usually mutated to X before meiosis occurred. An allele of x, termed x_1, identified in Mummy, a nonrogue producing variety, was shown to mutate to X very rarely in Xx_1 plants.

LABILITY IN SOMATIC CELLS OF A LOCUS IN MALVA PARVIFLORA Lilienfeld (11) discovered a mutant form with deeply incised leaves and smaller flowers, called laciniata (*lac*) among plants grown from *Malva parviflora* seed collected in the wild. Laciniata and two mutants derived from it, incisa (*inc*) and normal-2 (*no-2*) behaved as recessives to wild type. It was observed, however, that *no-2/lac* and *no-2/inc* heterozygotes, initially intermediate in form between the respective parental types, changed progressively, as development proceeded, in the direction of the *no-2* homozygote. Progeny tests using seed from different branches showed that parallel with the changes in external morphology of the plant the *lac* and *inc* genes became more and more like their *no-2* partner in action. Lilienfeld postulated continuously varying states of these mutant genes in heterozygotes with respect to effect on leaf form and lability.

SOMATIC CONVERSION (PARAMUTATION) IN OENOTHERA HETEROZYGOTES Renner (2) early observed in his classical studies on complex heterozygotes in Oenothera that the cruciata character (cross-shaped flowers of reduced size) displayed an unusual genetic phenomenon which he termed somatic conversion. The results of investigations of cruciata, extending over some 20 years, were summarized by Renner (12–14) in three articles in which references to the several original reports are listed.

Renner found that cruciata (*cr*) and normal (*Cr*) plants differed at a single locus. The offspring are normal when a homozygous normal plant is selfed, or is crossed with another normal. Similarly, cruciata × cruciata, or cruciata selfed, yields only cruciata progeny. When normal is crossed with cruciata, of whatever lineage, however, the *Cr cr* heterozygotes are always inconstant. The F_1 plants are sometimes visibly sectored, and Renner found, as he notes deVries had earlier reported, that the F_2 progeny in such cases vary in accordance with the form of the flowers that produced the seed, namely, cruciata, intermediate, or normal. It was concluded that in the somatic cells of *Cr cr* plants, *Cr* often is heritably changed into *cr*, and *cr* occasionally is changed into *Cr*. *Cr* and *cr* were considered as collective symbols, however, each representing an allelic series. Renner (12–14) conjectured that *Cr* and *cr* may each consist of subunits that can vary in number, and so in expression, as a result of some kind of physical exchange between the alleles in *Cr cr* heterozygotes, in the resting or prophase nuclei of somatic cells.

Paramutation in the Tomato (*Lycopersicon*)

Following X-ray treatment of seed of the Lukullis tomato variety, Hagemann (15) isolated a mutant allele termed sulfurea (*sulf*) that was subsequently found to be paramutagenic. The other *sulf* alleles used in Hagemann's studies occurred by spontaneous mutation, or arose by paramutation in *sulf*/+ heterozygotes. Two phenotypic classes of mutants, both recessive to normal, were established (*a*) *sulf-pura*, which conditioned yellow cotyledons and yellow foliage leaves, and (*b*) sulf-variegata (*sulf-vag*) which causes green flecks on a yellow background in both cotyledons and foliage leaves. Homozygous *sulf-pura* shoots bear flowers and fruits if grafted on green plants. Hagemann (4) has presented a succinct, critical comparison of the tomato results with those from other organisms.

The *sulf*$^{+}$ allele is stable in homozygotes, and is always unstable in +/*sulf-vag* and +/*sulf-pura* plants. The heterozygotes form green cotyledons but become variegated, with a variety of leaf patterns, as plant development proceeds. Green branches on +/*sulf* heterozygotes, following selfing, give green and variegated progeny in a 3:1 ratio. Aberrant proportions occur, however, among the sexual offspring of variegated branches on the same plant; the greater the extent of variegation on the parent branch the higher the proportion of *sulf* offspring. Pure yellow branches yield only yellow (*sulf*/*sulf*) progeny, that ordinarily die as seedlings. The extent of variegation is the same following reciprocal crosses between normal and *sulf* plants, and chromosome number and meiosis are normal in +/*sulf* plants. Hagemann concluded that in the vegetative cells of *sulf*/+ heterozygotes the $sulf^{+}$ allele becomes unstable in response to its *sulf* partner and mutates with varying frequencies to *sulf*. It was noted that +/+ cells, which would accompany *sulf*/*sulf* cells if mitotic crossing over in the 4-strand stage underlies the observed variegation, were lacking. Paramutation was regularly unidirectional, from +/*sulf* to *sulf*/*sulf* (4, 15–20).

Hagemann (16) observed that the *sulf-pura* groups comprised alleles that varied in grade of paramutagenicity from 0.5% to 100%. The grade reflects the percentage of plants heterozygous for the *sulf* allele in question that show variegated foliage. The maximum paramutagenicity of the *sulf-vag* alleles tested was 10%. The level of paramutagenicity of an allele may be either increased or decreased by mutation, but no *sulf* allele was found that was devoid of paramutagenicity. Paramutagenicity, and effect on plant phenotype in homozygous condition, of a newly arisen *sulf* allele derived from a given +/*sulf* plant, do not necessarily correspond in these respects to the *sulf* allele in the parent plant. A +/*sulf-pura* heterozygote whose *sulf-pura* allele scored 90% on the scale of paramutagenicity, for example, yielded new *sulf* paramutants of both the *pura* and *vag* classes, and these varied in grade of paramutagenicity. Hagemann (17, 18) notes that this evidence proves that somatic conversion in the tomato is a different phenomenon than conversion in Neurospora, Aspergillus, and Saccharomyces, in which cases the recovered allele is identical with the allele originally present. No $sulf^{+}$ allele was found among diverse stocks of *Lycopersicon esculentum* and *L. pimpinellifolium* that was insensitive to paramutation in +/*sulf* heterozygotes. Paramutagenicity appeared to be dependent only on the level of activity of the *sulf* allele employed. The residual heredity was found to affect paramutation, however, in *L. esculentum* hybrids with *L. hirsutum* and *Solanum pennellii*. Paramutation was reduced in the F_1 interspecific hybrids, and segregation for amount of change was observed in subsequent generations (16, 19, 20.)

Hagemann (4) found that the *sulf* locus maps close to, or possibly within, the heterochromatic part of chromosome 2. Both the centromere and the nucleolar organizer are in this region. He suggested that *sulf* may have become paramutagenic as a result of heterochromatization of this site (cf. Prokofyeva-Belgovskaya, 21). The functional activity of the locus, with respect to chlorophyll formation, was reduced concurrently.

Paramutation, as exhibited by the formation of variegated sectors, occurred in about 60% of +/*sulf*/*sulf* trisomic plants. In contrast, +/+/ *sulf* individuals remained green throughout, even when vegetatively propagated. Hagemann (4) concluded that within the nucleus of heterozygotes a threshold determined by the dosage of paramutable and paramutagenic alleles must be exceeded in order for paramutation to occur. Once the threshold is exceeded the amount of paramutation that results is a function of the activity of the paramutagenic allele present. Hagemann notes that although trisomics in Oenothera were not studied, Oehlkers (22) and Renner (23) reported that cruciata flowers often appeared on *Cr/cr/cr* triploid plants, but not at all on triploid *Cr/Cr/cr* individuals.

Recently Ecochard (106) has reported the occurrence, among the immediate sporophytic descendants of irradiated Lycopersicon female gametophytes, of three additional examples of paramutation, termed C6, C11, and C12, and expressed as mottling for chlorophyll pigmentation. The breeding behavior of the C6 paramutant paralleled that of the *sulf* allele described by Hagemann (loc. cit.). Paramutagenicity of C11, on the other hand, was transitory. The stable, nonparamutagenic allele to which the C11 paramutant invariably gave rise, conditioned virescent yellow foliage, without variegation. C12 was intermediate between C6 and C11 with respect to stability for paramutagenicity. The gametes derived from most of the sporogenous cells in the foundation, heterozygous plant transmitted a stable, nonparamutagenic chlorophyll deficient, recessive allele causing lethality at the seedling stage. A few sporogenous cells, however, retained C12 in paramutagenic form. Ecochard showed that chlorophyll forming activity of the Cl l paramutant could be increased, with low frequency, by treatment with gamma rays or ethyl methane sulfonate.

Paramutation at the R Locus in Maize

EARLY INVESTIGATIONS Reviews of *R* paramutation in maize were published by Brink (24) and by Brink, Styles, & Axtell (25). The reader is referred to these articles for summaries of the evidence on which the conclusions arrived at in the earlier studies, and briefly noted below, were based.

The standard R^r allele, which normally conditions dark aleurone mottling in single dose ($R^r rr$), colored seedlings and anthers, is heritably changed to a weakly pigmenting form, $R^{r'}$, following passage through a heterozygote with *R*-stippled (R^{st}), a factor that conditions a fine, but irregularly distributed purple spotting of the aleurone. Paramutant $R^{r'}$ (designated $R^{r'}$) tends to revert toward the level of action of standard R^r in the $R^{r'}$ $R^{r'}$ offspring of R^r R^{st} individuals selfed (26). R-marbled (R^{mb}) which gives coarse aleurone spotting also is paramutagenic (27). R^r becomes paramutagenic on passage through an $R^r R^{st}$ plant, although weakly so as compared with R^{st} (28). The paramutagenic alleles, R^{st} and R^{mb}, are not altered with respect to either aleurone phenotype or paramutagenicity in heterozygotes with R^r (26, 27).

Direct evidence was obtained that the paramutant R' phenotype is *R*-locus dependent. Also tests of the hypothesis that *R* paramutation is attributable to either an autonomous cytoplasmic element or a particle produced and released by R^{st} and then incorporated at the *R* locus in $R^r R^{st}$ plants, changing R^r to $R^{r'}$, gave negative results throughout (29, 30). Brown (31) observed that the capacity to become paramutagenic is conserved when R^r mutates to R^g (colored aleurone, green anther) but not to r^r (colorless aleurone, red anther). When R^{st} mutates to self-colored aleurone (R^{sc}) paramutagenicity was unchanged, abolished, or altered in degree, in different instances (32). Paramutant $R^{r'}$ alleles derived from a given $R^r R^{st}$ plant are heterogeneous in pigmenting potential, and also are mestastable (24, 33). Change of $R^{r'}$ in level of pigmenting action can occur autonomously, i.e., in $R^{r'}$ hemizygotes (33,34). Paramutation is progressive in successive generations of heterozygotes (32,35). Phenotypic changes associated with R^r paramutation are manifested in the plant as well as in the seed (36). *R* paramutation is, at least mainly, a somatic cell phenomenon (37–39).

Insertion of standard R^r into a reciprocally translocated chromosome involving a break in chromosome 10L either proximal or distal to *R* and not necessarily close to the *R* locus, or into a K10 chromosome which carries a large heterochromatic segment of unknown origin at the end of 10L and at least 35 map units distal to *R*, reduces the sensitivity of R^r to paramutation in R^{st} heterozygotes in a manner that may persist for at least one generation after return of the allele to a structurally normal chromosome (40–46). R^r is subject to both heritable repression and heritable enhancement of its expression in the aleurone, as a result of paramutation.

Paramutability is an intrinsic property of the R^r allele; the capacity to undergo directed, heritable change is not necessarily dependent upon prior association of R^r with a paramutagenic partner in a heterozygote. Occurrence of heritable enhancement in R^r or $R^{r'}$ aleurone pigmenting potential in hemizygotes, i.e., when the allele is in opposition to a deficiency for the *R* locus, excludes an interpretation of *R* paramutation in terms of an interallelic transfer of a particle, or particles (34, 47).

Changing R^r from the standard to paramutant forms reduces the frequency of mutation of R^r to r^r from 12.7×10^{-4} to 7.43×10^{-4} (48). The R^r allele is extremely sensitive to change in aleurone pigmenting potential when $R^r R^{st}$ seed, but not R^r pollen, is treated with the alkylating agents diethyl sulfate or ethyl methane sulfonate (49). Pollen transmissible differences in *R* aleurone expression were observed among the offspring of RR^{st} heterozygotes given unlike photoperiodic treatments (50). Sensitivity of R^r to paramutation may be changed by pretreating R^r pollen with gamma rays or X-rays (51 –54). Aleurone pigmenting potential of R^r alleles in strongly repressed form as a result of paramutation often was raised by treatment of seed with X-rays (53).

R factors derived from maize races indigenous to the Andean region of South America are characteristically insensitive to the paramutagenic action of R^{st}. Many *R* alleles from this region, however, are paramutagenic (55, 56). All R^r and R^g alleles tested from United States, Canada, Mexico, and Central America were sensitive to R^{st} action, as were a few from South America outside the Andean region. Occasional *R* alleles are paragenetically amorphic (56). Mutants from R^{st} termed smoky and nearly colorless, conditioning unusual aleurone spotting patterns differed significantly from each other and from standard R^{st} in frequency of mutation to self-colored aleurone. Preliminary tests suggest changes in paramutagenicity also. The mutants did not carry *Mp*, *Spm*, or *Dt*, elements that are known to be involved in other mutable allele systems,at the *R* locus (57, 58).

THE GENETIC FINE STRUCTURE OF R-STIPPLED R^{st} involves two spatially related but functionally quite unlike classes of components. One class affects anthocyanin synthesis and distribution and the other, paramutagenicity. The initial evidence for this dualism was obtained by McWhirter & Brink (32) in a study of paramutagenicity of self-colored(R^{sc}) mutations from R^{st}. Knowledge of the organization of R^{st}, and other *R* alleles, has been greatly extended since by Ashman (59–61) and by Kermicle (62) and his students Satyanarayana (63) and W. Williams (64). The new evidence on genetic fine structure clearly shows that the elements underlying paramutagenicity are distinctive in kind, vary quantitatively, and lie within the locus in close association with components affecting anthocyanin formation.

The stippled allele, as the name implies, gives a fine, but irregular, purple spotting of the aleurone, the outer cell layer of the endosperm. Gavazzi (65) and W. Williams (64) have described variant forms. Small streaks of anthocyanin also may occur in the scutellum of R^{st} embryos and in the coleoptile of seedlings, but not elsewhere in the plant. R^{st} is highly unstable with respect to anthocyanin pigmentation. Mutations to R^{sc} are recoverable through eggs of $R^{st}R^{st}$ plants with a frequency of about 2×10^{-3} (66) and from sperm at the rate of about 6 $\times 10^{-3}$ (62). Ashman (66) identified a modifier of the R^{st} phenotype, termed M^{st}, 5.7 map units distal to *R*, that markedly affects the intensity of aleurone spotting but not the frequency of germinally transmissible R^{st} to R^{sc} mutations. McWhirter & Brink (32) found that a random sample of 83 R^{sc} mutants varied continuously in paramutagenic potential from the high level characteristic of the parent R^{st} allele to zero. Mutation of R^{st} to R^{sc} without change in paramutagenicity showed that the latter property is genetically separable from the stippled phenotype and also that a high level of paramutagenicity may be associated with an *R* allele either stable (R^{sc}) or unstable (R^{st}) for aleurone pigmentation.

McWhirter & Brink (32) postulated that R^{st} comprised an R^{g} gene, ordinarily conditioning self-colored aleurone in one, two or three doses, and green anthers, but which, as an R^{st} component, is immediately conjoined with an inhibitor(I^{R}), that suppresses R^{g} action. They inferred that R^{st} to R^{sc} mutations could occur by transposition of (I^{R}) from its initial position immediately adjacent to R^{g} to another site. The paramutagenic component was assumed to be unaffected by (I^{R}) in an R^{st} chromosome, and so is potentially fully active there. It was postulated that when (I^{R}) transposed, the new site would often be that of the nearby paramutagenic component (67). It was thought that in such case, (I^{R}) would act as a suppressor, not of aleurone color, but of paramutagenicity, in one or another degree. This latter notion, however, has failed to gain experimental support.

Subsequent studies, reviewed below, have validated the idea that R^{st} contains a basic anthocyanin factor, now designated (*Sc*), potentially capable of giving self-colored aleurone, and also a closely associated, transposable inhibitor of aleurone color, (I^R). The occurrence of a third component, (*Nc*), affecting aleurone pigmentation, also has been established among the mutant derivatives of R^{st}. At least two distinct processes have been shown to underlie mutants from R^{st} that vary with respect to (*Sc*), (I^R), and (*Nc*), namely, intralocus crossing over and transposition.

Ashman (59, 60) observed that R^{st} mutates to near-colorless (*Nc*) as well as to self-colored (*Sc*), aleurone. The term, near-colorless, signifies aleurone mottling at a level much lower that that conditioned by R^{st}. Ashman found three kinds of mutations for reduced aleurone pigmentation among 37 recombinant offspring of R^{st} plants heterozygous for R^r (colored seed, red anther) in which the *R* locus was flanked by marker genes. These were (class I) near-colorless aleurone, green anther, (class II) near-colorless aleurone, red anther, and (class III) colorless aleurone, red anther. Ten of 11 near-colorless, green anther mutants were as paramutagenic as R^{st} when tested in heterozygotes with R^r, and one was more paramutagenic. All 12 near-colorless aleurone, red anther mutants displayed paramutagenic action, but at lower levels than R^{st}. The 12 mutants also varied significantly in paramutagenicity among themselves. (Paramutagenicity of an allele (R^x) ordinarily is measured by scoring the R^r kernels resulting from $r^g r^g$ ♀ × standard-R^r/R^x ♂ matings for level of aleurone pigmentation against a graded set of seeds.) The 13 colorless aleurone, red anther mutants were nonparamutagenic throughout. Ashman concluded that (*a*) mutation of R^{st} to near-colorless (*Nc*) in $R^r R^{st}$ heterozygotes usually was accompanied by recombination between outside markers, (*b*) paramutagenicity and (*Nc*) were closely associated with each other at the *R* locus, and (*c*) paramutagenicity was partitionable by crossing over.

Composition of the unstable paramutagenic, *R*-stippled allele has been further elucidated in independent, fine structure studies by Gavazzi (65), Ashman (61), and Kermicle (62).

Ashman (61) verified the compound nature of R^{st} by resynthesis of the allele in heterozygotes carrying certain R^{st} mutants. A particular near-colorless aleurone mutant, originally of crossover origin, r^g (*Nc*)1-3, stable for colorless seed, unstable for anther color, and strongly paramutagenic, when entered in heterozygotes with any one of three R^{sc} mutants (either paramutagenic or nonparamutagenic) from R^{st}, for example, gave 19 reconstituted R^{st} alleles among 118,397 progeny. The distribution of outside markers showed that the reconstitution, in each case, involved recombination, and was that expected if R^{sc} and r^g(*Nc*)1-3 contributed, respectively, (*Sc*) and (I^R), the two components of the R^{st} complex established by Kermicle (62). Qualitative tests for paramutagenicity of four of the resynthesized R^{st} mutants gave positive results. Resynthesized R^{st} alleles were recovered also from one kind of near-colorless heterozygote, namely, r^g (*Nc*)1-3/r^g (*Nc*)1-2, but not from the respective homozygotes. All the evidence supported the conclusion that the reconstitution of R^{st} involved restoration of (*Sc*) and (I^R) to the *cis* configuration.

Kermicle (62) obtained critical evidence that meiotic crossing over was an important source of mutations from R-stippled. He isolated 15 mutations to a Navajo-stippled compound allele ($R^{nj:st}$) among 31,410 progeny of $R^{nj}R^{st}$ heterozygotes, all of which were recombinant for outside markers. This is about one mutant per 2000 gametes, a high value. Heterozygotes between the derived allele and R-stippled ($R^{nj:st}/R^{st}$) gave rise to self-colored (R_{sc}) mutants and Navajo revertants (R^{nj}) with about equal frequency and with the outside markers in the relationships expected if the two kinds of mutants were complementary crossovers.

The significant additional observation was made by Kermicle (62) that (*Sc*) is separable from (I^{R}), slightly distal, by meiotic recombination in $R^{st}R^{st}$ homozygotes. He pointed out that this fact implies an unequal exchange in a tandemly duplicated chromosome region. Loss of the instability component (I^{R}), rather than (*Sc*), by crossing over in $R^{st}R^{st}$ plants led to the conclusion that (I^{R}), but not necessarily (*Sc*), was associated with a differential region of the duplication.

The observed frequencies of kernels showing self-colored aleurone that gave R^{sc} progeny following pollination of $r^{g}r^{g}$ ♀♀ by $R^{st}r^{g}$ and $R^{st}R^{st}$ ♂♂ were 13.8 and 58.2×10^{-4}, respectively. Significantly, however, this wide disparity did not hold when the frequencies of R^{sc} offspring from kernels with stippled aleurone and self-colored embryos (i.e., nonconcordant seeds) from the same two matings were compared with each other. The observed values in this instance were 54.6 and 61.2×10^{-4}. Kermicle concluded that R^{st} is highly unstable at the second mitosis in the male gametophyte and that this instability is unaffected by heterozygosity of the parent sporophyte. He estimated that about one pollen grain in 86 from R^{st} microspores, from either source, carried one R^{st} and one R^{sc} sperm. The marked effect of heterozygosity for R^{st} ($R^{st}r^{g}$ versus $R^{st}R^{st}$) on frequency of sexually transmissible R^{sc} mutations was related to some stage in the plant cycle, probably meiosis, between the beginning of tassel differentiation and the first microspore nuclear division.

Kermicle (62) demonstrated that all cases of relatively high R^{sc} mutation rates, however, could not be accounted for in terms of intralocus recombination. An unequivocal example involved a genotype carrying R^{st} on a normal chromosome 10 and a 10^{B} chromosome that lacked the distal two-thirds of 10L, and hence the R locus. Plants of this constitution yielded R^{sc} mutations with a frequency of 39×10^{-4}, a value approximately twice as high as those for $R^{nj}R^{st}$, $R^{st}/r^{r}(I^{R})$, and $R^{st}/R^{nj:st}$ heterozygotes, and also for $R^{st}R^{st}$ homozygotes, in each of which genotypes meiotic crossing over is a major source of mutations. Kermicle pointed out that the mechanism of R^{st} to R^{sc} mutations in $R^{st}/10^{B}$ hemizygotes, in which recombination at the R locus is precluded, could be transposition of (I^{R}) to a different chromosome site or possibly separation of (*Sc*) and (I^{R}) by unequal exchange between R^{st} sister chromatids.

Analysis by Satyanarayana & Kermicle (68) of R^{sc} mutants from $R^{st}R^{st}$ plants originating during meiosis and recombinant for outside markers showed that usually paramutagenicity was reduced. Paramutagenicity was similarly reduced in crossover R^{sc} mutants from $R^{nj}R^{st}$ and $R^{st}/R^{nj:st}$ individuals. Post-meiotic

R^{sc} mutants from these genotypes, in contrast, were paramutagenic at the same level as R^{st}. Also changes in paramutagenicity in R^{sc} mutants from R^{st} hemizygotes were infrequent and small. It was concluded that the paramutagenic component of R^{st} was being fractionated in crossovers in the R^{st} complex that separated (*Sc*) from (I^R). A further sample of crossover derivatives from $R^{st}R^r$ plants belonging to the three classes earlier noted by Ashman (59, 60) were evaluated in terms of the structure of R^{st}. Class I plants were shown to be crossovers between (*Sc*) and (I^R) in the R^{st} chromosome. They contain the red anther component (*P*) of R^r and near-colorless (*Nc*) from R^{st}. Class II and class III derivatives were found to be crossovers distal to (I^R) in the R^{st} chromosome. Since class II derivatives carry near-colorless aleurone, (*Nc*) must be distal to (I^R). Class III offspring resulted from crossing over distal to (*Nc*) in the R^{st} chromosome. This evidence confirmed the order of the three components affecting aleurone color as centromere –(*Sc*) –(I^R) –(*Nc*).

Further studies led Satyanarayana & Kermicle (68, 69) to the conclusions that paramutagenic potential was distributed more or less uniformly along the chromosome between (*Sc*) and (*Nc*), and that R^{sc} mutants vary in length of this segment. Heterozygotes for R^r and various medium to strongly paramutagenic R^{sc} alleles yielded variants belonging to classes II and III but not to class I, thereby confirming the absence of (I^R) in R^{sc}. Class II frequency approximately equaled the sum of classes I and II from $R^{st}R^r$; class III frequencies were equal. On the other hand, one R^r heterozygote carrying a very weakly paramutagenic R^{sc}, and another carrying a nonparamutagenic R^{sc} mutant, yielded only class III crossovers, and at about one-half the rate for the three classes from $R^{st}R^r$ plants. Among R^r heterozygotes carrying eight different R^{sc} mutants that yielded both class III and class II crossovers the more strongly paramutagenic R^{sc} alleles gave more class II than class III progeny.

In a study by W. Williams (64) of 10 *R* alleles from maize races indigenous to the Andean region of South America and resembling standard R-stippled in phenotype, nine were found to carry the centromere-(*Sc*)-(I^R)-(*Nc*) complex characteristic of R^{st}. They varied, however, in intralocus recombination values, intensity of aleurone spotting, seedling pigmentation of derived R^{sc} mutants, and paramutagenicity. The remaining allele appeared to be distinct. It was nonparamutagenic, carried (*Sc*) and a repressor of aleurone color, but not (*Nc*).

W. Williams obtained evidence which suggested that the basis of paramutagenicity in R^{st} was a series of repeating units activated by a "controlling center" located closely proximal to (*Nc*). A nonparamutagenic allele, termed R^{st}: Laughnan, was thought to carry repeating units, but to lack the controlling center.

Satyanarayana (63) inferred transposability of (I^R) from instances of mitotic R^{st} to R^{sc} mutation in which the resulting genome was found to carry a modifier of R^{st} expression. It was subsequently observed by W. Williams (64) that (I^R) in trans heterozygotes with light stippled (R^{st}, but no M^{st} modifier present) gives an aleurone spotting phenotype approaching that of R^{st} linked with M^{st}. He found also that (I^R) frequently transposes to a new site when R^{st} mutates to R^{sc}

at the second pollen grain mitosis. (I^{R}) in the new position is then identifiable as a modifier of the R-stippled phenotype. Sometimes in such cases a transposed (I^{R}) is recoverable from both sperm in a pollen grain, thus indicating over-replication of the element in a single mitotic cycle. W. Williams (64) also confirmed Kermicles' observation that hemizygosity for the distal part of 10L, carrying the R locus, markedly increased the frequency of R^{st} to R^{sc} mutations relative to that for $R^{st}r^{g}$ plants.

THE COMPOSITION OF OTHER R ALLELES IN MAIZE The R^{r} allele, conditioning anthocyanin formation in seed and plant, and highly sensitive to paramutation, also is a compound structure. Stadler (70, 71) visualized R^{r}: Cornell, a representative of this class and indistinguishable by phenotypic or recombinational criteria from the standard R^{r} allele used in the Wisconsin paramutation studies, as containing a plant pigmenting determiner (*P*) and a seed pigmenting determiner (*S*) that mutated independently of each other. Mutations to colorless seed, red anther (r^{r}) appearing among the offspring of $R^{r}R^{r}$ homozygotes were observed often to involve recombination between markers flanking the R locus. This fact led Stadler & Nuffer (72) to suggest that (*P*) and (*S*) were carried in a tandem duplication the components of which retained synaptic homology, and so were capable of oblique pairing at meiosis. Linkage tests placed (*P*) proximal to (*S*). Chromatographic analysis by Gavazzi & Zannini (57) showed that the anthocyanins in the seed and vegetative parts of $R^{r}R^{r}$ plants are qualitatively alike.

Dooner & Kermicle (73), using standard R^{r}, confirmed the findings of Stadler & Nuffer. Recombination data gave an estimated length of the duplicate segment of 0.16 map units. Noting that recombination following oblique pairing in R^{r} duplication homozygotes would lead to loss of either (*P*) or (*S*), but not both, and that the indicated position of (*P*) and (*S*) would be a function of the ratio of the frequencies with which these two kinds of events occurred, Dooner & Kermicle showed that the pigment determiners mapped close to the proximal end of the respective segments. They estimated from the frequencies with which r^{r} mutations appear among the offspring of R^{r} plants heterozygous, respectively, for 1-element ($-S$) and 2-element (pS) R^{g} mutants from R^{r}, that oblique pairing in duplication homozygotes occurred with a frequency of about 0.5, where the theoretical maximum is 2.0.

Nine R^{g} mutants from R^{r}, when tested in R^{st} heterozygotes by Brown (31), were alike in paramutability, and also were indistinguishable in this respect from the parent R^{r} factor. Using a crossing over test devised by Stadler & Emmerling (74) Bray & Brink (48) proved that R^{g}_{4}, one of the R^{g} mutants Brown used, was of the 1-element ($-S$) and another, R^{g}_{6}, was of the 2-element (pS) type. The mean aleurone color scores in the paramutability tests in the two cases were 2.29 and 2.43, respectively. It is apparent, therefore, that mode of origin of an R^{g} mutant from R^{r}, either by loss of (*P*) or mutation of (*P*) to (*p*) does not necessarily affect sensitivity to paramutation. Twenty colorless aleurone, red anther (r^{r}) mutants from R^{r}, on the other hand, were found by Brown to be insensitive to R^{st} action when tested in $R^r r^r$ plants for acquisition of paramutagenicity following passage through $R^{st} r^r$ heterozygotes. Loss of the (S) component from R^r appeared to abolish paramutability.

Maize plants carrying a paramutant form of R^r in heterozygotes with r^g (colorless aleurone, green anther) often show marked reduction in anther color, as compared with $R^r r^g$ controls. Brink & Mikula (36) showed that a parallel difference in pigmentation of the coleoptile and leaf sheath, but not of the roots, appeared in paramutant $R^g r^g$ seedlings grown under appropriate test conditions. It is not known whether the paramutant expression of (P), as well as of (S) in these cases is due to a "spreading effect" to (P) of the paramutable component known to be closely associated with (S) or the presence at both (P) and (S) of paramutable elements. The fact, however, that F_1 $R^r R^{st}$, but not F_1 $r^r R^{st}$, plants (31) occasionally show tassel sectors with clearly reduced anther pigmentation suggests that (P), when alone, lacks a paramutable element.

The R-cherry allele (R^{ch}) in maize conditions anthocyanin formation in seed, various vegetative plant parts and, when the Pl gene also is present, in the pericarp. The results of mutation experiments led Sastry (75) to postulate that in addition to (P) and (S) two R components underlie this unusually broad spectrum of pigmentation, (Si) affecting silk, and (Ch) affecting pericarp color. Validation of the view that R^{ch} comprises four such elements must await the results of tests for intralocus recombination in R^{ch} heterozygotes. Sastry's evidence does not exclude the possibility that R^{ch} is a unitary allele with multiple pigmenting effects. R^{ch} was found to be paramutable in $R^{ch} R^{st}$ plants with respect to aleurone, but not to pericarp, pigmentation.

Evidence from a variety of sources shows that paramutability of R^r, and also of R^g mutants from it, varies in continuous fashion. McWhirter & Brink (32) found that the pigmenting potential of R^r was altered in varying degrees following passage through heterozygotes carrying R^{sc} mutants from R^{st} characteristically different in paramutagenic strength. R^r pigmentation level declined progressively in successive generations of R^r plants kept heterozygous with a weakly paramutagenic R^{sc} allele. Kermicle (33) showed that R^r pigmenting action is changed to varying levels in different instances on passage of R^r through a single $R^r R^{st}$ plant. Evidence of these kinds led Brink (24) and Sastry, Cooper & Brink (39) to suggest that the basis of R paramutability was a heterochromatic segment at the R locus that repressed action of the associated anthocyanin determiners. The repressor segment was assumed to contain a unit, termed a metamere, that varied in number. Degree of R^r repression was considered to be a function of metamere number. Paramutation, on this hypothesis, results in change in number of metameres in the repressor segment.

The mechanism whereby differences in number of repeating units in paramutable R alleles are supposed to arise has not been established. Brink & E. Williams (76) have suggested, however, that the underlying processes are similar to those involved in the transposition of repressor segments associated with other unstable loci in maize. They assumed that nucleotide sequence in the regions adjacent to a transposable element determines the frequency with which a form of micro-nondisjunction occurs whereby a transposable element is released from a donor site. Transposition to a new site was interpreted in terms of a chromosome model that invokes nicking, or single strand breaks, occurring throughout the genome as a prerequisite to unwinding, strand separation, and replication, of the DNA double helix. Brink & E. Williams postulated that increases in metamere number occur as a replication fork passes through the repressor segment associated with an R^r allele. A repressor segment that has undergone micro-nondisjunction just behind the fork is inserted, following overlap, at a nick immediately ahead of the fork, where it is again replicated. The result is one strand of the DNA double helix resembling the donor strand except for a small shift in position of the repressor segment, and a complementary strand carrying two, instead of one, repressor segments. Segments containing various multiples of the repressor element would be built up by repetition of this process in different somatic mitoses. Reduction in number of metameres in the repressor segment could result from incomplete copying of the elements during chromosome replication.

ALEURONE MOTTLING IN RELATION TO TRANSMISSION THROUGH MALE AND FEMALE GAMETOPHYTES Kermicle (77) has shown that *R* alleles (including standard R^r) that give mottled aleurone in single dose (*Rrr*) are transmitted to the endosperm at different levels of anthocyanin determining activity through male and female gametophytes. It has long been known that reciprocal crosses between colorless (*rr*) individuals and plants carrying an *R* allele of the kind in question give unlike phenotypes. The aleurone in *Rrr* seeds resulting from *rr* ♀ × *RR* ♂ matings is mottled, whereas that in *RRr* seeds formed after *RR* ♀ × *rr* ♂ crosses is solidly colored. The view previously prevailed that the difference in pigmentation reflected merely the *R* dosage difference between *Rrr* and *RRr* seeds. Utilizing Roman's (78) observation that a chromosome segment translocated to the centromeric portion of an accessory *B* chromosome in maize is often transmitted by the male gametophyte in duplicate, Kermicle was enabled to vary *R* dosage in the endosperm independently of parental origin. Contrary to expectation on the dosage hypothesis it was found that introduction into the endosperm via the pollen of two doses of *R* on the TB-10a translocated chromosome regularly gave mottled, rather than solidly colored, seeds. Dosage balance between the *R* and *r* alleles also was excluded as the basis of the difference in aleurone phenotype. A comparison of the two respective (♀-derived/♂-derived) genotypes, namely, *rr/RR* and *RR/rr*, obtained by crossing appropriately marked TB-10a and normal chromosome 10 stocks, afforded critical evidence on this point. Two *R* alleles of male origin in such tests regularly yielded mottled seeds, whereas two *R* alleles of female origin gave solidly colored kernels. Kermicle concluded that the *R*-mottled phenotype is a gametophyte effect, dependent on mode of sexual transmission.

Using paramutable *R* alleles that exert a unitary genetic effect on both endosperm and seedling color, Brink, Kermicle & Ziebur (79) subsequently showed that the gametophyte effect on level of *R* pigmenting action was not exhibited in embryo and young seedlings. They concluded that it was peculiar to the endosperm and rests on a derepression of *R*-pigmenting potential which is restricted to polar nuclei of the female gametophyte (which enter into formation of the primary endosperm nucleus) during the period before fertilization.

All *R* alleles that have given a positive test for paramutability also show the gametophyte effect. No *R* alleles that have proved entirely insensitive to paramutation, on the other hand, exhibit the phenomenon. It appears likely, therefore, that the gametophyte effect is an aspect of *R* paramutation.

R PARAMUTATION DOSAGE EFFECTS Styles (80) found that R^r aleurone testcross scores did not differ significantly when disomic R^rR^x and trisomic $R^rR^xr^g$ and $R^rR^rR^x$ individuals from $R^rR^rr^g \times R^xR^x$ matings (A series) were crossed on to r^gr^g ♀♀ (where R^r is paramutable, R^x is paramutagenic, and r^g is paragenetically amorphic). Similarly disomic R^rR^x and trisomic $R^rR^rR^x$ individuals from $R^rR^rR^r \times R^xR^x$ matings (B series) testcrossed on r^gr^g ♀♀ gave nearly equivalent paramutant R^r scores. Thus R^r is just as paramutable when present twice as when present only once, in a heterozygote with a paramutagenic allele. By the same token, the paramutagenic allele R^x is equally effective in causing change in R^r when the latter is present twice in the nucleus as when present only once. The R^r testcross scores from the B series of matings above (involving $R^rR^rR^r$ plants) were lower than those from the A series (involving $R^rR^rr^g$ plants). Styles points out that this result is in accord with expectation based on the outcome of earlier tests of the effect on sensitivity to paramutation of R^r dosage. R^r propagated in R^rR^r plants (two doses) retains its sensitivity to a paramutagenic allele, or becomes more sensitive to paramutation. R^r propagated in an *r* heterozygote (one dose) on the other hand, is less sensitive relative to R^r from an R^rR^r homozygote (34, 81, 82).

Paramutation at the B Locus in Maize

Paramutation (termed "conversion-type" inheritance) was reported by Coe (83) at the *B* locus on chromosome 2 in 1959. *B* paramutation corresponds to that of *R*, on chromosome 10, in the essential respect that, in certain heterozygotes, one allele regularly is heritably changed in response to presence of the other allele. It is differently manifested at the two loci, however, in certain other ways. A brief summary of Coe's findings follows. A full account of Coe's work on *B* paramutation and a critique of paramutation in general are presented in Coe's 1966 article (84).

The *B* alleles used in Coe's investigations may be characterized as follows: *B* = intense anthocyanin formation in many plant parts; changes infrequently to B', in *BB* individuals; paramutable. B^b = weak pigmentation, except for glume base; paragenetically amorphic, i.e., insensitive to change in B^bB' heterozygotes. b = colorless; paragenetically amorphic. B^v = unstable, colorless allele that frequently mutates in somatic cells to *B*, and occasionally to *b*; paramutable. $b-v$ = colorless mutant from B^v; becomes paramutagenic in $B'/b-v$ heterozygotes. B', $B^{v'}$, and $b'-v$ are the paramutant (and paramutagenic) forms of these respective alleles. The B' phenotype resembles that of B^b; B^v can form anthocyanin only in cells in which B^v has mutated to B; $b' - v$ forms no anthocyanin. B', B^v, and $b' - v$ are paramutagenic in heterozygotes with B.

The first observed instances of B paramutation related to two weakly colored individuals among 140 otherwise darkly colored offspring, one from each of two selfed ears on a single BB plant. The aberrant individuals bred as $B'B'$, i.e., as homozygous paramutant, through three generations of selfing. Outcrosses of $B'B'$, used as either female or male parent, to four different BB strains likewise yielded only $B'B'$ offspring. Among the offspring of $B'b$ plants, however, the B' and b phenotypes were normally distributed. These results suggested that B occasionally changes to B' in BB homozygotes, and that in BB' heterozygotes B invariably is altered to B' (83).

The B' mutation appeared initially as a well defined light sector on the darker background characteristic of BB plants. Thus the primary change from B to B' occurs in vegetative cells (84). The B alleles in all sublines of the foundation BB strain were found to be subject to change to B'. Furthermore, all B alleles tested from other sources proved highly sensitive to paramutation in BB' heterozygotes. B mutates much more frequently to B' in BB, than in Bb, plants. It is not certain, in fact, that B ever changes to B' in Bb individuals (84, 85). The initial change from B to B' in BB plants, and also the paramutation of B to B' in BB' individuals, appear to be one-step events. B' is stable. No change in pigmentation was observed through three generations of selection for either more or less plant color in $B'B'$ plants, nor was the level of B' paramutagenicity altered by repeated passage of B' through $B'B$ individuals (84). Linked marker genes on either side of B continue to assort normally after B changes to B'. B' is distributed in conventional fashion among the offspring of $B'b$ plants. These are the results expected if an alteration at the B locus underlies the B to B' phenotypic change (83, 84). The unstable colorless allele, B^v, becomes paramutant and paramutagenic following passage through a $B'B^v$ plant, just as do B (fully pigmented) and b (nonpigmented) mutants from B^v. It is evident, therefore, that mutability and paramutability are distinct properties of the unstable B^v allele, and also that functional expression of B, in terms of plant pigmentation, can vary independently of paramutation (84).

The paramutagenic property of B' is not infectious. Maternal haploid plants from BB ♀ × $B'B'$ ♂ matings were B in phenotype, even though the endosperm immediately associated with the haploid embryo is a fertilization product (BBB'). Furthermore, attempts to transfer the B' paramutagenic property by inoculation of B seedlings with macerated B' tissue gave negative results (84, 87).

The fact that BB' plants are darker in color than $B'B'$ or $B'b$ individuals was interpreted by Coe (84, 86) to mean that B' tends to suppress the pigmenting action of B, as such, in BB' heterozygotes. He inferred that paramutation of B to B' in BB' individuals does not take place until very late in the life cycle, probably at meiosis. It was observed that the tissue uncovered by deletion of a marked chromosome segment bearing B' in BB' heterozygotes, following X-irradiation at the 10-leaf stage, displayed the B phenotype. The occurrence of B

sectors on cobs, a late developing structure, of X-irradiated *BB′* plants also was construed as evidence for persistence of *B* in unchanged form through most of sporophytic stage (84).

Coe (84) postulated that the paramutant form of *B* is a *B* coding unit firmly combined in the chromosome with a distinct element *E*, and symbolized by *B.E.* He suggested that *E* could arise by somatic mutation of a regular chromosome constituent, *E*°, or it could be a kind of episome, specific for the *B* locus, phenotypically detectable only after integration at that site. In the somatic cells of *B.E/B* heterozygotes *B* pigmenting function tends to be suppressed in the *B.E* compound and also in the *B* allele carried by the homologous chromosome. *E* is assumed to replicate in phase with *B* in somatic mitosis. Late in the development of *B.E/B* plants, however, probably during chromosome synapsis in meiosis, *E* undergoes an extra replication. The supernumerary copy of *E* then combines with the *B* allele in the homologous chromosome, so that all gametes formed by plants of the *B.E/B* genotype are of the paramutant type, *B.E.*

Styles (88,89) obtained evidence pointing to the conclusion that the *B* locus on chromosome 2 and the *R* locus on chromosome 10 are of common origin. The *B* (booster) gene was originally reported by Emerson (90) as an intensifier of anthocyanin pigmentation of the plant. No *B* alleles were then known that affected seed pigmentation. More recently, however, Styles isolated a factor from a stock designated Peru 1497 that gave colored aleurone in $r^g r^g$ plants, that is, in individuals that carried a null allele for anthocyanin pigmentation at the *R* locus on chromosome 10. The newly recognized gene was carried by chromosome 2, and mapped at the *B* locus. The *B* locus, therefore, could appropriately be designated R_2. Styles found that the Peru 1497 gene gave solidly colored aleurone in single dose. No reduction in seed pigmentation was observed after it had been in the same genome with a paramutagenic R_1 (chromosome 10) gene. Furthermore, appropriate testcrosses showed that the Peru 1497 gene did not reduce the pigmenting action of an R_1 gene known to be paramutable.

The occurrence of paramutability at both *B* and *R* in maize evidently is not a coincidence but is associated with a common derivation of the two loci. Each locus affects anthocyanin formation in both seed and plant and also is the site of elements that underlie directed, heritable change in appropriate heterozygotes. The two loci are not congruent, however, in either of these respects. Coe (84) has summarized in detail the experimental evidence on paramutation from them and has juxtaposed the conclusions which the different investigators, working independently on *B* and *R*, have advanced. The interpretations in the two cases differ in basic respects. Coe's studies led him to the view that paramutation in *B B′* plants probably occurs only during chromosome pairing at meiosis and involves the transfer of a distinctive particle generated by *B′* from *B′* to *B* into which it then becomes incorporated (84). This interpretation contrasts fundamentally with that applied to the *R* data. *R* paramutation occurs in vegetative cells and rests on a property inherent in the paramutable allele rather than on acquisition by the latter of an element from a paramutagenic partner (34).

Magnification of Genes for Ribosomal RNA in Drosophila

A phenomenon simulating paramutation in other species appears to occur at the bobbed (*bb*) locus in the X and Y chromosomes of *Drosophila melanogaster*. This locus conditions the production of ribosomal RNA, and consists of multiple copies of a gene that acts additively on bristle size over a wide range. Wild type results if the X and Y chromosomes together contain about 150 copies of the gene. Progressively fewer than this number gives the bobbed phenotype in correspondingly increased degree (94).

Ritossa (91), Atwood (92), and Ritossa & Scala (93) observed that alleles giving a pronounced bobbed phenotype may revert gradually to wild type, a change that was termed "magnification". Henderson & Ritossa (94) and also Tartoff (95) found that reversion of a magnified bobbed allele (bb^{m}) is differentially affected by the kind of partner with which the allele is associated in a heterozygote.

Anomalous Genetic Phenomena in which Paramutation may be Involved

Bud failure in almonds, crinkle and variegation in sweet cherry, and strawberry June yellows are hereditary disorders in which virus-like symptoms are displayed but in which conventional tests fail to disclose an infectious agent. Aberrant Ratio in maize may be a related phenomenon. The genetic basis of the biological disturbances is unknown in these cases. There are indications, however, that paramutation is involved.

NONINFECTIOUS BUD FAILURE IN THE ALMOND (PRUNUS AMYGDALUS) Bud failure (BF) in the almond appears to be due to a genetic instability that is characteristic of certain clonal varieties. Transmission does not occur following grafting of an affected scion on a healthy tree. BF is perpetuated, however, by vegetative propagation and it is also pollen and seed transmitted (96–98). The principal symptom of the disease is the failure of buds to grow out in the spring. Not all buds on a shoot are equally affected so that bizarre branching patterns in a tree result.

Instructive data on the inheritance of BF were obtained by Kester (98) from matings between three BF trees of the Nonpareil variety, used as seed parents, and a series of almond cultivars of diverse history and status in terms of the disease. Three classes of branches on the Nonpareil-BF trees with respect to BF phenotype, severe, moderate, and slight, were used separately in the crosses. The pollen parents included, on the one hand, the old varieties Jordonola-BF and Jubilee-BF, which were known to be affected by BF in some degree throughout the respective clones, and, on the other, Ne Plus Ultra, a major commercial variety with no known history of BF in the clone. The percentage of progeny exhibiting BF after four years of growth from seed was much higher in the families derived from severely affected male parents than from male parents with no known history of BF. Progenies grown from seeds borne on Nonpareil-BF branches that showed slight symptoms of the disease contained significantly fewer BF individuals than their counterparts based on branches with a moderate or a severe BF expression. Kester (98) concluded that gametically transmissible gradients in BF potential occur that parallel the levels of expression of the disorder in the parent plants. The more pronounced the BF expression in the parent, the greater is the number of affected offspring at a given stage, the earlier is the age of onset, and the greater is the average level of severity. Accompanying the phenotypic shift characteristic of the BF developmental pattern is a parallel shift in genotype. Kester (98) directed attention to the general correspondence between bud failure in the almond and paramutation at the *R* and *B* loci in maize.

JUNE YELLOWS IN THE STRAWBERRY (FRAGARIA) June yellows is a noninfectious disease of the strawberry that involves a progressive, deleterious leaf variegation. H. Williams (99) and also Wills (100) found that the capacity to develop June yellows is transmitted through both pollen and eggs, but differentially from normal and chlorophyll deficient branches within a clone. They postulated an altered plasmagene as the genetic basis of the disease. According to Wills, frequent transmission through the pollen indicates a close association of the plasmagene with the nucleus. Kester (98) notes that the evidence leaves open the possibility that the disease is a paramutation phenomenon.

CRINKLE AND VARIEGATION IN THE SWEET CHERRY (PRUNUS AVIUM) Crinkle and variegation are leaf abnormalities exhibited by certain clonally propagated varieties of sweet cherry that resemble those resulting from some virus infections. The disorders, however, are not transmissible to a healthy stock by grafting. Symptom expression is highly variable with respect to amount and distribution of affected areas within a tree and also age of plant. Inheritance studies led Kerr (101) to suggest that each of these maladies is conditioned primarily by a recessive gene for which varieties that developed symptoms are heterozygous. Symptoms appear in such heterozygotes if the dominant partner in a gene pair mutates in somatic tissue to the recessive condition. Lethality of recessive homozygotes and modifying genes were assumed to account for the marked variation observed in the frequency with which affected individuals appeared in segregating families. Another possible explanation, to which Kester (98) has alluded, is that paramutation underlies the irregularities in sexual and somatic cell heredity characteristic of these diseases.

ABERRANT RATIO IN MAIZE Sprague & McKinney (102) found that maize plants artificially infected with an RNA virus causing barley stripe mosaic (BSMV) and then used as the male parent in outcrosses occasionally gave abnormal ratios for marker genes. Active BSMV virus has not been established in maize pollen or seed. This phenomenon, termed Aberrant Ratio (AR), was manifested in similar form at each of the few diverse marker loci that were extensively tested. Essentially the same distribution of offspring results from the use of a given AR plant as a male or a female parent. Distortion of the ratio in a given case involves a single locus; the distribution of other genes on the same, or different, chromosomes is not altered. The amount of distortion varies widely from one instance to another. Reversals in phase within a given lineage occur, although infrequently.

Two general classes of AR variants were found by Sprague & McKinney (103) with low frequency among the F_2 descendants of a plant homozygous for a dominant allele, such as *A*, infected with the virus, and then outcrossed to a normal *aa* strain. One class, designated *A*a*, yields an excess of the dominant phenotype, *A*, in appropriate testcrosses. The other category, termed *Aa**, gives an excess of the recessive phenotype when comparably examined. Both *A*a* and *Aa** individuals give equal numbers of *A* and *a* plants when crossed with an *aa* control stock that has not been exposed to the virus. This shows that the two alleles in AR heterozygotes assort at meiosis in the conventional way. *A*a* and *Aa** plants mated reciprocally with *aa* individuals extracted from *A*a* and *Aa** parents on the other hand, usually, but not invariably, give ratios that are significantly distorted in one degree or another. No regular pattern of distribution was discernible among the various forms of AR plants disclosed by these inter se matings.

The mechanism underlying AR remains obscure. Sprague & McKinney (103) suggest, without elaboration, that paramutation may be involved. An element is postulated "which can duplicate independently from chromosome replication" and also can transpose from one chromosome site to another. A locus with which this element becomes associated then gives distorted genetic ratios. Sprague & McKinney (103) infer that the element in question may represent a nucleotide sequence from the virus genome, a maize sequence modified indirectly by the virus, or another, even less direct, product of virus action on the maize plant.

The above finding that the genetic apparatus in maize may be heritably altered locally following virus infection may have important implications for both virology and chromosome constitution in higher plants. It has long been recognized that the gross external symptoms induced by some viruses in flowering plants frequently are duplicated by the action of chromosomal genes conditioning mottling and striping of the foliage. The parallel appears to be more than coincidental, particularly since the wide occurrence of temperate phages in bacteria, having chromosomal and cytoplasmic expressions distinct from each other, has been established. No virus in seed plants, however, has yet been proved to have a chromosomal counterpart. Plastid mutations induced by genic action have been recorded in a few instances (104). It may be significant in the present relation that the *R* locus in maize which, as noted earlier, affects anthocyanin pigmentation and is paramutable, contains an element that suppresses chlorophyll striping conditioned by genes at other chromosome sites (105). The possibility has not been ruled out that this seemingly foreign inclusion in the *R* locus is significant for *R* paramutation.

Paramutable loci, although unique in certain respects, are like other unstable loci in possessing elements distinct from, but affecting the expression of structural genes with which they are physically associated. Much remains to be resolved concerning the origin, nature, and significance for chromosome organization of these seemingly disparate elements. The AR findings in maize suggest the possibility that the chromosome components in question correspond to nucleotide sequences of virus origin, an idea that is experimentally testable. Further exploration in this direction might yield new evidence significant for chromosome evolution and cell organization in higher plants.

ACKNOWLEDGMENTS

This article is No. 1613 from the Laboratory of Genetics, University of Wisconsin, Madison. The studies reported in which the writer participated were aided by grants from the research committee of the Graduate School of funds supplied by the Wisconsin Alumni Research Foundation; by grants from the National Science Foundation and the Atomic Energy Commission. The writer is indebted to his colleagues, J. L. Kermicle, H. Dooner, and O. E. Nelson, for helpful suggestions during preparation of the manuscript.

Literature Cited

1. Thomas, C.A., Jr. 1971. The genetic organization of chromosomes. *Ann. Rev. Genet.* 5:237-54
2. Renner, O. 192I. Über Oenothera atrovirens Sh. et Bartl. und über somatische Konversion im Erbgang des cruciata-Merkmals der Oenotheren. *Z. Indukt. Abstamm. Verebungsl.* 74:91–124
3. Mitchell, M. B. 1955. Further evidence of aberrant recombination in Neurospora. *Proc. Nat. Acad. Sci., USA* 41:935–37
4. Hagemann, R. 1969. Somatic conversion (paramutation) at the *sulfurea* locus of *Lycopersicom esculentum* Mill. III. Studies with trisomics. Appendix: characteristics of allele-induced heritable changes in eukaryotic organisms. *Can. J. Genet. Cytol.* 11:346–58
5. Bateson, W., Pellew, C. 1915. On the genetics of "rogues" among culinary peas (*Pisum sativum*). *J. Genet.* 5:15–36
6. Bateson, W., Pellew, C. 1916. Note on an orderly dissimilarity in inheritance from different parts of a plant. *Proc. Roy. Soc.* B 89:174–75
7 Bateson, W., Pellew, C. 1920. The genetics of "rogues" among culinary peas (*Pisum sativum*). *Proc. Roy. Soc.* B 91:186–95
8. Bateson, W. 1926. Segregation. *J. Genet.* 16:201–35
9. Brotherton, W., Jr. 1923. Further studies on the inheritance of "rogue" type in garden peas (*Pisum sativum* L.). *J. Agri. Res.* 24:815–52
10. Brotherton, W., Jr. 1924. Gamete production in certain crosses with "rogues" in peas. *J. Agr. Res.* 28:1247–52
11. Lilienfeld, F. A. 1929. Verebungsversuche mit schlitzblättrigen Sippen von *Malva parviflora*. I. Die *laciniata* Sippe. *Bibl. Genet.* 13:1–214
12. Renner, O. 1958 Über den Erbgang des *cruciata*-Merkmals der Oenotheren. VIII. Verbindungen der *Oenothera Hookeri*, *Oe. franciscana* und *Oe. purpurata*. *Flora* 145:339–73
13. Renner, O. 1958. Über den Erbgang des *cruciata*-Merkmals der Oenotheren. VIII. Mitteilung. Verbindungen der *Oenothera atrovirens*, und Rückblick. *Z. Verebungslehre* 377–96

14. Renner, O. 1959. Somatic conversion in the heredity of the *cruciata* character in Oenothera. *Heredity* 13:283–88
15. Hagemann, R. 1958. Somatische Konversion bei *Lycopersicon esculentum* Mill. Z. *Verebungslehre* 89:587–613
16. Hagemann, R. 1969. Somatische Konversion (Paramutation) am *sulfurea* Locus von *Lycopersicon esculentum* Mill. IV. Die genotypische Bestimmung der Konversionshäufigkeit. *Theor. Appl. Genet.* 39:295–305
17. Hagemann, R. 1961. Mitteilungen über somatische Konversion. 1. Ausschluss des Vorliegens von somatischem Austausch. *Biol. Zentralbl.* 80:477–78
18. Hagemann, R. 1961. Mitteilungen über somatische Konversion. 2. In welchem Ausmass ist die somatische Konversion gerichtet? *Biol. Zentralbl.* 80:549–50
19. Hagemann, R. 1961. Mitteilungen über somatische Konversion. 3. Die Konversionshäufigkeit in Bastarden zwischen *sulfurea* Homozygoten und verschiedenen Sippen des Subgenus *Eulycopersicon. Biol. Zentralbl.* 80:717–19
20. Hagemann, R. 1966. Somatische Konversion am *sulfurea* Locus von *Lycopersicon esculentum* Mill. II. Weitere Beweise für die somatische Konversion. *Kulturpflanze* 14:171–200
21. Prokofyeva-Belgovskaya, A.A. 1948. Heterochromatization as a change of chromosome cycle. *J. Genet.* 48:80–98
22. Oehlkers, F. 1935. Die Erblichkeit der Sepalodie bei Oenothera und Epilobium. Studien zum Problem der Polymerie und des multiple Allelomorphismus III. Z. *Bot.* 28:161–222
23. Renner, O. 1942. Beiträge zur Kentniss des *cruciata*-Merkmals der Oenotheren. IV. Gigas Bastarde. Labilität und Konversibilität der *Cr*-Gene. Z. *Verebungslehre* 80:590–611
24. Brink, R.A. 1964. Genetic repression of *R* action in maize. In *Role of the Chromosomes in Development*. Locke, M. Ed. pp. 183–230. Academic: New York
25. Brink, R.A., Styles, E.D., Axtell, J. D. 1968. Paramutation: directed genetic change. *Science* 159:161–70
26. Brink, R. A. 1956. A genetic change associated with the *R* locus in maize which is directed and potentially reversible. *Genetics* 41:872–89
27. Brink, R. A., Weyers, W. H. 1957. Invariable genetic change in maize plants heterozygous for marbled aleurone. *Proc. Nat. Acad. Sci., USA.* 43:1053–60
28. Brown, D. F., Brink, R. A. 1960. Paramutagenic action of paramutant R^{r} and R^{g} alleles in maize. *Genetics* 45:1313–16
29. Brink, R. A., Brown, D. F., Kermicle, J., Weyers, W. H. 1960 Locus dependence of the paramutant *R* phenotype in maize. *Genetics* 45:1297–1312
30. Brink, R. A., Kermicle, J. L., Brown, D. F. 1964. Tests for a gene-dependent cytoplasmic particle associated with *R* paramutation in maize. *Proc. Nat. Acad. Sci., USA* 51:1067–74
31. Brown, D.F. 1966. Paramutability of R^{g} and r^{r} mutant genes derived from an R^{r} allele in maize. *Genetics* 54:899–910
32. McWhirter, K. S., Brink, R. A. 1962. Continuous variation in level of paramutation at the *R* locus in maize. *Genetics* 47:1053–74
33. Kermicle, J. L. 1963. *Metastability of paramutant forms of the R gene in maize.* Ph.D Thesis, Univ. Wisconsin, Madison
34. Styles, E. D., Brink, R. A. 1969. The metastable nature of paramutable *R* alleles in maize. IV. Parallel enhancement of *R* action in heterozygotes with *r* and in hemizygotes. *Genetics* 61: 801–11
35. Mikula, B. C. 1961. Progressive conversion of *R*-locus expression in maize. *Proc. Nat. Acad. Sci., USA* 47:566–71
36. Brink, R. A., Mikula, B. 1958. Plant color effects of certain anomalous forms of the R^{r} allele in maize. Z. *Indukt. Abstamm. Verebungsl.* 89:94–102
37. Brink, R. A. 1959. Paramutation at the *R* locus in maize plants trisomic for chromosome 10. *Proc. Nat. Acad. Sci., USA* 45:819–27
38. McWhirter, K.S., Brink, R. A. 1963. Paramutation in maize during endosperm development. *Genetics* 47: 1053–74
39. Sastry, G. R. K., Cooper, H. B., Jr., Brink, R. A. 1965. Paramutation and somatic mosaichism in maize. *Genetics* 52:407–24
40. Brink, R. A., Blackwood, M. 1961. Persistent enhancement of R^{r} action in maize by structural alterations of chromosome 10. *Genetics* 46: 1185–1205

41. Brink, R. A. 1961. Relative insensitivity of enhanced R^r to paramutation in maize plants heterozygous for the stippled allele. *Geneties* 46:1207–21
42. Brink, R. A., Notani, N. K. 1961. Effect on R^r action in maize of a structural alteration distal to the R locus in chromosome 10. *Genetics* 46:1223–30
43. Brink, R. A., Weyers, W. H. 1960. Effect of an abnormal knob-carrying chromosome 10 on paramutation of R^r in maize. *Genetics* 45:1445–55
44. Brink, R. A. 1969. Abnormal chromosome 10 and R paramutation in maize. *Mutat. Res.* 8:285–302
45. Brink, R. A., Venkateswarlu, J. 1965. Effect of a translocation in maize on sensitivity to paramutation of R^r alleles from three geographic sources. *Genetics* 51:585–91
46. Kester, D. E., Brink, R. A. 1966. Variation in paramutability of R^r in maize following extraction of the allele from a linked reciprocal translocation. *Genetics* 54:1401–07
47. Styles, E. D., Brink, R. A. 1966. The metastable nature of paramutable R alleles in maize. I. Heritable enhancement in level of standard R^r action. *Genetics* 54:433–39
48. Bray, R. A., Brink, R. A. 1966. Mutation and paramutation at the R locus in maize. *Genetics* 54:137–49
49. Axtell, J. D., Brink, R. A. 1967. Chemically induced paramutation at the R locus in maize. *Proc. Nat. Acad. Sci., USA* 58:181–87
50. Mikula, G. 1967. Heritable changes in R-locus expression in maize in response to environment. *Genetics* 56:733–42
51. Linden, D. B. 1963. Radiation induced modification of paramutation expression. *Proc. XI Int. Congr. Genet.* 1:951 (abstr.)
52. Linden, D. B. 1963. Effects of radiation on paramutation. *Radiat. Res.* 19:184
53. Shih, K. L., Brink, R. A. 1969 Effects of X-irradiation on aleurone pigmenting potential of standard R^r and a paramutant form of R^r in maize. *Genetics* 61:167–77
54. Shih, K. L. 1969. Effects of pretreatment with Xrays on paramutability or paramutagenicity of certain R alleles in maize. *Genetics* 61:179–89
55. Linden, D. B., Rodriguez, V. 1965. Paramutagenic systems in some South American races of corn. *Genetics* 51:847–55
56. Van der Walt, W. J., Brink, R. A. 1969. Geographic distribution of paramutable and paramutagenic R alleles in maize. *Genetics* 61:677–95
57. Gavazzi, G., Zannini, N. 1966. Diversa sensibilita' alla paramutazione di due subunita' del locus R, in *Zea mays. Inst. Lombardo-Accad. Sci. Lett.* B 100:49–62
58. Gavazzi, G., Maldotti, C. 1967. Genetic instability of R^{st} in somatic and germinal tissues of maize. *Atti Assoc. Genet. Ital.* 12:385–400
59. Ashman, R. B. 1965. Mutants from maize plants heterozygous R^rR^{st} and their association with crossing over. *Genetics* 51:305–12
60. Ashman, R. B. 1965. Paramutagenic action of mutants from maize plants heterozygous R^rR^{st}. *Genetics* 52: 835–41
61. Ashman, R. B. 1970. The compound structure of the R^{st} allele in maize. *Genetics* 64:239–45
62. Kermicle, J. L. 1970. Somatic and meiotic instability of R-stippled, an aleurone spotting factor in maize. *Genetics* 64:247–58
63. Satyanarayana, K. V. 1970. *Organization of the pigmenting and paramutagenic determinants of the R-stippled gene in maize.* Ph.D. Thesis, Univ. Wisconsin, Madison, 96 pp
64. Williams, W. M. 1972. *Variability of the R-stippled gene in maize.* Ph.D. Thesis, Univ. Wisconsin, Madison, 108 pp
65. Gavazzi, G. 1966. Genetic analysis of some derivatives of R^{st} in maize. *Atti Assoc. Genet. Ital.*, Pavia 11:117–25
66. Ashman, R. B. 1960. Stippled aleurone in maize. *Genetics* 45:19–34
67. Van Schaik, N., Brink, R. A. 1959. Transpositions of Modulator, a component of the variegated pericarp allele in maize. *Genetics* 44:725–38
68. Satyanarayana, K. V., Kermicle, J. L. 1973. The R-stippled allele in maize. I. Relation of changes in paramutagenic action to recombinant origin of self-colored mutations. *Genetics* (in press)
69. Satyanarayana, K. V., Kermicle, J. L. 1973. The R-stippled allele in maize. II. Arrangement of the pigmenting and paramutagenic components. *Genetics* (in press)
70. Stadler, L. J. 1942. Some observations on gene variability and spontaneous mutation. *Spragg Memorial Lectures (3rd Ser.)*, 3–15. Michigan State Coll., East Lansing

71. Stadler, L. J. 1951. Spontaneous mutations in maize. *Cold Spring Harbor Symp. Quant. Biol.* 16:49–63
72. Stadler, L. J., Nuffer, M. G. 1953. Problems of gene structure. II. Separation of R^r elements (S) and (P) by unequal crossing over. *Science* 117: 471–72 (abstr.)
73. Dooner, H., Kermicle, J. L. 1971. Structure of the R^r tandem duplication in maize. *Genetics* 67:427–36
74. Stadler, L. J., Emmerling, M. H. 1956. Relation of unequal crossing over to the interdependence of R^r elements (P) and (S). *Genetics* 41:124–37
75. Sastry, G. R. K. 1970. Paramutation and mutation of R^{ch} in maize. *Theor. Appl. Genet.* 40:185–90
76. Brink, R. A., Williams, E. 1973. Mutable *R*-Navajo alleles of cyclic origin in maize. *Genetics* 73:273–96
77. Kermicle, J. L. 1970. Dependence of the *R*-mottled aleurone phenotype in maize on mode of sexual transmission. *Genetics* 66:69–85
78. Roman, H. 1947. Mitotic nondisjunction in the case of interchanges involving the B-type chromosomes in maize. *Genetics* 32:391–409
79. Brink, R. A., Kermicle, J. L., Ziebur, N. K. 1970. Derepression in the female gametophyte in relation to paramutant *R* expression in maize endosperms, embryos, and seedlings. *Genetics* 66:87–96
80. Styles, E. D. 1970. Paramutation of the *R* locus in maize trisomics and comparison of paramutation systems in maize and tomato. *Can. J. Genet. Cytol.* 12:941–46
81. Styles, E. D. 1967. The metastable nature of paramutable *R* alleles in maize. II. Characteristics of alleles differing in geographic origin. *Genetics* 55:399–409
82. Styles, E. D. 1967. The metastable nature of paramutable *R* alleles in maize. III. Heritable changes in level of *R* action in heterozygotes carrying different paramutable *R* alleles. *Genetics* 55:411–22
83. Coe, E. H., Jr. 1959. A regular and continuing conversion-type phenomenon at the *B* locus in maize. *Proc. Nat. Acad. Sci., USA* 45:828–32
84. Coe, E. H., Jr. 1966. The properties, origin, and mechanism of conversion-type inheritance at the *B* locus in maize. *Genetics* 53:1035–63
85. Coe, E. H., Jr. 1961. A test for somatic mutation in the origination of conversion-type inheritance at the *B* locus in maize. *Genetics* 46:707–10
86. Coe, E. H., Jr. 1968. Heritable repression due to paramutation in maize. *Science* 162:925
87. Coe, E. H., Jr. 1961. Some observations bearing on plasmid versus gene hypotheses for a conversion-type phenomenon. *Genetics* 46:719–26
88. Styles, E. D.. 1964. A duplicate *R* locus. *Maize Genet. Coop. News Lett.* 38:134–35
89. Styles, E. D. 1965. An aleurone color factor seemingly at the *B* locus. *Maize Genet. Coop. News Lett.* 39:172–73
90. Emerson, R. A. 1921. The genetic relations of plant colors in maize. *Cornell Univ. Agr. Exp. Sta. Memoir* 39:1–156
91. Ritossa, F. M. 1968. Unstable redundancy of genes for ribosomal RNA. *Proc. Nat. Acad. Sci., USA* 60:509–16
92. Atwood, K. C. 1969. Some aspects of the bobbed problem in Drosophila. *Genetics* 61 (Suppl. 1, pt. 2):319–27
93. Ritossa, F. M., Scala, G. 1969. Equilibrium variations in the redundancy of rDNA in *Drosophila melanogaster*. *Genetics* 61 (Suppl. 1, pt. 2):305–17
94. Henderson, A., Ritossa, F. 1970. On the inheritance of rDNA of magnified bobbed loci in *D. melanogaster*. *Genetics* 66:463–73
95. Tartof, K. D. 1971. Increasing the multiplicity of ribosomal RNA genes in *Drosophila melanogaster*. *Science* 171:294–97
96. Wilson, E. E., Schein, R. D. 1956. The nature and development of non-infectious bud-failure in almond. *Hilgardia* 24:519–42
97. Kester, D. 1968. Noninfectious bud-failure, a nontransmissible inherited disorder in almond. I. Pattern of phenotype inheritance. *Proc. Am. Soc. Hort. Sci.* 92:7–15
98. Kester, D.E. 1968. Noninfectious bud-failure, a nontransmissible inherited disorder in almond. II. Progeny tests for bud-failure. *Proc. Am. Soc. Hort. Sci.* 92:16–28
99. Williams, H. 1955. June yellows: a genetic disease of the strawberry. *J. Genet.* 53:232–43
100. Wills, A. B. 1962. Genetical aspects of strawberry June yellows. *Heredity* 17:361–72
101. Kerr, E. A. 1963. Inheritance of crinkle, variegation, and albinism in sweet cherry. *Can. J. Bot.* 41:1395–1404

102. Sprague, G. E., McKinney, H. H. 1966. Aberrant ratio: an anomaly in maize associated with virus infection. *Genetics* 54:1287–96
103. Sprague, G. F., McKinney, H. H. 1971. Further evidence on the genetic behavior of AR in maize. *Genetics* 67:533–42
104. Rhoades, M. M. 1955. Interaction of genic and non-genic hereditary units and the physiology of non-genic inheritance. *Handb. Pflanzenphysiologie* 1:2–57
105. Emerson, R. A., Beadle, G. W., Fraser, A. C. 1935. A summary of linkage studies in maize. *Cornell Univ. Agr. Exp. Sta. Memoir* 180:1–83
106. Ecochard, R. 1972. New cases of somatic conversion (paramutation) in tomato (*Lycopersicon esculentum*) Mill. *Theor. Appl. Genet.* 42:189–95

LONGITUDINAL DIFFERENTIATION OF CHROMOSOMES

3051

T. C. Hsu

University of Texas, M. D. Anderson Hospital & Tumor Institute, Houston, Texas

I. INTRODUCTION

It is well known that metaphase chromosomes do not possess many morphological characteristics which can be used to distinguish them within a complement. Only a few criteria can be employed to describe the chromosomes: the length, the position of centromere (hence the arm ratio), and the presence or absence of secondary constriction(s). It is possible to find a few species with low diploid numbers, e.g., *Crepis capillaris*, in which all chromosomes are morphologically distinct. However, even in these species, chromosome pairs with identical morphology are not uncommon. For example, in *Drosophila melanogaster* ($2n = 8$) Chromosomes II and III are nearly indistinguishable, and in *D. virilis* ($2n = 12$) five pairs of chromosomes are morphologically similar.

Cytologists have tried for decades to devise means for differentiating chromosomes longitudinally. Probably one of the earliest was the distribution of heterochromatin. Since the last century, microscopists have observed deeply staining chromatin masses, termed chromocenters, in interphase nuclei. The nature, the distribution, and the function of the chromocenters were completely unknown. During the late 1920s and early 1930s Heitz (1-3) made a series of careful cytological observations and discovered that chromosomes contain two types of chromatin, one condensing during mitosis and decondensing during interphase (euchromatin) and the other remaining condensed throughout the cell cycle (heterochromatin). The chromocenters are equivalent to the heterochromatin. Heterochromatin is distributed in special regions of the chromosomes, but identification of heterochromatic regions on metaphase chromosomes was difficult because at that stage, though best for discerning chromosome morphology, both euchromatin and heterochromatin are condensed, thus giving no differentiation in staining reaction. In interphase and prophase where the differentiation in staining reaction is best, recognition of individual chromosomes is not feasible. The only stage at which a staining differentiation is possible, and the chromosomes can be recognized, is the brief stage in mitosis known as prometaphase. In species with a low diploid number, such as *Drosophila melanogaster*, identification of heterochromatic segments can be performed with absolute certainty on prometaphasic complements. However, in species with a large number of chromosomes, especially those with numerous pairs possessing similar morphology, the prometaphase analysis again gives ambiguous data.

The localization of heterochromatin within the chromosomes requires no experimental procedure, merely a patient search for appropriate mitotic figures. The first case of experimentally induced differential staining along the chromosomes was probably the "nucleic acid starvation" of Darlington & La Cour (4). These investigators found that when root-tips of *Trillium* were treated for prolonged periods at 0°C, the chromosomes showed unstained segments or bands. However, this method never gained popularity, even though it was confirmed by other investigators (5, 6), because (*a*) not all metaphases show such characteristics and (*b*) the mitotic rate is, of course, extremely low at such severe temperatures.

In the hope of differentiating chromosomes longitudinally, cytologists have treated live tissues and cells with a variety of chemicals which can induce chromosome breaks to determine whether some of these agents cause breakage in specific regions. A few such compounds were indeed found (7), but the data must be tediously collected and statistically evaluated, so that this approach is at least impractical, if not meaningless. Furthermore, the responses of chromosomes to the drugs depend heavily on the stage of the cell cycle, thus complicating the experimental design.

Due to the lack of simple and reproducible procedure, or procedures, to differentiate ordinary metaphase chromosomes, cytologists turned their attention to special types of chromosomes, e.g., the polytene chromosomes, the lampbrush chromosomes, the pachytene bivalents, etc., for detailed analyses of many biological problems relating to chromosomes. Those who worked on the majority of species, including man, were obliged to deal with empirical systems and wait for a breakthrough.

The breakthrough did come. In fact, several significant advances in chromosome cytology came clustering within a short two-year period that so revolutionized the field of cytogenetics (as well as concepts of chromosome structure) that what was in vogue prior to this recent period has become the Stone Age approach. One no longer refers to a chromosome of a species as a member of, say, the 6–12+ X group because every chromosome of every species is recognizable. Each chromosome can be segmented into many identifiable regions or bands and tracing rearrangements is now feasible. In 1971 an ad hoc committee meeting on Standardization of Human Chromosomes was held in Paris to revise nomenclature systems, in the light of new techniques and new findings (8). In this report four different banding patterns were recognized (now known as Q-bands, C-bands, G-bands, and R-bands).

The present paper attempts to review briefly the subject of the "banded chromosomes", how the techniques were developed, what are the characteristics, what the bands might mean, and what are their applications to biology and to medicine. It is not the intention of this review to cover exhaustively all the references published to date. Further advances will undoubtedly be made, but what has already been accomplished should be regarded as an important milestone in the field of chromosome cytology.

II. DIFFERENTIAL FLUORESCENCE

The first real advance in differentiating metaphase chromosomes longitudinally came from the Institute for Medical Cell Research and Genetics, Karolinska Institutet at Stockholm, under the leadership of T. Caspersson. The original idea was, in a nutshell, something like this: Many fluorochromes have affinity with DNA, usually by intercalation. The nuclei and the chromosomes will show fluorescence under ultraviolet optics. If a fluorochrome molecule could be attached to an alkylating agent, e.g., a mustard, the latter might crosslink guanines and would therefore preferentially bind GC-rich regions of the chromosomes. If the chromosomes possess G-C base pairs in clusters, the fluorochromes may show differences in brightness along the metaphase chromosomes.

In a series of papers Caspersson and collaborators (9-11) discovered that quinacrine mustard (QM) indeed showed such a desirable quality. From a number of materials tested, including plants as well as animals, each chromosome displays, after QM staining and fluorescence in ultraviolet light, characteristic bright and dark bands or zones (Fig. 1). Applying the procedure to human chromosomes, this group of investigators (12-14) found that every chromosome pair of the human karyotype could be recognized by fluorescent characteristics. The fluorescent bands are of a variety of shades of brightness (or darkness). For example, the distal portion of the human Y chromosome is exceptionally bright when its fluorescence is compared with the remaining chromosomes of the complement. This characteristic enabled cytologists to identify the Y element even in interphase nuclei (15).

The combination of the position, the width and the brightness of the QM bands (now known as the Q-bands) is so unique for each chromosome that every element in the human complement can be recognized with relative ease and Q-band idiograms have been constructed (Fig. 2). One may also describe the banding pattern of each chromosome by performing densitometer tracings (16) and arriving at an idealized graphic representation of the karyotype (Fig. 3). Such advances caused the existing human cytogenetics to become obsolete almost overnight because numerous cases accumulated during the ten years prior to the discovery of differential fluorescence (1959-1969) would have to be reinvestigated.

At first it was thought that the mustard moiety was responsible for binding QM to DNA. The differential fluorescence by QM was therefore interpreted as the reflection of base composition along the chromosomes with GC-rich regions

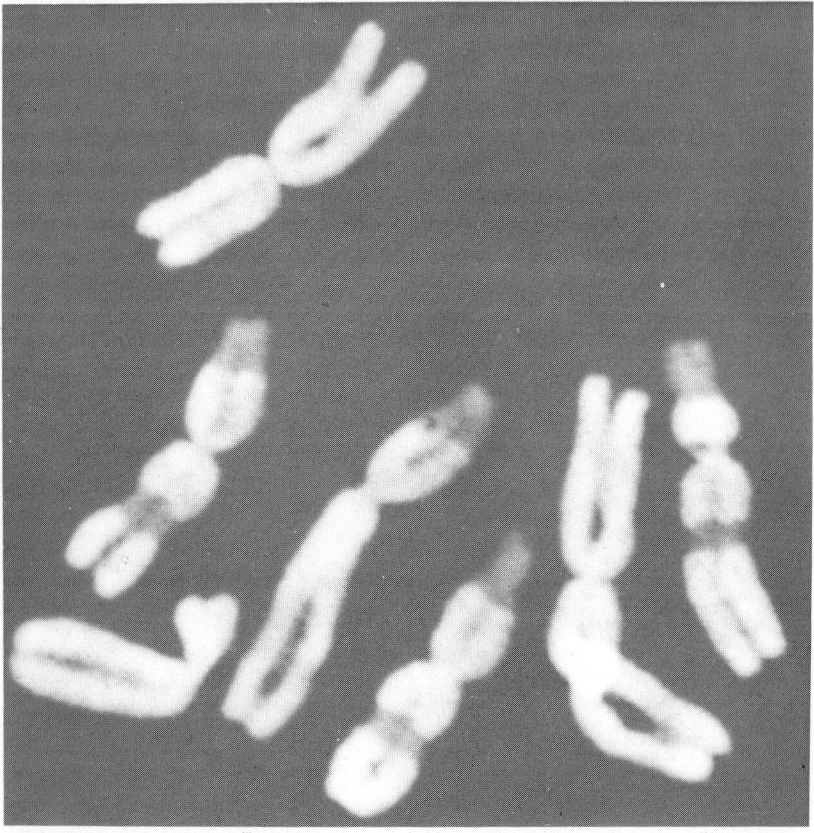

Figure 1. Metaphase chromosomes of *Scilla sibirica*, stained with quinacrine mustard, showing differential fluorescent segments (Q-bands) along the chromosomes. Courtesy Dr. Lore Zech.

binding more QM. However, several pieces of recent evidence appeared to throw considerable doubt on this hypothesis: (*a*) quinacrine dihydrochloride, which does not have an alkylating agent linked onto it, gives Q-band patterns identical with those of QM; (*b*) bright fluorescence was found on chromosomal regions known to be AT-rich (17); and (*c*) recent chemical studies indicate that AT-rich DNA tends to enhance the quinacrine fluorescence, while GC-rich DNA tends to quench it (18, 19). Thus it appears that QM has an affinity for AT-rich regions instead of GC-rich regions.

Although the mechanism for differential chromosomal fluorescence requires much more research, the discovery of Q-banding has had an immense impact not only on cytogenetics but on studies in chromosome organization as well. For

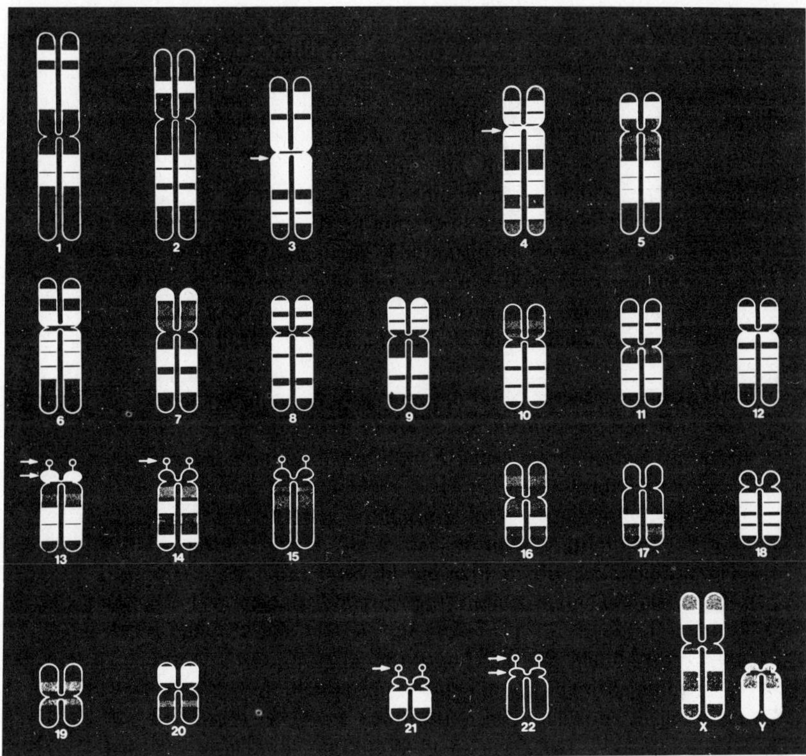

Figure 2. Q-band idiogram of human chromosomes. Courtesy Dr. C. C. Lin.

example, the random-folding-fiber model of chromosome structure (20) must be revised. If the "fibers" indeed fold back and forth along a metaphase chromosome, this would require a certain order by which DNA sequences with similar base composition are arranged side-by-side along segments of chromosomes in order to obtain banding (21).

It may be of interest to note in passing that Caspersson et al (11) found a peculiar behavior of ethidium bromide, viz., it produces a rather uniform fluorescence in the chromosomes of *Vicia faba* but a reversed brightness pattern from QM in those of *Scilla sibirica*. It also produces a uniform fluorescence of human chromosomes (L. Zech, personal communication). This contrast in fluorescence behavior is not explained. From biophysical data, ethidium does not appear to show base pair preference (22).

Differential fluorescence along metaphase chromosomes is not limited to the direct application of fluorochromes such as QM, quinacrine dihydrochloride, etc. Fluorochrome molecules can be conjugated to specific antibodies from patients

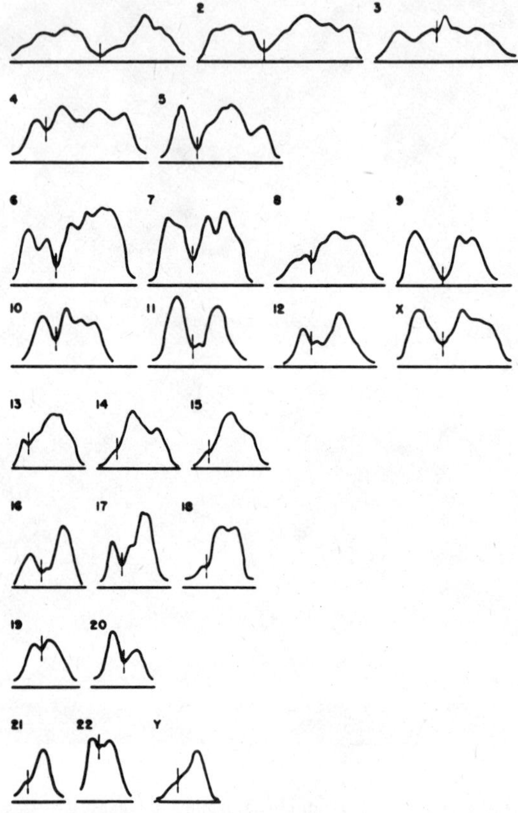

Figure 3. A graphic representation of fluorescent patterns of the 24 (23 autosomes + X + Y) human chromosomes. After Caspersson et al (16). Courtesy of Professor T. Caspersson.

with autoimmune diseases. As Dev et al (23) demonstrated, anti-adenine antibody gives a fluorescence pattern similar to that of Q-banding. If this fluorescence indeed indicates the anti-A reaction, the similarity between Q-banding and anti-A banding supports the notion that QM prefers AT-rich regions. A number of investigators also used fluorescent dyes (mainly acridine orange) or anti-single-stranded DNA to study chromosome fluorescence in conjunction with other treatments, but these experiments really belong to the subject on heterochromatin and will be discussed later.

The Q-banding technique, though extremely useful, has a few unfortunate shortcomings: (*a*) The equipment is expensive. Many laboratories cannot afford such a luxury but would like to be able to recognize chromosomes for routine cytogenetic analysis as well as for research work. (*b*) The fluorescence is not permanent. In fact, in most cases the slides must be studied and photographs must be taken immediately following staining or the fluorescence will fade. (*c*) The more recent Giemsa procedures (Sections III and IV), when the slides are properly prepared, reveal more detailed banding patterns than Q-banding. However, the Giemsa banding techniques still cannot completely replace Q-banding. For example, the extremely bright fluorescence of the heterochromatin in human Y and the polymorphic centromeric fluorescent marker on human Chromosome 3 make them easy targets to trace when such chromosomes are involved in rearrangements. The Giemsa techniques cannot unequivocally identify a particular heterochromatin or a particular band when extensive translocations and inversions upset its original spatial relationships with other bands. Furthermore, fluorescence microscopy, especially that in conjunction with immunological techniques, is expected to flourish in experimental cell research, including chromosome studies.

III. CONSTITUTIVE HETEROCHROMATIN

Historical

As mentioned previously, heterochromatin was one of the earliest criteria used to differentiate chromosomes longitudinally. It can be detected in late prophase, prometaphase, and meiotic prophase. In *Drosophila melanogaster*, the locations of heterochromatic segments are well-known: the centromeric areas of the autosomes, the proximal one-third of the X, and the entire Y. Numerous genetic studies demonstrated that heterochromatin contains very few, if any, structural genes (24), but heterochromatin may inhibit the phenotypic expression of structural genes in euchromatin when they are brought into juxtaposition (position effect).

Although the nature and the exact function of heterochromatin were not understood, it seemed highly suggestive, at least at first, that heterochromatin is intrinsically different from euchromatin. This conclusion was clouded as observational data accumulated. The X chromosome of the male grasshopper, for example, shows a strong heterochromatic staining behavior (heterpycnosis) in spermatocytes, but the same X chromosome does not appear heterochromatic in somatic cells of the same individual nor in the female cells.

More striking was the case of the mealy bug, *Planococcus citri*, in which an entire haploid set of chromosomes was found to assume a heterochromatic appearance in the male. All evidence indicated that this set of chromosomes is of paternal origin and has no genetic expression. Since this set of chromosomes is always discarded in gametogenesis, the male invariably received its functional chromosome set from the mother. Yet this same set, when passed to the sons in the next generation, becomes heterochromatic and functionless. Thus the situation existing in the grasshopper and the mealy bug definitely suggests that heterochromatin is the result of the responses of chromosomes to the physiological milieu instead of intrinsic difference.

Heterochromatin, therefore, appears to be of two basic types—one containing structural genes, but cytologically condensed and genetically inactivated depending upon physiological and developmental processes (the mealy bug type), and the other containing no structural genes (the original heterochromatin). To distinguish these two types, Brown & Nelson-Rees (25) and Brown (26) suggested two terms, *facultative heterochromatin* for the former and *constitutive heterochromatin* for the latter.

Heterochromatin of Mammalian Chromosomes

Prior to 1970, heterochromatin of mammalian chromosomes was more inferred than actually demonstrated by staining. Of course, there was no problem demonstrating chromocenters in interphase nuclei, but locating every piece of heterochromatin in metaphase chromosomes was not feasible. In species where an entire chromosome arm or an entire chromosome is made of heterochromatin, differences in chromosome condensation between such heterochromatic elements and euchromatic chromosomes can be observed in prophase and prometaphase (27). The best examples are the European field vole, *Microtus agrestis*, and the Chinese hamster, whose long arm of the X chromosome and the entire Y chromosome are heterochromatic. At metaphase, however, the distinction between euchromatin and heterochromatin becomes very subtle or nonexistent.

By allowing the cells to incorporate tritiated thymidine into their DNA and harvesting the cells sometime later, Taylor (28) discovered, in the autoradiographs of the Chinese hamster cells, that the long arm of the X and the entire Y replicate later than other chromosomes. This phenomenon was confirmed (29) and was later extended to the heterochromatin of *Microtus agrestis* (30).

During the same period, Lyon (31), in her genetic analysis of the mouse, discovered the phenomenon of inactivation of an X chromosome in female somatic cells. According to her genetic data, as well as those of Russell (32), either the paternal or the maternal X chromosome can be inactivated. This phenomenon was quickly correlated with the sex chromatin (33), a "chromocenter" found only in female somatic cells. Also, cytological examination of female somatic mitosis occasionally revealed one X chromosome to be more condensed than other chromosomes. It was implicit that one of the two X chromosomes, identical in their genetic content, became heteropycnotic during development, a situation analogous to that found in the mealy bugs.

The discovery of heterochromatization of an X chromosome suggested that in mammalian systems both constitutive and facultative heterochromatins exist. Trititiated thymidine autoradiography showed that the inactivated X chromosome is also late-replicating (34). Thus late-replication does not differentiate the two types of heterochromatin.

Since it was not feasible to identify heterochromatin positively in conventional preparations of mammalian cells, late DNA replication was used exclusively to identify heterochromatin. However, this method has several disadvantages: (*a*) the procedure is time-consuming, and proper cells must be selected at the right time of harvest; (*b*) it becomes uncertain when the heterochromatic piece is very small, because one or two silver grains would not provide unequivocal evidence;

(*c*) it does not differentiate the two types of heterochromatin; (*d*) when a complement contains many late replicating segments, one begins to wonder whether the data are correct. A good example of this last point is found in the work of Galton & Holt (35) on the DNA replication sequences of the Syrian hamster. They found numerous late replicating chromosome arms and they speculated that all these segments may be heterochromatic but could not confirm such hypothesis cytologically. Recently we (36) proved that they were correct.

The C-band Technique

In an experiment designed to determine the cytological location(s) of mouse satellite DNA by *in situ* nucleic acid hybridization, Pardue & Gall (37) noticed that the areas where the satellite DNA is localized (centromeric regions of all autosomes and the X chromosome) stained deeply with Giemsa whereas the remaining chromosome arms were lightly stained. This phenomenon had not been observed in the mouse cells. Apparently the series of treatments used in the *in situ* hybridization procedure had caused a differential response between chromosomal regions, viz., areas rich in satellite DNA content stained more deeply than areas without the satellite DNA. The differential staining does not depend on the presence of radioactive complementary RNA molecule. Pardue & Gall considered these areas to be heterochromatin.

In their experiments involving human repetitious DNA, Arrighi et al (38) also noted deeply stained areas (mostly centromeric) in human chromosomes. Several pertinent points from this study can be listed here: (*a*) These deeply stained areas represent constitutive heterochromatin of the human complement. (*b*) Each chromosome has its own characteristic amount and location of constitutive heterochromatin. For example, Chromosome 1 has a large piece on its long arm extending from the centromeric constriction, Chromosome 2 has a tiny piece at the centromeric constriction, and the Y chromosome has a large segment at the distal end of its long arm (corresponding to the segment displaying the bright Q-band). (*c*) Facultative heterochromatin is not revealed since the two X chromosomes behave similarly in staining property. Figure 4 presents a male human karyotype showing the distribution of constitutive heterochromatin.

The discovery of a procedure for staining constitutive heterochromatin, though a byproduct, was as important a contribution as the main purpose of the experiment, the *in situ* hybridization. It marked the first time that cytologists could rely on a simple and reproducible procedure for revealing constitutive heterochromatin at metaphase as well as other stages, and it helped to correlate heterochromatin and repeated DNA sequences. Furthermore, it cleared up a considerable amount of ambiguity in interpreting karyologic evolution and molecular evolution, and it led to the discovery of Giemsa banding of chromosomes to be discussed later.

The procedure (39), referred to as C-banding by the Paris Conference, is indeed very simple: Treat conventional preparations (squash or air-dried) with a brief exposure of .2N HCl (10-30 min), wash in deionized water, then treat the slides with NaOH (2X SSC with pH adjusted to 12) for not more than 2 min, rinse in ethanol several times, incubate in 6X SSC overnight at 60°C, and stain in Giemsa solution.

To resolve the question of whether C-bands are equivalent to the classic heterochromatin of Heitz, Hsu (40) applied the new procedure to neuroblast preparations of *Drosophila melanogaster* to determine whether it could produce C-bands on metaphase chromosomes and, if so, whether the C-bands corresponded to the heterochromatin segments known for that species. The affirmative results indicated the applicability of the term to all species whenever a heavily stained chromosomal segment appears following the treatments.

In the majority of chromosomes of the majority of mammalian species, heterochromatin is centromeric. Its amount on each particular chromosome appears to be characteristic, but polymorphism does occur among individuals, at least in man (41, 42). In some individuals, one or more chromosomes, e.g., 1, 9, 16, etc., may possess heterochromatin twice as long as that of its homolog. Such persons show no phenotypic abnormality that can be attributed to the variation in the amount of heterochromatin. This phenomenon is consistent with the classic concept that heterochromatin is genetically inert. As it is known that heterochromatin contains many repeated DNA sequences, it is conceivable that any sequence can synapse with any other sequence of the homolog during meiosis, thus increasing the frequency of unequal crossing over, which may account for the variability of the C-bands.

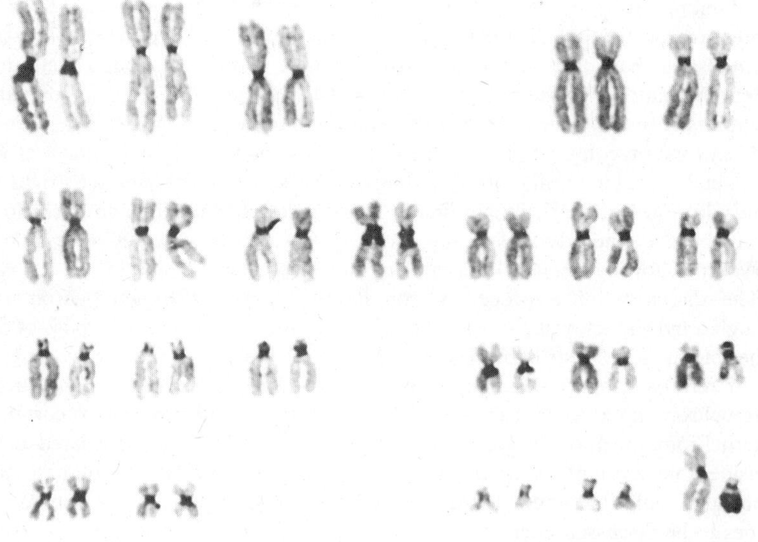

Figure 4. A human karyotype showing distribution of constitutive heterochromatin (C-bands). Courtesy Mrs. Ann P. Craig-Holmes and Dr. Margery W. Shaw.

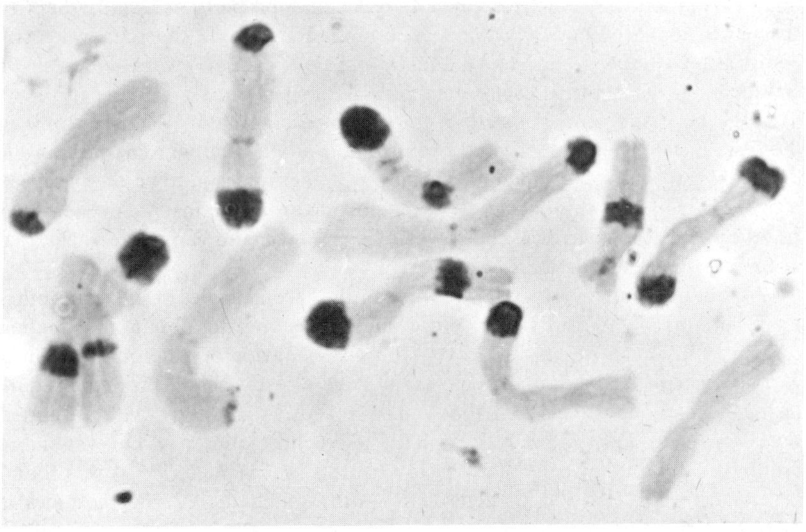

Figure 5. Heterochromatin patterns of metaphase chromosomes of *Scilla sibirica.* Courtesy Dr. C. G. Vosa.

Interstitial and terminal C-bands have been observed in some species. Possibly these originated from the centromeric regions but were displaced by inversions or translocations. In certain groups of animals, e.g., the hamster genus *Mesocricetus,* many chromosomal arms are totally heterochromatic (36, 43, 44). These correspond to the late-replicating arms found by Galton & Holt (35).

Heterochromatin can be found most abundantly in the sex chromosomes. As proposed some time ago by Ohno et al (45), the functional mammalian X represents a definite proportion (approximately 5%) of the genome. Those species, e.g., *Microtus agrestis,* possessing large X chromosomes invariably contain a large amount of constitutive heterochromatin in the X chromosomes. This suggestion found support from DNA replication studies (30) as well as from C-band staining (36, 46).

The Y chromosome of mammals is also interesting regarding heterochromatin distribution. In some mammals where the Y chromosome is a very tiny element (e.g., the marsupial *Didelphis albiventris* of the American tropics), it contains no heterochromatin. Presumably this chromosome contains only the functional genetic material for the Y chromosome. When the Y chromosome is slightly larger, e.g., over 2μ in length, it is invariably heterochromatic throughout its entire length. Without question, the functional portion is included in it, but this portion, being so small, is probably masked by the heterochromatic part. Thus the C-band technique is excellent for identifying the Y chromosome of mammals if uncertainty occurs in conventional preparations. For example, Kreizinger &

Shaw (47) laboriously employed the late-replication property to identify the Y of species in the deer mice, *Peromyscus*, and in some cases the data were still equivocal. With the C-band technique, Bradshaw & Hsu (48) were able to single out the Y chromosome in every metaphase by inspection.

In the bird complements, aside from the heterochromatic W chromosome in the female cells, the microchromosomes contain more heterochromatin than the macrochromosomes (49). In some species, e.g., the domestic fowl, the parakeet, etc., there is hardly any detectable heterochromatin in the macrochromosomes. In others, e.g., the sparrows, centromeric heterochromatin can be demonstrated in the macrochromosomes.

Without question, cytologists interested in the chromosomes of life forms other than mammals and birds will adopt the procedures to obtain information regarding the amount, the locations, and the behavior of heterochromatin in various systems. Indeed, plant cytologists have already begun this exploration. Using a modified C-band method, mainly replacing NaOH with saturated $Ba(OH)_2$, Vosa (50) showed that plant heterochromatin may be even more dramatic than animal heterochromatin (Fig. 5). In *Scilla sibirica*, it was found (51) that the distribution of heterochromatin among the 12 chromosomes is extremely polymorphic.

The Nature of the C-bands

Sufficient data are now available to correlate repeated DNA sequences (including satellite fractions) and constitutive heterochromatin (52). It is, therefore, tempting to interpret that highly repetitious DNA is responsible for C-band staining. Since the preparations are treated with acid, alkali, or elevated temperatures, cellular DNA is presumably denatured. The overnight incubation at 60°C in saline-citrate solution presumably renatures the DNA. Thus it appears, superficially at least, a reasonable proposition that highly repetitious DNA renatures under the prescribed conditions while low repetitious DNA and unique DNA do not, thereby resulting in the differential staining reaction.

However, the situation may not be so simple. Doubts still exist regarding the denaturation-renaturation scheme because DNA *in situ* may behave differently from DNA in solution in response to the various treatments. Some investigators (53, 54) attempted to use acridine orange fluorescence in combination with the treatments to obtain information regarding the strandedness of the DNA molecules in the chromosomes (single-stranded, red fluorescence; double-stranded, yellow-green fluorescence). Usually the chromosomes show yellow-green fluorescence, indicating a double-stranded nature.

When mouse cells are treated with acid, alkali, or high temperature, the chromosome fluorescence is red, but the centromeric regions (C-bands) quickly return to yellow-green fluorescence, suggesting that the DNA in these areas (repeated sequences) renatures rapidly.

Such explanation, though seemingly plausible, is probably oversimplified. Color changes with acridine orange fluorescence in relation to the DNA strandedness are not as reliable as some investigators would like to think.

However, no better hypothesis is available to account for the color changes that do take place. It is interesting to note that in *Microtus agrestis*, known C-bands will fluoresce strongly in QM after "denaturation-renaturation" treatments while the Q-bands will disappear (55). Another fluorochrome, a benzimidazole derivative, may prove to be useful because it shows Q-bands as well as C-bands (56).

An additional piece of evidence favoring the denaturation-renaturation scheme is found in the work of Mace et al (57) who prepared fluorescein-labeled antibody against single-stranded DNA originally obtained from the serum of patients with systemic lupus erythematosis. The antibody produced no fluorescence in control cytological preparations but gave excellent fluorescence when the preparations were "denatured". If the preparations were allowed to "denature" and to "renature" as in the C-band procedure, the C-band regions failed to show fluorescence.

Even though the observations on fluorescence can be accepted as evidence to indicate denaturation and renaturation of DNA in the chromosomes, a void still exists in explaining Giemsa staining. What does Giemsa stain—DNA, protein, or other components? Suppose Giemsa does react with DNA and, therefore, stains it. Why, then, does it stain heterochromatin more intensely following such simple treatments? Or does Giemsa stain the protein components of the chromosomes? This is possible, but how do we explain the heterochromatin reaction? Must the proteins responsible for the Giemsa stain associate with double-stranded DNA before it can take up the stain?

At least two groups of investigators (53, 58) found that the NaOH treatment followed by saline-citrate solution incubation induces a considerable amount of DNA loss from the fixed cells. Perhaps constitutive heterochromatin presents itself because it renatures quickly to prevent its being extracted.

IV. THE GIEMSA BANDS

Many cytologists have observed banded chromosomes which occasionally appeared in their preparations, but no one could reproduce these experimentally. Perhaps one of the earliest reports in which banded chromosomes were experimentally induced was the one of Stubblefield (59) who treated Chinese hamster cells with a prolonged exposure to colcemid. The mitotic cells, after a certain period of arrest, reverted to interphase without division, thereby giving rise to multinucleated, polyploid daughter cells. Some of these cells, when reentering mitosis, may show segmented chromosomes with characteristically condensed and stretched regions. Unfortunately, the procedure was not expanded or applied by others because this method does not yield a large number of cells with these desirable features.

The G-bands

In the middle of 1971 several papers appeared in the literature claiming successful induction of banded human chromosomes using Giemsa staining (60-63). All of these found their origin in the heterochromatin staining procedure

(39). Most human cytogenetics laboratories employ air-dried or flame-dried preparations of lymphocyte cultures (64) instead of squash preparations. When the C-band procedure or its modifications are applied to such preparations, a different set of crossbands appears along the arms of the chromosomes. These bands are now known as the G-bands.

These remarkable procedures were, within a few months, overshadowed by the trypsin procedure (65), a derivation of the pronase technique (66), which produces more detailed banding patterns within a matter of minutes. If conditions are standardized, the trypsin technique is highly reproducible. The cells in air-dried slides are treated with a diluted trypsin solution for not more than two or three minutes (depending upon the concentration), rinsed, and stained in Giemsa or similar stains. For flame-dried slides, both the concentration of trypsin and the treatment time should be increased (67).

For cytogeneticists, these G-band techniques are marvelous (Fig. 6). As in the case of Q-banding, practically every chromosome within a complement can be positively identified and rearrangements can be traced (68). Roughly, the G-bands correspond well with the Q-bands, i.e., the deeply stained Giemsa bands are equivalent to the bright fluorescent bands stained with quinacrine. However, disagreements do occur. For example, in Q-band preparations the human Y chromosome has a distinctly bright segment (dark C-band), but in G-band preparations this distinction is not clear. Likewise, C- and G-bandings do not

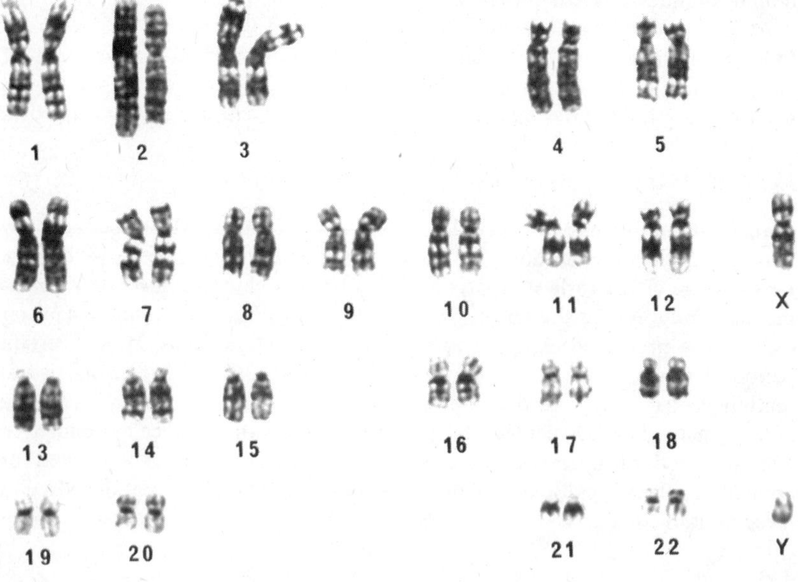

Figure 6. A human G-band karyotype. Trypsin treatment, Leishman stain. Courtesy Mrs. Marina Seabright.

always match either. In some cases, distinct C-bands are unstained in G-band preparations (e.g., human chromosome No. 9).

In practice, we believe that to gain maximum information the best way is to apply all three (Q, C, G) procedures. One can at least perform Q-banding and C-banding on the same cell (69), and it is even feasible to apply all three on the same preparations (70).

The Nature of the G-bands

No one knows exactly how G-bands are induced. At first it seemed easy to propose a mechanism similar to that proposed for the C-bands, viz., repetitious DNA denaturation and reassociation. However, the following data are against this type of interpretation: (*a*) some C-bands are negative in G-band staining, and (*b*) agents whose action do not involve DNA, e.g., trypsin, can produce G-bands.

A rather comprehensive study conducted by Kato & Moriwaki (71) showed that a variety of agents can induce G-bands. These investigators found that acids are ineffective to produce chromosome bands, but many salts are capable of producing bands if the pH is alkaline. Strong bases were also found to be potent agents in this regard, and some protein denaturants such as urea, guanidine-HCl, and several surface active compounds (e.g., sodium dodecyl sulfate) were also effective. Kato & Moriwaki considered that solubilization or extraction of some chromosomal proteins, probably of acidic nature, is the primary cause of the appearance of the banded structure in chromosome arms. More specific is the report of Utakoji (72) who induced G-bands by using the cupric sulfite reagent, which acts by breaking disulfide bonds of proteins.

Since we have no information regarding how different species of nuclear proteins are distributed along the chromosomes, it is indeed difficult to interpret the available cytochemical data with any degree of confidence. If the G-banding is purely the result of altering the proteins of the chromosomes, it would then imply that these species of chromosomal proteins are distributed along the chromosomes in clusters. An alternative speculation is that both DNA and proteins are involved in the cytochemical reactions. Several indirect evidences seem to favor this view: (*a*) The G-bands are comparable to the Q-bands. (*b*) A variety of compounds that are capable of inducing G-bands have chelating properties. Dev et al (73) maintained that trypsin at very low temperature or at low pH (away from its optimal proteolytic activity) still induces G-bands. Deaven (74) also obtained good G-band preparations with trypsin treatment at 0°C. Such data suggest that the action of trypsin in inducing G-bands may indeed be a chelation rather than protein digestion. (*c*) Urea is known to cleave proteins from DNA. (*d*) Adding agents such as Actinomycin D into cultures for a few hours before harvest can induce G-banding of metaphase chromosomes without any post-fixation treatment (75). Actinomycin D is not known to react with nuclear proteins, but it is known to bind DNA at the guanine moiety. Quite possibly the antibiotic molecules bind DNA at specific loci, thus interfering with protein binding during the G_2 phase, which is necessary for chromosome condensation in preparation for mitosis. If this is the likely explanation, the unstained and the lightly stained bands represent those responding to the treatments. Therefore, the G-banding induced by Actinomycin D prefixation treatment agrees with the notion that the dark G-bands (or bright Q-bands) represent AT-rich regions. (*e*) Recent (76) biochemical investigations suggest that arginine-rich histones preferentially bind GC-rich DNA. Is it possible that arginine-rich histone is involved in G-banding?

The R-bands

The R-bands (77) should be considered as a variation of the Giemsa bands. They show a pattern which is the "reverse" of the G-bands, i.e., lightly-stained G-bands become darkly stained in R-band preparations, and vice versa. The procedure is also a simple one: Air-dried slides are placed in phosphate buffer (pH 6.5) at 86-87°C for 10 min, rinsed in tap water and stained with Giemsa.

As far as utility is concerned, R-bands, in my opinion, are not as good as G-bands, particularly those produced by the trypsin procedure. In good G-band preparations, one can discern more detailed banding patterns than in the R-band preparations. However, R-banding may be an important tool in deciphering chromosome organization in the future, as it may represent the other side of the same coin. One must be able to explain R-banding as well as G-banding satisfactorily. Our preliminary experiments (75) showed that R-bands can be induced by adding certain chemical agents (e.g., ethidium bromide, nogalamycin) into the culture medium for a few hours and staining the chromosomes without post-fixation treatments; the opposite of the Actinomycin D-induced G-bands. Quite possibly such results suggest that the composition of DNA may be partially responsible for the appearance of banding, and R-banding, being the opposite of G-banding, should add equal weight to the explanation of longitudinal differentiation of chromosomes.

V. CONTRIBUTIONS OF BANDING TECHNIQUES TO BIOLOGY

Needless to say, human and mammalian cytogeneticists welcome these new tools with great enthusiasm. Although the methodologies have been in existence only for a short time, some significant contributions are already on hand.

Mammalian and Human Cytogenetics

Although the mammalian chromosome bands are not as detailed as those in the polytene chromosomes of *Drosophila* or *Chironomus*, they provide sufficient landmarks for recognizing chromosomes as well as for major zoning of each chromosome. They give cytogeneticists tools for positive identification of trisomics, monosomics, translocations, and at least major inversions and deletions.

Thus far, mouse genetics has received the most benefit from the banded chromosomes. The miserable mouse chromosomes suddenly became wonderful, and a nomenclature system was established (78). The tremendous amount of genetic information and the large number of available stocks, particularly

TABLE 1 Assignment of Chromosomes and Linkage Groups in *Mus musculus* (79)

Chromosome	Linkage Group	Chromosome	Linkage Group
1	XIII	11	?
2	V	12	?
3	?	13	XIV
4	VIII	14	III
5	XVII	15	?
6	XI	16	?
7	I	17	IX
8	XVIII	18	?
9	II	19	XII
10	X	X	XX

reciprocal translocations, could be immediately utilized to correlate genetic and cytological findings. The original banding pattern was made from QM-stained preparations, but G-band idiogram for the mouse chromosomes was also published recently (79).

The laboratory mouse has a diploid number of 40 and 20 linkage groups. Employing various stocks containing reciprocal translocations, investigators in the laboratory of O. J. Miller were able to identify the chromosomes involved in each translocation. The principle is simple. Suppose two different sets of translocations share the involvement with a common linkage group, say, linkage group V. Cytologically, one chromosome should show abnormality in both translocations. This chromosome, which happened to be Chromosome 2, must be equivalent to linkage group V. The investigators have been able to correlate most of the known linkage groups with specific chromosomes (Table 1) as well as to identify the approximate break points of these translocations (Fig. 7). For details of this area of research one should consult the review of Miller & Miller (80).

Human cytogenetics likewise has received a new lease on life from the banded chromosomes. One can identify with certainty not only total trisomics, but also partial trisomy resulting from reciprocal translocations of one of the parents. The analyses have just begun. We anticipate great advances within the next few years in clinical cytogenetics and in genetic counseling, based on more accurate information.

Somatic Cell Hybrids

It is well known that in interspecific somatic cell hybrids one set of component chromosomes will be progressively eliminated, leaving only a few chromosomes surviving with the complement of the other species. In selected mutants that must rely on the presence of certain genes to survive, the chromosome(s) carrying the particular gene(s) from the set that would ordinarily be lost must be selected

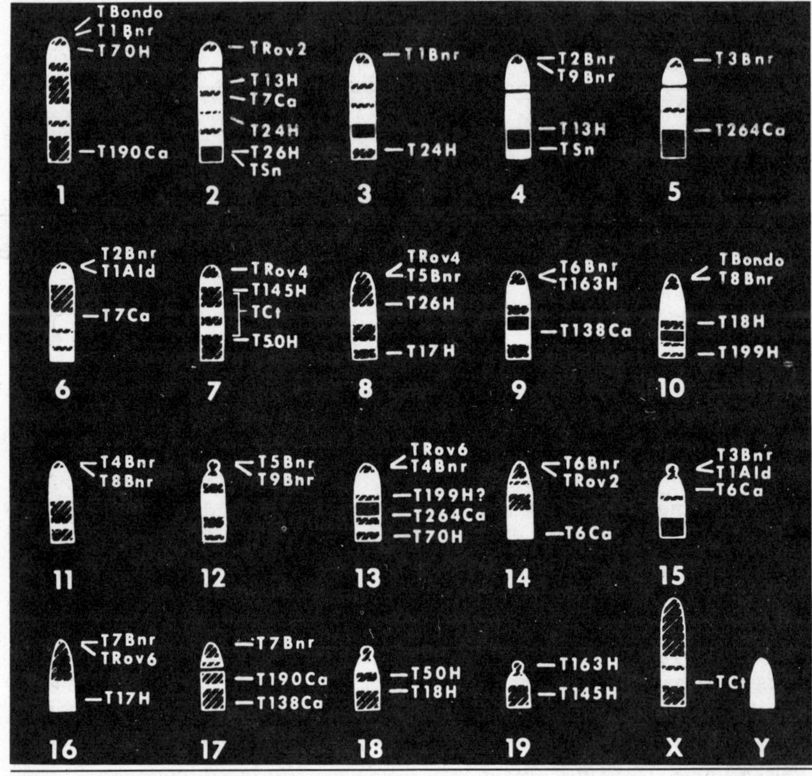

Figure 7. Q-band idiogram of the mouse, *Mus musculus,* with approximate breakage points of a number of reciprocal translocations. After Miller & Miller (79).

for and retained. Banding patterns (81, 82) enabled the identification of these chromosomes. More cases will undoubtedly accumulate in the future and will contribute significantly to the field of somatic cell genetics.

Chromosomes and Evolution

Obviously, being able to recognize individual chromosomes of any complement will greatly contribute to our understanding of phylogeny, population dynamics, and related subjects. Thus far, construction of banding idiograms has been mainly limited to single species (83-90), but a few attempts have been made to compare the chromosomes between related species. The best example is the identification of translocated (or fused) elements from the house mouse (*Mus musculus*) to the tabacco mouse (*Mus poschiavinus*), the latter having seven different sets of Robertsonian fusions (91). The banding patterns also help to verify inversions (92).

Probably the most striking report was the comparison between the karyotypes of man (Hominidae) and those of apes (Pongidae), particularly those of the chimpanzee (93). Figure 8 demonstrates that only one set of Robertsonian translocation and a few possible inversions occurred in the history of divergence between these two taxa belonging to two separate families. This report suggests that gene arrangements within chromosomes may be very conservative in the course of organic evolution. Such conservatism is also found in our own studies (94, 95) when we compare the chromosomes of various species of climbing rats, *Tylomys*, and also between two species of old world monkeys.

When considering evolutionary significance of karyotypic changes, one must not neglect the information regarding C-banding, because information obtained from Q-, G-, or R-banding without C-banding can be misleading. We have said that constitutive heterochromatin (C-bands) contains a large amount of highly repetitive DNA and is probably without structural genes in most, if not all, cases. Furthermore, as several investigators have postulated (96, 97), repetitive DNA may be formed by saltatory replication, i.e., a huge number of copies may be formed within a few generations. Thus an individual with many heterochromatic chromosome arms is genetically the same as another individual with no such arms. But C-bands do not always reveal themselves in G- or Q-band preparations. The most striking example is found in the cactus mouse, *Peromyscus*

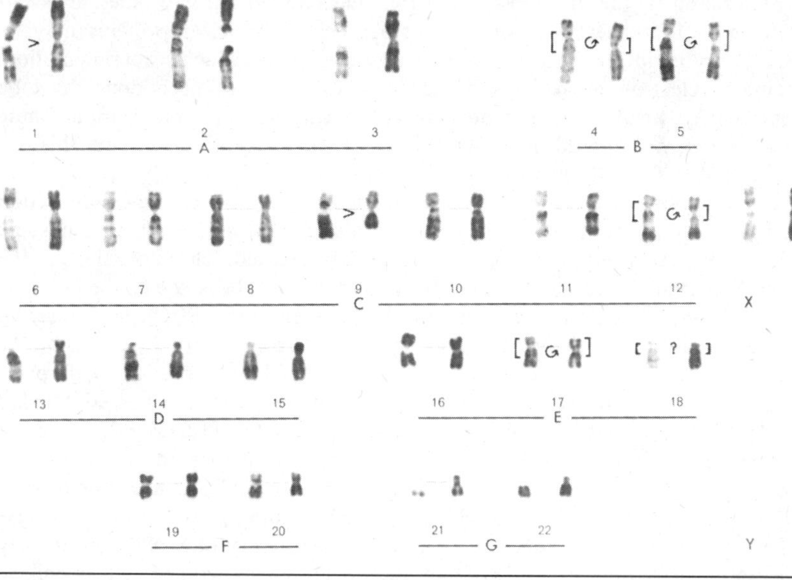

Figure 8. Comparison between human and chimpanzee karyotypes. R-bands. Nomenclature system follows human idiogram. For each pair, human chromosome at left, chimpanzee chromosome at right. Brackets, possible inversions; >, possible deletions or additions. Courtesy Dr. J. deGrouchy.

eremicus (unpublished data). This species possesses 48 biarmed chromosomes but each chromosome has one totally heterochromatic arm. In C-band preparations this distinction is very clear, but the same heterochromatic arms show various shades of staining (including nearly negative staining) in G-band preparations. Deleting the C-bands, the functional genetic material of *P. eremicus* (96 arms) is the same as that of *P. crinitus* (56 arms). Using G-banding or Q-banding alone, *eremicus* would appear to contain a good deal more genetic material than *crinitus*.

Chromosome Organization and Physiology

That chromosomes have stripes is, of course, not a new phenomenon—the polytene chromosomes are famous because of their stripes! What was surprising was that banded metaphase chromosomes are no more difficult to demonstrate than ordinary, unbanded chromosomes; and one wonders how cytologists of yesteryear (myself included) did not find such simple techniques earlier. As polytene chromosomes contributed significantly to our understanding of chromosome organization, the banded metaphase chromosomes should also prove to be extremely useful to this field in the future. We now have evidence, from *in situ* hybridization studies, to conclude that at least some repeated DNA sequences are clustered in the chromosomes. From the banded chromosomes, it is not outrageous to suspect that one or more species of nuclear proteins may be unevenly distributed along the chromosomes also. Much experimentation, including electron microscopy, is needed, but what has been done must be considered a breakthrough in our concept of chromosome organization. These new findings have posed more questions than they have answered, but this is a usual phenomenon in the progress of science.

Combinations of techniques for banding and for the premature condensation of chromosomes (98-100) should be useful also for the studies on chromosome physiology. As every student in biology has learned, chromosomes in the interphase nucleus were not discernible except in special cases, e.g., the polytene chromosomes. The premature chromosome condensation (PCC) procedure, an induction of unscheduled chromosome condensation in interphase by fusing the interphase cell with a mitotic cell, makes possible the observation of chromosomes at any stage of the cell cycle or cells out of the cycle (G_0). Indeed, Unakul et al (101) found that C- and G-bands can be induced in PCC as easily as those at metaphase. Thus it will be possible in the future to study a number of problems relating to chromosome physiology. For example, it should be profitable to find out the sequence of changes in the banding pattern from telophase when the chromosomes start to decondense to the entry of S phase when they are presumably fully extended. Once such a pattern is worked out, it will be possible to compare the chromosome banding patterns between cells in the cell cycle and cells not in the cell cycle (differentiating and differentiated cells). It will also be possible to estimate the rate of chromosome duplication during S phase, especially if suitable materials are used. Immediate effects of chemicals, radiation and viruses on interphase chromosomes can be visualized also with such techniques.

Literature Cited

1. Heitz, E. 1928. Des Heterochromatin der Moose. I. *Jahrb. Wiss. Bot.* 69:762-818
2. Heitz, E. 1933. Die Somatische Heteropyknose bei *Drosophila melanogaster* und ihre genetische Bedeutung. *Z. Zellforch.* 20:237–87
3. Heitz, E. 1934. Über α-und β-Heterochromatin sowie Konstanz und Bau Chromomeren bei *Drosophila. Biol. Zentralbl.* 54:588–609
4. Darlington, C. D., La Cour, L.F. 1940. Nucleic acid starvation of chromosomes in *Trillium. J. Genet.* 40:185–213
5. Callan, H. G. 1942. Heterochromatin in *Triton. Proc. Roy. Soc. London, B* 130:324–35
6. Wilson, G. B., Boothroyd, E. R. 1944. Temperature-induced differential contraction in the somatic chromosomes of *Trillium erectum* L. *Can. J. Res. C* 22:105–19
7. Kihlman, B. A. 1966. *Actions of chemicals on dividing cells.* Prentice-Hall, Englewood Cliffs, N. J.
8. Paris Conference. 1971. Standardization in Human Cytogenetics. 1972. *Cytogenetics* 11:317–62
9. Caspersson, T., Farber, S., Foley, G. E., Kudynowski, J., Modest, E. J., et al. 1968. Chemical differentiation along metaphase chromosomes. *Exp. Cell Res.* 49:219–22
10. Caspersson, T., Zech, L., Modest, E. J., Foley, G. E., Wagh, U., et al. 1969. Chemical differentiation with fluorescent alkylating agents in *Vicia faba* metaphase chromosomes. *Exp. Cell Res.* 58:128–40
11. Caspersson, T., Zech, L., Modest, E. J., Foley, G. E., Wagh, U., et al 1969. DNA-binding fluorochromes for the study of the organization of the metaphase nucleus. *Exp. Cell Res.* 58:141–52
12. Caspersson, T., Zech, L., Johansson, C. 1970. Differential binding of alkylating fluorochromes in human chromosomes. *Exp. Cell Res.* 60:315–19
13. Caspersson, T., Zech, L., Johansson, C. 1970. Analysis of the human metaphase chromosome set by aid of DNA-binding fluorescent agents. *Exp. Cell Res.* 62:490–92
14. Caspersson, T., Zech, L., Johansson, C., Modest, E. J. 1970. Identification of human chromosomes by DNA-binding fluorescing agents. *Chromosoma* 30:215–27
15. Pearson, P. L., Bobrow, M., Vosa, C. G. 1970. Technique for identifying Y chromosomes in human interphase nuclei. *Nature* 226:78–80
16. Caspersson, T., Lomakka, G., Zech, L. 1971. The 24 fluorescence patterns of the human metaphase chromosomes—distinguishing characters and variability. *Hereditas* 67:89–102
17. Ellison, J. R., Barr, H. J. 1972. Quinacrine fluorescence of specific chromosome regions: Late replication and high A:T content in *Samoaia leonensis. Chromosoma* 36:375–90
18. Pachmann, U., Rigler, R. 1972. Quantum yield of acridine interacting with DNA of defined base sequence. *Exp. Cell Res.* 72:602–08
19. Weisblum, B., DeHaseth, P. 1972. Quinacrine—a chromosome stain specific for deoxyadenylate-deoxythymidylate-rich regions in DNA. *Proc. Nat. Acad. Sci. USA* 69:629–32
20. DuPraw, E. J. 1968. *Cell and Molecular Biology*. Academic: N. Y.
21. Comings, D. E. 1973. The structure of human chromosomes. In *The Cell Nucleus*, ed H. Busch. New York: Academic. In press
22. Waring, M. J. 1965. Complex formation between ethidium bromide and nucleic acids. *J. Mol. Biol.* 13:269–82
23. Dev, V. G., Warburton, D., Miller, O. J., Miller, D. A., Erlanger, B. F., et al. 1972. Consistent pattern of binding of antiadenosine antibodies to human metaphase chromosomes. *Exp. Cell Res.* 74:288–93
24. Hannah, A. 1951. Localization and function of heterochromatin in *Drosophila melanogaster. Advan. Genet.* 4:87–127
25. Brown, S. W., Nelson-Rees, W. A. 1961. Radiation analysis of a Leganoid genetic system. *Genetics* 46:983–1007
26. Brown, S. W. 1966. Heterochromatin. *Science* 151:417–25
27. Lee, J. C., Yunis, J. J. 1971. A developmental study of constitutive heterochromatin in *Microtus agrestis. Chromosoma* 32:237–50
28. Taylor, J. H. 1960. Asynchronous duplication of chromosomes in cultured cells of Chinese hamster. *J. Biophys. Biochem. Cytol.* 7:455–64
29. Hsu, T. C. 1964. Mammalian chromosomes *in vitro* XVIII. DNA replication sequence in the Chinese hamster. *J. Cell Biol.* 23:53–62
30. Schmid, W. 1967. Heterochromatin in mammals. *Arch. Julius Klaus-Stift. Vererbungsforsch,* 42:1–60

31. Lyon, M. F. 1961. Gene action in the X-chromosome of the mouse (*Mus musculus L.*). *Nature* 190:372–73
32. Russell, L. B. 1961. Genetics of mammalian sex chromosomes. *Science* 133:1795–1803
33. Barr, M. L., Bertram, E. G. 1949. A morphological distinction between neurones of the male and female, and the behavior of the nucleolar satellite during accelerated nucleoprotein synthesis. *Nature* 163:676–77
34. Morishima, A., Grumbach, M. M., Taylor, J. H. 1962. Asynchronous duplication of human chromosomes and the origin of sex chromatin. *Proc. Nat. Acad. Sci. USA* 48:756–63
35. Galton, M., Holt, S. F. 1964. DNA replication patterns of the sex chromosomes in somatic cells of the Syrian hamster. *Cytogenetics* 3:97–111
36. Hsu, T. C., Arrighi, F. E. 1971. Distribution of constitutive heterochromatin in mammalian chromosomes. *Chromosoma* 34:243–53
37. Pardue, M. L., Gall, J. G. 1970. Chromosomal localization of mouse satellite DNA. *Science* 168:1356–58
38. Arrighi, F. E., Saunders, P. O., Saunders, G. F., Hsu, T. C. 1971. Distribution of repetitious DNA in human chromosomes. *Experientia* 27:964–66
39. Arrighi, F. E., Hsu, T. C. 1971. Localization of heterochromatin in human chromosomes. *Cytogenetics* 10:81–86
40. Hsu, T. C. 1971. Heterochromatin pattern in metaphase chromosomes of *Drosophila melanogaster*. *J. Hered.* 62:285–87
41. Craig-Holmes, A. P., Shaw, M. W. 1971. Polymorphism of human constitutive heterochromatin. *Science* 174:702–04
42. Craig-Holmes, A. P., Moore, F. B., Shaw, M. W. 1973. Polymorphism of human C-band heterochromatin. I. Frequency of variants. *Am. J. Human Genet.* 25:181–92
43. Voiculescu, I., Vogel, W., Wolf, U. 1972. Karyotyp und Heterochromatinmuster des rumänischen hamsters (*Mesocricetus newtoni*). *Chromosoma* 39:215–24
44. Popescu, N. C., DiPaolo, J. A. 1972. Identification of Syrian hamster chromosomes by acetic-saline-Giemsa (ASG) and trypsin techniques. *Cytogenetics* 11:500–07
45. Ohno, S., Beçak, W., Beçak, M. L. 1964. X-autosome ratio and the behavior pattern of individual X-chromosomes in placental mammals. *Chromosoma* 15:14–30
46. Arrighi, F. E., Hsu, T. C., Saunders, P. P., Saunders, G. F. 1970. Localization of repetitive DNA in the chromosomes of *Microtus agrestis* by means of *in situ* hybridization. *Chromosoma* 32:244–36
47. Kreizinger, J. D., Shaw, M. W. 1970. Chromosomes of *Peromyscus* (Rodentia, Cricetidae). II. The Y chromosomes of *Peromyscus maniculatus*. *Cytogenetics* 9:52–7
48. Bradshaw, W. N., Hsu, T. C. 1972. Chromosomes of *Peromyscus* (Rodentia, Cricetidae). III. Polymorphism in *Peromyscus maniculatus*. *Cytogenetics* 11:436–51
49. Stefos, K., Arrighi, F. E. 1971. Heterochromatic nature of W chromosome in birds. *Exp. Cell Res.* 68:228–31
50. Vosa, C. G., Marchi, P. 1972. Quinacrine fluorescence and Giemsa staining in plants. *Nature New Biol.* 237: 191–92
51. Vosa, C. G. Personal communication
52. Arrighi, F. E., Saunders, G. F. 1973. The relationship between repetitious DNA and constitutive heterochromatin with special reference to man. In "*Modern Aspects of Cytogenetics: Constitutive Heterochromatin in Man*," Symposia Medica Hoechst. In press
53. Comings, D. E., Avelino, E., Okada, T. A., Wyandt, H. E. 1973. The mechanism of C- and G-banding of chromosomes. *Exp. Cell Res.* 77:469–93
54. Lubs, H. A., McKenzie, W. H., Merrick, S. 1973. Comparative methodology and mechanism of banding. In "*Chromosome Identification*" Nobel Symposia Monograph 23: In press
55. de la Chapelle, A., Schröder, J., Selander, R. K. 1971. Repetitious DNA in mammalian chromosomes. *Hereditas* 69:149–53
56. Hilwig, I., Gropp, A. 1972. Staining of constitutive heterochromatin in mammalian chromosomes with a new fluorochrome. *Exp. Cell Res.* 75:122–26
57. Mace, M. L., Jr., Tevethia, S. S., Brinkley, B. R. 1972. Differential immunofluorescent labeling of chromosomes with antisera specific for single-strand DNA. *Exp. Cell Res.* 75:521–23
58. Hsu, T. C. 1973. Service of *in situ* nucleic acid hybridization to biology. In "*Chromosome Identification*," Nobel Symposia Monograph No. 23: In press
59. Stubblefield, E. 1966. Mammalian chromosomes *in vitro* XIX. Chromosomes of Don-C, a Chinese hamster fibroblast strain with a part of autosome 1b translocated to the Y chro-

mosome. *J. Nat. Cancer Inst.* 37:799–817
60. Sumner, A. T., Evans, H. J., Buckland, R. A. 1971. A new technique for distinguishing between human chromosomes. *Nature New Biol.* 232:31–32
61. Drets, M. E., Shaw, M. W. 1971. Specific banding patterns of human chromosomes. *Proc. Nat. Acad. Sci. USA* 68:2073–77
62. Patil, S. R., Merrick, S., Lubs, H. A. 1971. Identification of each human chromosome with a modified Giemsa stain. *Science* 173:821–22
63. Schnedl, W. 1971. Analysis of the human karyotype using a reassociation technique. *Chromosoma* 34:448–54
64. Moorhead, P. S., Nowell, P. C., Mellman, W. J., Battips, D. M., Hungerford, D. A. 1960. Chromosome preparations of leucocytes cultured from human peripheral blood. *Exp. Cell Res.* 20:613–16
65. Seabright, M. 1971. A rapid banding technique for human chromosomes. *Lancet* ii:971–72
66. Dutrillaux, B., de Grouchy, J., Finaz, C., Lejeune, J. 1971. Mise en évidence de la structure fine des chromosomes humains par digestion enzymatique (pronase en particulier). *C. R. Acad. Sci. Paris* 273:587–88
67. Wang, H. C., Fedoroff, S. 1972. Banding in human chromosomes treated with trypsin. *Nature New Biol.* 235:52–53
68. Seabright, M. 1972. The use of proteolytic enzymes for the mapping of structural rearrangements in the chromosomes of man. *Chromosoma* 36:204–10
69. Gagné, R., Tanguay, R., Laberge, C. 1971. Differential staining patterns of heterochromatin in man. *Nature New Biol.* 232:29–30
70. Lubs, H. A. Personal communication
71. Kato, H., Moriwaki, K. 1972. Factors involved in the production of banded structures in mammalian chromosomes. *Chromosoma* 38:105–20
72. Utakoji, T. 1973. Differential staining of human chromosomes treated with potassium permanganate and its blocking by organic mercurials. *Chromosomes Today* 4: In press
73. Dev. V. G., Warburton, D., Miller, O. J. 1972. Giemsa banding of chromosomes. *Lancet* i:1285
74. Deaven, L. L., Petersen, D. F. 1973. The chromosomes of CHO, an aneuploid Chinese hamster cell line: G-band, C-band, and autoradiographic analysis. *Chromosoma* 41:129–44
75. Hsu, T. C., Pathak, S., Shafer, D. A. 1973. Induction of chromosome crossbanding by treating cells with chemical agents before fixation. *Exp. Cell Res.* In press
76. Clark, R. J., Felsenfeld, G. 1972. Association of arginine-rich histones with GC-rich regions of DNA in chromatin. *Nature New Biol.* 240:226–29
77. Dutrillaux, B., Lejeune, J. 1971. Sur une nouvelle technique d'analyse du caryotype humain. *C. R. Acad. Sci., Paris* 272:2638–40
78. Committee on Standardized Genetic Nomenclature for Mice: Standard karyotype of the mouse, *Mus musculus*. 1972. *J. Hered.* 63:69–72
79. Wurster, D. H. 1972. Mouse chromosomes identified by trypsin-Giemsa (T-G) banding .*Cytogenetics* 11: 379–87
80. Miller, D. A., Miller, O. J. 1972. Chromosome mapping in the mouse. *Science* 178:949–55
81. Chen, T. R., Ruddle, F. H. 1971. Karyotype analysis utilizing differentially stained constitutive heterochromatin of human and murine chromosomes. *Chromosoma* 34:51–72
82. Miller, O. J., Allderdice, P. W., Miller, D. A., Breg, W. R., Migeon, B. R. 1971. Human thymidine kinase gene locus: Assignment to chromosome 17 in a hybrid of man and mouse cells. *Science* 173:244–45
83. Schnedl, W., Schnedl, M. 1972. Banding patterns in rat chromosomes (*Rattus norvegicus*). *Cytogenetics* 11:188–96
84. Schnedl, W. 1972. Giemsa banding, quinacrine fluorescence and DNA-replication in chromosomes of cattle (*Bos taurus*). *Chromosome* 38:319–28
85. Wolman, S. R., Phillips, T. F., Becker, F. F. 1972 Fluorescent banding patterns of rat chromosomes in normal cells and primary hepatocellular carcinomas. *Science* 175:1267–69
86. Unakul, W., Hsu, T. C. 1972. The C- and G-banding patterns of *Rattus norvegicus* chromosomes. *J. Nat. Cancer Inst.* 49:1425–31
87. Fredga, K. 1971. Idiogram and fluorescence pattern of the chromosomes of the Indian muntjac. *Hereditas* 68:332–37
88. Gustavsson, I., Hageltorn, M., Johansson, C., Zech, L. 1972. Identification of pig chromosomes by quinacrine mustard fluorescence technique. *Exp. Cell Res.* 70:471–74
89. Kato, H., Yosida, T. H. 1972. Banding patterns of Chinese hamster chromosomes revealed by new techniques.

Chromosoma 36:272–80
90. Cooper, J. E. K., Hsu, T. C. 1972. The C-band and G-band patterns of *Microtus agrestis* chromosomes. *Cytogenetics* 11:295–304
91. Zech, L., Evans, E. P., Ford, C. E., Gropp, A. 1972. Banding patterns in mitotic chromosomes of tobacco mouse. *Exp. Cell Res.* 70:263–68
92- Yosida, T. H., Sagai, T. 1972. Banding pattern analysis of polymorphic karyotypes in the black rat by a new differential staining technique. *Chromosoma* 37:387–94
93. Turleau, C., de Grouchy, J. 1972. Caryotypes de l'homme et du chimpanzé. Comparison de la topographie des bandes. *C. R. Acad. Sci., Paris* 274:2355–57
94. Pathak, S., Hsu, T. C., Helm, J. D., III. 1973. Chromosome homology in the climbing rats, genus *Tylomys* (Rodentia:Cricetidae). *Chromosoma* In press
95. Stock, A. D., Hsu, T. C. 1973. Evolutionary conservatism in arrangement of genetic material: A comparative analysis of chromosome banding between the Rhesus macaque (2n=42, 84 arms) and the African green monkey (2n=60, 120 arms). *Chromosoma.* In press
96. Walker, P. M. B. 1971. "Repetitive" DNA in higher organisms. *Progr. Biophys. Mol. Biol.* 23:67–101
97. Britten, R. J., Davidson, E. H. 1971. Repetitive and non-repetitive DNA sequences and a speculation on the origins of evolutionary novelty. *Quart. Rev. Biol.* 46:111–38
98. Johnson, R. T., Rao, P. N., Hughes, S. D. 1970. Mammalian cell fusion III. A HeLa cell inducer of premature chromosome condensation active in cells from a variety of animal species. *J. Cell Physiol.* 76:151–57
99. Johnson, R. T., Rao, P. N. 1971. Nucleo-cytoplasmic interactions in the achievement of nuclear synchrony in DNA synthesis and mitosis in multinucleate cells. *Biol. Rev.* 46:97–155
100. Stenman, S., Saksela, E. 1971. The relationship of Sendai virus-induced chromosome pulverization to cell cyclus in HeLa cells. *Hereditas* 69:1–14
101. Unakul, W., Hsu, T. C., Rao, P. N., Johnson, R. T. 1973. Giemsa banding in prematurely condensed chromosomes obtained by cell fusion. *Nature New Biol.* 242:106–07

DNA OF *DROSOPHILA* CHROMOSOMES 3052

Charles D. Laird

Department of Zoology, University of Washington, Seattle, Washington

1. Introduction

A remarkable combination of molecular, chemical, and cytological techniques is being applied to understanding the structure and function of *Drosophila* chromosomes. This has resulted in resolutions to several long-standing questions of chromosome and DNA organization. Because of the intensive activity in this area, this review has become, in part, a discussion of current experimental results. Many people have generously communicated unpublished and in-press information that provides a focus for future experimental analysis of *Drosophila* chromosomes.

The general direction of the review is from chromosomes of many different species of *Drosophila* to DNA of a few species and to DNA of chromomeres. It is my hope that the reader will be able to separate hypotheses from well-established facts, and in so doing be able to generate more appropriate models and experimental ideas.

2. The Genus Drosophila Includes Diverse Species

The genus *Drosophila* was established by Fallen in 1823. By 1952, Patterson & Stone (78) listed over 700 species, and, with the addition of the Hawaiian *Drosophila* species (25, 128) the number is now well over 1200. By contrast, there are 567 species and 97 genera in the rodent family Cricetidae (2). Considerable species diversity occurs within the genus *Drosophila*. Chromosome patterns show the extent of variation usually observed in broader taxons, although the fundamental number (91) of chromosome arms (5 rods and 1 dot) and gene linkage relationships (106) show the constancy expected for related organisms. The variation in the genus of haploid chromosome number (from 3–7) is comparable to that found in many mammalian orders (Fig. 1a). Comparable variation is observed in the mammalian orders Primates and Artiodactyla and occasionally within families (Fig. 1b, ref. 50, 51). Studies on DNA homologies have confirmed that *Drosophila* species exhibit considerable divergence in nucleotide sequences (34, 64, 90). Measurements have been extended to nonrepetitive DNA by solution renaturation (see Laird, McConaughy & McCarthy 66). These data are in qualitative agreement with those obtained with repetitive sequences (Table 1). Thermal stability measurements indicate that the genus *Drosophila* contains species whose DNA sequences are more different from each other than are those of members of different families (Bovidae and Suidae) in the order Artiodactyla (66). Extensive differences in density satellite DNAs (discussed below, see Rae 86) are also comparable to those observed in broader mammalian taxons (124). While this diversity is of taxonomic interest it is especially important to note that conclusions made for one species may be applied only very cautiously to another. For example genetic data from *D. melanogaster* and cytochemical results from *D. hydei* may often be at odds because of very important biological differences, rather than for reasons of experimental inconsistency. To compare data on these two species is equivalent to comparing data on the genomes of cow and pig, or of human and galago (59).

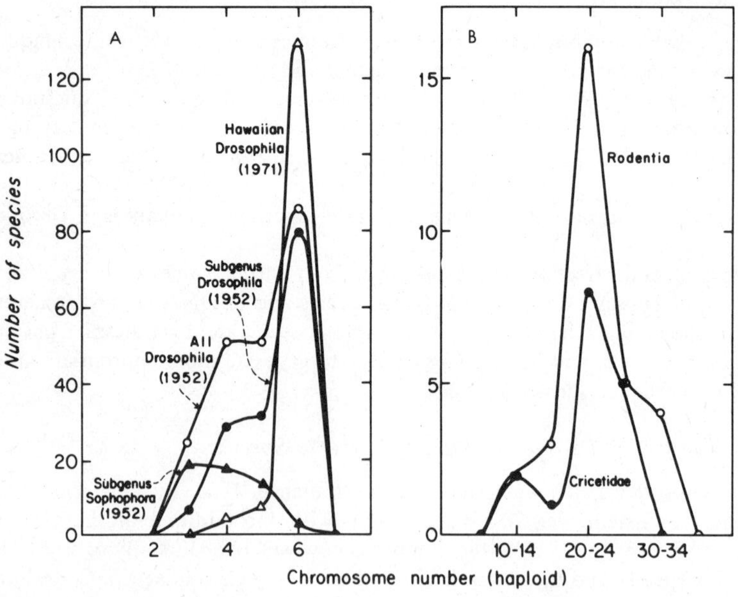

Figure 1. Chromosome number distributions in the genus *Drosophila* and in the order Rodentia. *Drosophila* data are taken from Patterson & Stone (78) and from Clayton (25). The subgenus *Drosophila* includes, for example, *D. virilis*, *D. hydei*, and *D. americana*; *Sophophora* includes *D. melanogaster* and *D. simulans*. For comparison, chromosome numbers of Rodentia, and the family Cricetidae (taken from refs. 50 and 51), indicate a twofold range, comparable to the variation observed in the genus *Drosophila*.

TABLE 1 Interspecific nucleic acid reactions

Species compared[4]	*competition*[1]			*direct binding*[2]			ΔT_m^3	
			H^3 cRNA	H^3 DNA filter (slow)	H^3 DNA filter (fast)	H^3 DNA solution (unique)	filter (fast)	solution (unique)
	relative percent							
melanogaster - simulans	81[c]	80[a]	34 - 60[b]	79[c]	70[c]	87[d]	3°[a]	3°[d]
simulans - melanogaster			36 - 40[b]					
melanogaster - virilis	43[c]			40[c]				
melanogaster - hydei						60[d]		11°[d]
hydei - melanogaster	30[c]							
melanogaster - mulleri						26[d]		11°[d]
melanogaster - busckii						18[d]		15°[d]
melanogaster - funebris	34[c]	25[a]	10[b]	24[c]	35[c]	17[d]	13°[a]	17°[d]
melanogaster - salmon						9[d]		19°[d]

a. Laird & McCarthy (64)
b. Robertson et al (90)
c. Entingh (34)
d. Laird & Dickson, unpublished results
1. Competition curves for filter reactions
2. Direct binding of labeled cRNA, or H^3 DNA, to filters, or in solution with H^3 -labeled unique DNA [see Laird et al (66)].
3. Thermal stabilities were determined by denaturing duplexes off filters or off hydroxyapatite columns.
4. Labeled nucleic acids were from the first species of the pair.

3. *Haploid Genome Sizes of Different Species Vary at Least Twofold*

The size of the haploid genome of *Drosophila melanogaster* has been estimated accurately by Feulgen staining and quantitative cytochemistry to be 11×10^{10} daltons, assuming that chicken erythrocyte nuclei, used as a cytophotometric standard, have a DNA content of 2.5×10^{12} daltons (87). Kinetic data of DNA renaturation give an estimate of 8.5×10^{10} daltons (62), which is in reasonable agreement with the Feulgen values. This reflects a relatively small genome only 1/20 that of humans and two to three times larger than that of slime mold, *Dictyostelium discoideum* (37). This small size is of practical use for nucleic acid hybridization or renaturation (reaction times are shorter). However, genome size is not invariant in the genus *Drosophila*. Renaturation kinetics suggest that *D. hydei* and *D. funebris* genomes are respectively 1.4 and 2.0 times larger than that of *D. melanogaster* (32, 62, 65). Quantitative cytochemistry also indicates differences in DNA content in sperm from different species (Table 2)(Rasch, pers. comm.). Thus, *D. virilis* has about twice as much DNA per sperm as does *D. melanogaster*.

A third estimate of genome size can be derived from physical measurements of the size of DNA molecules in different species. Kavenoff & Zimm (55) used viscoelastic retardation times of *Drosophila* cell lysates in an attempt to determine whether DNA molecules could be found which were long enough to account for all of the DNA in a unineme chromatid of the largest chromosome. Methods for these molecular weight calculations were verified with phage and bacterial DNAs (58). Based on (*a*) the percentage of the genome represented by the largest chromatid, and (*b*) on the size of the longest DNA molecule, their genome size estimates for *D. melanogaster*, *D. hydei*, *D. virilis*, and *D. americana* fit closely with the cytochemical data obtained by E. Rasch (Table 2).

TABLE 2 Haploid genome sizes (daltons) of some *Drosophila* species

Species	Feulgen	kinetic	Extrapolation from molecular weights of DNA molecules
D. melanogaster	11×10^{10}(a)	8.5×10^{10}(d)	11×10^{10}(g)
D. simulans		8.5×10^{10}(e)	
D. hydei	15×10^{10}(b)	14×10^{10}(f)	12×10^{10}(g)
	12×10^{10}(c)		
D. virilis	20.4×10^{10}(♂)(c)		24×10^{10}(g)
	23.5×10^{10}(♀)(c)		
D. americana			20×10^{10}(g)
D. funebris		17.0×10^{10}(e)	

(a) Rasch, Barr & Rasch, (87)
(b) Mulder et al (73)
(c) Rasch, personal communication
(d) Laird (62)
(e) Laird & McCarthy (65) (relative to *D. melanogaster*)
(f) Dickson, Boyd & Laird, (32)
(g) Kavenoff & Zimm (55)

The corollary of these molecular weight measurements (55) is that DNA of a single chromatid is one continuous molecule extending through the centromere. These high molecular weight molecules were unaffected by pronase, suggesting that protein linkers are not necessary for the structural continuity of chromosomal DNA. The conclusion that DNA is molecularly continuous through *Drosophila* centromeres was further strengthened by showing that a pericentric inversion of the longest chromosome of *D. melanogaster* (changing chromosome 3 from a metacentric to an acrocentric), did not increase the molecular weight of the longest DNA molecules. Higher molecular weights could be obtained, however, from *D. melanogaster* strain T (1;3) OR 38. In this case, chromosome 3 had been lengthened about 35% by an X to 3 translocation. An even greater increase in molecular weight was observed in comparing lysates of *D. americana* to those of *D. virilis*. The karyotypes of these species differ in that *D. americana* contains two large metacentric chromosomes that are homologous to four acrocentric chromosomes of *D. virilis*. Presumably, Robertsonian centric fusion (91) has generated chromosomes twice as large in *D. americana*, and these contain DNA molecules twice as long as those of *D. virilis*.

The variation in genome size among *Drosophila* species is of particular interest because of the evidence that chromomeres, corresponding to bands in polytene chromosomes, are units of function (6). Thus *D. hydei* is thought to have about 2000 bands in contrast to about 5000 for *D. melanogaster* (7, 13). Indications are that the *D. hydei* genome is about 10–40% larger than that of *D. melanogaster* (Table 2), so it seems likely that individual chromomeres in *D. hydei* are more than three times larger than those of *D. melanogaster* (73). Similarly, *Chironomus tentans* chromomeres contain about 60×10^6 daltons of DNA (compared with 13×10^6 daltons for *D. melanogaster*), as a consequence of having 1900 bands and a haploid genome of 12×10^{10} daltons (31). Do genes correspond to chromomeres in these other diptera, as they appear to in *D. melanogaster* (6)? It would indeed be remarkable to find that gene size and number could vary by over threefold in different diptera. However, a solid conclusion on this point must wait for (*a*) genetic data in other species, (*b*) confirmation of band numbers, (*c*) data on the fraction of the genome that replicates to form the bands in polytene chromosomes [values for *D. hydei* and *D. melanogaster* are comparable —about 75% (73, 92), although the fraction for *D. virilis* is probably much lower —about 50% (26)], and (*d*) a demonstration that DNA in bands of polytene chromosomes is an accurate representation of DNA sequences in diploid cells [such as has been found for *D. hydei* (32)].

4. *DNA Replication*

If it is true that *Drosophila* DNA molecules are of chromosomal length, as mentioned above (55), then the replication of these molecules must be via multiple replication points, as interkinesis (presumably the S period) during early embryonic cleavage divisions is only about 3 –4 minutes long (97). For example, *D. melanogaster* has a haploid DNA content of 11×10^{10} daltons, or about 40 times that of *Escherichia coli*. As 40 times more DNA must be replicated in one-fifth the time of *E. coli* DNA replication, it was plausible that at least 200 replication sites are distributed along the approximate 50,000 μm of DNA (81). This would result in replication bubbles (52) every 250 μm. Multiple bubbles are in fact observed, but they are even closer (between 1 and 10 μm) than expected if DNA replication proceeds at the same rate in *Drosophila melanogaster* as it does in *E. coli*. [Fig. 2, from Wolstenholme (132)]. The mean center-to-center distance reported by Wolstenholme (132) was about 4 μm. Hogness (49a) finds a similar distribution of DNA replication forks, and concludes that origins of replication are spaced at an average distance of 6000 nucleotide pairs (about 2 μm). These spacings of 2-4 μm may be compared with an average of 7 μm of DNA per chromomere in *D. melanogaster*.[1] It is reasonable to expect Hogness' and Wolstenholme's values of 2 –4 μm to be a slight underestimate of the size of embryonic DNA replication units, because longer adjacent units would be more difficult to find than would smaller adjacent units. It seems tenable, therefore, that during early embryonic cleavage divisions, chromomeres represent 1-4 units of DNA replication.

An intriguing problem exists concerning DNA synthesis during formation of polytene chromosomes (133). Several lines of evidence indicate that there is unequal replication of euchromatic and heterochromatic sequences (see Section 5 D, nuclear DNAs). If DNA molecules are initially continuous in *Drosophila*

---

[1] From Rudkin's data (92) it appears that only 75% of the *Drosophila melanogaster* genome replicates to form polytene chromosomes in salivary glands. Thus 50,000 μm DNA $\times$ 0.75 $\div$ 5000 chromomeres = 7.5 μm DNA/chromomere. The other 25% of the DNA, primarily in the centric heterochromatin, may also represent chromomeres. These would not be observed as bands in polytene chromosomes, however, and hence they should not be included in the calculation of DNA per chromomere.

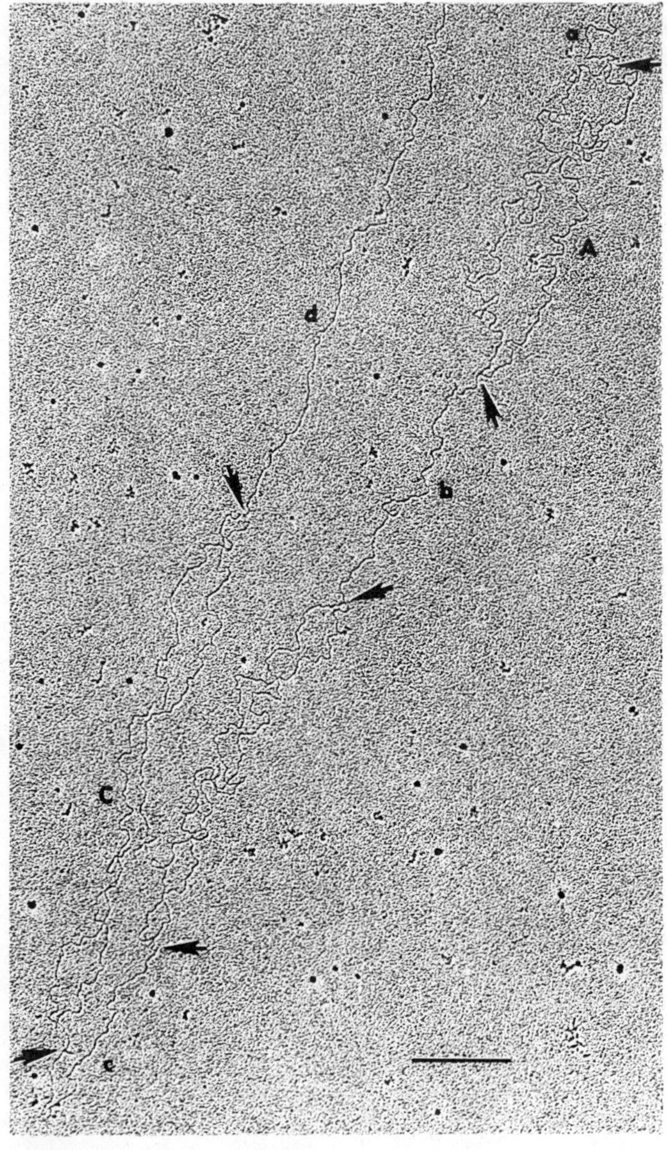

Figure 2. A replicating DNA molecule from a *D. melanogaster* embryo. This electron micrograph, kindly provided by D. Wolstenholme (132), shows a DNA molecule over 30 μm long which contains three replication bubbles of length 3.8 μm (A); 3.3 μm (B); and 3.6 μm (C). A molecule of this length represents the DNA content of about four average-sized chromomeres. Molecules such as these are present in cleavage-stage *D. melanogaster* embryos (0 –120 min. after fertilization) in which a complete round of DNA replication occurs in less than five minutes (97). The bar represents 0.5 μm.

unit chromatids (55), and if some heterochromatic sequences (density satellite) are not replicated and other sequences are under-replicated (ribosomal DNA)(49) then either polynucleotide discontinuities or multifork sites must be generated during polytenization (Fig. 3).

The control of such a replication pattern must be very precise, as deficiencies in ribosomal genes are apparently compensated for by over-replication in polyneme or polytene tissues (101, 102). These estimates were made by hybridizing ribosomal RNA to DNA of diploid and nondiploid tissue of *D. melanogaster* strains carrying from either one or two nucleolar organizer regions. This replication control may be the cause of "reversion" of bobbed mutants (putative rDNA deficiencies) (45, 88, 89) as well as the restoration to normal levels of rDNA in adult flies which have one (instead of two) nucleolar organizers (113, 114). Thus under this hypothesis the change would occur in the control of DNA replication, rather than in the primary number of rRNA genes in gametes (102). With bobbed mutant "reversions" (45, 88), the inheritance of a control locus for rDNA replication would be difficult to distinguish from an increase of rDNA in gametes except by rRNA hybridization with haploid, or possibly diploid, cell DNA.

5. *Base Composition, Buoyant Density Data*

A. TOTAL DNA AND SOME SPECIFIC GENES Kirby (56) and Mead (72) analyzed chemically the base composition of DNA from *D. melanogaster* and observed a guanine-cytosine (G+C) content of 39–43%, which is characteristic of metazoan DNAs. Base compositions for several specific loci have also been reported. tRNA and 5S RNA genes have G+C contents of 59% and 52% respectively (115). The ribosomal RNA cistrons (18S and 28S) have a relatively low G+C content of 41 and 38%, as compared with 65% for mammalian rDNA. The G+C content is 31% if one includes the nonconserved low G+C region in the transcription units of *D. melanogaster* (80). The ribosomal transcription unit in *D. melanogaster* is about 2.8×10^6 daltons, of which 72% is reported to represent the 18S and 28S components. This biochemical estimate has been confirmed by electron microscope visualization of ribosomal DNA transcription in *D. melanogaster* (43a). Spacer lengths of at least 0.5 μm were also observed between transcription units. These data compare with a ribosomal transcription unit of 4.1×10^6 daltons in mouse, of which 56% represents 28S and 18S sequences (80). Unfortunately, its base composition makes difficult the fractionation of rDNA in *Drosophila* as the usual methods used for other metazoan rDNA utilize large differences in G+C content between rDNA and total DNA (17). It is thus likely that transcription inhibitors that act preferentially on high G+C DNA, such as actinomycin D, will be relatively ineffective in *Drosophila* for specific inhibition of rRNA synthesis. This may be only a temporary disadvantage, however, because of the increasing suitability of electron microscope techniques for studies of the transcription of ribosomal, nonribosomal, and mitochondrial genes (24a, 43a, 62a, 72a).

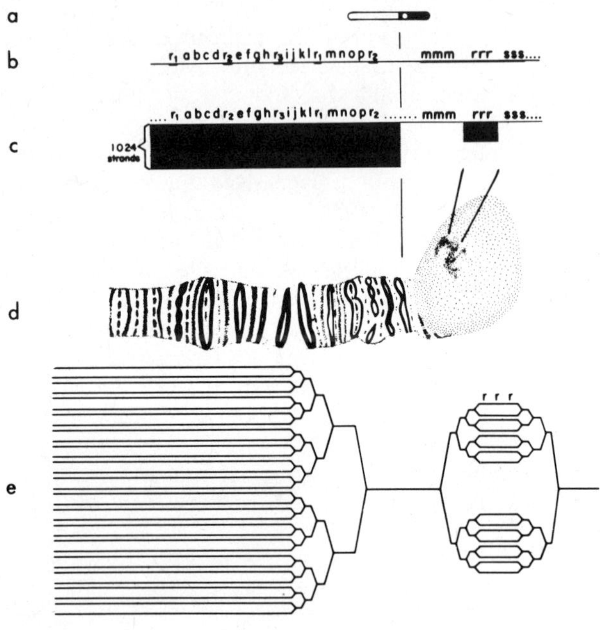

Figure 3. Schematic representation of the X chromosome of *D. hydei* : Relationship between a unit chromatid and its polytene chromosome. (a) X chromosome as a unit chromatid at metaphase. Centromere (circle), heterochromatin (shaded) and euchromatin (unshaded) are indicated. (b) A portion of extended DNA from an X chromosome. r_x's represent short repeat sequences; nonrepeat sequences are indicated by a,b,c, etc. Clustered repeats in the heterochromatin (mmm, rrr, sss) refer to, for example, density satellite DNAs (sss), or ribosomal DNA (rrr). (c) Replication of DNA (133) leads to polytene chromosomes, which appear to contain most of the unique sequences (32) and euchromatic repeat sequences (r_1, r_2, r_3). Heterochromatic repeats mmm and sss are not replicated (32, 38) whereas ribosomal DNA (rrr) is partially replicated (49, 102). (d) The base of the X chromosome is illustrated with an attached nucleolus (dotted) containing replicated ribosomal DNA (76). The bands presumably result from a specific packing of DNA into 100-250 Å fibers, and a folding of these fibers to give densely packed DNA (bands) and extended fibers (interbands) (33, 98). (e) A model of how a continuous DNA molecule (55), from a chromatid, could be regionally replicated as in the X polytene chromosome. Each line represents a DNA double-helix. Branch points are structurally analogous to DNA replication forks. One suggestion as to the origin of these structures (D. Nash, pers. communication) is that they result from a blockage of DNA replication. Chromomeres distal to heterochromatin (or from a central r, outward) may replicate DNA until such replication is blocked near the α heterochromatin, and on each side of the ribosomal DNA. Regions of putative multi-forks could result in peculiar banding patterns, such as observed in β heterochromatin, or at telomeres (not shown). Translocations of chromomeres to under-replicated telomere regions might be observed genetically, but not cytologically, ie. there would be a loss of bands in polytene chromosomes that is correlated with genetic translocation but not with genetic deficiency (Judd, pers. comm.).

B. CYTOPLASMIC DNAS The plethora of buoyant density satellites in *D. melanogaster* DNA preparations makes desirable a careful evaluation of possible cytoplasmic components. The only such component about which there is general (although not complete) agreement is mitochondrial DNA from *D. melanogaster*. Reports from two laboratories (20, 82) indicate that *D. melanogaster* mitochondrial DNA has the following properties: (*a*) It is circular, with a length of 5.5 μm, similar to that observed for other mitochondrial DNAs (111). [Wolstenholme (132) has measured the contour length of circular DNA, presumed to be mitochondrial DNA, in *D. melanogaster* embryos, using fd, ϕx 174 and λ DNAs as length standards. He obtained means of 6.1 μm and 5.0 μm for *D. melanogaster* and rat mitochondrial DNAs respectively.] The conclusion by Travaglini & Schultz (122) that *D. melanogaster* DNA does not contain circular mitochondrial DNA is thus not supported by other experiments. (*b*) It has an unusually low G+C content (22%) as judged from its buoyant density (1.681 gm/cm^3 relative to *E. coli* of 1.710 gm/cm^3.) Such a low G+C content for mitochondrial DNA has been reported only for *Tetrahymena pyriformis* (110, 111). (*c*) Mitochondrial DNA from *D. melanogaster* has an unusually heterogeneous base composition. A two (20) or three (82) step thermal melting transition indicates that about 72% of the mitochondrial genome has a mean G+C content of 24%, while the remainder has a mean G+C of 18%. Partial denaturation maps of *D. melanogaster* mitochondrial DNA could conceivably result in uniquely oriented molecules, a circumstance of some advantage in the mapping of mitochondrial genes, replication origins, and transcription sites.

Mitochondrial DNAs from other *Drosophila* species have been less well characterized. Travaglini et al (122) have isolated closed circular DNAs from the species *D. simulans*, *D. nasuta*, *D. virilis*, and *D. hydei*. These DNAs have buoyant densities ranging from 1.683 –1.692 gm/cm^3. Although the size and origin of these DNAs remain to be determined, it is tenable that this variation represents sequence changes in mitochondrial DNA.

Polan et al (82) and Bultmann & Laird (20) also report the presence of an unusual density satellite in preparations of DNA from unfertilized eggs and cell cultures of *D. melanogaster*. This component has a density of about 1.697 gm/cm^3, and comprises about 50% of the cytoplasmic DNA of unfertilized eggs (82), but only a minor fraction of cell culture DNA (20). It does not appear to be amplified ribosomal DNA, based on RNA/DNA hybridization assays (82). Its origin and function thus remain obscure.

C. DNAS OF UNCERTAIN ORIGIN The d(A-T)-rich density satellite (1.678 gm/cm^3) in *D. melanogaster*, usually less than 4% of the total DNA, has been reported to contain 83% alternating adenine and thymine (36). It is present in DNAs isolated from embryos and adults (9, 10) but not in DNA from salivary gland nuclei (quoted in 120). Blumenfeld & Forrest (9) claim, on the basis of DNAs isolated from strains with different numbers of Y chromosomes, that the d(A-T)-rich DNA in *D. melanogaster* is preferentially located on the Y chromosome. They prefer the hypothesis of nuclear origin for the d(A-T)-rich DNA because their isolation techniques included partial purification of nuclei. Since none of the published reports on the presence of this component in nonpolytene cells has demonstrated a high degree of nuclear purity, however, it is formally tenable that Y-linked genes determine the extent of replication of a cytoplasmic d(A-T)-rich component. This is a problem of some difficulty in *Drosophila* because of the heterogeneous cell populations, and because chromosome polytenization results in under-replication of a significant proportion of centric heterchromatin DNA (38, 73, 92). A resolution of this problem will probably await analysis of chromosomal DNA, either by chromosome isolation, or by *in situ* hybridization. With either a cytoplasmic, nuclear sap, or chromosomal origin, however, these components promise to be of considerable interest.

D. NUCLEAR DNAS *Main band.* The principal component of *D. melanogaster* DNA in CsCl gradients has a density of 1.702 gm/cm^3 corresponding to 42% G+C (65). Small differences in the densities of main band DNAs from different species have also been reported (65, 121). The range is about 0.004 gm/cm^3, indicating a likely G+C change of 4%. This variation is comparable to that found in the class Mammalia (3).

Nuclear satellites. Only recently has there been general agreement on whether some of the density satellites are nuclear in origin. It now seems likely that part or all of the component that bands at approximately 1.690 gm/cm^3 in *D. melanogaster* DNA is chromosomal. This conclusion comes from *in situ* hybridization studies with purified satellite DNA (38). Suggestions in previous reports that this satellite is cytoplasmic (12, 60, 65) were misleading. The occurrence of differential DNA replication in polytene cells, or differential DNA extraction (Botchan, pers. comm.), apparently led to variable proportions of this density satellite in DNA preparations. It is likely that there are two satellite sequences at this density position (38). One of these satellites could be cytoplasmic in origin, although it seems unlikely. An interesting and presently unresolved question concerns the release, after shearing, of 8% of main-band DNA into this density-satellite region (1.690 gm/cm^3) (60). If these results can be confirmed, they would indicate the interspersion of high and low G+C sequences either in the centromere region, as shown by *in situ* hybridization, or in euchromatin at a concentration too low to be detected with this technique. As these satellite sequences are reported to have a high degree of reiteration (17,000 copies of 260 nucleotides, and 450 copies of 800 nucleotides) (12), this suggestion of interspersion is of considerable importance in understanding repeat sequence function.

In addition to showing the chromosomal location (38, 85) of the 1.690 gm/cm^3 *D. melanogaster* satellite, *in situ* hybridization experiments indicate that some centromeric DNA is under-replicated during polytene chromosome formation. Satellite sequences hybridized about the same extent to centric heterochromatin DNA of polytene chromosomes as to DNA in nuclei of 2C or 4C cells (38). Renaturation kinetics and CsCl pycnography of polytene DNA from *D. hydei* also gave results consistent with the *in situ* hybridization experiments (32).

The similarity between the *D. melanogaster* and mouse satellite DNAs is striking. This is apparent from the proportion of the DNA involved (7–10%); the base composition (about 32% G+C); the centromeric localization (53, 75); the variability in related species (125); and the high degree of redundancy (12, 127). The availability of numerous *Drosophila* species makes likely the possibility of understanding the evolutionary steps in satellite DNA formation, [see also related work with rodent satellites (71, 109)]. Gall & Atherton (pers. comm.) and D. Parry (pers. comm.) have found considerable satellite variation in *D. virilis* and *D. nasuta* species groups. *D. hydei* satellites also seem to be species specific (48). It now seems likely that nucleotide sequencing will be feasible with *Drosophila* satellites. Gall (39a) has sequenced RNA transcribed from *D. virilis* density satellites. His results indicate that the three satellites have very simple and closely related nucleotide sequences. These satellites, with densities (in gm/cm^3) of 1.692 (I), 1.688 (II), and 1.671 (III), make up about 25%, 8%, and 8% of diploid nuclear DNA (38). The primary nucleotide sequence of satellite I is 5′ ACAAACT while that of II differs from I by the first C being T; III differs from I in that the last C is replaced by T. While these observations make the question of function only more enticing, they indicate the feasibility of experimental approaches more direct than renaturation kinetics and base composition analyses. For further consideration of repetitive *Drosophila* DNAs compared to those of other organisms, see Rae (86).

6. *Localization of Specific Nucleotide Sequences by* in situ *Hybridization*

The ability of *in situ* hybridization to couple chromosome morphology with biochemical fractionation techniques allows mapping of specific sequences (39). This technique must rank as one of the most fundamental innovations in cytogenetics. The density satellites and the 28S and 18S ribosomal genes have been localized in heterochromatin (12, 38, 48, 76). 5S RNA sequences have been localized to band 56EF (131); transfer RNA sequences seem to be scattered throughout the genome (104); histone genes are in region 39E–40A (77).

These localizations by *in situ* hybridization must be considered as representing the regions that contain the highest concentrations of sequences complementary to the labeled RNA or DNA. If, for example, each chromomere contained a short nucleotide sequence identical to the density satellite sequence, such euchromatic localization would be difficult to detect because of the regionally low concentration of hybridized nucleic acid.

The conclusion that histone genes are clustered in one site is of considerable significance for chromomere structure. The histone gene redundancy has been estimated to be 60 for each of 5 histone mRNAs (8a). This was estimated by measuring the rate of hybridization of sea urchin histone RNA to *Drosophila melanogaster* DNA filters. A total of 300 cistrons, each making a protein of average molecular weight of 10,000, would result in a gene cluster encompassing 9×10^4 nucleotides (300 nucleotides/cistron $\times$ 300 cistrons). An average *D.*

melanogaster chromomere contains, at its haploid level, only about 2×10^4 nucleotides (65). It is not yet known whether all five histones mRNAs are hybridizing to this site. It seems likely, however, from filter hybridization experiments with fractionated histone mRNA, that at least three of the five mRNAs are involved in the *in situ* hybridization (8a). It is also unclear how many polytene chromosome bands contain sequences complementary to histone mRNA. Chromosomes that are stretched in regions 39E–40A have labeled histone mRNA over two large bands plus the interband region. This interband may in fact contain several very faint bands (8a). To summarize the limits of these experiments, 200–300 cistrons of 3–5 different sequences are distributed among 1–5 bands. Some of these bands are likely to contain DNA that is predominantly responsible for specifying mRNA sequences. In addition, either a single band contains sequences for different mRNAs, or else adjacent bands are involved in specifying different histones. If the former is true, it would be an exception to the one band-one function conclusion (6,54). If the latter holds, this would represent a very interesting clustering of genes with related function.

Other data from sequence mapping may be considered similarly: the 5S RNA sequences are about 200-fold redundant (115). If each sequence is 120 nucleotides, then the 5S cluster must represent 24,000 nucleotides. This is close to an average size chromomere, but it does not account for possible spacer sequences which in *Xenopus laevis* are six times longer than the 5S RNA sequences (18). Thus an informationally saturated and poly-complementation chromomere seems likely here too. A similar conclusion can be made for the BR II band in *Chironomus* (30, 67).

One exception to the saturated-chromomere conclusion made for histone and 5S RNA cistrons might be the tRNA loci. Steffenson & Wimber (104) hybridized, *in situ*, H^3-labeled tRNA to *Drosophila melanogaster* polytene chromosomes. Based on extensive statistical analysis of autoradiographic data for the X and the second chromosomes they predicted that between 130 and 140 sites exist in the complete genome for tRNA genes. From saturation hybridization of *Drosophila melanogaster* DNA with tRNA, it is known that there are a total of 750 genes, or an average of 12 genes for each of the possible 62 different tRNAs (115), assuming two termination codons without tRNAs in *Drosophila* (22). Thus each tRNA site would be expected to represent, on the average, about 5 or 6 tRNA genes. The amount of information for tRNA at each site is about 5×80 nucleotides, or 2% of the average chromomere DNA. The possibility of additional loci existing in these chromomeres is of great interest, and deficiency mapping of these particular sites may be useful (69).

7. *Renaturation Kinetics of DNA*

Complete renaturation kinetics have been done with DNAs from *D. melanogaster*, *D. simulans*, *D. funebris*, and *D. hydei* (32, 47, 62, 65, 134). For these species, several generalizations can be made: about 75 –85% of the DNA renatures with kinetics expected for sequences present once per haploid cell. From this it may be concluded that *Drosophila* chromatids are unineme (32, 62).[2] If such sequences are nonrepetitive in a lateral sense, so must they be nonrepetitive in a tandem sense. Thus most of these sequences do not participate in a master-slave chromomere organization (21, 117). It is also likely that these nonrepetitive sequences are not functionally repetitive at the level of specifying identical proteins by virtue of having only 3rd position codon changes (28). This conclusion is based on the measurement of kinetics and stabilities of *D. melanogaster* DNA at different temperatures of renaturation. The dependence of the second order rate constant k_2 on temperature was similar for T4, *Bacillus subtilis*, and *Drosophila melanogaster* nonrepetitive DNA (63). Renaturation kinetics of *D. hydei* salivary gland DNA indicate that most unique sequences are replicated during polytenization (32). This is inferred from the observation that DNA from polytene chromosomes contains essentially the same sequence diversity as DNA from diploid cells. A more conclusive experiment utilized radioactively labeled unique-sequence DNA from nonpolytene cells. This was renatured with a large excess of DNA from polytene chromosomes (62a). The extent and kinetics of the reaction are consistent only with the conclusion that there is uniform replication of more than 80% of unique sequence DNA during formation of salivary gland polytene chromosomes in *D. hydei*. Cytogenetic data derived from analyses of meiosis should therefore be applicable to considerations of DNA in polytene chromosomes (6). Hennig (47) has presented data in conflict (by a factor of five) with this conclusion, but the absence of control renaturation curves in his experiments makes the alternative data (32, 62a) more reliable at this time.

Approximately 5–10% of nuclear DNA from diploid cells is very fast-renaturing ($C_o t_{1/2}$ less than 0.1 M sec).[3] This value may be as high as 40 –50%, as in the case of *D. virilis* (26, 38). These sequences are predominantly localized in centric heterochromatin, as was discussed in the section on *in situ* hybridization. The conclusion that some satellite sequences are fast-renaturing because of sequence reiteration is firmly established with rodent density satellite sequences (99, 100, 108, 127) and more recently, with satellites of *D. virilis* (Gall, 39a). Such nucleotide sequencing can be used to infer the sequence of the basic ancestral repeat, whereas renaturation kinetics are influenced substantially by divergence within the repeat. In cases where considerable sequence divergence has occurred, such as in the α satellite of guinea pig, these methods may give quite different results (100, 108).

---

[2] Obviously it is true that *Drosophila* chromatids or chromosomes can on occasion be bineme, multineme, polyneme, or polytene, when cell DNA contents are 8C or higher. For example, the 8C or 16C brain ganglion chromosomes must be bineme or tetraneme (40).

[3] Possible explanations as to why earlier experiments (65) did not detect this component are that there was preferential loss during DNA extraction of the 1.690 gm/cm^3 satellite, or that renaturation occurred too rapidly to measure at the high DNA concentrations used. The use of hypochromicity for detecting minor fractions is technically difficult. More recent estimates (12, 63) utilized fractionated DNA.

The absence of information on the functional aspects of such reiterated DNAs has fostered speculation ranging from that of a "housekeeping" (126) role to that of "bulk" centromere DNA (125). These suggestions were, respectively, that reiterated DNAs are involved in a sequence specific way with chromatin folding at the centromere region, or that reiterated DNAs of any nucleotide sequence were located at centromere regions in order to keep such sequences functionally innocuous. A definitive test of function would involve asking whether cells without such centromeric repetitive sequences have altered properties. An increase in nondisjunction of the X and Y chromosomes in *D. melanogaster* is observed with an X chromosome that is deficient for most of the centric heterochromatin (In[1]$sc^{4L}sc^{8R}$, ref 79a). More precise localization of repetitive sequences, and a more detailed analysis of nonrepetitive sequences in centric heterochromatin, will be useful in understanding the significance of this observation. It should be possible to construct other *D. melanogaster* chromosomes that are deficient for centromeric heterochromatin by use of the appropriate inversion and translocation stocks (69).

A striking reduction, to less than 1%, of tandemly organized repeat sequences has been observed in DNA from a very different type of insect than *Drosophila*—from an hemipterous insect with diffuse contromeres, *Oncopeltus fasciatus* (61). This observation suggests that spindle fibers may attach to chromosomes [at least polycentric chromosomes (27)] in the absence of long tandem repeat sequences. It is perhaps more consistent with the proposal that centromeres act as a storage location for tandem repeats and that the absence of discrete centromeres in *Oncopeltus* puts limits on the amount of such repeats that can be accommodated by this organism. The possible participation of these sequences in speciation or in the evolution of new genes has been discussed extensively (15, 71).

There is a strong suggestion that a second class of very rapidly renaturing DNA may be explicable in an alternative way. Single-stranded DNA with inverted repeats should self-nature rapidly. This has been observed for many eukaryotic DNAs using hydroxyapatite to fractionate base-paired molecules immediately after denaturation. *D. melanogaster* DNA seems to have about one inverted repeat every 80×10^6 daltons of DNA, or one every four chromomeres (129, 130). Such sequences in other organisms are about 500 –900 nucleotides long, although data have not yet been reported for *Drosophila*.

"Moderately" fast renaturing *Drosophila* DNA is observed between a $C_o t_{1/2}$ of 0.1 and 10 (32, 47, 62, 65, 134). Since euchromatic but not heterochromatic DNA is replicated during formation of polytene chromosomes, the presence of this component in salivary gland DNA suggests that these sequences are euchromatic in location. Circumstantial evidence from *in situ* hybridization suggests a similar conclusion (38, 85). The amount (5–15%)[4] and family size (an average of 40–100 for *D. melanogaster*) (65, 134) indicate that there is an average of about 3000 such

[4] This amount is probably species specific. For *D. hydei*, Dickson et al (32) report about 5%; Hennig (47) reports 5–10%. For *D. melanogaster*, this value could be as high as 15% (134), although the evidence is indirect.

families. Wu, Hurn & Bonner (134) have examined this DNA with electron microscope techniques. They report that repeats are about 150 nucleotides long, and are separated by nonrepetitive DNA of 750 nucleotides. Thus an average chromomere of 20,000 base pairs could have 20 –25 short repeat sequences.

8. *Formation of Rings from Linear DNA Molecules*

The suggestion that some genomes contain repeat sequences originated from genetic tests with bacteriophage. The hypothesis was advanced that some phage T4 heterozygotes could be rationalized in terms of the existence of a population of linear genomes whose ends represent cuttings of a headful of DNA from a genome polymer (94, 103, 105). As the headful unit was slightly larger than the sum of individual phage genes, terminal genes would be expected to be present twice. Which sequence was repeated would depend on where the terminal cuts were made.

An elegant test of this model was made by Thomas and colleagues in the closely-related phage, T2 (70, 119). Exonuclease treatment was used to hydrolyze several hundred nucleotides from the 3′ end of each strand, a process termed resection (118). Individual molecules were then observed to form circular structures after a short incubation under conditions of DNA renaturation, indicating that the single stranded ends that had been exposed by resection had complementary sequences. [For a review of concepts and data on terminal repetition and sequence permutation in phage DNA, see Thomas (116)]. From a consideration of the T2 experiments, it is apparent that molecular length, as provided by intact phage genomes, and resection distance are important in circle formation.

Thomas and his colleagues have applied these techniques to DNAs of eukaryotes as an approach toward understanding sequence organization (118). This transition to eukaryotic genomes requires fragmentation of eukaryotic DNA from chromosomal sized molecules of over 10^4 μm into pieces of 1-3 μm. This is necessary to obtain optimum ring formation. A surprisingly high percentage of circles was observed after exonuclease resection for fragments of DNA from several eukaryotes. The mouse satellite DNA, which has by kinetic estimates a reiteration length of 150 –350 nucleotides (108, 127), formed 20% circles before exonuclease III resection, and up to 70% circles after resection. This is not surprising, as sequence repetition has been observed by both direct sequencing and by renaturation kinetics. However *Drosophila hydei* salivary gland DNA, which shows only about 5% repeat sequences by renaturation kinetics (32, 47) formed about 6 –7.5% rings, and between 3 –5% other circular structures (lariats and polycircles) (Lee, pers. comm. and 68). At least half of these circular structures had high thermal stabilities, indicative of ring closure regions of greater than 50–100 nucleotides (8, 66, 117a). Indeed, up to 15% circular structures have been reported by Hennig (47), compared to 35% with DNA from *D. hydei* adults. Even 6% rings is, however, too high to reconcile with renaturation kinetics if the repeat sequences are distributed randomly. This immediately raises the possibility that repeat sequences are not located at random, but that they have instead a very regular distribution in *Drosophila* DNA. This possibility will be discussed further in the next section.

More recent data have revealed that the degree of resection and the fragment size are important. For *D. hydei* DNA, the highest frequency of rings was obtained with 1.5 μm fragments and resection of 500 –1000 nucleotides (68). Closure lengths with *D. hydei* rings were about 350 nucleotides (8) as measured by the distance between single-strand "whiskers" on double-stranded circles. Experiments with DNA from *D. virilis* adults also showed optimum ring formation with fragments between 1.5–2 μm. A marked drop (from 17%–4%) was observed as fragments decreased in size from 1.8–0.4 μm (5300–1200 nucleotides). Because ring formation with nonpolytene DNAs from *D. melanogaster* is dominated by density satellite sequences (Hogness, 49a), it seems likely that a large fraction of *D. virilis* rings are a result of cyclization of the satellite DNAs that contain the heptanucleotide tandem repeat sequences (39a). It is not clear, however, why the frequency of rings would decrease with fragments of 0.4 μm (about 1200 nucleotides) since these would contain over 150 copies of the heptanucleotide repeat. Because α heterochromatin, which contains the satellite DNAs, is under-replicated in the polytene chromosomes, it is possible to circumvent the satellite contribution to ring formation by using salivary gland chromosome DNA. Ony limited data on short fragments (<1.5 μm) of salivary gland DNA have been reported. It is clear however, that the efficiency of circle formation decreases drastically for fragments greater than 2 μm for salivary gland DNAs from *D. virilis*, *D. melanogaster*, and *D. hydei*. For the latter two DNAs, the decrease is considerably greater than for adult DNAs. This observation will be important in the model discussed below.

Another uncertainty that surrounds these circle formation experiments is whether the double-strand breaks that lead to fragments are introduced at random (118). To clarify this situation, an attempt was made to break DNA in a manner different from hydrodynamic shear, ie. by nuclease fragmentation (68). DNA from *D. virilis* adults gave comparable values of rings (17%) when fragments were formed by either endonuclease I or R·H, or with shear forces. This suggests that ring formation does not depend on unique break points. Data on salivary gland DNA fragmented by endonucleases are not yet available. While it might be argued that the distribution of endonuclease sensitive sites could correspond to the distribution of weak points (gaps, or single-strand breaks), this seems an unlikely possibility. The following models will hence be based on the assumption that fragments are formed by random breakage.

9. *Models of Sequence Arrangement in Drosophila DNA*

Data from renaturation kinetics (32, 65), circle formation (8, 68), and electron microscope visualization of repeats (134) are put together here in an attempt to rationalize these apparently contradictory data (Fig. 4). These models have been considered previously by Thomas et al (119a) with somewhat different emphasis.

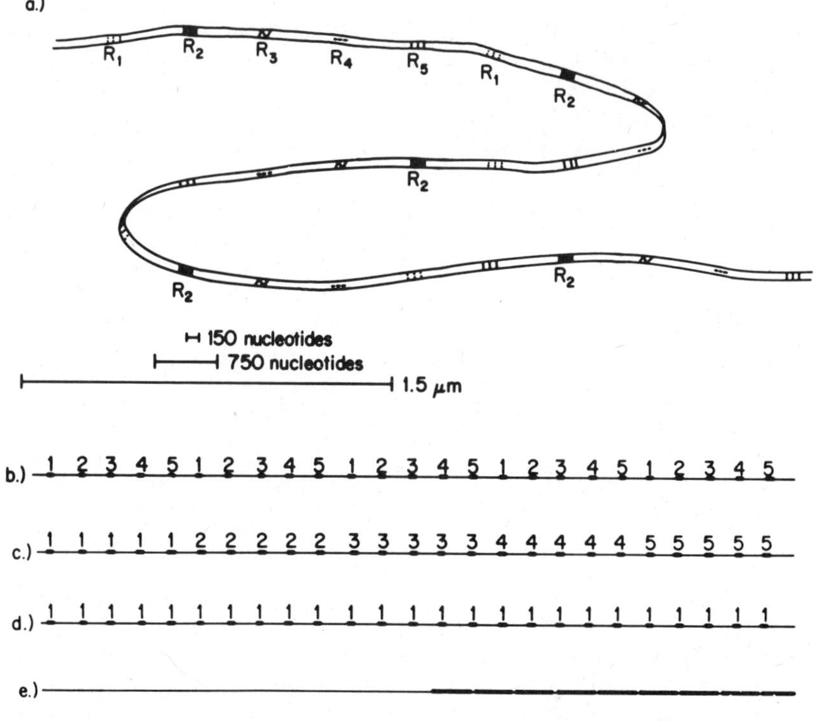

Figure 4. Some models of sequence organization in DNA of a *D. melanogaster* chromomere. Scales of 150 and 750 nucleotides, and 1.5 μm, refer to part (a). Each long line in (b) through (e) represents about 7 μm DNA. Heavy line segments in (b) through (d) indicate short repeats (150 nucleotides) of sequence indicated by numbers 1-5. Then regions indicate nonrepeat sequences.

Model (a) is an expanded form of (b), indicating the 1.5 μm hypothetical separation of the *same* repeat (R_2, for example), and the 750 nucleotide unique sequences that separate *different* repeats R_1, R_2 etc. Model (d) proposed by Bonner & Wu (11), shows all of the repeats in a chromomere as being similar. This does not seem to fit the data of Lee & Thomas (68) for polytene chromosome DNA. Model (e) is that favored by Bick, Huang & Thomas (8), based on the paucity of single-stranded regions (gaps) near ring closure sites. However, with this model it is difficult to arrange repeat sequences (heavy lines) such that ring frequencies decrease with small fragments and renaturation kinetic data are not contradicted.

These models are attempts to rationalize kinetic, electron microscope, and ring formation data. Critical tests in one species are necessary to demonstrate the validity of any or all of these possibilities. A unique and specific prediction of model (a), for example, is that ring frequencies with salivary gland DNA will have peaks at multiples of the 1.5 μm repeat separation length, i.e. 3.0, 4.5 μm, etc (62a). As we have no information on the function of these short repeat sequences, there is no logical *a priori* reason to discount any of the models. Indeed, different chromomeres may have quite different sequence organization.

Model (a) is an attempt to illustrate a regular alternation of short repeat sequences of 150 nucleotide pairs and longer unique sequences of 750 nucleotide pairs. Four different repeats and five different unique sequences separate the same repeat, for example R_2. Model (b) is equivalent to (a) except that the scale is reduced for comparison with models (c)–(e). Model (c) is a variant of (d) in the sense that (c) illutrates five similar repeats in succession, whereas (d) is essentially that proposed by Bonner & Wu (11), with 25 similar repeats separated by unique sequences. Model (e) is favored by Bick et al (8), with all repeats clustered into about half the DNA. Each line represents the length of DNA helix in an average *D. melanogaster* chromomere, ie. 7 μm. The important feature of model (a) is that two similar or identical repeat sequences have a spacing of 5000–6000 nucleotides. Nonrepetitive sequences (750 nucleotides) separate non-related repeats. Thus unrelated repeats occur every 900–1000 nucleotides, and related repeats every 5000–6000 nucleotides (Fig. 4). These features give the following crucial properties with respect to circle formation:

(*a*) A break induced at random in DNA would be an average of about 400 nucleotides from a repeat sequence. Half of all such fragments should have breaks *less than* 400 nucleotides from a repeat. Hence resection to the optimum 500 nucleotides should expose repeats on a substantial proportion of fragments.

(*b*) Fragments much greater or shorter than 5000–6000 nucleotides would not have the second related repeat within resection distance of the other end. Hence circle formation should be optimal with fragments a few hundred nucleotides longer than the spacing distance between related repeats.

The size dependence of ring formation would also distinguish between models (b) and (c)-(d). A sharp decrease in frequency of rings would be expected with 0.8 –1.2 μm fragments under model (b), but not under models (c) or (d). Only limited data on this point exist for salivary gland DNAs. However, the data (68) for *D. hydei* DNA suggests a peak in ring frequency for fragments of 1.2 μm. Thus model (b) would seem a reasonable fit to both kinetic and ring formation data in several *Drosophila* species.

Model (d) was favored by Bonner & Wu (11) because of a good fit between their calculations and the observed frequency of rings (68). However, their graph from Lee & Thomas (68) appears to be an incorrect plot of adult DNA data. The decrease in rings with larger fragments of salivary gland DNA is much more marked than for adult DNAs, and it therefore seems likely that model (d) is incorrect in detail, although the spacing suggestion is useful. A more quantitative treatment of these models occurs in ref. 62a.

(*c*) Another prediction of model (b)-(d) is that single strand regions would be expected to occur in some rings. These regions would represent nonrepeated sequences in single-strands exposed by resection, internal to the terminal repeat sequences. Few such rings were observed (8). However, while the paucity of these rings is inconsistent with models (b)-(d), such structures are difficult to resolve unambiguously with the electron microscope.

Model (e), favored by Bick, Huang & Thomas (8) has the advantage that it does not predict rings with internal single strand regions. But it is more difficult to reconcile with the sharp decrease in rings with small fragments. It is also less consistent with renaturation kinetics of salivary gland DNA (32) since it predicts 50% fast-renaturing DNA, in contrast to the observed 5%. Although Lee & Thomas (68) have suggested that renaturation kinetic data are not precise enough to distinguish between a genome with 95% single copy DNA and one with 50% single copy and 50% repeat sequences of fivefold redundancy, close examination of the data does not support their contention (62a, 63).

Thus I favor model (b) because it results in fewer inconsistencies with published data. It also leads to the interesting prediction that a small number of chromomeres share some repeat sequences. This prediction can be derived as follows: *D. melanogaster* chromomeres are represented by 7–8 μm DNA. Within this DNA, model (b) predicts the presence of 20-25 regions of nonrepeat sequences, separated by short (150 nucleotide) repeats. These repeats would represent 5 different sequences, each present 5 times (in one chromomere) at a spacing of about 1.5 μm. These repeats are likely to be of moderate repetition frequency, with an average family size of 50–100 (65, 85, 134). Thus 50–100 repeats distributed 5 per chromomere would result in 10–20 chromomeres sharing the same repeat. An individual chromomere could thereby be interconnected with 5 different groups of 10–20 chromomeres. It is not clear how uniform this proposed sequence distribution is, or whether the precise spacing is species or locus specific. It is reasonable to expect that in addition to the intermittent repeats, some euchromatic chromosome regions represent clustered repeats such as the loci for 5S RNA, tRNA, and histone mRNA. For the reasons discussed above, however, the fraction of the *Drosophila melanogaster* euchromatic genome so organized is probably small—perhaps less than 0.05. The possibility is also open that a substantial proportion of *D. melanogaster* DNA, perhaps 50%, represents long regions of nonrepeated sequences as has been reported for *Xenopus* (31a). Additional experimental data and approaches are necessary to resolve these questions.

The possible function of this type of repeat sequence distribution has been the subject of several recent papers. One suggestion is that repeat sequences are involved in the control of transcriptional activity (14, 42). Another proposal is that such sequences are useful in packing DNA into chromatin fibers and for the folding of these fibers into chromosomes (29, 79, 107). Recognition of sequences could be mediated by double-stranded DNA, single-stranded regions, or bifunctional proteins. These suggestions are not mutually exclusive, as the control of transcriptional activity must be mediated at some level by the state of chromatin condensation (44a, 71a).

The putative role of such repeat sequences in the processing of RNA is also testable. The presence of repeats in phage λ is a useful example. Here, the distribution of poly G binding sites is internal in transcription units. Champoux & Hogness (23) speculate that poly C regions act as divider sequences for transcription or post-transcriptional processing.

If chromosomal DNA is structurally continuous (55), as discussed in Section 3, it seems likely that the information for chromosome folding must lie in the DNA sequence itself rather than in, for example, protein linkers. Although a highly specific pattern of bands is apparent in polytene chromosomes (6), it is likely that only a minor fraction of chromomeres are functionally active in polytene tissues (4, 19). Thus chromomere structure, as reflected in polytene chromosome bands, is probably not designed to serve only the functional requirements of polytene cells. More plausible is the expectation that the band structure reflects an underlying organization of chromomeres in nonpolytene DNA in which case structure-determining nucleotide sequences would play an important role in interphase chromosomes as well. Band morphology mutants, for example, might manifest their major functional defects only in nonpolytene tissue, although a concomitant disruption in band structure would be observable at the level of the polytene chromosome.

Of related interest is the macronuclear DNA of the ciliated protozoan, *Stylonychia* (1, 83). Prescott and his colleagues have presented evidence that the metabolically functional macronucleus in *Stylonichia* contains only 5% of the sequences found in micronuclear DNA. Data are derived from CsCl pycnography and from renaturation kinetics of macronuclear DNA. This 5% occurs as small fragments (0.80 μm) with one end having an affinity for bacterial RNA polymerase. Cells whose micronuclei have been surgically removed continue to grow and divide, suggesting that the information remaining in the macronucleus is sufficient for cell growth and division. Meiosis does not occur, since this is a micronuclear function. Macronulei also do not have condensed chromosomes or organized separation of genetic material. This probably accounts for culture death, after about 500 divisions, due to gene deficiencies caused by random segregation of DNA (1).

These results with macronuclear DNA of *Stylonychia* suggest that such cells can survive with only 5% of their germ line DNA. It is not yet known whether RNA synthesis in these macronuclei occurs constitutively. However, *Stylonychia* offers an unusual opportunity to study several possible candidates for the function of the remaining sequences: meiosis,[5] chromosome structure, and the control of RNA synthesis.

10 Transcription

A. DIVERSITY OF RNA SEQUENCES It is becoming clear that RNA transcription is not limited to minor chromosomal regions. Turner & Laird (123) reported that an average of 30% of each chromomere (or the entirety of 30% of the chromomeres) is transcribed in embryos, larvae, pupae, and adults. Johnson (pers. comm.) observed similar values for a single cell type (*Drosophila* tissue culture cells). These quantitative estimates were obtained with nonrepeated DNA sequences, thus ensuring that specific RNA/DNA hybrids were measured (1, 6, 41, 43). These values assume that only one of the two complementary DNA strands is transcribed in vivo. In the experiments of Turner & Laird (123), analysis of the kinetics of hybridization indicate that most of these RNA sequences are rare, i.e., less than 10 per cell, on the average. These data indicate

[5] It is interesting that in *Drosophila*, meiotic mutants occur with relatively high frequency in natural populations [8%, compared to 40% for recessive lethals (93)], suggesting that a considerable fraction of the genome is involved in the meiotic process.

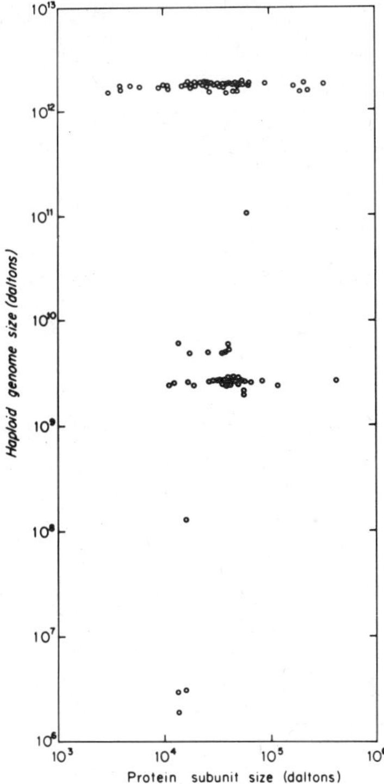

Figure 5. Size of protein subunits in organisms of different genome sizes. Data from Klotz & Darnall (57) and (5) are graphed with genome sizes of the appropriate organism (44). Major groups are vertebrates ($1 - 2 \times 10^{12}$ daltons); bacteria and fungi ($10^9 - 10^{10}$ daltons) and RNA virus ($10^6 - 10^7$ daltons).

that the structure of much of the DNA does not preclude RNA synthesis. It is not yet clear whether or not this high extent of transcription has functional consequences.

Other biochemical data are fragmentary for nonribosomal RNA in *Drosophila* (see Section 6). In part this stems from the difficulty in distinguishing nonribosomal from ribosomal RNA synthesis. In *Chironomus*, however, it appears from biochemical analysis of Balbiani Ring II RNA (30) that transcriptional units are very long—perhaps representing complete transcription of individual chromomeres. No information is yet available on spacings between transcription units, or whether adjacent units have the same polarity.

B. SIZE OF MESSENGER RNA It is likely *a priori* that large transcription units, if they are general, do not correspond to large protein subunits in *Drosophila*. The range in protein size is the same for viral and bacterial proteins as for more complex organisms (57 and Fig. 5). It is now known specifically that mRNA in *Drosophila melanogaster* has an average length of about 0.5 μm (24), which is sufficient to code for a protein of about 50,000 daltons. These measurements were made by isolating polyribosomes from embryos, and by estimating ribosome numbers and spacing with sucrose gradients and electron microscopy. Is one such mRNA derived from each chromomere? If so, what is the fate or function of the other RNA that is present (123)?

C. CHROMOMERES AND GENES The dilemma posed by chromomeres representing single functional genetic groups (6, 54) is that chromomeres are large (20,000 base pairs in *D. melanogaster*) compared to our concept of a gene in bacteria (about 1000 base pairs). If we accept the hypothesis that a one-to-one correlation does exist between chromomeres and genes, we are only at the beginning of the problem. We clearly need a more complete biochemical and genetic definition of such a gene.

It is plausible that the mutants that have led to presently recognized functional groups (6, 54) will include mutations in both control and structural genes. Suppose that a chromomere contains one structural locus, and ten control loci. If a mutation in any of the ten control loci prevents expression of the structural gene (cis-dominant) then this is equivalent to having a multi-functional control locus with a single structural gene. Under this model, all mutations could appear to be allelic in complementation tests (54, 95).

Alternatively, consider the model that each chromomere represents one control locus, with many structural genes. Here, mutations in different structural genes would be expected to complement each other. One rationalization for why this has not been found is that the probability of detecting control mutations may be considerably higher than for structural gene mutations. For example, if the respective relative probabilities are 20:1, then from the binomial distribution the probability that a set of 10 mutations represents two or more structural gene loci is 0.07. Only in this 7% of the complementation groups would complementation be observed among the 10 mutants. Such structural gene mutants would appear to be abnormal in that they would complement only with each other, but not with the rest of the alleles which would represent control loci. Thus one would expect only chromomeres with very large numbers of alleles to give rise to more than one complementing group.

Cytogenetic mapping suggests that some known loci correspond to restricted regions of chromomeres (6). This observation is consistent with either model, except that it would imply that mutations thus far observed at these loci are either all structural (one structural gene) or all control (one control gene). Analysis at the protein level will be necessary to resolve these two possibilities. Encouraging progress in isolating mutations in a specific protein is expected now that a double selection system is available (96).

11. Concluding Remarks

Several long-standing problems of chromosome structure are being resolved: it is likely that (*a*) chromatids in haploid and and diploid *Drosophila* cells are unineme; (*b*) DNA in each such chromatid is structurally continuous; (*c*) DNA in polytene chromosome bands is an accurate reflection of meiotic chromomere

DNA. The *in situ* hybridization technique has provided a powerful approach to specific cytogenetic mapping. Questions of gene location (histone, tRNA, and rRNA genes) and the nature of heterochromatic DNA are being answered. For example, the centric heterochromatin regions of the X chromosome and the major autosomes of *D. virilis* share the same (or very similar) heptanucleotide sequences.

Evidence is accumulating that *Drosophila* chromomeres represent units of function at many levels: Nucleotide sequence signals are thus likely to exist within chromomeres for DNA replication, chromomere (and chromosome) folding, and RNA synthesis and processing. Do control sequences also exist for chromosome pairing and recombination? Are there special integrative sequences (14, 42) for coordinating poly-chromomeric activities (4)?

On the biochemical side, regular spacings of repeat sequences (Fig. 4) indicate that there are isolable candidates for at least some of these putative functions. The use of electron microscope mapping of low G+C sequences (35) and of repeat sequences (134) may be useful in making these suggestions more precise. In bacteriophage, homopolymer binding has been useful in uncovering possible divider sequences within transcription units (23, 112). It seems plausible that similar techniques will be valuable in understanding chromomere organization in *Drosophila*.

However, the regular spacing of repeat sequences should not obscure the observation that the sequence diversity of DNA in *Drosophila* is enormous compared to bacteria. In *Drosophila melanogaster*, about 15% of euchromatic DNA is intermediate in repetition. The remaining 85% of euchromatic DNA renatures as unique sequences. Do these nonrepeated sequences represent information for proteins and/or control genes? Are they spacer sequences that have diverged rapidly after initial duplication because of an absence of selection? The extensive transcription (123), and the observations that unique sequences do not diverge at unusually high rates in mammals (59, 66, 74) or in *Drosophila* (Table 1) are perhaps more consistent with a functional role.

Thus, while the general features of gene and sequence arrangement in *Drosophila* are becoming clear, only vague clues exist as to the significance of this arrangement. It seems likely that these questions will be resolved by analysis of gene structure and action at specific loci.

Literature Cited

1. Ammermann, D. 1971. Morphology and development of the macronuclei of the ciliates *Stylonychia mytilus* and *Euplotes aediculatus*. *Chromosoma* 33:209–38
2. Anderson, S., Jones, J.K. 1967. *Recent Mammals of the World*. Ronald Press
3. Arrighi, F.E., Mandel, M. Bergendahl, J., Hsu, T.C. 1970. Buoyant densities of DNA of Mammals. *Biochem. Genet.* 4:367–76
4. Ashburner, M. 1970. Function and structure of polytene chromosomes during insect development. *Advan. Insect Phys.* 7:1–95
5. Atlas of protein sequence and structure, Vol. 4. 1969. M.O. Dayhoff, ed. *Nat. Biomed. Res. Found.*, Silver Spring, Maryland

6. Beermann, W. 1972. *Chromomeres and Genes*. In *Results and Problems - Cell Differentiation*, Vol. 4, Beerman, W., Reinert, J., H. Ursprung, ed. Springer-Verlag, Berlin
7. Berendes, H.D. 1963. The salivary gland chromosomes of *Drosophila hydei* Sturtevant. *Chromosoma* 14: 195–206
8. Bick, M.D., Huang, H.L., Thomas, C.A. Jr. 1973. The stability and fine structure of eukaryotic DNA rings in formamide. *J. Mol. Biol.* 77:75–84
8a. Birnstiel, M.L., Weinberg, E., Pardue, M.-L. 1973. 9S Polysomal RNA from Sea Urchin Embryos. In *Symposium on Molecular Cytogenetics*, B. Hamkalo, J. Papaconstantinou, editors. Plenum
9. Blumenfeld, M., Forrest, H.S. 1971. Is *Drosophila* dAT on the Y chromosome? *Proc. Nat. Acad. Sci. USA* 68:3145–49
10. Blumenfeld, M., Forrest H.S. 1972. Differential under-replication of satellite DNAs during *Drosophila* development. *Nature New Biol.* 239:170–72
11. Bonner, J., Wu, J.-R. 1973. A proposal for the structure of the Drosophila genome. *Proc. Nat. Acad. Sci. USA* 70:535–37
12. Botchan, M., Kram, R., Schmid, C.W., Hearst, J.E. 1971. Isolation and chromosomal localization of highly repeated DNA sequences in *Drosophila melanogaster. Proc. Nat. Acad. Sci. USA* 68:1125–29
13. Bridges, C.B. 1935. Salivary chromosome maps. *J. Hered.* 26:60–64
14. Britten, R.J., Davidson, E.H. 1969. Gene regulation for higher cells: A theory. *Science* 165:349–57
15. Britten, R.J., Kohne, D.E. 1968. Repeated sequences in DNA. *Science* 161:529–40
16. Brown, I.R., Church, R.B. 1971. RNA transcription from nonrepetitive DNA in the mouse. *Biochem. Biophys. Res. Comm.* 42:850–56
17. Brown, D.D., Dawid, I.B. 1968. Specific gene amplification in oocytes. *Science* 160:272–80
18. Brown, D.D., Wensink, P.C., Jordan, E. 1971. Purification and some characteristics of 5S DNA from *Xenopus laevis. Proc. Nat. Acad. Sci. USA* 68:3175–79
19. Bultmann, H., Clever, U. 1969. Chromosomal control of foot pad development in *Sarcophaga bullata. Chromosoma* 28:120–35
20. Bultmann, H., Laird, C.D. 1973. Mitochondrial DNA from *Drosophila melanogaster. Biochim. Biophys. Acta.* 299:196–209
21. Callan, H.G. 1967. The organization of genetic units in chromosomes. *J. Cell Sci.* 2:1–7
22. Caskey, C.T. 1970. The universal RNA genetic code. *Quart. Rev. Biophys.* 3:295–326
23. Champoux, J.J., Hogness, D.S. 1972. The topography of lambda DNA; polyriboguanylic acid binding sites and base composition. *J. Mol. Biol.* 71:383–405
24. Chooi, W.Y., Laird, C.D. Size of messenger RNA in *Drosophila melanogaster*. In preparation
24a. Chooi, W.Y., Laird, C.D. 1973. Transcription and translation in mitochondria of *Drosophila melanogaster*. In preparation
25. Clayton, Frances E. 1971. Additional Karyotypes of Hawaiian Drosophilidae In *Studies in Genetics*, VI, 171–81, M.R. Wheeler, ed., University of Texas publication 7103, Austin, Texas
26. Cohen, E. 1973. *Biochemical characterization and cytological localization of repetitive DNAs in Drosophila virilis*. Ph.D. thesis, Yale Univ.
27. Comings, D.E., Okada, T.A. 1972. Holocentric chromosomes in *Oncopeltus*: kinetochore plates are present in mitosis but absent in meiosis. *Chromosoma* 37:177–92
28. Crick, F.H.C. 1966. Codon-anticodon pairing: The wobble hypothesis. *J. Mol. Biol.* 19:548–55
29. Crick, F. 1971. General model for the chromosomes of higher organisms. *Nature* 234:25–27
30. Daneholt, B. 1972. Giant RNA transcript in a Balbiani ring. *Nature New Biol.* 240:229–32
31. Daneholt, B., Edstrom, J.-E. 1967. The content of deoxyribonucleic acid in individual polytene chromosomes of *Chironomus tentans. Cytogenetics* 6:350–56
31a. Davidson, E. 1973. Sequence organization of the eukaryotic genome and sequence representation in messenger and nuclear RNA. In *Symposium on Molecular Cytogenetics*, B. Hamkalo, J. Papaconstantinou, editors. Plenum
32. Dickson, E., Boyd, J., Laird, C.D. 1971. Sequence diversity of the polytene chromosome DNA from *Drosophila hydei. J. Mol. Biol.* 61:615–27
33. DuPraw, E.J., Rae, P.M.M. 1966. Polytene chromosome structure in relation to the "folded fibre" concept. *Nature* 212:598–600
34. Entingh, T.D. 1970. DNA hybridiza-

tion in the genus *Drosophila*. *Genetics* 66:55–68

35. Evenson, D.P., Mego, W.A., Taylor, J.H. 1972. Subunits of chromosomal DNA. I. Electron microscopic analysis of partially denatured DNA. *Chromosoma* 39:225–35
36. Fansler, B.S., Travaglini, E.C., Loeb, L.A., Schultz, J. 1970. Structure of *Drosophila melanogaster* dAT replicated in an *in vitro* system. *Biochem. Biophys. Res. Commun.* 40:1266–72
37. Firtel, R.A., Bonner, J. 1972. Characterization of the genome of the cellular slime mold *Dictyostelium discoideum*. *J. Mol. Biol.* 66:339–61
38. Gall, J.G., Cohen, E.H., Polan, M.L. 1971. Repetitive DNA sequences in *Drosophila*. *Chromosoma* 33:319–44
39. Gall, J.G., Pardue, M.-L. 1971. Nucleic acid hybridization in cytological preparations. In *Methods in Enzymology*, Vol. 21, p. 470–80 L. Grossman, K. Moldave, eds. Academic: N.Y.

39a. Gall, J.G. 1973. Repetitive DNA in Drosophila In *Symposium on Molecular Cytology*. B. Hamkalo, J. Papaconstantinou, editors, Plenum

40. Gay, H., Das, C.C., Forward, K., Kaufmann, B.P. 1970. DNA content of mitotically-active condensed chromosomes of *Drosophila melanogaster*. *Chromosoma* 32:213–23
41. Gelderman, A.H., Raku, A.V., Britten, R.J. 1971. Transcription of nonrepeated DNA in neonatal and fetal mice. *Proc. Nat. Acad. Sci. USA*. 68:172–76
42. Georgiev, G.P. 1969. On the structural organization of operon and the regulation of RNA synthesis in animal cells. *J. Theoret. Biol.* 25:473–90
43. Hahn, W., Laird, C.D. 1971. Transcription of nonrepeated DNA in mouse brain. *Science* 173:158–61

43a. Hamkalo, B.A., Miller, O.L., Jr. 1973. Ultrastructural Aspects of Genetic Activity. In *Symposium on Molecular Cytogenetics*, B. Hamkalo, J. Papaconstantinou, editors. Plenum

44. Handbook of Biochemistry, 2nd edition. 1970. H.A. Sober, ed. Chem. Rubber Co., Cleveland, Ohio

44a. Harris, H. 1970. *Nucleus and Cytoplasm*, 2nd edition. Oxford University Press

45. Henderson, A., Ritossa, F. 1970. On the inheritance of rDNA of magnified bobbed loci in *D. melanogaster*. *Genetics* 66:463–73
46. Hennig, W. 1972. Highly repetitive DNA sequences in the genome of *Drosophila hydei*. I. Preferential localization in the X chromosome heterochromatin. *J. Mol. Biol.* 71:407–17
47. Hennig, W. 1972. Highly repetitive DNA sequences in the genome of *Drosophila hydei*. II. Occurrence in polytene tissue. *J. Mol. Biol.* 71:419–31
48. Hennig, W., Hennig, I., Stein, H. 170. Repeated sequences in the DNA of *Drosophila* and their localization in giant chromosomes. *Chromosoma* 32:31–63
49. Hennig, W., Meer, B. 1971. Reduced polyteny of ribosomal RNA cistrons in giant chromosomes of *Drosophila hydei*. *Nature New Biol.* 233:70–72

49a. Hogness, D. 1973. The arrangement and replication of information in *Drosophila* chromosomes. In *Symposium on Molecular Cytogenetics*, B. Hamkalo, J. Papaconstantinou, editors. Plenum

50. Hsu, T.C., Benirschke, K. 1967. *An Atlas of Mammalian Chromosomes*. Vol. 1, Springer-Verlag: New York
51. Hsu, T.C., Benirschke, K. 1968. *An Atlas of Mammalian Chromosomes*. Vol. 2, Springer-Verlag, New York
52. Huberman, J.A., Riggs, A.D. 1968. On the mechanism of DNA replication in mammaliam chromosomes. *J. Mol. Biol.* 32:327–41
53. Jones, K. 1970. Chromosomal and nuclear localization of mouse satellite DNA in individual cells. *Nature*. 225:912–15
54. Judd, B.H., Shen, M.W., Kaufman, T.C. 1972. The anatomy and function of a segment of the X chromosome of *Drosophila melanogaster*. *Genetics* 71: 139–56
55. Kavenoff, R., Zimm, B.H. 1973. Chromosome sized DNA molecules from *Drosophila*. *Chromosoma* 41: 1–27
56. Kirby, K.S. 1962. Deoxyribonucleic acids. IV Preparation of deoxyribonucleic acid from *Drosophila* eggs. *Biochim. Biophys. Acta*. 55:382–84
57. Klotz, I.M., Darnall, D.W. 1969. Protein subunits: a table. 2nd ed. *Science* 166:126–28
58. Klotz, L.C., Zimm, B.H. 1972. Size of DNA determined by Viscoelastic measurements: Results on Bacteriophages, *Bacillus subtilis*, and Escherichia coli. *J. Mol. Biol.* 72:779—800
59. Kohne, D.E., 1970. Evolution of higher-organism DNA. *Quart. Rev. Biophys.* 3:327–75
60. Kram, R., Botchan, M., Hearst, J.E. 1972. Arrangement of the highly reiterated DNA sequences in the centric

heterochromatin of *Drosophila melanogaster*. Evidence for interspersed spacer DNA. *J. Mol. Biol.* 64:103–17

61. Lagowski, J., Yu, W. M.Y. Forrest, H., Laird, C.D. 1973 Dispersity of repeat sequences in *Oncopeltus fasciatus*, an organism with diffuse centromeres. *Chromosoma*. In press

62. Laird, C.D. 1971. Chromatid structure: relationship between DNA content and nucleotide sequence diversity. *Chromosoma* 32:378–406

62a. Laird, C.D., Chooi, W.Y., Cohen, E., Dickson, E., Hutchison, N., Turner, S. 1973. Organization of *Drosophila* DNA. *Cold Spring Harbor Symp. Quant. Biol.* 38

63. Laird, C.D., Dickson, E. 1973. Renaturation kinetics of fractionated *Drosophila melanogaster* DNA. *J. Mol. Biol.*

64. Laird, C.D., McCarthy, B.J. 1968. Magnitude of interspecific nucleotide sequence variability in *Drosophila. Genetics* 60:303–22

65. Laird, C.D., McCarthy, B.J. 1969. Molecular characterization of the *Drosophila* genome. *Genetics 63: 865–82*

66. Laird, C.D., McConaughy, B.L., McCarthy, B.J. 1969. Rate of fixation of nucleotide substitutions in evolution. *Nature* 224:149–54

67. Lambert, B. 1972. Repeated DNA sequences in a Balbiani ring. *J. Mol. Biol.* 72:65–75

68. Lee, C.S., Thomas, Jr. C.A. 1973. Formation of Rings from *Drosophila* DNA fragments. *J. Mol. Biol.* 77:25–42

69. Lindsley, D.L., Sandler, L., Baker, B.S., Carpenter, A.T.C., Denell, R.E., Hall, J.C., Jacobs, P.A., Miklos, G.L.G., Davis, B.K., Gethmann, R.C., Hardy, R.W., Hessler, A., Miller, S.M., Nozawa, H., Parry, D.M., Gould-Somera, M. 1972. Segmental aneuploidy and the genetic gross structure of the *Drosophila* genome. *Genetics* 71:157–84

70. MacHattie, L.A., Ritchie, D.A., Thomas, C.A., Jr., Richardson, C.C. 1967. Terminal repetition in permuted T2 bacteriophage DNA molecules. *J. Mol. Biol.* 23:355–63

71. Mazrimas, J.A., Hatch, F.J. 1972. A possible relationship between satellite DNA and the evolution of Kangaroo rat species (Genus *Dipodomys*). *Nature New Biol.* 240:102–05

71a. McConaughy, B.L., McCarthy, B.J. 1972. Fractionation of chromatin by thermal chromatography. *Biochemistry* 11:998–1003

72. Mead, C.G. 1964. A deoxyribonucleic acid-associated ribonucleic acid from *Drosophila melanogaster. J. Biol. Chem.* 239:550–54

72a. Miller, O.L., Jr. Bakken, A.H. 1972. Morphological Studies of Transcription. *Acta Endocrinol.* 168:155–77

73. Mulder, M.P., Duijn, P. van, Gloor, H.J. 1968. The replicative organization of DNA in polytene chromosomes of *Drosophila hydei. Genetica* 39:385–428

74. Ohta, T. 1972. Evolutionary rate of cistrons and DNA divergence. *J. Mol. Evol.* 1:150–57

75. Pardue, M.L., Gall, J.G. 1970. Chromosomal localization of mouse satellite DNA. *Science* 168:1356–58

76. Pardue, M.-L., Gerbi, S.A., Eckhardt, R.A., Gall, J.G. 1970. Cytological localization of DNA complementary to ribosomal RNA in polytene chromosomes of Diptera. *Chromosoma* 29: 268–90

77. Pardue, M.-L., Weinberg, E., Kedes, L.H., Birnstiel, M.L. 1972. Localization of sequences coding for histone messenger RNA in the chromosomes of *Drosophila melanogaster. J. Cell. Biol.* 55:199a

78. Patterson, J.T., Stone, W.S. 1952. *Evolution in the genus* Drosophila Macmillan Co: New York

79. Paul, J. 1972. General theory of chromosome structure and gene activation in eukaryotes. *Nature* 238:444–46

79a. Peacock, W.J. 1965. Nonrandom segregation of chromosomes in *Drosophila* males. *Genetics* 51:573–83

80. Perry, R.P., Cheng, T.-Y., Freed, J.J., Greenberg, J.R., Kelley, D.E., Tartof, K.D. 1970. Evolution of the transcriptional unit of ribosomal RNA *Proc. Nat. Acad. Sci. USA* 65:609–16

81. Plaut, W., Nash, D., Fanning, T. 1966. Ordered replication of DNA in polytene chromosomes of *Drosophila melanogaster. J. Mol. Biol.* 16:85–93

82. Polan, M.L., Friedman, S., Gall, J.G., Gehring, W. 1973. Isolation and characterization of mitochondrial DNA from *Drosophila melanogaster. J. Cell Biol.* 56:580–89

83. Prescott, D.M., Murti, K.G., Bostock, C.J. 1973. The genetic apparatus of *Stylonychia. Nature*. 242:576, 597–600

84. Pyeritz, R., Lee, C.S., Thomas, Jr., C.A. 1971. The cyclization of mouse satellite DNA. *Chromosoma* 33: 284–96

85. Rae, P.M.M. 1970. Chromosomal distribution of rapidly reannealing DNA in *Drosophila melanogaster. Proc. Nat. Acad. Sci. USA* 67:1018–25

86. Rae, P.M.M. 1972. The distribution of repetitive DNA sequences in chromosomes. In *Advances in Cell and Molecular Biology* Vol. II, 109–49. E.J. DuPraw, ed Academic: N.Y.
87. Rasch, E.M., Barr, H.J., Rasch, R.W. 1971. The DNA content of sperm of *Drosophila melanogaster. Chromosoma* 33:1–18
88. Ritossa, F.M. 1968. Unstable redundancy of genes for ribosomal RNA. *Proc. Nat. Acad. Sci. USA* 60:509–16
89. Ritossa, F.M., Atwood, K.C., Spiegelman S. 1966. A molecular explanation of the bobbed mutants of *Drosophila* as partial deficiencies of "ribosomal" DNA. *Genetics* 54:819–34
90. Robertson, F.W., Chipchase, M., Nguyen, T.M. 1969. The comparison of differences in reiterated sequences by RNA-DNA hybridization. *Genetics* 63:369–85
91. Robertson, W.R.B. 1916. Chromosome studies, I. Taxonomic relationships shown in the chromosomes of Tettigidae and Acrididae: V-shaped chromosomes and their significance in Acrididae, Locustidae, and Gryllidae: Chromosomes and variation. *J. Morphol.* 27:179–331
92. Rudkin, G.T. 1969. Non-replicating DNA in *Drosophila. Genetics* 61: Suppl. 227–38
93. Sandler, L., Lindsley, D.L., Nicoletti, B., Trippa, G. 1968. Mutants affecting meiosis in natural populations of *Drosophila melanogaster. Genetics* 60: 525–58
94. Séchaud, J., Streisinger, G., Emrich, J., Newton, J., Lanford, H., Reinhold, H., Stahl, M.M. 1965. Chromosome structure in phage T4. II. Terminal redundancy and heterozygosis. *Proc. Nat. Acad. Sci. USA* 54:1333–39
95. Shannon, M.P., Kaufman, T.C., Shen, M.W., Judd, B.H. 1972. Lethality patterns and morphology of selected lethal and semi-lethal mutations in the zeste-white region of *Drosophila melanogaster. Genetics* 72:615–38
96. Sofer, W.H., Hatkoff, M.A. 1972. Chemical selection of alcohol dehydrogenase negative mutants in *Drosophila. Genetics* 72:545–49
97. Sonnenblick, B.P. 1950. The early embryology of *Drosophila melanogaster.* In *The Biology of Drosophila* M. Demerec, ed. 62–163, Wiley: N.Y.
98. Sorsa, M., Sorsa, V. 1968. Electron microscopic studies on band regions in *Drosophila* salivary chromosomes. *Ann. Acad. Sci. Fenn. Ser.* A. IV 127:1–8
99. Southern, E.M. 1970. Base sequence and evolution of guinea-pig α satellite DNA *Nature* 227:794–98
100. Southern, E.M. 1971. Effects of sequence divergence on the reassociation properties of repetitive DNAs. *Nature New Biol.* 232:82–83
101. Spear, B.B. 1972. Tissue specific variations in the multiplicity of genes for ribosomal RNA in *Drosophila melanogaster. J. Cell Biol.* 55:246a
102. Spear, B.B., Gall, J.G. 1973. Independent control of ribosomal gene replication in polytene chromosomes of *Drosophila melanogaster. Proc. Nat. Acad. Sci. USA* 70:1359–63
103. Stahl, F.W., Steinberg, C.M. 1964. The theory of formal phage genetics for circular maps. *Genetics* 50:531–38
104. Steffensen, D.M., Wimber, D.E. 1971. Localization of tRNA genes in the salivary chromosomes of *Drosophila* by RNA:DNA hybridization. *Genetics* 69:163–78
105. Streisinger, G., Edgar R.S., Denhardt, G.H. 1964. Chromosome structure in phage T4, I. Circularity of the linkage map. *Proc. Nat. Acad. Sci. USA* 51:775–79
106. Sturtevant, A.H., Novitski, E. 1941. The homologies of the chromosome elements in the genus *Drosophila. Genetics* 26:517–41
107. Sutton, W.D. 1972. Chromatin packing, repeated DNA sequences, and gene control. *Nature New Biol.* 237:70–71
108. Sutton, W.D., McCallum, M. 1971. Mismatching and the reassociation rate of mouse satellite DNA. *Nature New Biol.* 232:83–85
109. Sutton, W.D., McCallum, M. 1972. Related Satellite DNAs in the Genus *Mus. J. Mol. Biol.* 71:633–56
110. Suyama, J., Preer, J.R., Jr. 1965. Mitochondrial DNA from protozoa. *Genetics* 52:1051–58
111. Swift, H., Wolstenholme, D.R. 1969. Mitochondria and chloroplasts: nucleic acids and the problem of biogenesis (genetics and biology) Frontiers of Biology, 15:972–1046. In *Handbook of Molecular Cytology*. A. Lima-de-Faria, ed. North Holland Publ. Co: Amsterdam
112. Szybalski, W., Kubinski, R., Sheldrick, P. 1966. Pyrimidine clusters on the transcribing strand of DNA and their possible role in the initiation of RNA synthesis. *Cold Spring Harbor Symp. Quant. Biol.* 31:123–27
113. Tartof, K.D. 1971. Increasing the multiplicity of ribosomal RNA genes

in *Drosophila melanogaster. Science* 171:294–97

114. Tartof, K.D. 1973. Regulation of ribosomal RNA gene multiplicity in *Drosophila melanogaster. Genetics* 73:57–71
115. Tartof, K.D., Perry, R.P. 1970. The 5S RNA genes of *Drosophila melanogaster. J. Mol. Biol.* 51:171–83
116. Thomas, C.A., Jr. 1967. The rule of the Ring. *J. Cell. Physiol.* Supp. I to Vol. 70:13–33
117. Thomas, C.A., Jr. 1970. *The theory of the master gene.* Neurosciences: Second Study Program. F.O. Schmitt ed. Rockefeller Univ. Press. N.Y.
117a. Thomas, C.A., Jr. Dancis, B.M. 1973. Ring Stability. *J. Mol. Biol.* 77:43–56
118. Thomas, C.A., Jr., Hamkalo, B.A. Misra, D.N., Lee, C.S. 1970. Cyclization of eukaryotic deoxyribonucleic acid fragments. *J. Mol. Biol.* 51:621–32
119. Thomas, C.A., Jr., MacHattie, L.A. 1964. Circular T2 DNA molecules. *Proc. Nat. Acad. Sci. USA* 52: 1297–1301
119a. Thomas, C.A., Zimm, B.H., Dancis, B.M. 1973. Ring Theory. *J. Bol. Biol.* 77:85–100
120. Travaglini, E.C., Petrovic, J., Schultz, J. 1972 . Characterization of the DNA in *Drosophila melanogaster. Genetics* 72:419–30
121. Travaglini, E.C., Petrovic, J., Schultz, J. 1972. Satellite DNAs in the embryos of various species of the genus *Drosophila. Genetics* 72:431–39
122. Travaglini, E.C., Schultz, J. 1972. Circular DNA molecules in the genus *Drosophila. Genetics* 72:441–50
123. Turner, S.H., Laird, C.D. 1973. Diversity of RNA sequences in *Drosophila melanogaster.* Biochem. Genet. In press
124. Walker, P.M.B. 1968. How different are the DNAs from related animals? *Nature,* 219:228–32
125. Walker, P.M.B. 1971. Origin of satellite DNA. *Nature* 229:306–08
126. Walker, P.M.B., Flamm, W.G., McClaren, A. 1969. Highly repetitive DNA in rodents. In *Frontiers of Biology,* 15:52–66. Handbook of Molecular Cytology. Lima-de-Faria, A., ed. North Holland: Amsterdam
127. Waring, M. Britten, R.J. 1966. Nucleotide sequence repetition: a rapidly reassociating fraction of mouse DNA. *Science* 154:791–94
128. Wheeler, M., Hamilton, N. 1972. Catalogue of *Drosophila* species names. *Studies in Genetics* 7:257–68 University of Texas publ. 7213
129. Wilson, D.A. 1972. *Palindromes in chromosomes.* Ph.D. thesis, Harvard University
130. Wilson, D.A., Bick, M.D., Thomas, C.A. Jr. 1972. Eukaryotic DNA contains inverted repetitions. *J. Cell Biol.* 55:283a
131. Wimber, D.E., Steffensen, D.M. 1970. Localization of 5S RNA genes on *Drosophila* chromosomes by RNA-DNA hybridization. *Science* 170: 639–41
132. Wolstenholme, D.R. 1973. Replicating DNA molecules from eggs of *Drosophila melanogaster. Chromosoma* In press
133. Woods, P.S., Gay, H., Sengün, A. 1961. Organization of the salivary-gland chromosome as revealed by the pattern of incorporation of H^3-thymidine. *Proc. Nat. Acad. Sci. USA* 47:1486—93
134. Wu, J.-R., Hurn, J., Bonner, J. 1972. Size and distribution of the repetitive segments of the *Drosophila* genome. *J. Mol. Biol.* 64:211–19

LOCALIZATION OF GENE FUNCTION

3053

Donald E. Wimber and Dale M. Steffensen

Department of Biology, University of Oregon, Eugene, and Department of Botany, University of Illinois, Urbana

Introduction

Cytogenetic methods are traditionally used to estimate the location of genes on chromosomes. By using cytogenetic procedures in some favorable cases, the control of, or the appearance of, a specific protein or RNA can be ascribed to a distinct chromosome region. Recently a very powerful method, which circumvents traditional genetic procedures, has become available for locating specific genes. The development of in situ RNA–DNA hybridization techniques has made it possible to observe the sites of production of specific classes of RNA on cytological preparations. In this methodology particular species of radioactive RNA or DNA are hybridized to denatured DNA in the chromosomes of standard cytological preparations. As complementary nucleic acid sequences anneal specifically to each other, autoradiographs of these cytological hybridizations reveal a radioactive site on the chromosome, which presumably marks the location of the DNA coding for the RNA species (1–3). In theory, the site of production on any RNA species sufficiently different from other forms can be located on chromosomes. In the last few years these methods have revealed the sites of production in certain organisms of 18 and 28S ribosomal RNA, transfer RNAs, histone message, and possibly hemoglobin message. In addition, in situ annealing methods have shown the location of a number of different satellite or reiterated DNAs on chromosomes. This review will deal mainly with in situ hybridization, practice, theory, results, and future potential.

The methods for in situ nucleic acid hybridization have been largely derived from those developed for gel or filter annealing techniques. The in situ methods allow one to determine the site of particular cistrons along a chromosome, but at present are only crude quantitative measures of cistronic redundancy. On the other hand, the filter annealing techniques are mainly quantitative measures of nucleic acid complementarity. The in situ methods are still undergoing change;

they have by no means been perfected. The two main areas that deserve further work are the refinement of quantitative methods and the development of techniques that will enhance the sensitivity of the system (1, 4).

Methods and Theory

The methods as they have developed up to now generally involve the production and purification of nucleic acids that have been labeled to a very high degree with tritium (^{3}H) or ^{125}I. Cytological preparations are made and are treated to remove the histones and RNA and then the DNA is denatured by heat, high or low pH, or formamide. The radioactive nucleic acid of one or a few species in a salt solution is placed over the denatured cytological preparation at a temperature 10–20° below the average melting temperature for the DNA. Molecules of the nucleic acid in the salt solution anneal to the complementary DNA sequence of the cytological preparation and if enough radioactivity is present, an autoradiographic image can be produced that will reveal the site of the binding (1, 4, 5).

In theory, any gene can be localized by this method, but in practice, there are several limiting parameters. These are gene redundancy, specific activity of the nucleic acid, annealing efficiency, and autoradiographic efficiency. These will be discussed in that order.

GENE REDUNDANCY The DNA sequences in higher organisms are present in the unduplicated haploid genome in numbers ranging from only one to several thousand. Some of the genes of known function are reiterated while others are thought to be present only once. Multiple copies of the 18 and 28S ribosomal RNA cistrons (100–100,000) are present in all eukaryotic organisms where experimental evidence is available (6–8). The 5S ribosomal cistrons and the transfer RNA genes are also reiterated (9–12). Among the genes that are known to code for proteins, only the histone cistrons are clearly reiterated (13). Others, such as the hemoglobin, ovalbumin, crystallin, etc. genes are probably present but once or at best a few times (14–16). The evidence gathered by Laird and others on DNA renaturation kinetics strongly upholds the idea that a substantial portion of the genome of higher organisms (the slowly renaturing part) is present as single copies (17, 18). This supports the so-called unineme theory (one DNA double helix per chromatid) of chromosome structure. On the other hand, there remains a substantial amount of evidence that supports a polyneme (two or more DNA double helices per chromatid) structure, at least in some organisms (19). There is no unanimous agreement; indeed, different types of strandedness among chromosomes may exist between organisms or even within an organism (20). We need not come to any conclusion about polynemy or uninemy except to say that polynemy would simplify the localization of genes on chromosomes, because it would provide more binding sites for individual RNA species.

For in situ hybridization, the chances of detecting a hybrid between a nucleic acid species in solution and a DNA sequence on a chromosome is greatly enhanced in cases where there is more than one copy of the cistron. In the haploid or diploid state the numbers of cistrons that are reiterated is probably fairly limited. On the other hand, the multistranded, polytene chromosomes are ideal for such studies. In the dipteran insects many polytene chromosomes show a lateral redundancy of several hundred to several thousand and cytological preparations are easily made. Cistrons that are present but once in the haploid state would be present 1024 times in a fully developed polytene chromosome of *Drosophila melanogaster*. Polytene chromosomes are also present in some ciliated protozoans and in the suspensor cells of some higher plants; these can be exploited in some cases (21, 22), but their preparation in usable numbers is difficult. Among higher plants and animals, endoreduplicated chromosomes can sometimes be found; this multichromatid condition can be produced by a variety of chemical treatments and could be used to increase the number of chromatid copies within a cell to a small degree. This would be advantageous in cases where only a small increase in cistronic redundancy would make the difference between a viable and inviable experiment. A major breakthrough in methodology would be the development of an in vitro system where polytene chromosomes could be produced in cells from organisms that ordinarily contain only diploid chromosomes.

SPECIFIC ACTIVITY OF THE NUCLEIC ACID Very high specific activity is a necessity for in situ hybridization procedures. With tritium-labeled nucleic acids, the long half-life (12.26 years) means that a large fraction of the bases need to be labeled in order that the radioactivity bound to a chromosomal site can be detected. One disintegration per month per molecule would give a nucleic acid with a specific activity of 1.4×10^{13} dpm/μmole. A site having a redundancy of 100 with a 10-20% hybridization efficiency and a 10% autoradiographic efficiency would yield only one or two autoradiographic silver grains per month. This is barely at the level of experimental detection (4). An increase in the specific activity by an order of magnitude clearly places the experiment well within the workable limits, i.e., 10-20 silver grains per month. Nucleic acids of this specific activity are difficult to produce in vivo, as radiation-induced lethality of the organism or cells often occurs. We have successfully produced *Drosophila melanogaster* ribosomal ^3H-RNA having a specific activity of 4×10^{12} dpm/μmole in vivo (23). This gives about 0.29 disintegrations per month per molecule. In the *D. melanogaster* system this is usable because the ribosomal cistrons are reiterated about 130 times in the haploid genome, and with a salivary gland nucleus having moderate polyteny, several thousand cistrons are present (24). Thus, 50-200 autoradiographic silver grains would be realized within a month. Gall and co-workers (2) have produced ^3H-RNA of a somewhat higher specific activity from *Xenopus laevis* tissue cultures. RNA of even higher specific activity can be produced by using the *Escherichia coli* RNA polymerase system. Here highly radioactive RNA is synthesized in vitro from satellite or total DNA by supplying the polymerase system with radioactive nucleotides. RNA having a specific activity an order of magnitude greater than that produced by in vivo systems can easily be synthesized, e.g. rRNA of 7×10^{13} dpm/μmole or about 5 d/month/molecule (25, 26).

Recently Prensky et al (5) have been able to iodinate RNA in vitro to specific activities heretofore only barely attainable with tritium, and there is every reason to believe that much higher specific activities can be reached. ^{125}I has a half life of about 60 days and produces autoradiographs with only slightly lower resolution than tritium (cf. Fig. la and lb). Prensky et al have produced ^{125}I – 5S RNA having a specific activity of 4.6×10^{12} dpm/μmole. If this were ribosomal RNA it would translate into about 8 d/month/molecule. Such high specific activities shorten considerably the time necessary for autoradiographic exposure. An order of magnitude increase in specific activity easily places the localization of the low or nonreiterated sequences in diploid cells within the realm of experimental attack.

ANNEALING EFFICIENCY In order to maintain adequate chromosome morphology for these RNA-DNA annealing procedures, the denatured but complementary DNA strands within the chromosome must remain in close proximity. During the hybridization period one would suppose that two competing reactions occur—that of DNA-DNA renaturation and that of RNA-DNA annealing. There is little information in the literature on the amount that DNA-DNA reannealing competes with the RNA-DNA reaction. Furthermore, it is difficult to find information on the efficiency on the initial denaturation. Generally, in situ hybridization procedures have evolved by empirical steps, and the quantitative details are unknown. With this in mind, we have performed some recent experiments (unpublished observations, Szabo & Steffensen) in which we preincubated denatured slides of *Drosophila melanogaster* chromosomes for times up to 10 hours under conditions of salt and temperature used in the annealing procedures; results with both 5S RNA and rRNA hybridized to the preincubated slides show no significant difference in autoradiographic grain numbers over chromosome region 56F (the 5S DNA site) or over the nucleolus (the rDNA site)

⟶

Figure la Autoradiograph of *Drosophila melanogaster* polytene chromosomes, hybridized to homologous ^{3}H-5S RNA and competed with excess nonradioactive transfer RNA. Radioactivity is at 56F, chromosome 2R. Exposure time, 18 months.

Figure lb Autoradiograph of *D. melanogaster* polytene chromosomes, hybridized with homologus ^{125}I-5S RNA. 56F shows radioactivity. Compare grain spread with Fig. la. Exposure time, 4 days.

Figure lc Two partly overlapping first meiotic anaphase divisions in pollen mother cells of *Tulbaghia violacea* (N=6). Hybridized to ^{125}I-5S RNA from *D. melanogaster*. A pair of chromosomes from each cell shows high annealing to the 5S RNA. Probable pairs indicated by arrows. Exposure time, 3 days.

Figure ld A portion of a lampbrush chromosome bivalent from *Taricha granulosa*. Hybridized to homologous ^{125}I-5S RNA. The two homologues are visible with a connecting chiasma (Xma). The touching ends of the homologues are indicated (T) between two axial granules. A region of high 5S RNA annealing is next to the chiasma. Exposure time, 7 days.

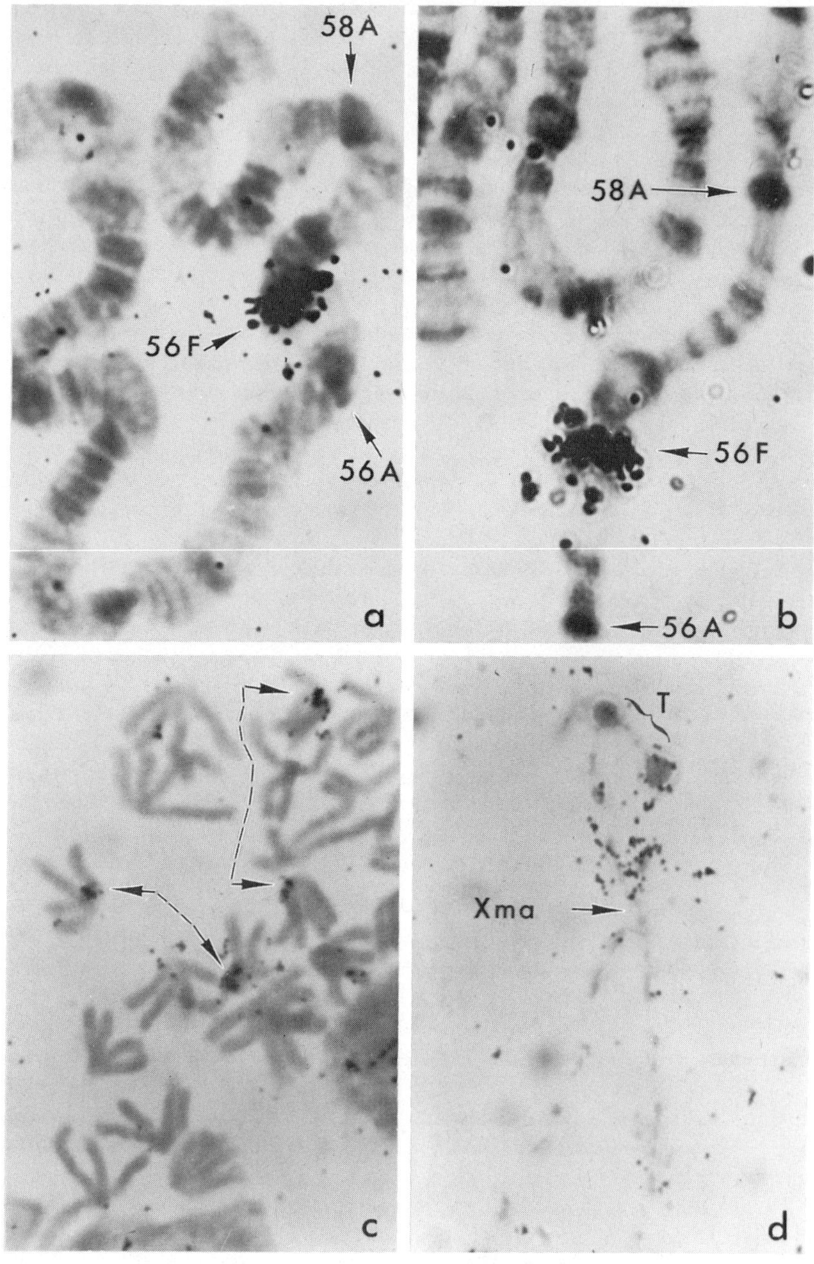
58A
56F
56A
a
58A
56F
56A
b
c
T
Xma
d

with or without preincubation. These results suggest that there is little renaturation of the DNA. Thus, the DNA in the cytological preparations behaves very much like the DNA immobilized on filters or gels. It seems to be prevented from reannealing by the presence of protein molecules or other physical constraints.

We have hybridized 5S RNA to both the extended lampbrush chromosomes from oocytes of *Taricha granulosa* and to the compact meiotic chromosomes from the male; identical procedures for denaturing and annealing were followed (unpublished observations). The results show at least an order of magnitude more grains over the extended lampbrush preparations than over the compact male meiotic material (Fig. ld). This suggests that the more diffuse and elongated condition of the lampbrush chromosomes enhances the degree of annealing to the DNA. There may be differences in autoradiographic efficiency between the two cases, but we feel that it is unlikely to be the main reason for the results. Whatever the ultimate cause, the main conclusion seems clear—the autoradiographic picture that one obtains is dependent upon the physical state of the chromosome prior to processing. Lampbrush chromosomes would seem to be prime targets for studies with single or low copy genes, since here, in contrast to the highly condensed mitotic or meiotic chromosomes, a higher annealing efficiency or autoradiographic efficiency or both is realized.

Annealing efficiency is also dependent upon the concentration of the RNA in the hybridizing solution, as well as the duration of the reaction. Although a multitude of experiments offer information on these points with filter hybridization, little information is available for in situ hybrids. The empirical methods that have evolved the present techniques have not yet given quantitative formulations. We and others have estimated an annealing efficiency of 5-10% (23, 27). If anything, this is a low estimate and may possibly be revised upward by a factor of two or three when quantitative methods are perfected. Given the differences in chromosomes, the degree of contraction, variation in nuclear proteins, and the vagaries of the hybridization procedure, we suspect that only rough quantitative measures will eventually be devised.

AUTORADIOGRAPHIC EFFICIENCY The efficiency with which (^3H) β-rays are detected by the common autoradiographic emulsions is usually between 5–10% (28). The very weak β-ray energies from ^{3}H generally preclude any higher efficiencies using standard techniques. This is an unfortunate drawback with these methods since most of the β-rays emitted from a site will not be detected. Some investigators using autoradiographic techniques have used phosphor coatings on slides or exposed slides directly in a scintillation fluid to increase the autoradiographic efficiency (29). Such methods should have substantial merit for in situ hybrids.

On the other hand, ^{125}iodine shows an autoradiographic efficiency of 20% or more (30). However, as some of the electrons are more energetic than those produced by tritium, a somewhat poorer resolution is found in the autoradiographs. This can be a drawback when small chromosomes are being examined.

Gene Localization

Through the use of in situ hybridization procedures it is theoretically possible to find the production site on the chromosomes of all of the multitude of species of RNA produced within a cell. Going further, it is possible to produce radioactive RNA from constitutive heterochromatin (that part of the genome that probably does not code for RNA species) and find the areas on the chromosomes carrying these base sequences. Until now, the sites of only a few genes have been identified using annealing methods and these have been mainly those cistrons concerned with the mechanics of protein synthesis. These are known to be reiterated in the genome, and thus afford relatively easy detection.

18 AND 28S RIBOSOMAL GENES A large amount of experimental evidence has shown the nucleolus organizers to be the chromosomal sites for the 18 and 28S ribosomal genes (24, 31). Further, it has been shown that from less than a hundred to several thousand copies of ribosomal cistrons exist in the haploid genome, depending on the species. Because of the degree of reiteration, this was a logical test system to use in the development of in situ hybridization techniques and indeed, it was used by three research groups who independently worked out successful methodology at nearly the same time (2, 3, 32). Gall & Pardue (2) used ^{3}H-rRNA produced in *Xenopus laevis* tissue cultures to demonstrate the localization of the ribosomal cistrons within the oocytes, oogonia, and somatic cells of *Xenopus*. The genes were located as expected in association with the nucleoli, and further, Gall & Pardue clearly demonstrated the amplification of the rDNA cistrons during meiotic prophase in the oocyte. An unexpected finding, which would have been difficult to discover by other means, was that some amplification occurs during the oogonial stages, hence, it is not totally a phenomenon confined to the oocytes. *Xenopus* ribosomal RNA was successfully hybridized to other distantly related species (25); three genera of flies showed specific localization of the silver grains over the nucleoli. Ribosomal cistrons (rDNA) were shown to be in some of the micronucleoli of *Rhynchosciara hollaenderi* and *Sciara coprophila* and at an unsuspected spot on the C-chromosome of *R. hollaenderi*. The finding of ribosomal cistrons at a chromosome site not associated with the cell's nucleolus suggests that there may be nucleolar organizers that function in some other tissue or some other time in ontogeny.

The nucleolus organizer of *D. melanogaster* is located in region 20 (proximal part of the X-chromosome) of the polytene chromosome map. However, in situ annealing of ribosomal RNA (rRNA) to squash preparations shows label confined to the nucleolus and not to any banded chromosome region, as if the rDNA were spun out of the chromosome or was reduplicated exogenous to the chromosome. The organization of the X-chromosome of *D. melanogaster* in this region has been the topic of some recent papers (33, 34). Polytene chromosomes are special cases of differential gene duplication, where parts are reduplicated and others are not. Thus, the nucleolus organizer seems to undergo gene duplication during polytenization while being buried deep within a heterochromatic segment that does not replicate (35, 36). We might speculate that the rDNA may undergo an amplification similar to that which occurs in many oocytes, and the amplified rDNA is actually free of the chromosome. That this may be the case is suggested by the frequent observation of fragmented nucleoli that contain rDNA and their attachment to specific chromosome regions, such as 56F (4) or 33-34 (37). The finding that the fragmented nucleoli are often but not always associated with specific nonnucleolar chromosome regions implies that there may be some specific functional relationship between these parts of the genome.

The proximal portion of the X-chromosome is interesting from yet another aspect of ribosome formation and function. Genetic data from *D. melanogaster* × *D. simulans* hybrids show that the structural genes for at least eight ribosomal proteins from both the 60 and 40S ribosomal subunits map within segment 19 to 20A (Steffensen, unpublished).

Tritium labeled rRNA from a human tissue culture has recently been used to locate the sites that code for the 18 and 28S rRNA on the human genome (27). As had been expected, the sites of the secondary constrictions on chromosomes 13, 14, 15, 21, and 22 were the annealing positions. One of the major unanswered questions about the coding sites for the ribosomal genes is why there are several pairs of chromosomes carrying them in some organisms as in man and only one pair in others.

Lampbrush chromosomes from animals offer some interesting possibilities for in situ hybridizations. They are very elongate bivalent chromosomes from the oocytes of animals. The isolation techniques are best known for the amphibians. Lateral loops of nucleoprotein spin out from the main chromosome axis and a number of landmarks such as spheres, large chromomeres, nucleoli, etc. can be used to identify individual chromosomes. The lampbrush chromosomes of *Triturus viridescens* have recently been hybridized with *Xenopus* rRNA (38). The radioactivity was found over the rDNA contained in the numerous micronucleoli, but no label was seen over the axis of the chromosome at the site of the nucleolus organizer. This observation is probably a result of a short autoradiographic exposure, as we have recently hybridized homologous ^{125}I – rRNA to lampbrush chromosomes of another salamander, *Taricha granulosa*, and clearly found label over the chromosome axis at the nucleolus organizer as well as the micronucleoli.

Not to be outdone, the botanists have utilized the polytene chromosomes found in the suspensor cells of the beans, *Phaseolus coccineus* and *P. vulgaris* for in situ hybridization studies (21, 22, 39). They were able to obtain ^{3}H – rRNA of a high enough specific activity by growing *Vicia faba* root tips in a ^{3}H-uridine solution or by organ culturing hypocotyls from germinating *Phaseolus* seeds in a radioactive medium. Brady & Clutter (21) found hybridization to the nucleoli of the two pair of chromosomes that carry the nucleolar organizers in *Phaseolus*. On the other hand, in the same species Avanzi et al (22, 39) obtained hybridization not only to the nucleolar organizers but also to two other chromosome regions and to some of the micronucleoli that are produced at one of the nucleolar organizers. There are several differences in technique between the two experiments so it is difficult to make any judgments about the additional annealing sites.

In some recent work at the University of Oregon, R. L. White (unpublished) has been able to anneal radioactive lily or onion rRNA to the pollen mother cells of both species. The nucleolus clearly showed label associated with the perinucleolar chromatin. However, a more intriguing phenomenon was that the tapetal cells of both lily and onion showed a very large amount of rRNA annealing to the nuclei, estimated to be perhaps 100 times that found over the pollen mother cells. Although tapetal nuclei are often found to be polyploid, the degree of polyploidy would not account for the number of silver grains found over these nuclei. Therefore, it is assumed that the tapetal tissue of these species and perhaps all Angiosperms, undergoes an amplification of the ribosomal cistrons similar to that found within the oocytes of many vertebrate and invertebrate animals (40, 41). Amplification of higher plant rDNA has been looked for before (6, 42), but the studies have been by filter hybridization of rRNA to DNA extracted from entire plants or gross plant parts at different developmental stages. No clear-cut amplification has been observed. Amplification of genes in restricted cells or tissues could easily be missed when using the filter hybridization assay, simply because it is often too difficult to obtain a large enough quantity of DNA from the specialized cells to carry out experiments. In such cases, in situ hybridization is of high merit.

5S Ribosomal Cistrons A few years ago it was considered likely that the 5S ribosomal genes would also be located at the nucleolus along with the 18 and 28S cistrons. However, evidence accumulated in 3 systems, *Xenopus laevis*, *Drosophila melanogaster*, and HeLa cells, that the 5S genes were not contiguous with the other ribosomal genes (43-45). Then in 1970 we demonstrated that in *D. melanogaster*, the 5S genes were on another linkage group entirely, the right arm of chromosome two at 56F (Fig. la, lb, 2b) (23). We originally thought that 56E might also be involved but autoradiographs of finer resolution indicate that the label is localized over 56F. As far as we know, no mutations have been mapped within this region. However, Lindsley et al (46) have synthesized segmental aneuploids for band 56F. One might expect to recover *bobbed*-like mutants (*bobbed* mutants are deletions of 18-28S ribosomal cistrons) when there are fewer copies of the 5S DNA than in wild type, but the hemizygotes show no unusual phenotypic effect with just one dose of 5S DNA (D. Lindsley, personal communication).

It remains to be conclusively demonstrated how 56F interacts with the nucleoli on the X and Y chromosomes. We think that part of the communication problem within the nucleus may be solved by a physical association of the 5S region with the nucleolus. We often see a close contact of region 56F with the nucleolus in our squashes (Fig. 2b) (4). Amaldi & Buongiorno-Nardelli (11) examined the nuclear placement of both 5S and transfer RNA (tRNA) genes within the interphase nuclei of Chinese hamster cells by in situ methods. They found in both cases grain localization to the perinucleolar dense chromatin but not exclusively so. Unfortunately no metaphase chromosomes were examined. This may be a reflection of an interaction or cooperation of genes on several chromosomes clustered around the nucleolus, similar to the association that we sometimes found with 56F and the nucleolus in *D. melanogaster*. However, we feel that this observation on Chinese hamster nuclei needs confirmation in light of the difficulty that we have had with rRNA contamination of smaller RNA species. We found that even acrylamide gel purified 5S RNA still contained fragments of rRNA that specifically bind to the nucleolus (23). Thus, the clustering of grains around the nucleolus in the Chinese hamster may be a reflection of rRNA contamination in the annealing mixture.

These observations bear on the sensitivity of in situ hybridizing systems. RNA that by most criteria is "biochemically pure" may still contain cleavage products of high molecular weight species. Hence, in our work on *Drosophila*, tRNA contained degradation products of both 5S and rRNA that could be detected by in situ hybridization but not on acrylamide gels. The addition of an excess of unlabeled 5S and rRNA to the tRNA hybridizing medium reduced or eliminated the grains over 56F and the nucleolus but did not affect the tRNA hybridizing sites (4, 47). Such "competition" type experiments should routinely be carried out as part of the experimental protocol for most studies of gene localization.

Recently we have used ^{125}I-5S RNA from *D. melanogaster* in several heterologous hybridizations with higher plant chromosomes (unpublished). In all of the species we have tried, a few areas of the plant genome showed appreciable binding. For example, *Tulbaghia violacea* (society garlic) meiotic chromosomes showed one pair of chromosomes with a region of high annealing (Fig. lc). Although we have not yet performed competition experiments in which homologous unlabeled 5S RNA was in excess, it seems likely that these are regions coding for 5S RNA. The 5S DNA is conservative in an evolutionary sense (48) and contains substantial numbers of base sequences that are little changed from very ancient times. In a similar way *Xenopus laevis* 18 and 28S ribosomal RNA will form heterologous hybrids to several different genera of insects (25). The ease with which such heterologous hybridizations can be made will allow for convenient studies among numerous species and genera with respect to the evolution of the placement of these cistrons on the genome.

→

Figure 2a Proximal part of *D. malanogaster* polytene chromosome 2L. Hybridized to ^{3}H-9S messenger RNA (probable histone message) from a sea urchin (*Psammachinus milaris*). Competed with excess nonradioactive *Escherichia coli* RNA. The region showing radioactivity is probably 39DEF. Exposure time, 99 days. Photo provided by M. L. Pardue et al (52).

Figure 2b *D. melanogaster* polytene chromosomes hybridized to homologous ^{3}H-5S RNA containing 18 and 28S rRNA as an impurity. Nucleolus (N) is densely labeled. Region 56F on chromosome 2R shows the 5S label. Nucleolus and region 56F frequently show close relationships. Exposure time, 2 months.

Figure 2c and 2d Autoradiograph of *D. melanogaster* polytene chromosomes hybridized to homologous ^{3}H-transfer RNA. The proximal part of chromosome 3R, region 82F, is one of the sites of high annealing.

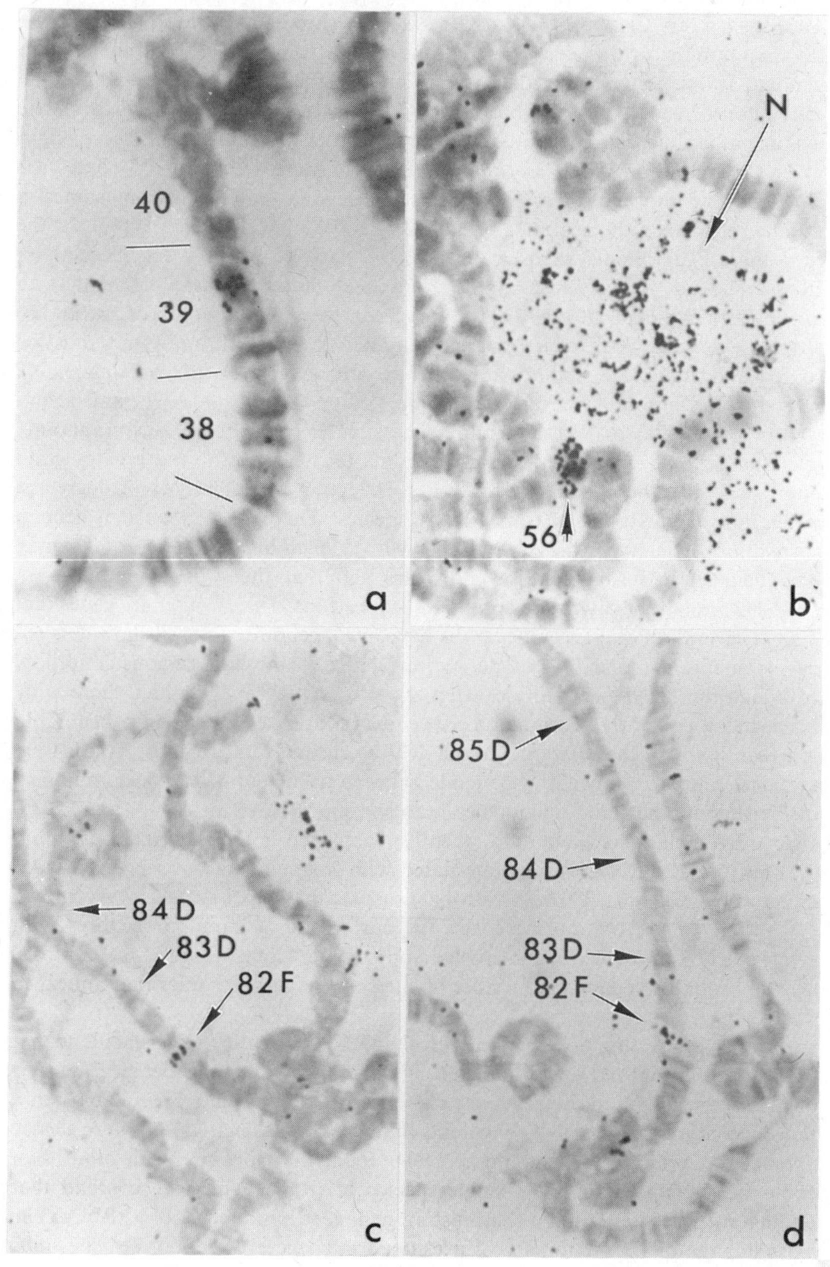

40
39
38
a
N
56
b
84 D
83 D
82 F
c
85 D
84 D
83 D
82 F
d

TRANSFER RNA CISTRONS Ritossa et al (10) showed that in *D. melanogaster* the total number of transfer RNA cistrons amounted to about 750 in the haploid genome. If one assumes that there are about 60 different types of tRNA, and the reiteration of each transfer DNA (tDNA) sequence is about equal, then on the average a haploid genome would contain 13 copies of each tDNA. These ought to be detectable on the *Drosophila* polytene chromosomes if the 13 copies are clustered and indeed, when ^{3}H-tRNA is annealed to polytene chromosomes, a number of sites on the genome preferentially show label (Fig. 2c, 2d) (47). At the time these experiments were carried out, we had a fairly low specific activity tRNA; indeed, it was close to the theoretical minimum usable for such an experiment. We expected only about one grain per site per month of autoradiographic exposure. With such low grain numbers, multiple chromosomes must be scored and statistical methods must be used to give levels of confidence. We scored the X-chromosome and three-quarters of chromosome two and found well over 100 sites of high annealing. The sites are distributed fairly randomly over the genome but there are some regions that clearly showed greater grain numbers than others. This probably is a reflection of differences in degree of reiteration or clustering of some tRNA species. There seemed to be a general correlation of many of the labeled sites with *Minute* loci on the chromosomes. Atwood (10) had suggested several years ago that the *Minute* mutations in *Drosophila* might be alterations in the tDNA cistrons. The *Minutes* are genes that map at several different regions on the *Drosophila* genome; they all show delayed development, a similar phenotype, and are lethal in the homozygous condition. Although the *Minutes* broadly overlap the sites of tRNA annealing, there is by no means a conclusive correlation. Recently Lindsley et al (46) have been able to assign many of the *Minutes* to more definite chromosome regions. At this time a critical comparison needs to be made between particular *Minute* loci and areas of the genome that bind unequivocally to specific tRNA species.

A problem that we have only recently recognized is the very real possibility that our transfer RNA was contaminated with fragments of messenger RNA or high turnover nuclear RNA. We strongly suspect that when we take every known precaution to prevent nuclease activity during the isolation procedures with transfer RNA and when we introduce appropriate "competition" type controls, we will probably reduce the number of sites on the chromosomes that anneal to tRNA.

Another interesting observation with the tRNA annealing experiments is that puffs are often but not always labeled. Puffs are generally considered to be sites of message production, however the presence of label at puffs is probably not a reflection of messenger RNA fragments in our tRNA, as many puffs were clearly unlabeled. Certainly the amount of DNA found in a puff is much more than could be accounted for by a single species of tRNA. Thus, we contend that puffing at these sites is more than just an enhanced production of a tRNA. The close linkage of tDNA to sites of presumed message production, i.e., the puffs (49), may be a reflection of an operon that includes a minor tRNA species that is necessary for the message translation.

A more difficult task than identifying the sites of all of the tRNA genes is the localization of individual amino acid accepting forms of tRNAs. One possible route to take in this problem is to separate the total tRNA into various fractions by chromatographic or electrophoretic methods. Various fractions will be enriched for certain tRNA species and may bind to a limited number of sites on the genome. On the other hand, we do not know how specific will be the annealing; a great deal of cross hybridizing may occur so that it may be very difficult to pinpoint individual species of tRNA genes. Perhaps a surer method would be to charge a mixture of tRNA with a particular radioactive amino acid and then use the total mixture as the annealing solution. The competition provided by the nonradioactive forms would probably cut down on the cross-hybridizations. Methods are now available to produce peptide chains on charged tRNAs. Thus, one could transfer several radioactive amino acids onto the charged tRNA and increase the specific activity by an order of magnitude or more (50).

HISTONE CISTRONS Kedes & Birnstiel and others (13) have shown that the cistrons that code for histone message are reiterated in the sea urchin and are very active early in embryogeny when a large fraction of the total mRNA is coding for histone. A substantial amount of work on the histones among a large number of organisms has shown that the amino acid sequences are very similar between taxonomically diverse groups (51). With this background, Pardue et al (52) prepared tritium-labeled histone message from sea urchin blastulae (*Psammechinus milaris*) and challenged polytene chromosomes of *Drosophila melanogaster* with this message. A region, 39E-40A, of the left arm of chromosome 2 showed distinct label (Fig. 2a). Several mutants are known to exist within this general region of chromosome 2 but nothing is yet known about this exact locus. It is not known whether all of the histone messages are produced at this site or whether this is a reflection of but one or two binding to the region.

HEMOGLOBIN CISTRONS There is a substantial amount of disagreement in the published information about the reiteration of the hemoglobin cistrons in man as well as other animals (14, 53, 54). It seems likely that the redundancy of the locus is low, viz. less than 10 cistrons (54). Considering this low multiplicity it seems unlikely that the cistrons could be located on the human genome using hybridizing techniques with moderately radioactive hemoglobin message. Price et al (55) produced a hemoglobin message having a specific activity of about 100 dpm/μg. They hybridized the ^{3}H-hemoglobin messenger RNA to human chromosomes and reported high binding regions on chromosomes 2 and 4 or 5. As pointed out by Bishop & Jones (56), this observation is very likely in error, because it would be virtually impossible to obtain an autoradiograph with RNA having the specific activity reported by Price et al. On the other hand, with the specific activities now attainable with ^{125}I-RNA, the localization of the hemoglobin cistrons as well as numerous other messages should be possible.

Other Messenger Cistrons Not a great many highly purified messages have been isolated and of those that have, the reiteration of their cistrons within the genome is generally low or not known. Lambert et al (49, 57) have published work that strongly suggests that the Balbiani rings (BR) in *Chironomus tentans* produce messenger RNA. The BR of *C. tentans* appear in full development concomitantly with the appearance of specific protein fractions in the salivary secretions. Lambert et al (57) have been able to label the RNA of the cells of the salivary gland to a high degree by organ culture in a medium containing ^{3}H-labeled nucleosides. One of the major BR of the genome can be isolated by microdissection, and its associated RNA can be isolated and hybridized back to cytological preparations. The radioactivity specifically binds in great abundance to the BR from whence it was isolated. Similar high binding of nuclear sap RNA and peripheral cytoplasm RNA to the same BR was also found. Other sites of lower label were found, to be sure, but the highest annealing region was the BR. The evidence clearly shows that there is a substantial amount of RNA in the cytoplasm and nucleoplasm that is complementary to the BR DNA. However, it is yet to be shown that this RNA is associated with polysomes in the cytoplasm and translated into protein (58).

The purification of specific messenger RNAs can be a difficult procedure. Most purified messenger RNAs have come from differentiated systems that are mainly involved in the production of one protein in abundance, e.g. immunoglobin from plasma cell tumors, ovalbumin in the chick oviduct, etc. (15, 59). It may be possible in the future to separate particular messenger RNAs from polysomal mixtures in enough abundance to perform an ^{125}I annealing experiment. The development of convenient and reproducible assay systems for such messages is a necessity in order that the protein for which the message is coding may be accurately identified. A start on the development of a dependable assay system has been made with the oocyte translation system evolved in Gurdon's laboratory. When heterologous messenger RNA is injected into living oocytes of *Xenopus laevis* it is translated and novel polypeptides appear, which can be isolated and analyzed (60, 61).

Recent work suggests that most messenger RNAs have a region of polyadenylic acid which is added to the 3′ end during the maturation of the message (62). This condition could provide a convenient labeling method for any message. Very high specific activity ^{125}I- or ^{3}H-polyuridylic acid or polythymidylic acid could be added to the hybridization mixture along with the messenger RNA. The messenger RNA would anneal to the chromosome at the complementary site and the polyuridylic acid would presumably anneal to the polyadenylic acid tails. This would eliminate the necessity for labeling each message as it is purified; radioactive polyuridylic or polythymidylic acid could simply be kept "on the shelf" as a convenient tag for any message.

Satellite and Other Repetitive DNAs We will not attempt to cover the literature under this extensive topic. Several review articles have recently appeared (63–65). In situ hybridizing techniques have been valuable adjuncts to the study of satellite and repetitive DNA. Satellite DNAs can be separated from bulk nuclear DNA by isopycnic centrifugation in neutral or alkaline CsCl, $Ag^+ - Cs_2SO_4$, or $Hg^+ - Cs_2SO_4$. Jones (66) and Pardue & Gall (67) were first to use in situ hybridizing methods to localize mouse satellite DNA. The mouse satellite fraction is about 10% of the total nuclear DNA; it can be separated from bulk DNA, incubated with RNA polymerase and a highly radioactive RNA complement can be prepared. In situ annealing revealed that the satellite DNA is part of the centromeric heterochromatin of all of the chromosomes except for perhaps the Y. In a similar way satellite DNA was isolated from the salamander, *Plethodon cinereus*, and its complementary RNA was found to hybridize to centromeric heterochromatin (68). In *Rhynchosciara hollaenderi* hybridizations showed the satellites to be present in the centromere area as well as in some of the telomere regions (69). On the other hand one of the satellite DNAs from *D. pseudoneohydei* (70) showed localization to a large number of sites throughout the genome, and no particular region could be singled out as being more heavily labeled than the others. It seems obvious from these studies that the satellite DNAs need not necessarily be localized at particular chromosome regions. The fact that they are highly localized in some organisms and of a more general pattern in others would indicate that they may be functioning in manifestly diverse ways.

Repetitive but nonsatellite DNA has also been isolated from bulk DNA by the use of denaturation:renaturation methods developed by Britten & Kohne and others (71). Many workers (64, 72–74) have used in situ annealing procedures with RNA produced from rapidly reassociating fractions of denatured DNA. The DNA most rapidly reannealing is often fairly pure satellite fractions and these can show localization as in the case of *Microtus agrestis* where the sex chromosomes are highly enriched for the satellite fraction (72). However, much of the repetitive, nonsatellite DNA isolated by renaturation methods shows no obvious localization on the genome. A finding that may become axiomatic within a few years is that most satellite and a substantial amount of other repetitive DNA is never transcribed into RNA, or if it is transcribed it may never be translated (65, 75). There is some transcription to be sure, as in the case of the reiterated ribosomal or transfer RNA cistrons but the bulk of it seemingly never serves as a template.

Unfortunately, little is known about the function of satellite and nontranscribed or translated repetitive DNA. A number of possibilities have been proposed that range from maintaining centromere strength, acting as nontranslated spacers between cistrons, to possible control elements.

Conclusions

In situ hybridization methods offer a valuable adjunct to traditional cytogenetic techniques. The primary gene product, RNA, will anneal to the site of complementarity on the denatured chromosome, and the region of annealing can be identified by autoradiography if the RNA is sufficiently radioactive. The technique works well with cistrons that have a moderate to high reiteration on the genome. For this reason, the laterally redundant polytene chromosomes are ideal for gene localization studies. Highly radioactive RNA is a necessary prerequisite for the method. While moderately radioactive RNA can be produced in vivo and is often radioactive enough for studies with polytene chromosomes, a highly radioactive product is necessary for locating genes of low to moderate reiteration. Thus, the newly developed method for iodinating RNA to very high specific activities in vitro offers a substantial advance in the technology of gene localization. Using ^{125}I-RNA one can work with haploid or diploid cells. Genes that have a moderate to high level of reiteration are easily located on the genome and those that have a low reiteration or are present but once are now within experimental grasp.

ACKNOWLEDGMENTS

The preparation of this review and the work of the authors cited herein was supported by NIH research grant GM 18829 from the Institute of General Medical Sciences (to DEW) and National Science Foundation research grant GB 29603X (to DMS). We thank Dr. P. A. Duffey, Mr. P. Szabo, and Ms. D. R. Wimber for their critical comments on the manuscript.

Literature Cited

1. Gall, J., Pardue, M. 1971. Nucleic acid hybridization in cytological preparations. *Methods Enzymol.* 21:470–80
2. Gall, J., Pardue, M. 1969. The formation and detection of RNA-DNA hybrid molecules in cytological preparations. *Proc. Nat. Acad. Sci. USA* 63:378–83
3. John, H., Birnstiel, M., Jones, K. 1969. RNA-DNA hybrids at the cytological level. *Nature* 223:582–87
4. Steffensen, D., Wimber, D. 1972. Hybridization of nucleic acids to chromosomes. *Results Prob. Cell Diff.* 3:47–63
5. Prensky, W., Steffensen, D., Hughes, W. 1973. The use of iodinated RNA for gene localization. *Proc. Nat. Acad. Sci. USA*. In press
6. Ingle, J., Sinclair, J. 1972. Ribosomal RNA genes and plant development. *Nature* 235:30–32
7. Hotta, Y., Miksche, J. 1973. Molecular hybridization of coniferous ribosomal RNA to DNA. *Proc. 8th Cent. State Forest Tree Improv. Conf.* In press
8. Bostock, C. 1971. Repetitious DNA. *Advan. Cell Biol.* 2:153–223
9. Tartof, K., Perry, R. 1970. The 5S RNA genes of *Drosophila melanogaster*. *J. Mol. Biol.* 51:171–83
10. Ritossa, F., Atwood, K., Spiegelman, S. 1966. On the redundancy of DNA complementary to amino acid transfer RNA and its absence from the nucleolar organizer region of *Drosophila melanogaster*. *Genetics* 54:663–76
11. Amaldi, F., Buongiorno-Nardelli, M. 1971. Molecular hybridization of Chinese hamster 5S, 4S and "pulse labelled" RNA in cytological preparations. *Exp. Cell Res.* 65:329–34
12. Hatlen, L., Attardi, G. 1971. Properties of the HeLa cell genome complementary to tRNA and 5S RNA. *J. Mol. Biol.* 56:535–53
13. Kedes, L., Birnstiel, M. 1971. Reiteration and clustering of DNA sequences complementary to histone messenger RNA. *Nature N. B.* 230:165–69

14. Harrison, P., Hell, A., Birnie, G., Paul, J. 1972. Evidence for single copies of globin genes in the mouse genome. *Nature* 239:219–20
15. Means, A., Comstock, J., Rosenfeld, G., O'Malley, B. 1972. Ovalbumin messenger RNA of chick oviduct: partial characterization, estrogen dependence, and translation in vitro. *Proc. Nat. Acad. Sci. USA* 69:1146–50
16. Berns, A., van Kraaikamp, M., Bloemendal, H., Lane, C. 1972. Calf crystallin synthesis in frog cells: the translation of lens-cell 14S RNA in oocytes. *Proc. Nat. Acad. Sci. USA* 69:1606–9
17. Dickson, E., Boyd, J., Laird, C. 1971. Sequence diversity of polytene chromosome DNA for *Drosophila hydeii*. *J. Mol. Biol.* 61:615–27
18. Laird, C. 1971. Relationship between DNA content and nucleotide sequence diversity. *Chromosoma* 32:378–406
19. Wolff, S. 1969. Strandedness of chromosomes. *Int. Rev. Cytol.* 25:279–96
20. Gay, H., Das, C., Forward, D., Kaufmann, B. 1970. DNA content of mitotically-active condensed chromosomes of *Drosophila melanogaster*. *Chromosoma* 32:213–23
21. Brady, T., Clutter, M. 1972. Cytolocalization of ribosomal cistrons in plant polytene chromosomes. *J. Cell Biol.* 53:827–32
22. Avanzi, S., Durante, M., Cionini, P., D'Amoto, F. 1972. Cytological localization of ribosomal cistrons in polytene chromosomes of *Phaseolus coccineus*. *Chromosoma* 39:191–203
23. Wimber, D., Steffensen, D. 1970. Localization of 5S RNA genes on *Drosophila* chromosomes by RNA-DNA hybridization. *Science* 170:639–41
24. Ritossa, F., Spiegelman, S. 1965. Localization of DNA complementary to ribosomal RNA in the nucleolus organizer region of *Drosophila melanogaster*. *Proc. Nat. Acad. Sci. USA* 53:737–45
25. Pardue, M., Gerbi, S., Eckhardt, R., Gall, J. 1970. Cytological localization of DNA complementary to ribosomal RNA in polytene chromosomes of Diptera. *Chromosoma* 29:268–90
26. Brown, J., Jones, K. 1972. Localization of satellite DNA in the quail. *Chromosoma* 38:313–18
27. Henderson, A., Warburton, D., Atwood, K. 1972. Location of ribosomal DNA in human chromosome complement. *Proc. Nat. Acad. Sci. USA* 69:3394–98
28. Cleaver, J. 1967 *Thymidine metabolism and cell kinetics*. Amsterdam: North-Holland
29. Sade, R., Folkman, J., Cotran, R. 1972. DNA synthesis in endothelium of aortic segments in vitro. *Exp. Cell Res.* 74:297–306
30. Ada, G., Humphrey, J., Askonas, B., McDevitt, H., Nossal, G. 1966. Correlation of grain counts with radioactivity (^{125}I and tritium) in autoradiography. *Exp. Cell Res.* 41:557–72
31. Wallace, H., Birnstiel, M. 1966. Ribosomal RNA cistrons and the nucleolar organizer. *Biochim. Biophys. Acta* 114:296–310
32. Buongiorno-Nardelli, M., Amaldi, F. 1970. Autoradiographic detection of molecular hybrids between rRNA and DNA in tissue sections. *Nature* 225:946–48
33. Viinikka, Y., Hannah-Alava, A., Arajärvi, P. 1971. A reinvestigation of the nucleolus-organizing regions in the salivary gland nuclei of *Drosophila melanogaster*. *Chromosoma* 36:34–45
34. Schalet, A., Lefevre, G. The localization of "ordinary" sex-linked genes in section 20 of the polytene X chromosome of *Drosophila melanogaster*. *Chromosoma*. In press
35. Blumenfeld, M., Forrest, H. 1972. Differential under-replication of satellite DNAs during *Drosophila* development. *Nature N. B.* 239:170–72
36. Rudkin, G. 1969. Nonreplicating DNA in *Drosophila*. *Genetics (Suppl.)* 61:227–38
37. Hannah-Alava, A. Personal communication
38. Barsacchi, G., Gall, J. 1972. Chromosomal localization of repetitive DNA in the newt, *Triturus*. *J. Cell Biol.* 54:580–91
39. Avanzi, S., Buongiorno-Nardelli, M., Cionini, P., D'Amoto, F. 1971. Cytological localization of molecular hybrids between rRNA and DNA in the embryo suspensor cells of *Phaseolus coccineus*. *Acad. Naz. Lincei. Rendic. Cl. Sci. Fis. Mat. Nat.*, Ser. VIII 50:357–61
40. Brown, D., Dawid, I. 1968. Specific gene amplification in oocytes. *Science* 160:272–80
41. Cave, M. 1972. Localization of ribosomal DNA within oocytes of the house cricket, *Acheta domesticus* (Orthoptera: Gryllidae). *J. Cell Biol.* 55:310–21
42. Chen, D., Osborne, D. 1970. Ribosomal genes and DNA replication in

germinating wheat embros. *Nature* 225:336–40
43. Aloni, Y., Halten, L., Attardi, G. 1971. Studies of fractionated HeLa cell metaphase chromosomes II. Chromosomal distribution of sites for transfer RNA and 5S RA. *J. Mol. Biol.* 56:555–63
44. Brown, D., Weber, C. 1968. Gene linkage by RNA-DNA hybridization I. Unique DNA sequences homologous to 4S RNA, 5S RNA and ribosomal RNA. *J. Mol. Biol.* 34:661–80
45. Tartof, K., Perry, R. 1970. The 5S RNA genes of *Drosophila melanogaster*. *J. Mol. Biol.* 51:171–83
46. Lindsley, D. et al. 1972. Segmental aneuploidy and the genetic gross structure of the *Drosophila* genome. *Genetics* 71:157–84
47. Steffensen, D., Wimber, D. 1971. Localization of tRNA genes in the salivary chromosomes of *Drosophila* by RNA-DNA hybridization. *Genetics* 69:163–78
48. Forget, B., Weissman, S. 1969. The nucleotide sequence of ribosomal 5S ribonucleic acid from KB cells. *J. Biol. Chem. 244:3148–65*
49. Grossbach, U. 1969. Chromosomen-Aktivität und biochemische Zell-differenzierung in den Speicheldrüsen von *Camptochironomus*. *Chromosoma* 28:136–87
50. Menninger, J., Mulholland, M., Stirewalt, W. 1970. Peptidyl-tRNA hydrolase and protein chain termination. *Biochim. Biophys. Acta* 217:496–511
51. DeLange, R., Smith, E. 1971. Histones: structure and function. *Ann. Rev. Biochem.* 40:279–314
52. Pardue, M., Weinberg, E., Kedes, L., Birnstiel, M. 1972. Localization of sequences coding for histone messenger RNA in the chromosomes of *Drosophila melanogaster*. *J. Cell Biol.* 55:199a
53. Bishop, J., Pemberton, R., Baglioni, C. 1972. Reiteration frequency of hemoglobin genes in the duck. *Nature N.B.* 235:231–34
54. Ostertag, W., von Ehrenstein, G., Charache, S. 1972. Duplicated α-chain genes in Hopkins-2 haemoglobin of man and evidence for unequal crossing over between them. *Nature N.B.* 237:90–94
55. Price, P., Conover, J., Hirschhorn, K. 1972. Chromosomal localization of human hemoglobin structural genes. *Nature* 237:340–42
56. Bishop, J., Jones, K. 1972. Chromosomal localization of human hemoglobin structural genes. *Nature* 240:149–50
57. Lambert, B., Wieslander, L., Daneholt, B., Egyhazi, E., Ringborg, U. 1972. In situ demonstration of DNA hybridizing with chromosomal and nuclear sap RNA in *Chironomus tentans*. *J. Cell Biol.* 53:407–18
58. Lambert, B. 1973. Tracing of RNA from a puff in the polytene chromosomes to the cytoplasm in *Chironomus tentans* salivary gland cells. *Nature*. 242:51
59. Stavnezer, J., Huang, R. 1971. Synthesis of a mouse immunoglobulin light chain in rabbit reticulocyte cell-free system. *Nature N.B.* 230:172–76
60. Gurdon, J., Lane, C., Woodland, H., Marbaix, G. 1971. Use of frog eggs and oocytes for the study of messenger RNA and its translation in living cells. *Nature* 233:177–82
61. Laskey, R., Gurdon, J., Crawford, L. 1972. Translation of encephalomyocarditis viral RNA in oocytes of *Xenopus laevis*. *Proc. Nat. Acad. Sci. USA* 69:3665–69
62. Darnell, J., Philipson, L., Wall, R., Adesnik, M. 1971. Polyadenylic acid sequences: role in conversion of nuclear RNA into messenger RNA. *Science* 174:1507–10
63. Eckhardt, R. 1972. Chromosomal localization of repetitive DNA, pp. 271–92. In: *Evolution of genetic systems*, Ed. Smith, H., New York: Gordon & Breach
64. Rae, P. 1972. The distribution of repetitive DNA sequences in chromosomes. *Advan. Cell Mol. Biol.* 2:109–49
65. Walker, P. 1971. "Repetitive" DNA in higher organisms. *Progr. Biophys. Mol. Biol.* 23:145–90
66. Jones, K. 1970. Chromosomal and nuclear location of mouse satellite DNA in individual cells. *Nature* 225:912–15
67. Pardue, M., Gall, J. 1970. Chromosomal localization of mouse satellite DNA. *Science* 168:1356–58
68. Macgregor, H., Kezer, J. 1971. The chromosomal localization of a heavy satellite DNA in the testis of *Plethodon c. cinereus*. *Chromosoma* 33:167–82
69. Eckhardt, R., Gall, J. 1971. Satellite DNA associated with heterochromatin in *Rhynchosciara*. *Chromosoma* 32:407–27
70. Hennig, W., Hennig, I., Stein, H. 1970. Repeated sequences in the DNA of *Drosophila* and their localization in

giant chromosomes. *Chromosoma* 32:31–63
71. Britten, R., Kohne, D. 1968. Repeated sequences in DNA. *Science* 161:529–40
72. Arrighi, F., Hsu, T., Saunders, P., Saunders, G. 1970. Localization of repetitive DNA in the chromosomes of *Microtus agrestis* by means of in situ hybridization. *Chromosoma* 32:224–36
73. Botchan, M., Kram, R., Schmid, C., Hearst, J. 1971. Isolation and chromosomal localization of highly repeated DNA sequences in *Drosophila melanogaster*. *Proc. Nat. Acad. Sci. USA* 68:1125–29
74. Saunders, G., Shirakowa, S., Saunders, P., Arrighi, F., Hsu, T. 1972. Populations of repeated DNA sequences in the human genome. *J. Mol. Biol. 63:323–34*
75. Darnell, J. 1968. Ribonucleic acids from animal cells. *Bacteriol. Rev.* 32:262–90

DOSAGE COMPENSATION IN *DROSOPHILA*[1] 3054

John C. Lucchesi

Department of Zoology, University of North Carolina, Chapel Hill, North Carolina

Introduction

The regulation of gene activity, *sensu strictu*, can occur only at the level of transcription. The onset of transcription, the rate at which it proceeds, and its cessation are regulatory functions common to all living organisms. The program that orders these functions for a given gene is the first level where notable evolutionary divergence may occur. Increasingly complex integrative systems of individual gene programs are the basis for cellular and tissue differentiation in higher forms. Therefore, it is probable that, in order to understand properly the process of retrieval of pertinent genetic information during the development of multicellular organisms, one should study those regulatory mechanisms that are a consequence of the level of complexity exhibited by such organisms and are thereby relatively restricted in their occurrence. An example of this type of regulation is provided by the phenomenon of "dosage compensation".

Many species of higher organisms possess a regulatory mechanism which compensates for differences in the number of given genes in males and females. For example, in species where the male is the heterogametic sex, the phenotypes produced by many sex-linked genes are identical in males (with one X chromosome and, therefore, one dose of such genes) and in females (with two X chromosomes and, therefore, double the dose of such genes). This phenomenon was first described by Bridges (1) in *Drosophila melanogaster* and termed "dosage compensation" by Muller (Muller et al 2). Its evolutionary implications were first considered by Stern (3) and then by Muller who reviewed two decades of experimental and theoretical considerations in his famous Harvey Lecture on the subject of genetic adaptation (4). In another famous lecture, Stern provided the last comprehensive discussion of dosage compensation in *Drosophila* to date (5).

---

[1] This paper is dedicated with affection and gratitude to Professor Curt Stern, in honor of his seventieth birthday.

The purpose of the present review is to retrace as succinctly as possible the historical development of the concept of dosage compensation in *Drosophila*, to review new experimental evidence accumulated over the past dozen years, and to discuss various hypotheses and models that have been proposed to account for the compensation. Since most of the analysis has been performed with *D. melanogaster*, all statements made during the course of this review will, therefore, refer to this species unless otherwise indicated. The survey of the literature for this review was concluded in February, 1973.

X-Linked Gene Activity in Diploid Males and Females

PARAMETERS The original studies on dosage compensation made use of mutant alleles of X-linked genes, especially hypomorphic (leaky) mutants of genes responsible for the pigmentation of the eyes. Phenotypic measurements, showing equality between homozygous females and hemizygous males, consisted of visual estimation of eye color intensity (1,4,5) or of spectrophotometric determination of extracted pigment (Smith & Lucchesi 6). Such spectrophotometric measurements were performed on flies with wildtype genotypes (4,6) and showed that dosage compensation applies to normal alleles as well.

More recently, dosage compensation has been established for activity levels of enzymes whose structural genes are located on the X chromosome. Komma (7) and Seecof et al (8) reported the same level of 6-phosphogluconate dehydrogenase [6PGD, locus: 0.9 (9)] and of glucose-6-phosphate dehydrogenase [G6PD, locus: 63 (10)] activity in males and females. Similar results were obtained with respect to tryptophan pyrrolase [locus: 33 (11)] by Tobler et al (12) and by Baillie & Chovnick (13). Finally, equivalent fumarase specific activity levels were reported for males and females by Whitney & Lucchesi (14); the structural gene of this enzyme has recently been mapped at 19.9 on the X chromosome (15).

The characteristic relationships involved in dosage compensation are illustrated by the following generalized scheme where *g* is an allele of the X-linked gene, *Df*(*g*) represents an X chromosome deficient for the locus of *g*, *Dp*(*g*) represents a duplication for *g* (either a small chromosomal segment attached to its own centromere or inserted in an otherwise normal chromosome). A *g*/*Df*(*g*) female has a level of phenotypic expression for the gene in question which is lower than that of a *g*/Y male. The latter's phenotype is equal to that of a homozygous *g*/ *g* female. In a male, though, two doses of the allele (*g*/*Dp*(*g*)/Y) result in a greater phenotypic expression than that of *g*/Y or *g*/*g*; the same is true of three doses of the allele in a female [*g*/*g*/*Dp*(*g*)]. These relationships are summarized in the following graded series where genotypes are arranged according to amount of gene product: $g/Df(g) < g/g \cong g/\mathrm{Y} < g/g/Dp(g) < g/Dp(g)/\mathrm{Y}$. An example of X-linked enzyme activity measurements is provided in Table 1; the data are from Seecof et al (8).

Dosage compensation is not restricted to adult flies but occurs in larval stages, as well. Lucchesi & Rawls (16) have verified that all of the relationships outlined

TABLE 1. 6PGD activities for various gene dosages.[a]

Genotype	Description	No. 6PGD genes[b]	Enzyme activity[c]
g/Df(g)	Deficiency female	1	2.81± 0.04
g/g	Normal female	2	4.66± 0.03
g/Y	Normal male	1	4.65± 0.10
g/Dp(g)/Y	Duplication male	2	6.26± 0.26

[a] After Seecof et al (8).
[b] Number of 6PGD structural genes per diploid genome.
[c] Expressed as μ moles $\times 10^3$ of NADP reduced/ ml/ min/ mg live weight ± S.E.

above are applicable to third instar larvae with respect to 6PGD and G6PD activities. This is particularly relevant in light of the cytological observations performed on larval salivary gland chromosomes to be discussed below.

ACTIVITY OF BOTH X CHROMOSOMES IN FEMALE SOMA In contrast to mammals where dosage compensation appears to be mediated by the heterochromatization and inactivation of one of the two X chromosomes present in somatic cells of females [see Lyon (17) for a recent review], there is in *Drosophila* good evidence that both X chromosomes function in all cells of the female soma. It has been known for some time that females heterozygous for X-linked recessive mutants such as *y* (yellow hypodermis, bristles and hair), *w* (white eyes), *f* (forked bristles and hair), etc., fail to exhibit any visible mosaicism for these characters and appear uniformly wildtype. This argues against the inactivation in some cells, at some time during development, of the X chromosome bearing the wildtype allele of the mutants in question.

Additional evidence has been provided by Kazazian et al (18). Making use of *Drosophila* strains exhibiting either a fast or a slow electrophoretic variant of 6PGD, these workers obtained heterozygous females with an intermediate band in addition to the two parental bands. Such hybrid enzyme can best be explained by the dimeric nature of the enzyme molecules and the functioning of both alleles in the same cells of heterozygous females. Analogous conclusions were reached by Steele et al (19) with respect to G6PD.

CYTOLOGICAL MANIFESTATIONS OF DOSAGE COMPENSATION A sex difference in the morphology of the X chromosome is visible in certain tissues of *Drosophila*. This difference was first reported by Offermann (20) who observed that while the diameter of the paired X chromosomes in larval salivary gland nuclei of the female is comparable to that of the paired autosomes, in the male the width of the single X is comparable to the width of an autosomal pair. Making use of certain interspecific hybrids (*D. insularis* x *D. tropicalis*) in which somatic pairing of polytene chromosomes in the larval salivary gland nuclei does not take place, Dobzhansky (21) was able to compare a unipartite X chromosome in a female genome with one in a male genome, confirming the sexual dimorphism reported by Offermann. Using feulgen or ultra-violet microspectrophotometry, Rudkin (Aronson et al 22, Rudkin 23) determined that the single X chromosome in the male had the same amount of DNA as either of the two X chromosomes in the female, in spite of the fact that its volume seemed to be equivalent to that of both X chromosomes. Rudkin also reported that the single X contains approximately 10% more protein (low in tryptophan and phenylalanine) than either of the two X's in the female.

The functional significance of the cytological observations just described was established by Mukherjee & Beermann (24) and by Mukherjee (25). Following short pulse exposure of larval salivary glands to ^{3}H-uridine these investigators monitored the level of chromosomal RNA synthesis along the polytene chromosomes by means of autoradiography. The relative distribution of silver grains over the two paired X chromosomes and a control region of paired autosomes in female cells was equal to the relative distribution over the single X and the comparable region of paired autosomes in males. Mukherjee & Beermann noted that ^{3}H-uridine incorporation over occasional unpaired regions of X chromosomes sensibly exceeded half of the value obtained over the same regions in instances where the latter were paired. Nevertheless they emphasized that even if the male X grain counts were corrected for its unpaired condition, this chromosome's transcriptional activity would remain significantly greater than that of either of the two female X's. Holmquist (26), using the same type of experimental material, derived two independent methods to correct for the geometrical differences between a paired and an unpaired chromosome. This author calculated that the actual rate of RNA synthesis along the single male X was 43% greater than the rate along either of the two X's in a female. Holmquist underscored the fact that such a limited excess in the amount of chromosomal RNA of the male X is surprising in view of the 100% increase in phenotypic products of male X-linked genes.

The observations just discussed relate to whole chromosomes or segments of chromosomes spanning several polytene bands. Similar results are obtained, with few exceptions, if one compares the activity of single puffs, or the incorporation of ^{3}H-uridine over single chromomeres in males and females. Korge (27,28) studied RNA synthesis along eleven active loci at the tip of the X chromosome. Although it was necessary to pool the grain counts over certain adjacent loci, excellent equivalence in RNA synthesis of the resulting five groups was found in males and females. Similar results were obtained by Holmquist (26) with seven X chromosome intervals, each containing only one major band and interband. Chatterjee & Mukherjee (29) found that four or five out of a total of 17 puffing sites studied along the X chromosome of *D. hydei* showed significantly greater ^{3}H-uridine incorporation in paired female X's than in males, concluding that the activity of the majority of X chromomeres is compensated in this species as well.

CELLULAR AUTONOMY OF THE LEVEL OF X CHROMOSOME ACTIVITY Making use of an unstable ring X chromosome, Lakhotia & Mukherjee (30) obtained larval gynandromorphs whose tissues, including the salivary glands, were mosaics of female (ring-X/rod-X) and male (rod-X/0) cells. They found that the morphological appearance (width) and RNA synthetic activity of the single X chromosome in such larvae was the same as in normal males and equivalent to the width and activity of the double X in female cells.

Lakhotia & Mukherjee (30) also reported some preliminary results using the X-linked eye pigmentation hypomorphic mutant allele w^a (white-apricot). Adult gynandromorphs, of chromosomal constitution similar to the ones just discussed, occasionally exhibited a whole eye made up of male tissue (with one X and, therefore, a single dose of the w^a allele) while the other eye was made up of female tissue (with two X's and double the dose of the w^a allele). Such eyes were identical phenotypically, indicating that dosage compensation is autonomous at this level of gene activity as well.

DOSAGE COMPENSATION IN X-AUTOSOME TRANSLOCATIONS The activity of X chromosome genes translocated to an autosome has already been suggested, in the section entitled "parameters". Barring position effects (which usually affect most seriously the activity of those genes closest to the breakpoints of the rearrangement), X chromosome genes behave as such, regardless of their location in the genome. A special stock in which a small segment of the X chromosome bearing the structural gene for 6PGD is inserted into chromosome 3 was used by Seecof et al (8) and by Gvozdev et al (31) to produce females with one, two, or three doses and males with one or two doses of the structural gene in question. The direct correlation between 6PGD activity levels and structural gene dosage in each sex (see Table 1) indicates that genes on an X chromosome interval exhibit dosage compensation regardless of their location in the genome. Similar results were obtained for tryptophan pyrrolase activity by Tobler et al (12) using an X-3 and an X-Y translocation.

This is also true at the level of RNA synthesis. By monitoring ^{3}H-uridine incorporation along larval salivary gland polytene chromosomes, Holmquist (26) compared the transcription rate of an X chromosome interval translocated into the left arm of chromosome 3, with that of the same interval on a normal X chromosome. In male cells, the translocated segment exhibited the same enlarged and pale appearance and the same rate of RNA synthesis per unit DNA as those of the homologous interval on the normal X chromosome. In female cells, the translocated segment's activity per unit DNA was the same as that of the homologous interval of the bipartite X; its cross-sectional area was about half that of the bipartite chromosome.

Lakhotia (32) reported on the "functional morphology" of an autosomal segment (from the right arm of chromosome 3) inserted into an X chromosome. In salivary gland nuclei of male larvae this X chromosome was morphologically similar (enlarged, pale) to a normal X in control males; the inserted autosomal segment, on the other hand, was similar to the homologous region in a normal chromosome 3.

ROLE OF SEXUALITY IN DOSAGE COMPENSATION Since males and females may be held to represent inherently different physiological systems, the possibility that dosage compensation is brought about by this difference and that sex physiology, per se, plays a role in the regulatory mechanism, was originally evoked by Goldschmidt (33). Evidence in support of this contention was obtained by Komma (7). This author made use of two autosomal mutants, transformer (*tra*) and doublesex (*dsx*) which affect sex differentiation. The *tra* mutant alters the development of putative females towards extreme pseudo-maleness but has no apparent effect on the development of male embryos. Male and female embryos homozygous for *dsx*, on the other hand, develop as intersexes. By assuming that there is a gradient of physiological milieu from "femaleness" to "maleness" and that the relative "femaleness" or "maleness" of the milieu acts to decrease or increase, respectively, gene action, Komma predicted that two-X pseudo-males and intersexes should have a higher gene activity than their normal sisters (since they would be more male-like) while single-X intersexes should have a lower activity than their normal brothers (since they would be more female-like). Measuring G6PD activity in normal and abnormal flies, Komma verified the expectation for two-X pseudo-males and intersexes although he failed to substantiate it for single-X intersexes. Furthermore, he could not demonstrate any effect of sexuality on 6PGD activity.

On the other hand studies by Muller (4) on the eye pigmentation of two-X individuals homozygous for an X-linked eye color hypomorphic mutant and for *tra* showed that such extreme male-like intersexes were phenotypically identical to their sexually normal sisters. Smith & Lucchesi (6) reexamined more fully the possibility that sex physiology plays a role in dosage compensation. Using *dsx* to alter sex physiology, the spectrophotometric determination of red pigment as a measure of gene product, and a controlled uniform genetic background to reduce inherent variability among the different types of flies compared, these authors concluded that sex differences between males and females cannot account for dosage compensation.

Mention must be made of the fact that the Y chromosome appears to play no role in the dosage compensation of X-linked genes. This was validated at the level of a terminal phenotypic product (such as an eye pigment) by Muller (4), at the level of enzyme activity by Seecof et al (8) and by Lucchesi & Rawls (34), and at the level of RNA synthesis along larval salivary gland chromosomes by Kaplan & Plaut (35).

X-Linked Gene Activity in Heteroploids

In this section, measurements of X-linked gene activity performed in genotypes with differing numbers of structural genes, of X chromosomes, and of sets of autosomes per nucleus, will be discussed.

TRIPLOIDS All of the evidence reviewed to this point supports the contention that the activity of each X-linked gene dose in a two-X diploid genome is half that of the single gene dose in a single-X diploid genome. Lucchesi & Rawls (34) investigated whether this relationship extends to the activity of X-linked genes in a three-X triploid genome, i.e. whether a further reduction in activity per X-linked gene dose (and a concomitant proportional reduction in autosomal gene activity, necessary to maintain the normal balance between X-linked and autosomal gene products) occurs in a triploid. X-linked (6PGD and G6PD) and autosomal [α-glycerophosphate dehydrogenase, GPDH, locus: 20.5 on chromosome 2 (36); NADP-dependent isocitrate dehydrogenase, IDH, locus: 27.1 on chromosome 3 (37)] enzyme activity levels per unit of DNA were found to be identical in diploid and triploid females, leading to the conclusion that the contribution of each dose of a given gene to the level of enzyme activity is equal in diploid and triploid cells. (The presence of an extra genome in the latter leads to a proportional increase in cell constituents resulting in a proportional increase in cell size.)

TRIPLOID INTERSEXES Triploid intersexes (bearing two X chromosomes and three sets of autosomes) normally occur among the offspring of triploid females. Although a certain range of phenotypic sexuality is possible, the relative "maleness" or "femaleness" of intersexes produced is characteristic of the particular triploid line used. Lucchesi & Rawls (38) measured autosomal (IDH) and X-linked (6PGD) gene activity in phenotypically male-like intersexes produced by the same triploid females used in the experiments just discussed. They found that enzyme activity levels are equivalent in such intersexes and in triploid females and, therefore, that X-linked gene activity expressed on a *per gene* basis is greater in these triploid intersexes than in triploid or diploid females. This conclusion is consistent with results of measurements of X chromosome activity performed by autoradiographic monitoring of ^{3}H-uridine incorporation along larval salivary gland polytene chromosomes by Maroni & Plaut (39, 40). These authors found the same relative level of grain counts over the bipartite X of intersexes' nuclei and over the tripartite X of triploids, indicating that the rate of RNA synthesis per X chromosome is greater in the former than in the later. Maroni & Plaut (41) also reported corroborating data for G6PD activity in triploid intersexes.

METAFEMALES These females have three X chromosomes and two sets of autosomes. Stern (5) reported that such females whose X chromosomes are homozygous for an eye pigment hypomorphic mutant allele have an intensity of eye color closely resembling that of normal diploid females and males. Although accurate enzyme activity or cytogenetic measurements are not yet available to substantiate this contention, Stern's observation suggests that each dose of an X-linked gene is less active in a metafemale than in a diploid female.

Table 2. Localization of structural genes and enzyme activity levels in males and females of different *Drosophila* species

Enzymes	*D. melanogaster*		*D. pseudoobscura*		*D. willistoni*	
G6PD	X	(10)	XL or R	(47)	—	
male/female[a]	1.00	(8)	1.04	(42)	—	
Esterase[b]	3L	(45)	XR	(42)	—	
male/female	1	(46)	1.05	(42)	—	
IDH	3L	(37)	—		XR	(48)
male/female	1.17	(38)	—		1.00	(42)

[a] Ratio of specific activities measured in crude extracts of 1-24 hr old adults.
[b] Esterase-6 in *D. melanogaster*, esterase-5 in *D. pseudoobscura*; the two enzymes are probably homologous.

Dosage Compensation in Other Drosophila Species

Investigations of X-linked gene activity in other species of *Drosophila* may serve the purpose of determining whether the phenomenon of dosage compensation is widespread throughout the genus. Such a study is that of Chatterjee & Mukherjee (29) with *D. hydei*. As previously mentioned, these authors reported that single X chromosome activity in males is largely equivalent to that of the double X in females. Muller (4) pointed out that an understanding of the evolution of the regulatory mechanism may be sought in species whose X chromosome consists of two parts or arms, one descended from the original rod-shaped X (characteristic of species such as *D. melanogaster*), and the other corresponding to an original autosomal arm. In fact, Muller attempted a preliminary comparison of phenotypic levels mediated by X-linked hypomorphic mutants in males and females of *D. pseudoobscura* and concluded that further evidence involving exact dosage studies would be most valuable. Factual information of this nature has been obtained by Abraham & Lucchesi (42) in *D. pseudoobscura* and *D. willistoni*. The X chromosome in both species is metacentric with one arm homologous to the X and the other homologous to the left arm of chromosome 3 of *D. melanogaster* (43, 44). Abraham & Lucchesi measured levels of male and female activity of enzymes whose structural genes are located on those elements of the two selected species' genomes which are homologous to the *D. melanogaster* X and 3L. Their results, suggesting dosage compensation for all X-linked loci, are summarized in Table 2.

Discussion

Prior to considering various general aspects of dosage compensation emerging from the body of experimental evidence outlined in the previous sections, some of the limitations inherent in the nature of this evidence must be underscored. Although it is clear that by monitoring ^{3}H-uridine incorporation along polytene chromosomes one measures chromosomal RNA synthesis, i.e. transcription, it is also quite probable that only a small fraction of this RNA finds its way to the cytoplasm as *bona fide* messenger RNA to be translated into gene product (49,50). One must, therefore, conclude that differential synthesis of m-RNA in males and females has not been directly demonstrated. Similarly, although X-linked enzyme activity levels are equivalent in males and females, an equality in number of enzyme molecules synthesized has never been directly demonstrated. Lastly, phenotypic products such as eye pigments are the end result of long chains of biochemical reactions controlled by numerous autosomal and X-linked genes; such products, therefore, may be considered too remote to serve as a measure of specific genetic regulation. These considerations notwithstanding, the congruity of the observations made at the levels of terminal phenotypic product, enzyme activity and chromosomal RNA synthesis provides reasonable justification for proceeding with the discussion.

SUMMARY OF FEATURES The salient features of the phenomenon of dosage compensation appear to be the following:

(*a*) An X-linked gene is twice as active in a male as it is in a female. This is true whether one or two doses of the gene in question are present in the male, and whether the female is a diploid with one, two, or three doses or a triploid with two or three doses per genome.

(*b*) X chromosome function can occur at a level that is intermediate between the normal male and female levels (viz. gene activity in triploid intersexes; the

Table 3. Summary of X-linked gene activity measurements in diploid and heteroploid genotypes

	Genotype	Description	No. structural[a] genes	Gene product[b] per cell	Gene product[b] per gene dose	Documentation phenogenetic[c]	Documentation cytological[d]
a.	g/Y;AA	Male	1	2	2	(7,8,12-14,31)	(26-29)
b.	g/Dp(g)/Y;AA	Male-duplicated	2	4		(8,12,31)	(26-28)
c.	g/Df(g);AA	Female-deficient	1	1		(8,12,31)	(27,28)
d.	g/g;AA	Female	2	2		(7,8,12-14,31)	(26-29)
e.	g/g/Dp(g);AA	Female-duplicated	3	3	1	(8,12,31)	(26-28)
f.	g/g/g/;AAA	Female-triploid	3	3		(34,41)	(39,40)[f]
g.	g/g/Df(g);AAA	Female-triploid, deficient	2	2		(34,41)	—
h.	g/g;AAA	Intersex-male like	2	3	1.5	(38,41)	(39,40)[f]
i.	g/g/Dp(g);AAA	Intersex-male like, duplicated	3	4.5		(38)	—

[a] Number of doses of an X-linked gene under study per genome.
[b] The gene product per cell in a normal diploid female is set at 2, and the product per gene at 1; all other values are relative to these.
[c] Restricted to measurements of X-linked enzyme activities.
[d] Measurements of ^{3}H-uridine incorporation in larval salivary gland nuclei.
[e] The only available data here is unexpectedly at odds with phenogenetic observations: females heterozygous for a small deficiency produced nearly as much RNA as normal females, in the region spanned by the deficiency.
[f] Total X chromosome activity was assured rather than that of single chromomeres.

characteristic intermediate activity level per gene is maintained whether there are two or three doses of the structural gene present in the genome).

The above statements are represented in Table 3. Some additional features of dosage compensation are:

(*c*) The regulatory mechanism appears to affect the level of X chromosome transcription.

(*d*) The activity of the genes on a segment of X chromosome is not affected by the relocation of the segment within the genome and is directly correlated to the relative number of X chromosomes/sets of autosomes in the genome.

MODELS Dosage compensation can be attained by (*a*) a decrease of X-linked gene activity such that the amount of cellular product resulting from two gene doses in a female is reduced to the level resulting from a single gene dose in a male; (*b*) an increase of X-linked gene activity such that the amount of cellular product resulting from a single gene dose in a male is augmented to the level resulting from a double gene dose in a female; or (*c*) a combination of both mechanisms. Much of the phenogenetic data has been interpreted as evidence for the first type of mechanism (4, 8, 51). Cytological evidence, on the other hand, suggests that compensation may be achieved by the second type of mechanism (26,52), although exception has been taken to this conclusion by Muller & Kaplan (51). Mention must be made that Mukherjee et al (52) and Lakhotia (32) offer the observation that the X chromosome in male larval salivary gland nuclei replicates faster than the paired X's in female nuclei or the autosomes (53) as additional cytological evidence in support of a regulatory mechanism operating in the male on the single dose of X-linked genes. It must, nevertheless, be emphasized that there is, to date, no definitive empirical basis for distinguishing between the two views; therefore models which are based on either of them must be considered.

A reduction of gene activity in females to a level characteristic of males can be rationalized by assuming that there exists on the X chromosome a gene whose activity is itself not dosage compensated and whose product has a repressive effect on the transcription of all X-linked genes. If the probability of transcription for an X-linked gene were inversely proportional to the concentration of the regulatory molecule and directly proportional to the number of doses of the gene, the levels of product per cell registered for genotypes a–e of Table 3 would be expected. In the case of genotypes f–i, one must consider the lower specific activity of the regulatory molecule due to the proportionately larger size of triploid cells in comparison to diploid cells with the same number of X chromosomes and, therefore, doses of the compensator locus. A reduction of the repressive effect by one third yields the levels of products per cell expected for these genotypes.

Equalization of X-linked gene products in males and females by a regulatory mechanism which enhances gene activity (positive regulation) can be explained by postulating that there exists an autosomal gene whose activity is dosage-dependent and whose product is necessary for the transcription of all X-linked genes. Assuming a low number of regulatory molecules produced per compensator locus and numerous sites of action for such molecules along the X chromosome (one for each transcriptional unit), an increase in the number of X chromosomes per genome would effectively titrate the intranuclear concentration of these molecules. The probability of transcription of an X-linked gene would be directly proportional to the concentration of regulatory molecules and the number of doses of the gene, and inversely proportional to the number of X chromosomes in the genome. This simple relationship, comprising the essential features of a model proposed by Maroni & Plaut (40, 41), is sufficient to account for all levels of product per cell and activity per gene listed in Table 3.

EVOLUTIONARY CONSIDERATIONS The existence, in species such as *D. pseudoobscura* and *D. willistoni*, of dosage compensation for that segment of the X chromosome that corresponds to an autosomal arm in some ancestral form argues in favor of a regulatory mechanism that increases X-linked gene activity in flies. The rationale is based (*a*) on the fact that, in a species like *D. melanogaster*, the levels of various gene products mediated by the structural genes on an autosomal arm (such as 3L) are equivalent in males and females that possess equal doses of these structural genes, and (*b*) on the reasonable assumption that these levels of gene products are optimal with respect to those of all other genes in the genome. When, during the course of evolution, an ancestral autosomal arm becomes part of the X chromosome [as in the case of *D. pseudoobscura* and *D. willistoni* (54, 55)] and is represented only once in a male genome, the normal optimal balance of its gene products to other gene products can only be maintained in males if the activity of the single dose of genes on this formerly autosomal arm is appropriately increased.

ACKNOWLEDGMENTS

I wish to express my gratitude to the following colleagues who have provided me with encouragement, stimulating exchanges, and constructive criticisms during the course of my work on dosage compensation: Drs. W. Beermann, B.I. Kiefer, G. Maroni, P.D. Smith, and J.B. Whitney. The same acknowledgment is due to student associates: I. Abraham and G.C. Bewley. I am most particularly indebted to John M. Rawls for his indefatigable contribution to the experimental side of the study of compensation and for patiently apprising me of valuable conceptual insights into the basis of this regulatory phenomenon. Thanks are due to Drs. D.P. Costello and C.A. Hutchison for critically reading this manuscript prior to its submission.

Research in my laboratory has been supported by Research Grant GM-15691 from the National Institutes of Health.

Literature Cited

1. Bridges, C.B. 1922. The origin of variations in sexual and sex-limited characters. *Am. Natur.* 56:51–63
2. Muller, H.J., League, B.B., Offermann, C.A. 1931. Effects of dosage changes of sex-linked genes, and the compensatory effect of other gene differences between male and female. *Anat. Rec.* 51 (suppl.):110
3. Stern, C. 1929. Über die additive Wirkung multipler Allele. *Biolog. Zentralbl.* 49:261–90
4. Muller, H.J. 1950. Evidence of the precision of genetic adaptation. *Harv. Lect.* Ser. XLIII:165–229
5. Stern, C. 1960. Dosage compensation - development of a concept and new facts. *Can. J. Genet. Cytol.* 2:105–18
6. Smith, P.D., Lucchesi, J.C. 1969. The role of sexuality in dosage compensation in Drosophila. *Genetics* 61: 607–18
7. Komma, D.J. 1966. Effect of sex transformation genes on glucose-6-phosphate dehydrogenase activity in *Drosophila melanogaster. Genetics* 54: 497–503
8. Seecof, R.L., Kaplan, W.D., Futch, D.G. 1969. Dosage compensation of enzyme activities in *Drosophila melanogaster. Proc. Nat. Acad. Sci. USA* 62: 528–35
9. Young. W.J. 1966. X-linked electrophoretic variation in 6-phosphogluconate dehydrogenase in *Drosophila melanogaster. J. Hered.* 57:58–60
10. Young, W.J., Porter, J.E., Childs, B. 1964. Glucose-6-phosphate dehydrogenase in *Drosophila*: X-linked electrophoretic variants. *Science* 143: 140–41
11. Baglioni, C. 1959. Genetic control of tryptophan peroxidase-oxidase in *Drosophila melanogaster. Nature* 184: 1084–85
12. Tobler, J., Bowman, J.T., Simmons, J.R. 1971. Gene modulation in *Drosophila*: dosage compensation and relocated v^+ genes. *Biochem. Genet.* 5: 111–17
13. Baillie, D.L., Chovnick, A. 1971. Studies on the genetic control of tryptophan pyrrolase in *Drosophila melanogaster. Mol. Gen. Genet.* 112:341–53
14. Whitney, J.B., III, Lucchesi, J.C. 1972. Ontogenetic expression of fumarase activity in *Drosophila melanogaster. Insect Biochem.* 2:367–70
15. Madhavan, K., Ursprung, H. 1973. The genetic control of fumarate hydratase (fumarase) in *Drosophila melanogaster. Mol. Gen. Genet.* 120: 379–80
16. Lucchesi, J.C., Rawls, J.M. 1973. Unpublished observations
17. Lyon, M.F. 1972. X-chromosome inactivation and developmental patterns in mammals. *Biol. Rev.* 47:1–35
18. Kazazian, H.H., Jr., Young, W.J., Childs, B. 1965. X-linked 6-phosphogluconate dehydrogenase in *Drosophila*: subunit associations. *Science* 150:1601–02
19. Steele, M.W., Young, W.J., Childs, B. 1969. Genetic regulation of glucose-6-phosphate dehydrogenase activity in *Drosophila melanogaster. Biochem. Genet.* 3:359–70
20. Offermann, C.A. 1936. Branched chromosomes as symmetrical duplications. *J. Genet.* 32:103–16
21. Dobzhansky, Th. 1957. The X-chromosome in the larval salivary glands of hybrids *Drosophila insularis* x *Drosophila tropicalis. Chromosoma* 8: 691–98
22. Aronson, J.F., Rudkin, G.T., Schultz, J. 1954. A comparison of giant X-chromosomes in male and female *Drosophila melanogaster* by cytophotometry in the ultraviolet. *J. Histochem. Cytochem.* 2:458–59
23. Rudkin, G.T. 1964. The proteins of polytene chromosomes. *The Nucleohistones*, eds. Bonner & Ts'o, 184–92. San Francisco: Holden Day
24. Mukherjee, A.S., Beermann, W. 1965. Synthesis of RNA by the X-chromosomes of *Drosophila melanogaster* and the problem of dosage compensation. *Nature* 207:785–86
25. Mukherjee, A.S. 1966. Dosage compensation in *Drosophila*: an autoradiographic study. *The Nucleus* 9: 83–96
26. Holmquist, G. 1972. Transcription rates of individual polytene chromosome bands: effects of gene dose and sex in *Drosophila. Chromosoma* 36: 413–52
27. Korge, G. 1970. Dosage compensation and effect for RNA synthesis in chromosome puffs of *Drosophila melanogaster. Nature* 225:386–88
28. Korge, G. 1970. Dosiskompensation und Dosiseffekt für RNS-Synthese in Chromosomen-Puffs von *Drosophila melanogaster. Chromosoma* 30:430–64
29. Chatterjee, S.N., Mukherjee, A.S. 1971. Chromosomal basis of dosage compensation in *Drosophila*. V. Puff-

wise analysis of gene activity in the X-chromosome of male and female *Drosophila hydei. Chromosoma* 36: 46–59

30. Lakhotia, S.C., Mukherjee, A.S. 1969. Chromosomal basis of dosage compensation in *Drosophila*. I. Cellular autonomy of hyperactivity of the male-X chromosome in salivary gland and sex differentiation. *Genet. Res.* 14:137–50
31. Gvozdev, V.A., Birstein, V.J., Faizullin, L.Z. 1970. Gene dependent regulation of 6-phosphogluconate dehydrogenase activity of *D. melanogaster. Drosophila Inform. Serv.* 45:163
32. Lakhotia, S.C. 1970. Chromosomal basis of dosage compensation in *Drosophila*. II. DNA replication patterns in an autosome-X insertion in *D. melanogaster. Genet. Res.* 15:301–07
33. Goldschmidt, R.B. 1954. Different philosophies of genetics. *Science* 119: 703–10
34. Lucchesi, J.C., Rawls, J.M. 1973. Regulation of gene function: a comparison of enzyme activity levels in relation to gene dosage in diploids and triploids of *Drosophila melanogaster. Biochem. Genet.* 9:41–51
35. Kaplan, R.A., Plaut, W. 1968. A radioautographic study of dosage compensation in *Drosophila melanogaster. J. Cell Biol.* 39:7la
36. Grell, E.H. 1967. Electrophoretic variants of α-glycerophosphate dehydrogenase in *Drosophila melanogaster. Science* 158:1319–20
37. Fox, D.J. 1970. The soluble citric acid cycle enzymes of *Drosophila melanogaster*. I. Genetics and ontogeny of NADP-linked isocitrate dehydrogenase. *Biochem. Genet.* 5:69–80
38. Lucchesi, J.C., Rawls, J.M. 1973. Regulation of gene function: a comparison of X-linked enzyme activity levels in normal and intersexual triploids of *Drosophila melanogaster. Genetics.* 73:459–64
39. Maroni, G., Plaut, W. 1972. A cytological analsyis of dosage compensation in *D. melanogaster* triploids. *Genetics* 71:s37
40. Maroni, G., Plaut, W. 1973. Dosage compensation in *Drosophila melanogaster* triploids. I. Autoradiographic study. *Chromosoma* 40:361–77
41. Maroni, G., Plaut, W. 1973. Dosage compensation in *Drosophila melanogaster* triploids. II. Glucose-6-phosphate dehydrogenase activity. *Genetics*. In press
42. Abraham, I., Lucchesi, J.C. 1973. Dosage compensation in *Drosophila pseudoobscura* and *D. willistoni*. In preparation
43. Sturtevant, A.H., Tan, C. C. 1937. The comparative genetics of *Drosophila pseudoobscura* and *Drosophila melanogaster. J. Genet.* 34:415–32
44. Sturtevant, A.H., Novitski, E. 1941. The homologies of the chromosomal elements in the genus Drosophila. *Genetics* 26:517
45. Wright, T.R.F. 1963. The genetics of an esterase in *D. melanogaster. Genetics* 48:787–801
46. Aronstam, A.A., Kusin, B.A., Korochkin, L.I. 1973. Effect of actinomycin D and cycloheximide on the activity of esterase-6 in *D. melanogaster. Isozyme Bull.* 6:24
47. Prakash, S., Lewontin, R. C., Hubby J. L. 1969. A molecular approach to the study of genic heterozygosity in natural populations. IV. Patterns of genic variation in central, marginal and isolated populations of *Drosophila pseudoobscura. Genetics* 61:841–55
48. Lakovaara, S., Saura, A. 1972. Location of enzyme loci in chromosomes of *Drosophila willistoni. Experientia* 28: 355
49. Darnell, J.E. 1968. Ribonucleic acids from animal cells. *Bact. Rev.* 32: 262–90
50. Armelin, H.A., Margues, N. 1972. Transcription and processing of ribonucleic acid in *Rhynchosciara* salivary glands . I. Rapidly labeled ribonucleic acid. *Biochemistry* 11:3663–72
51. Muller, H.J., Kaplan, W.D. 1966. The dosage compensation of Drosophila and mammals as showing the accuracy of the normal type. *Genet. Res.* 8:41–59
52. Mukherjee, A.S., Lakhotia, S.C., Chatterjee, S.N. 1968. On the molecular and chromosomal basis of dosage compensation in *Drosophila. Nucleus*, Suppl.:161–73
53. Berendes, H.D. 1966. Salivary gland function and chromosome puffing patterns in *Drosophila hydei. Chromosoma* 17:35–77
54. Patterson, J.T., Stone, W.S. 1952. *Evolution in the Genus Drosophila*, 167–70. New York: MacMillan. 610 pp.
55. Buzzati-Traverso, A.A., Scossiroli, R.E. 1955. The "Obscura group" of the genus Drosophila. *Advan. Genet.* 7: 47–92

GENETICS OF RESISTANCE TO ENVIRONMENTAL STRESSES IN *DROSOPHILA* POPULATIONS 3055

P. A. Parsons[1]

Department of Genetics and Human Variation, La Trobe University, Bundoora, Victoria, Australia

Introduction

Recently there has been an increasing awareness of the need to study the genus *Drosophila* in the wild, to assess the various ecological and behavioral parameters determining its distribution (1,2). It must be stressed that the classic work of Dobzhansky and his collaborators has over the years provided much information (3,4) especially in *D. pseudoobscura* and closely related species such as *D. persimilis*. It is one of the curiosities of this sort of scientific endeavor, that the species *D. melanogaster*, perhaps best known genetically of all species in the genus, has been studied extensively under laboratory conditions for over 60 years, with little thought given to the situation in the wild. In this review, some environmental stresses, natural and artificial, for which genetic variability has been found within and between species of *Drosophila* will be examined, as they are important in determining the distribution of species of the genus in the wild. Stresses to be considered are temperature extremes, desiccation, anoxia as assessed by resistance to long-term exposure to CO_2, ^{60}Co γ-radiation, various chemical stresses (some of importance in nature because of the activities of man), and ethyl alcohol.

Genetic heterogeneity for many metrical traits in wild populations of *D. melanogaster* has been detected by setting up cultures from single inseminated females collected in the wild, and following the trait over a number of generations. A good example is the incidence of flies with more than the normal

[1] The work of the author has been supported by the Australian Research Grants Committee, and the Australian Institute for Nuclear Science and Engineering (Section 4). I am indebted to Dr. Lee Ehrman (State University of New York at Purchase) for her helpful criticism of the manuscript.

4 scutellar chaetae (termed additional chaetae for this purpose), which was followed for 45 generations in 3 strains collected at Leslie Manor, Victoria (referred to as LM strains) in late 1963. Each strain had a characteristic incidence of additional chaetae over this time, suggesting that the differences between strains were the result of genetic differences between gravid founder females (5). Diallel cross experiments indicated that the strains differed in additive genes. Further experiments with 15 LM strains collected in late 1965 showed that each strain maintained a characteristic incidence of additional chaetae over a period of 9 generations. As the initial founder females differed in scutellar chaeta number genes, the wild population must be polymorphic for the genes (6). These conclusions agree with those of Lewontin and Hubby (7) and other authors (8-10), in arguing for a high level of polymorphism in natural populations of outbred species. Similar results have been presented for other morphological traits such as sternopleural and abdominal chaeta numbers (11, 12), for behavioral traits such as mating speed and duration of copulation (13), and for the various environmental stresses to be discussed in this review.

Genetic heterogeneity can also be studied by carrying out selection experiments, but the time scale is longer. However, the selection experiment approach is useful for very specific chemical stresses such as insecticides (14). More powerful is the technique of ascertaining which strains set up from single inseminated females are extreme for a trait, and then basing directional selection on them (15). For scutellar chaeta number, this procedure led to extremely rapid responses to selection (16). Following selection, the genetic architecture of a trait can be studied in detail, using special stocks available in some species of *Drosophila*, especially *D. melanogaster*. In some cases this has led to the actual localization of genes controlling quantitative traits (15, 17, 18 for references).

2. *Environmental Stresses in Nature*

TEMPERATURE Ogaki & Nakashima-Tanaka (19) found that 2 wild strains of *D. melanogaster* were heat tolerant, and 2 mutant strains were heat susceptible. The genetic basis of heat resistance in one of the wild strains, Hikone-H, appeared to be dominant and involved chromosome 3. The ability to survive high temperature was studied (20) in *D. melanogaster* in the three LM strains referred to above by exposing adult flies aged 7 days to a high temperature shock (33.5°C for 24 hours). After the shock, the flies were returned to normal culture temperature, 25°C, to obtain percentage mortalities 7 days later. This was done between the 7th and 40th generations in the laboratory, and it was found that ability to withstand a high temperature shock varied among strains. This was consistent over the testing period of 33 generations. It can be argued that the differences between strains are genetic in origin, and arise from genetic differences between founder females. An analogous result came from a study of the ability of newly hatched larvae to emerge as adults at 30.5°C (21). The strain most sensitive to the temperature shock of 33.5°C applied to adults showed the poorest percentage emergence when larvae were set up at 30.5°C. Many high temperature sensitive mutants have been isolated in *D. melanogaster* after treatment with mutagens (22, 23). Suzuki (22) considers that they resemble the known missense basis for temperature sensitivity in microorganisms. The most common of these were induced by the alkylating agent ethyl methane sulphonate, and less commonly by mitomycin-C and γ-rays. Cold sensitive as well as heat sensitive mutants were found. Most mutants had a temperature-sensitive or lethal phase at one or two intervals during development, either the embryonic egg stage or the interval from the late third larval instar through the pupal stage. These represent two times of great morphogenetic activity and undoubtedly involve the activation of large numbers of developmental genes. Adult flies that do develop at permissive temperatures seem immune to temperature effects. Clearly much more work is needed on strains from natural populations on relative sensitivities at different stages of the life cycle.

Looking at flies from natural populations from a different point of view, Tantawy & Mallah (24) measured the percentage of eggs yielding progeny for *D. melanogaster* and its sibling species *D. simulans* in Uganda, an area of high temperature. Not surprisingly, superiority was found at high temperatures. The temperatures used by Hosgood & Parsons (20) are extreme, but they occur at Leslie Manor, Victoria, which is characterized by hot summers. Therefore it is reasonable to find some genetic control of the trait, even though flies presumably can avoid extreme temperatures by behavioral means. It is unlikely that temperature genes would be fixed in the population, but would be polymorphic to permit rapid adjustments to environmental changes as they occur. In *D. funebris*, Timofeeff-Ressovsky (25) studied the relative viability of flies from regions covering the major climatic zones of Europe, northern Africa, and into Asiatic Russia at 15°C, 22°C, and 28°C. The northern populations were more resistant to cold temperatures, and the southern to high temperature. Eastern populations, in contrast, showed resistance to both high and low temperatures, which could be accounted for in terms of the more "continental" climate of these regions with very low winter and very high summer temperatures. One must conclude that there are "temperature races" differing by various temperature genes.

The importance of temperature as an environmental factor has been stressed in *D. pseudoobscura*, where the third chromosomes are characteristically polymorphic for inversions in nature. In population cages at 25°C there are characteristic stable equilibria for pairs of inversions, clearly mediated by heterokaryotype advantage. Thus the two sequences Standard (*ST*) and Chiricahua (*CH*) gave at this temperature about 70% *ST* and 30% *CH* (26). This temperature can be regarded as fairly extreme for *D. pseudoobscura*, as it is difficult to culture the species at much higher temperatures in the laboratory, and one must often give this species temporary relief from the stress temperature or risk loss of the line entirely (27). Heterokaryotype advantage has been shown (28-33) to involve a number of fitness factors: (*a*) Selective elimination of homokaryotypes in the larval stages under high competition; (*b*) Heterokaryotype advantage for mating speed; and (*c*) Higher innate capacity for increase of numbers, higher productivity and higher population size of polymorphic compared with monomorphic populations. This argument for the last point is less direct, as it pertains to comparisons between monomorphic and polymorphic populations compared with direct comparison of homokaryotypes and heterokaryotypes.

In contrast, at 16.5°C little change occurred in the population cages (26). This is a temperature at which it is easy to breed the species in the laboratory. At the intermediate temperature of 22°C, polymorphism occurred in some cases, and in others monomorphism for populations containing Arrowhead (*AR*) and *CH* (34). Of those fitness traits referred to above that have been studied at temperatures in the range of 16.5°C–20°C, the advantage of heterokaryotypes or of polymorphic populations is reduced to low levels and in some cases disappears.

Therefore, the *D. pseudoobscura* data indicate greater heterokaryotype advantage under the stress of high temperature. Hence, it is not surprising that cyclical annual changes in inversion frequencies have been observed in nature, and that temperature may be one, but by no means the only, ecological factor involved (35). Altitudinal and geographical changes in inversion frequencies have often been reported in North America (36-38). In *D. persimilis*, a sibling species of *D. pseudoobscura*, altitudinal and geographical clines have been reported (39). In the South American species *D. flavopilosa* (40, 41) altitudinal clines are associated with fitness differences between karyotypes for two temperatures at which the larvae were permitted to develop in the laboratory. In *D. funebris* (42, 43) the incidence of inversions varies seasonally. From laboratory experiments where flies were exposed to +3°C and −2°C, Dubinin & Tiniakov hypothesized that certain inversions are favored during the warm period, but in the cold season this trend is reversed.

Brncic (44) found no seasonal or geographical variations in *D. pavani*. Thus in some species the chromosomal polymorphism is flexible, leading to varying frequencies in different environments due to variations in the selective values of homo- and heterokaryotypes. In other species the chromosomal polymorphism is more rigid, presumably because of less variation in selective values. Further examples of rigid species include *D. subobscura* and *D. willistoni*, and of flexible species, *D. robusta* (44, 45). However, the fact that differences have not been found does not mean that they do not exist. The type of laboratory experiments that have picked up genetic heterogeneity under extreme environments have probably not been widely carried out.

Tolerance to cold has been studied in its own right a few times (apart from indications of variations to cold tolerance already mentioned). It is a relevant stress variable in regions of cold winters. Substantial heterozygote advantage for cold temperature resistance was found in *D. pseudoobscura* (46) in that survival of chromosome III homozygotes was 50% less than that of the heterozygotes, and a much smaller differential was found for chromosome II, although in the same direction. This seems to parallel the heterozygote advantage found at high temperatures already discussed. Similarly (47) the longevity of adults in the presence of food at low temperatures (0^0 - 4°C) is greater for *ST/CH* heterokaryotypes than for the corresponding homokaryotypes. Not much other work has been done on the genetic basis of resistance of cold stress, although there are scattered reports on various overwintering mechanisms (2).

One of the main effects of seasonal cold stress is to reduce populations to very low sizes. Hoenigsberg et al (48) compared *D. melanogaster* populations in European (Hungarian) and neotropical (Colombian) environments, and found that the frequencies of drastic variants (lethals and semilethals having a fitness up to 10 and 50% of normals as homozygotes) were far higher in Colombia (about 35%) than in Hungary (about 10%). The Hungarian ecosystem is such that the population size of *D. melanogaster* undergoes serious periodic shrinkage, whereas the neotropical areas can support very large populations all the year. Other data cited (48) indicate generally a lower frequency of drastics in the more northern populations. Ives (49) studied populations of *D. melanogaster* in South Amherst, Massachusetts where winter temperatures are severe. Overwintering of larvae was found in a rotten apple pile to which apples were continuously added during the winter. On emergence in June and July, many of the adults contained chromosomes with allelic lethals, presumably because they were derived from a few founder individuals. Later in the season, the proportion of individuals carrying allelic lethals fell, and the proportion of nonallelic lethal chromosomes increased, presumably by the mixing of populations, and from mutations giving rise to new lethals. Thus cold-temperature stress, by causing large population shrinkages, can affect the makeup of the gene pool. It is likely that any situation of ecological marginality would have a lower proportion of lethals and semilethals than more central (and presumably ecologically less marginal) populations. The evidence from *D. pseudoobscura* agrees (50).

In relation to environmental stresses of ecological significance comparisons of sibling species are of interest. Tantawy & Mallah (24) found that the percentage emergence of flies from eggs was higher for *D. simulans* than *D. melanogaster* over the range 18°C–25°C, whereas over the entire range studied, 10°C–31°C, *D. melanogaster* had higher average survival. Thus *D. simulans* might be expected to outnumber *D. melanogaster* in regions where temperature extremes are not encountered, but in regions with wide temperature fluctuations, the reverse might be expected. The distribution data in the U.S.A. agree with this hypothesis (51), as *D. simulans* is more common in the south, where the climate is equable throughout the year, and *D. melanogaster* is more common in the north, where winters are severe and wide daily fluctuations in temperature occur. Observations in Australia are restricted, but *D. melanogaster* has been found to outnumber *D. simulans* at Leslie Manor where temperature fluctuations are wide (McKenzie & Parsons, unpublished). In Melbourne where temperature fluctuations are narrower *D. simulans* mainly predominates. The latter situation is also found in subtropical Brisbane, where temperature fluctuations are not wide. Hosgood & Parsons (52) set up 4 strains of *D. melanogaster* and 3 of *D. simulans*, each derived from single inseminated females collected in the wild. All strains were set up at 5 temperatures, 29.5°C, 27.5°C, 25°C, 20°C, and 15°C. After 5 generations, all strains of *D. melanogaster* were living at all temperatures, whereas for *D.*

simulans the 3 strains at 20°C were living and only one at 25°C (which in fact survived for 24 generations). At 29.5°C and 15°C all the *D. simulans* strains had died by the second generation, and at 27.5°C by the third generation. Extrapolating from the laboratory, *D. simulans* is much more restricted in its tolerance to diverse temperatures than *D. melanogaster* as found in nature.

Fluctuating, as opposed to fixed, temperatures have been shown to have effects on overall population fitness. Long (53) divided a *D. melanogaster* population into 4 groups, one of which was kept at a constant temperature of 25°C, and the 3 others under temperature cycles fluctuating between 20°C and 30°C with periods of 1 day, 1 month, and 3 months respectively, while in all cases maintaining an overall average of 25°C. Even though 30°C is extreme for *D. melanogaster*, various tests showed overall population fitnesses to be greater under the regimes of environmental variability. Oshima (54) found that for a diurnally varying temperature of 25° ± 5°C with a mean of 25°C, compared with a constant 25°C, fitnesses of heterozygotes for lethal, semilethal, and quasinormal genes (i.e. those that are not lethal or semilethal as homozygotes) were greater in the fluctuating environment. Additional evidence is provided by Beardmore (55).

Emerging from these studies is the conclusion that in fluctuating environments in *D. melanogaster* (25°C ± 5°C), the additive genetic variance is higher (56) than in a constant environment, and probably the genetic diversity is also higher. This was recently demonstrated (57) in experimental populations of *D. willistoni* using (*a*) a constant environment, (*b*) an environment where the temperature alternated weekly at 25°C and 19°C, or where 2 different yeasts were used, or where 2 different media were used, and (*c*) where the 3 pairs of factors above were varied all together. After 45 weeks, the populations were assayed by gel electrophoresis for polymorphisms at 22 loci, and it was found that the average heterozygosity per individual and the average number of alleles per locus were in the sequence (3)>(2)>(1), showing greatest genetic diversity in the most heterogeneous of the environments.

If the maximum temperature were extreme, say well above 30°C in *D. melanogaster*, this conclusion might not hold. Band (58) found genetic diversity as measured by frequency of lethals and semilethals, to be higher in a diurnal environment of 25°C ± 8°C than a constant environment of 25°C or a wide-ranging diurnal environment at 25°C ± 14°C. The 25°C ± 8°C system approximates the situation in the natural environment from which the flies were taken. Some conclusions are emerging from this work within the species *D. melanogaster* in relation to the effect of temperature fluctuations on fitness, but no such work has been reported on *D. simulans*. To understand more of the ecology of the two species in relation to temperature stress, further experiments on both species are needed.

DESICCATION Kalmus (59) and Waddington, Woolf & Perry (60) described differences between strains of *D. melanogaster* for preferences for environments with different humidities. Generally, females are more resistant than males, at least in *D. persimilis*, *D. pseudoobscura*, and *D. melanogaster* (61-63). In *D. pseudoobscura*, Heuts (47) found that relative humidity during the pupal stage at 25°C differentially affected the carriers of different gene arrangements. Pupae homozygous for *AR* had higher hatching percentages at low humidities, *CH* homozygotes at high humidities, and *ST* homozygotes were intermediate. Spassky (64) and Levine (65) found certain karyotypes to be affected differentially by certain combinations of temperature and humidity. Thomson (66) desiccated *ST/ST*, *ST/CH*, and *CH/CH* karyotypes and showed heterokaryotype advantage for survival 24–42 hours after the commencement of desiccation, such that at 42 hours the only survivors were heterokaryotypes. The heterokaryotypes were heavier than the homokaryotypes, suggesting a possible relationship between ability to survive and body weight. This result perhaps parallels the high levels of heterozygote advantage under temperature (hot and cold) stress, and suggests the possibility of enhanced heterosis in extreme environments (1, 67).

Parsons (68) studied resistance to desiccation in *D. melanogaster* in 18 LM strains, as measured by mortalities after 16 hours in a dry environment. Differences were found between strains, indicating genetic variability in wild populations for ability to withstand desiccation. Those strains with high wet and dry weights lost water relatively less rapidly and had lower mortalities than strains with lower wet and dry weights, as shown in Table 1, where correlation coefficients are given for several variables tested for the 18 strains. This agrees with data (66) in *D. pseudoobscura* above. Quite high heritabilities (in the broad sense) were found for all traits, as assessed from experiments on variability within and between five inbred strains.

TABLE 1 Correlation coefficients between various parameters in the desiccation experiments for female data (All correlations are based on 16 d.f.) (After Parsons 68)

	Mean weights after desiccation B	B/ A	Mean dry weights	Numbers dead (angularly transformed)
Mean net weights A	0.784***	0.597***	0.791***	-0.579*
Mean weights after desiccation B		0.961***	0.705**	-0902***
B/ A			0.578*	-0.926***
Dry weights				-0.436

*P< 0.05, **P< 0.01, ***P< 0.001

Assuming that levels of resistance to desiccation influence the distribution of species in the wild, it is reasonable that there should be genes segregating in populations for resistance to desiccation just as for temperature stresses. This is because populations, at least in some environments, would be exposed to desiccation stresses during some parts of the year and not others, therefore genetic variability for rapid adaption would be at a premium. It is relevant that intraspecific variation for resistance to desiccation has been found in the mosquitoes *Aedes aegypti* and *A. atropalpus* (69). Mosquitoes have more stringent humidity requirements than many other species, and susceptibility to water loss probably plays a major role in limiting the distribution of many species. In the region of Australia from which the 18 LM strains of *D. melanogaster* were extracted, which is characterized by long hot summers, the expected correlation between the two stresses of high temperature and desiccation was found (68), but this would not always be expected, as *Drosophila* is found in areas of high temperature and high humidity. More work would be rewarding, especially if body weight can be regarded as an adaptation to desiccation stress. Even so, it will be worth further considering body weight in flies from hot dry areas compared with those from hot wet areas (2, 70).

DISCUSSION Levins (70) studied acclimation to dry heat (a combination of high temperature and low humidity), and found that a process of physiological acclimation tended to occur especially in broader niched species such as *D. melanogaster*, and tended not to occur in narrow-niched species, e.g. *D. prosaltans*. This suggests a degree of individual flexibility enabling immediate adaptation to heat stress, as well as the genetic variation already discussed. Extrapolating to the wild, acclimatization may permit immediate feeding in dry places without immediate desiccation. At the behavioral level there is a further complication, as *D. willistoni*, and no doubt other species, tend to avoid stress by behavioral means. *D. willistoni* is in fact a moderately broad-niched species showing little acclimation or genetic variation for heat resistance in laboratory experiments, hence behavioral avoidance mechanisms could well be significant. If behavioral avoidance of environmental stresses is common, then Levins' conclusion of a negative correlation between strong habitat selection and individual (physiological) plasticity may be reasonable for the species studied, but the two cannot be regarded as mutually exclusive.

In *D. melanogaster* a behavioral mechanism is likely, because in extremely hot dry environments this species is difficult to trap in the middle of the day, and easier to collect in the early morning and evening. This may imply that when the external environment is extreme, flies tend to avoid it by seeking cooler niches, emerging under more favorable conditions. *D. melanogaster* taken at midday from exposed traps in Puerto Rico were 5% larger than those collected in the morning or late afternoon. In protected sites there was no difference (70). The implication is that only those flies that are genetically most resistant to desiccation by virtue of their large size appear at midday (68); those genetically less resistant do not. In brief, the species *D. melanogaster* shows genetic heterogeneity, physiological plasticity, and behavioral mechanisms in coping with hot dry environments. In other species of *Drosophila* the relative importance of these three mechanisms may vary in different ways, as will no doubt be elaborated by future work.

Data on extreme high and low temperatures and desiccation show large differences between heterokaryotypes (and heterozygotes) and homokaryotypes (and homozygotes) under conditions of extreme stress. Published data suggest that the more extreme the stress, the greater the level of heterokaryotype (or heterozygote) advantage (67). In more optimal environments heterokaryotype advantage is minimal. This provides an additional explanation for the high level of genetic polymorphism in natural populations being associated with a not excessive genetic load (2, 67). The need to explain a high level of genetic polymorphism in natural populations arose with the work of Lewontin & Hubby (7) and others (8-10). Lewontin & Hubby concluded that the average individual in *D. pseudoobscura* is heterozygous for at least 12% of genes in the entire genome based on studies of enzyme and protein variants, and it has been estimated that the proportion of polymorphic loci to total loci is in the range 30–70%. If polymorphism were due only to heterozygote advantage, the genetic load would be enormous. Various hypotheses, such as density-dependent selection, have been put forward to avoid this dilemma. Part of the problem resides in the definition of genetic load (10). Briefly, if genetic load is defined in relation to individuals with optimal genotypes, then it would be enormous; however, such genotypes are sufficiently rare in the population that they play little part in determining the average selective advantage at individual loci. Therefore, many selectively balanced polymorphisms may exist even if the fitnesses of optimal genotypes do not greatly exceed the population mean.

In nature, populations may be subjected to a number of environmental stresses during their existence, during which time heterozygote advantage for certain loci or inversions might be favored according to the stress, so enabling more ready adaptation to it. If such a high level of heterozygote advantage were restricted to stress environments, which would probably form a minor part of the population, the overall genetic load would not be high even with a high level of genetic polymorphism, under the assumption of a multiplicity of extreme ecological niches in which different heterozygotes would be advantageous.

There is little evidence for overdominance for enzyme polymorphisms, although Richmond & Powell (71) reported overdominance for a sex-linked enzyme locus, tetrazolium oxidase in *D. paulistorum*. Assuming the form of extreme-environment heterosis as postulated above, overdominance may not be expected in most laboratory experiments, which would normally have been carried out in nonextreme environments. Probably the next few years will provide much evidence on selective differentials for enzyme and protein variants in different environments, and already a number of reports are appearing on this issue (2). Although too early to generalize, the results often do show some dependence on environment.

In *D. pseudoobscura*, loci controlling enzyme and protein variants associated with different inversions could explain the associations of inversion frequencies with season, altitude, and other environmental factors. This possibility is enhanced if a molecular basis, partly or wholly, is accepted for extreme-environment heterosis. Even though no direct studies of environmental variables were carried out, it is possible that loci controlling enzyme and protein variants are associated with environmental factors. This is more likely if there is a molecular basis for extreme-environment heterosis, as suggested elsewhere (67, 72).

3. CO_2 *and Anoxia*

Sort-term sensitivity to CO_2 has been known for a number of years. L'Héritier & Teissier (73) found flies sensitive to even a 30 second exposure to CO_2, due to a cytoplasmic particle with virus-like properties. Such strains are common in *D. melanogaster* from Europe and some areas of South America (74). The virus, referred to as sigma, can grow normally and produce sensitivity after inoculation into several other species of *Drosophila* but not other genera (75, 76). In *D. affinis* and *D. athabasca*, cytoplasmic CO_2 sensitivity seems to follow the *D. melanogaster* pattern (77). This situation has been fairly extensively reviewed (78).

Resistance to long-term exposure to CO_2 has been less investigated genetically (79, 80). Using 18 LM strains in *D. melanogaster*, variation in mortalities was shown following exposure to a CO_2 atmosphere for $4\frac{1}{2}$ hours in a plastic desiccator flushed with CO_2, in which a high level of humidity was maintained. As for the stress traits already discussed, the founder females differed genetically, so that the population is polymorphic for genes conferring resistance to long-term exposure to CO_2.

Genetic analysis showed large additive genetic differences between inbred strains, and between those set up from single inseminated females collected in the wild. Nonadditive and reciprocal effects were generally small. The nonsignificance of the reciprocal effects contrasts with results expected from cytoplasmic CO_2 sensitivity. Chromosomal analyses revealed genetic activity, which was largely additive in a number of regions of the X, 2, and 3 chromosomes. Also, strains were found to differ from each other in resistance or sensitivity. The genetic architecture is such that there is not a multiplicity of polygenes each having a small effect on CO_2 resistance scattered throughout the genome, but rather genes or gene complexes with differing but reasonably large additive effects. The results agree with postulates (81, 82) that different wild strains should carry combinations of genes producing a relatively intermediate net effect.

Considerable literature on mechanisms of resistance to CO_2 revolves around whether its anaesthetic effect is specific to CO_2, or a generalized anoxia effect (83, 84). During anoxia, certain changes in metabolism normally result in the building up of an oxygen debt (85). This occurs by switching to metabolic pathways that are not oxygen dependent, but are usually less efficient in energy production, so enabling survival for a time. The survival time depends on the energy requirements of particular organs in the individual concerned, and these may be measured in a general way as the overall metabolic rate before anoxia. The higher the metabolic rate, the greater the energy requirement during anoxia, and the shorter the length of time the organism would be expected to survive. Assuming an anoxia effect, it would be expected that exposure to the inert gas N_2 should mimic CO_2, whereas if CO_2 acted as a direct poison this would not necessarily be so. Various lines of evidence (80) support the anoxia hypothesis: (*a*) Percentage mortalities plotted against time after CO_2 or N_2 exposure were almost identical, and the equations of the regression lines very similar. (*b*) Exposing 15 strains to CO_2 and N_2 respectively for $4\frac{1}{2}$ hours, gave correlation coefficients of 0.623 in females and 0.744 in males for angularly transformed mortalities, the former being significantly different from 0 at the 5% level and the latter at the 1% level. Therefore, over a series of strains, many similarities were found for the effects of the two gases. (*c*) Exposing four strains to CO_2 for 1-5 hours followed by N_2 so as to make a total of 5 hours gave similar results.

Therefore, anoxia seems more likely than a specific poisoning effect. If so, a relationship between metabolic rate and mortality would be expected. Fly size thus is expected to be relevant, since metabolic rate M as assessed by oxygen consumption per unit time, and body weight W are expected to be related in insects by $M = kW^b$, where k and b are constants (86, 87). The constant b is obtained from the slope of a plot of log oxygen consumption against log weight. If $b = 1$, metabolism is proportional to size, but in most organisms $b < 1$ in the vicinity of 0.73 (86, 87). Although only 18 strains were used, making estimates of b subject to considerable random error, in females $b = 0.75 \pm 0.37$ and in males 0.34 ± 0.55, which fits in with the general picture. Clearly, genetic differences in body weight ought to be associated with differences in metabolic rate, and hence resistance to anoxia. If CO_2 kills by anoxia, mortality should be related to body weight. In Table 2, correlation coefficients over the 18 LM strains are given between body weight, mortality under exposure to CO_2, and metabolic rate. Positive correlations were found between metabolic rates and CO_2 mortalities as predicted for an anoxia effect, and negative correlations occurred between body weights and both CO_2 mortalities and metabolic rates as also predicted. It seems fairly clear that in total the evidence favors CO_2 acting to produce anoxia.

TABLE 2 Correlation coefficients over 18 strains between body weights, mortality on exposure to CO_2 for 4½ hours, and metabolic rate assayed by O_2 uptake under standardized techniques (After Matheson & Parsons 80)

	Females	Males
Body weight - CO_2 mortality	-0.52*	-0.52*
Body weight - metabolic rate	-0.43	-0.66**
CO_2 - metabolic rate	+0.48*	+0.62**

*$P < 0.05$, **$P < 0.01$ for deviation from 0

Body weight is presumably a trait under stabilizing selection, as an optimum value would be favored. It is intriguing that so many other parameters showing genetic variability in natural populations are correlated with body weight. These include positive correlations between body weight, and resistance to desiccation and anoxia with CO_2 and N_2, and a negative correlation between body weight and metabolic rate. Therefore, genetic differences in body weights clearly are associated with a number of physiological variables, some of which have significance when considering the fitness of genotypes in the wild, as shown in the discussion on desiccation. Variations in resistance to anoxia are therefore associated with correlated factors, some of which are important in determining the distribution and survival of the species in different environments.

4. ^{60}Cobalt γ*-Radiation*

Ogaki & Nakashima-Tanaka (19) found one or more genes for radioresistance in *D. melanogaster* on chromosome 3, as measured by mortalities after high doses of ^{60}Co γ-rays. Parsons, MacBean & Lee (88) found that 18 LM strains differed in radioresistance to very high doses of ^{60}Co γ-rays, as measured by percentage mortalities after irradiation. A 4 x 4 diallel cross between 2 of the most resistant and 2 of the most sensitive strains showed that the differences between the 4 strains were mainly additive, and subsequent detailed genetic analyses revealed additive genes for resistance and sensitivity on various parts of chromosomes 2 and 3. In this sense, the genetic architecture corresponds to that for body weight showing high additivity, and so the trait is presumably largely under stabilizing selection.

Westerman & Parsons (89) examined the association between longevity and resistance to ^{60}Co γ-rays in 4 inbred strains for unirradiated flies, and for flies irradiated at 40 krads and then in 10 krad steps to 130 krads. Surprisingly, resistance as measured by the slope of the regression lines varied between strains. In other words, the longevity of unirradiated flies varies between strains, but is not necessarily associated with longevity after irradiation. In one sense, this was unfortunate, for it was hoped that there would be an association from the point of view of carrying out biochemical and physiological studies of the trait, which could be extrapolated from unirradiated to irradiated flies. To investigate further the genetic architecture of longevity after exposure to various doses of ^{60}Co γ-rays, longevity was examined after different doses, namely 0, 40, 60, 80, 100, and 120 krads for 4 inbred strains and all the possible hybrids between them (90). Just as for the inbred strains themselves, the regression lines of mean mortality on dose varied (Table 3). Thus the inbred strain OR and hybrids with OR generally, had the steepest slopes in both sexes, as compared with the other 3 inbred strains and their hybrids. This emphasizes the problems of extrapolating physiological or biochemical studies carried out at one dose to another dose.

The data made up a series of 4 x 4 diallel crosses from which analyses of

TABLE 3 Linear regression equations (Y) of mean mortalities on dose (after Westerman & Parsons 90)

	Females	Males
OR	38.74 - 0.31 x	39.60 - 0.34 x
OR x N4	55.20 - 0.41 x	42.80 - 0.36 x
OR x Y2	51.15 - 0.35 x	35.89 - 0.27 x
OR x Y4	53.12 - 0.40 x	41.59 - 0.35 x
N4	17.99 - 0.11 x	30.53 - 0.24 x
N4 x Y2	32.77 - 0.17 x	32.14 - 0.22 x
N4 x Y4	38.14 - 0.24 x	37.23 - 0.30 x
Y2	38.24 - 0.25 x	33.31 - 0.25 x
Y2 x Y4	37.24 - 0.20 x	39.19 - 0.30 x
Y4	24.21 - 0.15 x	20.90 - 0.16 x

variance are possible enabling estimation of the relative importance of additive genetic variation amd nonadditive (dominance) variation over a series of doses. At 120 krads, as expected from the earlier studies on strains set up from single inseminated females, additive genetic control was highly significant ($P<0.001$), and the nonadditive genetic component, while still significant ($P<0.001$), was minor by comparison. At 100 krads, the additive genetic effect was just significant ($P<0.05$), but the nonadditive genetic effect very significant ($P<0.001$). For the remaining doses, 40, 60, and 80 krads, the nonadditive effects only were significant ($P<0.001$, 0.01, and 0.001 respectively). The control data showed some additive genetic control ($P<0.01$) which was minor by comparison with the 120 krad data, and some nonadditive genetic control significant only at $P<0.05$. Hence the genetic architecture varies according to the dose of ^{60}Co γ-rays.

High temperature stress, like irradiation, reduces longevity. Parsons (91) examined longevity of adults in 4 inbred strains and their hybrids at 2 temperatures, 25°C and 29.5°C, using techniques identical to the γ-radiation studies. As 29.5°C is extreme for *D. melanogaster* this temperature stress was compared with the irradiation stress, and the data were reanalysed so that comparisons could be made (90). At 25°C no significant effects were found. This is in reasonable agreement with the data for untreated flies for which the significant additive and nonadditive effects were small by comparison with those obtained after irradiation, but at 29.5°C, longevity was controlled by nonadditive effects ($P<0.001$). Therefore the mode of inheritance is similar to that obtained for doses of 40–80 krads. Furthermore, mean longevity at 29.5°C was intermediate between the mean longevity at 60–80 krads. The evidence thus suggests that a temperature of 29.5°C and irradiation doses of 40–80 krads form similar stress environments with similar genetic architectures. Conveniently too, there is some very limited evidence (21) for an association (based on 3 LM strains) between resistance to high levels of γ-radiation (110 and 120 krads) and acute temperature shocks (flies exposed to 33.5°C for 24 hours) with mortalities corresponding to those obtained for high levels of γ-radiation. Furthermore, Ogaki & Nakashima-Tanaka (19) found that 2 radioresistant strains were heat tolerant, and 2 sensitive strains were heat sensitive. These problems are clearly in need of further investigation.

The genetic architecture of longevity under irradiation depends on the dose, and the same may well be true for various levels of temperature stress. Commencing with the control data, additive and nonadditive effects are small, which is perhaps reasonable, as longevity is probably not as directly related to fitness as traits such as viability or hatchability. Even so there is presumably some selection in nature for increased longevity insofar as this would increase fecundity, so that small additive and nonadditive effects seem reasonable (92, 93). At the lower dose rates, 40-80 krads, substantial and significant nonadditive effects occur. Here we are dealing with suboptimal artificial environments. With high and acute doses, 120 krads in particular, additive genetic control predominates in agreement with the acute anoxia stress just discussed, and the same form of genetic architecture will be shown for various acute chemical stresses in the next section. All of these stresses at the doses used lead to quite rapid death. Because of the correlation of anoxia and desiccation with body weight a similar genetic architecture can be assumed for desiccation. Thus resistance to acute stresses in *Drosophila* populations can be regarded as generally under additive control, indicating traits under stabilizing selection. In the case of γ-radiation, the transition to additive control at high doses from nonadditive control at lower doses is intriguing.

5. Specific Chemical Stresses

INTRODUCTION The reason for considering specific chemical stresses comes from their widespread use in the environment. The effects of insecticides on insect genomes are of prime importance. These environmental factors will undoubtedly assume progressively more importance with time. Even though most examples of evolutionary change due to man-made changes in the environment have come from species of more economic significance than *Drosophila*, the study of such phenomena in some of the species of *Drosophila* is instructive. The basic problem is that a rapid build-up of resistance to insecticides may occur, and this means increasing levels of pollution, or perhaps the change to other insecticides whereupon the process is likely to be repeated. To some extent the problem *may* be sidetracked by using biological control methods such as the release of sterile males, or artificial strains that lead to high levels of genetic deaths in natural populations. However, many of these techniques are only in the exploratory stage (94). Returning to chemical stresses, insecticides need not be used to enunciate general principles. The genetic architecture of resistance to ether will therefore be briefly considered first.

ETHER (AND CHLOROFORM) RESISTANCE Although of no known ecological effect, the anesthetic ether has been studied (95) from the point of view of genetic variability in *D. melanogaster* using techniques analogous to those described for resistance to CO_2 and ^{60}Co γ-radiation. It should be noted that Rasmusson (96) described a cytoplasmic effect for resistance to ether, while Ogaki, Nakashima-Tanaka & Murakami (97) found no such evidence, but an ether resistant strain was found to have major genetic activity at map locus 61 on the third chromosome, and minor activity on the X and fourth chromosomes.

In brief, Deery & Parsons' (95) results showed: (*a*) Heterogeneity between LM strains. (*b*) Additive genetic control between 2 resistant and 2 sensitive LM strains from a 4 x 4 diallel cross between them. Some smaller nonadditive effects were found between strains. This can be explained by the dominance of ether resistance over sensitivity, which is in partial agreement with Ogaki, Nakashima-Tanaka & Murakami (97) who found resistance to be completely dominant. Thus in natural populations ether resistance is mainly controlled by additive genes but with some dominance of resistance over sensitivity. (*c*) A 5 x 5 diallel cross between 5 highly inbred strains gave high heritabilities with large additive genetic effects and smaller but significant nonadditive effects because of the dominance of resistance over sensitivity. (*d*) Using two extreme strains, one sensitive and one resistant, the control of genetic activity was found on chromosome 3 and to a lesser extent on chromosome 2. Relative to the marker stock used, on chromosome 2 there was genetic activity for sensitivity at the proximal end, and on chromosome 3 a region of weak resistance proximally and strong sensitivity distally. Hence, based on these two strains alone, there are indications of genes controlling variations in resistance and sensitivity in various parts of chromosomes 2 and 3. As with the other traits discussed, if other strains were taken, no doubt different regions of genetic activity would be found. (*e*) Resistance to chloroform was investigated in less detail but indicated a situation similar to that for ether. Genetic control was mainly additive, but there was little or no correlation between strains for sensitivities to the two anesthetics. This is predictable because the anesthetics differ chemically and so presumably cause death by interacting with different metabolic pathways. Furthermore, the mortality curves after exposure to the anesthetics differed, as etherized flies usually die rapidly under the anesthetic, there being little difference in the number dead at 1 hour or 24 hours after etherization. For chloroform, death occurred progressively over the 24 hour period so that for the doses used, almost all flies were dead at 24 hours, but 4 hours after exposure, large differences between strains were found. (*f*) Unlike some of the stresses discussed no associations with body weight were found. This is not surprising because stresses of a more specific kind presumably interact with specific biochemical pathways not associated with body weight.

INSECTICIDES AND RELATED COMPOUNDS Prior to 1940 few insects were known to have developed resistance to insecticides, but after the introduction and use of DDT and other synthetic organic insecticides, the number of resistant strains began a sharp upward trend (98). Insecticides apparently form powerful selective agents favoring resistant mutants that were initially present in low frequencies in the original populations. A substantial literature on insecticide resistance has appeared, some dealing with genetical aspects in a number of species (14, 98-101). Here we will concentrate on *D. melanogaster* even though there is much evidence for the development of high levels of resistance in other insects such as house flies (102). Crow (14) considered that *D. melanogaster* had by 1957 not developed the high level of resistance of house flies, which is presumably still true, and is not unreasonable, as *Drosophila* would not have been a "target" species in thc same sense as house flies. More recently, at least in those countries where the use of DDT is restricted or banned, a fall in resistance would be expected as in relaxed selection.

In *D. melanogaster* larvae, Tsukamoto & Ogaki (103) and Tsukamoto (104) found large differences between various laboratory and wild strains for resistance to DDT due to a single dominant gene for resistance at about map locus 66 on chromosome 2. Tests showed the same region to be resistant to BHC (benzene hexachloride), parathion, and PU (phenylurea) but sensitive to PTU (phenylthiourea). In addition, the resistant Hikone wild strain larvae were resistant to nicotine sulphate, but in this case most of the resistance was due to a gene at 49 or 50 on chromosome 3. This third chromosome locus was associated with PTU and PU resistance (105, 106). Georghiou (98) reported that the only example so far of cross-resistance between a chlorinated hydrocarbon (DDT) and an organophosphate (parathion) is in *D. melanogaster*, and has not been found in insects such as house flies and mosquitoes.

Turning to adults, Crow (107) found resistance to DDT in *D. melanogaster* to be polygenic based on the development of a resistant strain by selection. Each of the major chromosomes was involved. Oshima (108) obtained results similar in principle. Other studies cited by Crow (14) agree with a polygenic interpretation, as is also the case for adult resistance to BHC. King & Sømme (100) produced over 50 generations of selection, 2 lines about 20 times as resistant as the control, and genetic activity for resistance involved the 3 major chromosomes. Crosses between the two selection lines after 20 generations of selection gave an F_1 population with the same resistance as the parental lines, but the F_2 resistance was significantly lower with a higher variance of mortality distributions. The two lines achieved resistance by consolidating different combinations of genes for resistance, again in agreement with polygenic control. To show the possible effects of DDT on the gene pool in nature, Merrell's (109) experiments in which resistance to DDT had been gradually increased over 10 years of selection by 100–300 times the unchanged control resistance are relevant.

It is not appropriate here to discuss the actual detoxication mechanisms involved in resistant lines, although there has been a considerable amount of research on this topic in the house fly (102). Research on insecticide resistance in an active *Drosophila* laboratory is probably uncommon, because of the problem of general contamination with insecticides. Even so, the need to discover the actual mechanism of insecticide resistance is emphasized by observations (105) that strains of *D. melanogaster* resistant to DDT and other insecticides as larvae are susceptible to PTU. Ogita (105) in fact reported differences in percentages of emergence of larvae grown on PTU medium, mainly controlled by genes on chromosomes 2 and 3 (see above), with a small contribution from an incompletely dominant gene on the X chromosome.

OTHER CHEMICAL STRESSES PTU is a chemical stress. Various mutants are known that are sensitive to PTU as larvae; in particular, ebony (110). A likely reason is that PTU acts as a dopa-oxidase inhibitor, ebony homozygotes having less of the enzyme than normals. Ebony homozygotes normally survive on food containing 0.04% PTU, whereas wild type larvae survive up to 0.12% PTU, and by selection survival on food containing 0.40% PTU has been found (Parsons, unpublished). There are also ebony alleles segregating in natural populations with little or no effect on body color, but which are as sensitive to PTU as ebony homozygotes. This polymorphism seems widespread in nature (111, 112).

There are without doubt a vast number of chemicals that affect different genotypes differentially, leading to genome changes if applied for adequate periods of time. Insecticides have been stressed, but with the increasing awareness of artificial environmental contaminants the list of chemicals will no doubt increase rapidly. A little work has been done on drugs. In *D. melanogaster* Duke & Glassman (113) located a recessive gene for streptomycin sensitivity and a semi-dominant gene for fluorouracil resistance, both on chromosome 3. Over 20 years ago, Herskowitz (114) gave a comprehensive list based on 314 papers of chemicals that had been reported to have genotypic or phenotypic effects on *Drosophila*, some of which may be important in the environment today. To show the effect of contaminants appearing in the environment is the report of Trout & Hanson (115) that exposure to Los Angeles smog reduces the longevity of the behavioral mutant Hyperkinetic (Hk^1) very substantially, and even the normal Canton-S stock was found to have its longevity abbreviated a little.

Chemical stresses may also arise from chemicals normally present in the environment of *Drosophila*, but in abnormal quantities. For example, sodium salts such as NaCl are normally present in the diet of *D. melanogaster*, but in the presence of 1.0M NaCl in the culture medium, considerable differences in emergence rates between various mutant and wild type strains occur. Excessive NaCl causes inhibition of growth and eventual death, but tolerance to NaCl can be increased by continuous selection (116). In one strain emergence was increased from 50.2% in 1M NaCl to 90.6% at the 9th generation of selection. Waddington (117) also developed highly resistant strains by selection, and found that the anal papillae of larvae resistant to high concentrations of NaCl were larger than those of susceptible larvae. Although the function of this organ is not precisely known, it is supposed to be concerned with osmotic regulation. Miyoshi (116) on the other hand found no such morphological changes.

Any extreme variation in concentration of a normally supplied nutrient may affect developmental rates and mortality, as documented by Sang (118) in his search for a suitable synthetic medium for *D. melanogaster*, and by Robertson (119) on variations in larval growth according to nutritional levels. While these form stresses, it is difficult to extrapolate to natural environments, although Robertson considers that variations in larval growth rates according to nutrition provide a flexible system for adjustment to different ecological conditions. For some of the more esoteric nutritional requirements of some specialized *Drosophila* species, without which such species apparently cannot survive, see (2, 120).

CONCLUSIONS Because man-made environmental factors of a chemical nature will undoubtedly increase in importance with time, it is important to consider genetic architectures under chemical stresses. Specific chemicals such as the insecticides and anesthetics discussed, may directly affect various biosynthetic pathways. They are controlled mainly by additive genes, probably of reasonably large effect. This is convenient for detailed genetic analysis, and is an architecture similar to that found for the other acute environmental stresses discussed. When artificial directional selection is practiced for such traits, the likely result is a rapid response to selection due to rearrangement of the additive genes. From the applied point of view it means that for specific chemicals, the rapid build-up of resistant strains is likely. To be effective, a new insecticide must be a chemical to which an organism has not been exposed. Unfortunately, a few resistants will usually survive an initial application, leading to the rapid build-up of resistant strains over a few generations. Then the insecticide becomes ineffective.

6. *Ethyl Alcohol*

Ethyl alcohol, as an environmental stress, falls into a special category for the sibling species *D. melanogaster* and *D. simulans*, because for the former it is hard to regard naturally occurring levels of alcohol as an environmental stress, whereas it can be so regarded for the latter (121). Although little is known of the ecology of these cosmopolitan species, it is likely that the interaction between them is finely balanced (1, 2 for reviews). Laboratory experiments have shown differential utilization of available resources for oviposition sites (122, 123), larval medium utilization sites, and pupation sites (122) by the two species. The outcome of competition experiments in the laboratory depends on both the environment in which they are conducted, and the precise strains used (124, 125). Extrapolation of these experiments to the wild presents enormous difficulties.

Collections at a vineyard and maturation cellar near Melbourne revealed that both species were present in the vineyard, but only *D. melanogaster* was collected within the cellar. This led to the inference that alcohol may be a factor in nature involved in the relative success of the two species. This was confirmed by a series of laboratory and field experiments (121) as follows: (*a*) Mean percentage survivals for adults at 20°C are given for 3 strains for each of the 2 species in Figure 1, after exposure to food containing 0, 3, 6, and 9% ethyl alcohol by volume for 6 days. The high sensitivity of *D. simulans* to the two highest alcohol levels compared with *D. melanogaster* is obvious. (Analogous results were obtained at 15°C and 25°C.) (*b*) Mean percentages of larvae emerging as adults at 20°C are given in Figure 1 for the same 3 strains in both species. For all strains of both species there is a decrease in emergence with increasing alcohol percentage, but generally *D. melanogaster* larvae are more tolerant. Thus the ratio of the angularly transformed percentages emerging on 9% alcohol medium to standard medium was 0.31, 0.34, and 0.27 in the three *D. melanogaster* strains, and 0.36, 0.43, and 0.44 in the three *D. simulans* strains (Analogous results were obtained at 15°C and 25°C.) (*c*) Significant heterogeneity for alcohol tolerance occurred between strains for both adult survival and emergence, as expected from results obtained for the other traits discussed, except for adult survival for *D. melanogaster*, but as only 3 strains were examined, the lack of significance is reasonable. (*d*) Taking normal media and media with 9% alcohol, and giving 2 strains of each species a choice, revealed that *D. melanogaster* had a slight preference for the media with alcohol, while *D. simulans* strongly rejected the alcoholic medium. (*e*) Field experiments revealed that outside the maturation cellar *D. simulans* was the more common species, while within the cellar only one fly of *D. simulans* was ever found, the population being exclusively *D. melanogaster*. During vintage, when fermentation occurs directly above the cellars, only *D.*

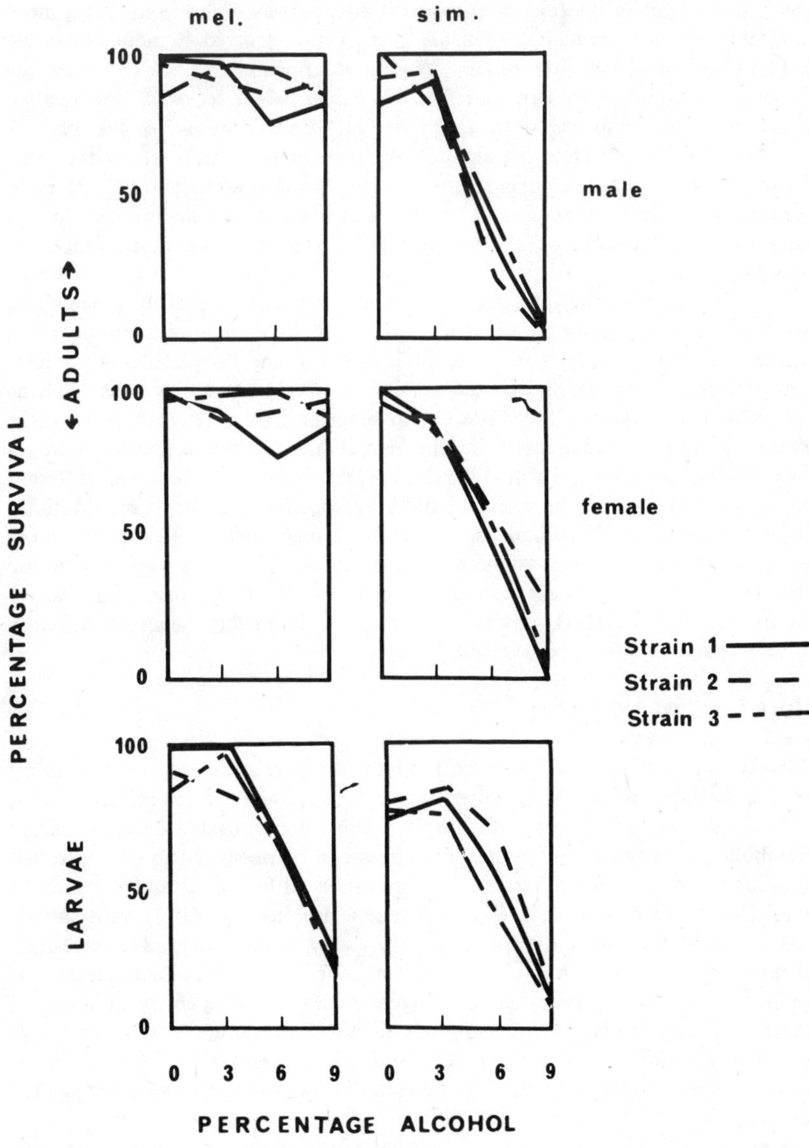

Figure 1 Mean percentage survival of adults after 6 days, and the percentage development from larvae to adults for 3 strains each of *D. melanogaster* and *D. simulans* on 0, 3, 6, and 9% alcohol (after McKenzie & Parsons 121)

melanogaster was in the fermentation area. A cline was found such that a more normal vineyard *D. simulans*: *D. melanogaster* ratio occurred 40 meters from the fermentation area, but after vintage the cline disappeared.

The distribution data can therefore be explained in terms of the relative tolerances of the 2 species, or of a preferential dispersal by one or the other of the species with respect to the alcohol resource. In the warm part of the year, which includes vintage, the temperature inside the maturation cellar would be in the range 16°–20°C, a temperature to which *D. simulans* is well-adapted, so that the presence of alcohol makes this ostensibly suitable environment unsuitable for *D. simulans*.

No work has been done on the genetic architecture of alcohol tolerance along the lines of the other stress traits discussed, but the likelihood of additivity must be assumed for high levels of stress, although this may be considerably higher than 9% alcohol for *D. melanogaster*. Certainly 100% alcohol is lethal, and as expected, *D. simulans* is the more sensitive species. In an environment where alcohol is common, adaptation may involve changes in the alcohol dehydrogenase (ADH) system and its modifier loci. ADH alleles extracted from different populations of *D. melanogaster* had differing specific activities (126). Modifier genes may alter the dehydrogenase activity in *D. melanogaster* (127). Jacobson et al (128) described a series of ADH configurations with differing electrostatic charges and heat stabilities, representing changes in protein formation. Clearly the further study of ADH activities and the potential of the system for variation in both species would be of interest.

Discussion and Summary

GENETIC ARCHITECTURES For very acute stresses, additive genetic control predominates, as shown by the discussion of anoxia, acute doses of ^{60}Co γ-rays, anaesthetics, DDT, and probably most if not all chemicals affecting specific metabolic pathways. Acute desiccation stress can be presumed to be controlled by additive genes because of its high correlation with anoxia, and the same is likely for acute temperature stress. Such genetic architectures imply traits subject to stabilizing selection, and, especially for the very specific chemical stresses, the likelihood of the rapid build up of resistant strains if directional selection is applied. This would occur via rearrangement and homozygosity of areas of genetic activity probably involving all major chromosomes, because where adequately studied, genetic activity for responses to stress was found on all major chromosomes. Possible problems in the study of less acute stresses are presented in Section 4.

A frequently discussed question is: what is the nature of polygenic inheritance? Certain stress traits may have advantages over the morphological traits more conventionally studied, such as chaeta number. Especially for chemical stresses, there is the possibility after locating areas of genetic activity, of finding out the precise function of the genes. For less precise environmental stresses such as temperature and desiccation, gene location could well reveal genes controlling rates of reactions as well. Approaching this level of sophistication for morphological traits may take longer, even though various morphogenetic substances have been suggested as their basis (129). On the other hand, stress traits have a disadvantage in analysis, as the assessment of a stress is based on populations of individuals rather than on individuals themselves, hence unstressed sibs will be important for the study of future generations.

THE COLONIZATION OF NEW HABITATS In the colonization of oceanic islands from the mainland, for example, the population sizes of the colonizers may be so small that some alleles may be fixed or absent. This is likely for most if not all quantitative traits, including the stress traits under discussion. If the founder population is small its influence may persist genetically for many generations, and even under a variety of environments as shown (130) for scutellar chaeta number under a number of temperature regimes in the laboratory. This stresses the importance of the founder principle (131). Especially for environmental stresses such as desiccation and temperature, the precise founder genotypes may well help to determine the likelihood of success of a colonizing population. This brings to light yet again the controversy as to the relative importance of chance variation (i.e. drift) and selection in evolution.

In some ways, colonization of new habitats has analogies with the directional selection experiment. Many traits discussed in this review are controlled by additive genes with high heritabilities. This means that in a new environment there could be fairly rapid adaptation for such traits, especially if an initial stress is fairly severe and there is adequate available additive genetic variation. But whether adaptation does occur is likely to depend on the precise genes present in the founder populations. According to chance some colonizers may survive and some may not. Generally, it may be expected that some such traits will be subject to periods of strong directional selection when a rapid response to selection would be desirable, while in a more fixed environment they would be under stabilizing selection. However, for traits such as temperature stress, it is likely in some environments that in one season directional selection would be in one direction (i.e. for resistance to high temperatures), and in another season in the reverse direction (i.e. for resistance to cold temperatures). Such oscillating directional selection could well simulate stabilizing selection, and lead to the high level of additive genetic control found for many stress traits.

ECOLOGICAL CONSEQUENCES For closely related species of the genus *Drosophila*, especially sibling species, the mechanisms isolating them from each other, i.e. preventing gene exchange between them, depend on a complex of behavioral and ecological factors. Some of these imply variations in resistance to various environmental stresses (1, 2). Thus *D. persimilis*, a sibling species of *D. pseudoobscura*, prefers a cooler niche than the latter species, and the degree of isolation between the two species under laboratory conditions is temperature-dependent

(132). As discussed, *D. melanogaster* can tolerate a greater range of temperatures than its sibling species *D. simulans*; but for these two species the most dramatic is the difference in alcohol tolerances of the two species (121).

Unfortunately the relevance of stresses in the wild is little understood, even though we have some laboratory information. The culmination of the ecological issue occurs when the extraordinarily diverse Hawaiian *Drosophila* fauna (133) are considered. Apart from a dependency on various plant species, the distribution of the various *Drosophila* species in the Hawaiian forest depends on factors such as wind intensity, humidity, temperature, light intensity, and acceptable ovipositional sites. Humidities below 90% and temperatures above about 21°C seem generally to be avoided. During periods of heavy overcast weather when relative humidities approach 100%, especially if misty rain is falling, the flies tend to move upwards into the vegetation of the forest, but on cloudless sunny days when humidity falls, they rapidly disappear, presumably seeking out small poorly lit areas of high humidity. Temperature is thought to be the most important factor. There is therefore a behavioral mechanism whereby flies move to suitable micro-environments. This could be important in Hawaii where seasonal changes are not as great as in the regions where most strains of *D. melanogaster* discussed in this review were derived. Even so, behavioral mechanisms are probably important in *D. melanogaster*. Extrapolation to the field from the laboratory must be attempted, and work on Hawaiian species is of clear importance in bridging this difficult gap.

It is not surprising that distribution of the Hawaiian flies varies according to forest environment since the fitness of genotypes depends on whole environment. We must consider all the factors contributing to fitness, and the effects of environmental stresses, natural and artificial, are important. I conclude by quoting from Fisher (134): "If therefore an organism be really in any high degree adapted to the place it fills in its environment, this adaptation will be constantly menaced by any undirected agencies liable to cause changes to either party in the adaptation". Environmental stresses of a man-made type are going to assume progressively more prominence with time, especially as they can be regarded as rather more "directed" than Fisher's "undirected agencies".

Literature Cited

1. Parsons, P.A., McKenzie, J.A. 1972. The ecological genetics of *Drosophila*. *Evol. Bio.* 5:87-132
2. Parsons, P.A. 1973. *Behavioural and Ecological Genetics*; *A Study in* Drosophila. Oxford: Clarendon Press
3. Dobzhansky, Th., Epling, C. 1944. Contributions to the genetics, taxonomy, and ecology of *Drosophila pseudoobscura* and its relatives. *Carnegie Inst. Wash. Publ.* 554:1-183
4. Dobzhansky, Th. 1951. *Genetics and the Origin of Species*. 3rd. ed. rev. New York: Columbia University Press
5. Parsons, P.A., Hosgood, S.M.W. 1967. Genetic heterogeneity among the

founders of laboratory populations of *Drosophila*. I. Scutellar chaetae. *Genetica* 38:328-39
6. Parsons, P.A., Hosgood, S.M.W., Lee, B.T.O. 1967. Polygenes and polymorphism. *Mol. Gen. Genet*. 99:165-70
7. Lewontin, R.C., Hubby, J.L. 1966. A molecular approach to the study of genic heterozygosity in natural populations. II. Amount of variation and degree of heterozygosity in natural populations of *Drosophila pseudoobscura*. *Genetics* 54:595-609
8. Harris, H. 1966. Enzyme polymorphisms in man. *Proc. Roy. Soc. Lond. Ser. B*. 164:298-310
9. Wallace, B. 1958. The average effect of radiation-induced mutations on viability in *Drosophila melanogaster*. *Evolution* 12:532-56
10. Sved, J.A., Reed, T.E., Bodmer, W.F. 1967. The number of balanced polymorphisms that can be maintained in a natural population. *Genetics* 55: 469-81
11. Parsons, P.A. 1968. Genetic heterogeneity among the founders of laboratory populations of *Drosophila melanogaster*. III. Sternopleural chaetae. *Aust. J. Biol. Sci*. 21:297-302
12. Parsons, P.A. 1970. Genetic heterogeneity among the founders of laboratory populations of *Drosophila melanogaster*. V. Sternopleural and abdominal chaetae in the same strains. *Theor. Appl. Genet*. 40:337-40
13. Hosgood, S.M.W., Parsons, P.A. 1967. Genetic heterogeneity among the founders of laboratory populations of *Drosophila melanogaster*. II. Mating behaviour. *Aust. J. Biol. Sci*. 20:1193-203
14. Crow, J.F. 1957. Genetics of insect resistance to chemicals. *Ann. Rev. Entomol*. 2:227-46
15. Lee, B.T.O., Parsons, P.A. 1968. Selection, prediction and response. *Biol. Rev*. 43:139-74
16. Hosgood, S.M.W., MacBean, I.T., Parsons, P. A. 1968. Genetic heterogeneity and accelerated responses to directional selection in *Drosophila*. *Mol. Gen. Genet*. 101:217-26
17. Thoday, J.M. 1961. The location of polygenes. *Nature* 191:368-70
18. MacBean, I.T., McKenzie, J.A., Parsons, P.A., 1971. A pair of closely linked genes controlling high scutellar chaeta number in *Drosophila*. *Theor. Appl. Genet*. 41:227-35
19. Ogaki, M., Nakashima-Tanaka, E. 1966. Inheritance of radioresistance in *Drosophila*. *Mutat. Res*. 3:438-43
20. Hosgood, S.M.W., Parsons, P.A. 1968. Polymorphism in natural populations of *Drosophila* for the ability to withstand temperature shocks. *Experientia* 24:727-28
21. Parsons, P.A. 1969. A correlation between the ability to withstand high temperatures and radioresistance in *Drosophila melanogaster*. *Experientia* 25:1000
22. Suzuki, D.T. 1970. Temperature-sensitive mutations in *Drosophila melanogaster*. *Science* 170:695-706
23. Fattig, W.D., Rickoll, W.L. 1972. Isolation of recessive third-chromosome temperature-sensitive mutants in *Drosophila melanogaster*. *Genetics* 71:309-13
24. Tantawy, A.O., Mallah, G.S. 1961. Studies on natural populations of *Drosophila*. I. Heat resistance and geographical variation in *Drosophila melanogaster* and *D. simulans*. *Evolution* 15:1-14
25. Timofeeff-Ressovsky, N.W. 1940. Mutations and geographical variation. In *The New Systematics*. Ed. J. Huxley 73-136. Oxford: Clarendon Press
26. Wright, S., Dobzhansky, Th. 1946. Genetics of natural populations. XII. Experimental reproduction of some of the changes caused by natural selection in certain populations of *Drosophila pseudoobscura*. *Genetics* 31:125-56
27. Ehrman, L. 1964. Genetic divergence in M. Vetukhiv's experimental populations of *Drosophila pseudoobscura*. I. Rudiments of sexual isolation. *Genet. Res*. 5:150-57
28. Beardmore, J.A., Dobzhansky, Th., Pavlovsky, O.A. 1960. An attempt to compare the fitness of polymorphic and monomorphic experimental populations of *Drosophila pseudoobscura*. *Heredity* 14:19-33
29. Dobzhansky, Th. 1947. Genetics of natural populations. XIV. A response of certain gene arrangements in the third chromosome of *Drosophila pseudoobscura* to natural selection. *Genetics* 32:142-60
30. Dobzhansky, Th., Lewontin, R.C., Pavlovsky, O. 1964. The capacity for increase in chromosomally polymorphic and monomorphic populations of *Drosophila pseudoobscura*. *Heredity* 19:597-614
31. Dobzhansky, Th., Pavlovsky, O. 1961. A further study of fitness of chromosomally polymorphic and monomorphic populations of *Drosophila pseu-*

doobscura. *Heredity* 16:169-79

32. Parsons, P. A., Kaul, D. 1966. Mating speed and duration of copulation in *Drosophila pseudoobscura*. *Heredity* 21:219-25
33. Spiess, E.B. 1970. Mating propensity and its genetic basis in *Drosophila*. In *Essays in Evolution and Genetics in honor of Theodosius Dobzhansky*. Eds. M.K. Hecht, W.C. Steere 315-79. New York: Appleton- Century-Crofts
34. van Valen, L., Levine, L., Beardmore, J.A. 1962. Temperature sensitivity of chromosomal polymorphism in *Drosophila pseudoobscura*. *Genetics* 33: 113-27
35. Birch, L.C. 1955. Selection in *Drosophila pseudoobscura* in relation to crowding. *Evolution* 9:389-99
36. Dobzhansky, Th. 1948. Genetics of natural populations. XVI. Altitudinal and seasonal changes produced by natural selection in certain populations of *Drosophila pseudoobscura* and *Drosophila persimilis*. *Genetics* 33: 158-76
37. Dobzhansky, Th., Anderson, W.W., Pavlovsky, O., Spassky, B., Wills, C.J. 1964. Genetics of natural populations. XXXV. A progress report on genetic changes in populations of *Drosophila pseudoobscura* in the American southwest. *Evolution* 18:164-76
38. Mayhew, S.H., Kato, S.K., Ball, F.M., Epling, C. 1966. Comparative studies of arrangements within and between populations of *Drosophila pseudoobscura*. *Evolution* 20:646-62
39. Spiess, E.B., Spiess, L.D. 1967. Mating propensity, chromosomal polymorphism, and dependent conditions in *Drosophila persimilis*. *Evolution* 21: 672-78
40. Brncic, D. 1962. Chromosomal structure of populations of *Drosophila flavopilosa* studied in larvae collected in their natural breeding sites. *Chromosoma* 13:183-95
41. Brncic, D. 1968. The effects of temperature on chromosomal polymorphism of *Drosophila flavopilosa* larvae. *Genetics* 59:427-32
42. Dubinin, N.P., Tiniakov, G.G. 1946. Inversion gradients and natural selection in ecological races of *Drosophila funebris*. *Genetics* 31:537-45
43. Dubinin, N.P., Tiniakov, G.G. 1947. Inversion gradients and selection in ecological races of *Drosophila funebris*. *Am. Natur.* 81:148-53
44. Brncic, D. 1969. Long-term changes in chromosomally polymorphic laboratory stocks of *Drosophila pavani*. *Evolution* 23:502-08
45. Krimbas, C.B. 1967. The genetics of *Drosophila subobscura* populations. III. Inversion polymorphism and climatic factors. *Mol. Gen. Genet.* 99:133-50
46. Marinkovic, D., Crumpacker, D.W., Salceda, V.M. 1969. Genetic loads and cold temperature resistance in *Drosophila pseudoobscura*. *Am. Natur.* 103: 235-46
47. Heuts, M.J. 1948. Adaptive properties of carriers of certain gene arrangements in *Drosophila pseudoobscura*. *Heredity* 2:63-75
48. Hoenigsberg, H.F., Castro, L.E., Granobles, L.A., Idrobo, J.M. 1969. Population genetics in the American tropics. II. The comparative genetics of *Drosophila* in European and neo-tropical environments. *Genetica* 40:43-60
49. Ives, P.T. 1970. Further genetic studies of the South Amherst population of *Drosophila melanogaster*. *Evolution* 24:507-18
50. Dobzhansky, Th., Hunter, A.S., Pavlovsky, O., Spassky, B., Wallace, B. 1963. Genetics of natural populations. XXXI. Genetics of an isolated marginal population of *Drosophila pseudoobscura*. *Genetics* 48:91-103
51. Wallace, B. 1968. *Topics in Population Genetics*. New York: W.W. Norton.
52. Hosgood, S.M.W., Parsons, P.A. 1966. Differences between *D. simulans* and *D. melanogaster* in tolerances to laboratory temperatures. *Drosophila Inform. Serv.* 41:176
53. Long, T. 1970. Genetic effects of fluctuating temperature in populations of *Drosophila melanogaster*. *Genetics* 66: 401-16
54. Oshima, C. 1969. Persistence of some recessive lethal genes in natural populations of *Drosophila melanogaster*. *Japan J. Genet.* 44:Suppl. 1, 209-16
55. Beardmore, J.A. 1970. Ecological factors and the variability of gene-pools in *Drosophila*. In *Essays in Evolution and Genetics in honor of Theodosius Dobzhansky*. Eds. M.K. Hecht, W.C. Steere 299-314. New York: Appleton-Century-Crofts
56. Beardmore, J.A. 1961. Diurnal temperature fluctuation and genetic variance in *Drosophila* populations. *Nature* 189:162-63
57. Powell, J.R. 1971. Genetic polymorphisms in varied environments. *Science* 174:1035-36

58. Band, H.T. 1963. Genetic structure of populations. II. Viabilities and variances of heterozygotes in constant and fluctuating environments. *Evolution* 17:307-19
59. Kalmus, H. 1945. Adaptive and selective responses of a population of *Drosophila melanogaster* containing *e* and *e+* to differences in temperature, humidity and to selection for developmental speed. *J. Genet.* 47:58-63
60. Waddington, C.H., Woolf, B., Perry, M.M. 1954. Environmental selection by *Drosophila* mutants. *Evolution* 8: 89-96
61. Kalmus, H. 1941. The resistance to desiccation of *Drosophila* mutants affecting body colour. *Proc. Roy. Soc. Lond. Ser. B.* 130:185-201
62. Perttunen, V., Salmi, H. 1956. The responses of *Drosophila melanogaster* (Dipt. Drosophilidae) to the relative humidity of the air. *Ann. Entomol. Fenn.* 22:36-45
63. Pittendrigh, C.S. 1958. Adaption, natural selection, and behavior. In *Behavior and Evolution*. Eds. A. Roe, G.G. Simpson 390-416. New Haven: Yale University Press
64. Spassky, B. 1951. Effect of temperature and moisture content of the nutrient medium on the viability of chromosomal types in *Drosophila pseudoobscura*. *Am. Natur.* 85:177-80
65. Levine, R.P.. 1952. Adaptive responses of some third chromosome types of *Drosophila pseudoobsura*. *Evolution* 6: 216-33
66. Thomson, J.A. 1971. Association of karyotype with body weight and resistance to desiccation in *Drosophila pseudoobsura*. *Can. J. Genet. Cytol.* 13: 63-69
67. Parsons, P.A. 1971. Extreme-environment heterosis and genetic loads. *Heredity* 26:479-83
68. Parsons, P. A. 1970. Genetic heterogeneity in natural populations of *Drosophila melanogaster* for ability to withstand dessication. *Theor. Appl. Genet.* 40:261-66
69. Machado-Allison, C.E., Craig, G.B. Jr. 1972. Geographic variation in resistance to desiccation in *Aedes aegypti* and *A. atropalpus* (Diptera: Culicidae). *Ann. Entomol. Soc. Am.* 65:542-47
70. Levins, R. 1969. Thermal acclimation and heat resistance in *Drosophila* species. *Am. Natur.* 103:483-99
71. Richmond, R.C., Powell, J.R. 1970. Evidence of heterosis associated with an enzyme locus in a natural population of *Drosophila. Proc. Nat. Acad. Sci. USA* 67:1264-67
72. Langridge, J. 1968. Thermal responses of mutant enzymes and temperature limits to growth. *Mol. Gen. Genet.* 103:116-26
73. L'Héritier, P., Teissier, G. 1937. Une anomalie physiologique hereditaire chez le Drosophile. *C. R. Hebd. Séances. Acad. Sci. Paris* 205:1099-1101
74. Kalmus, H., Kerridge, J., Tattersfield, F. 1954. Occurrence of susceptibility to carbon dioxide in *Drosophila melanogaster* from different countries. *Nature* 173:1101-02
75. L'Héritier, P. 1951. The CO_2 sensitivity problem in *Drosophila. Cold Spring Harbor Symp. Quant. Biol.* 16:99-112
76. L'Héritier, P. 1958. The hereditary virus of *Drosophila. Advan. Virus Res.* 5:195-245
77. Williamson, D.L. 1961. Carbon dioxide sensitivity in *Drosophila affinis* and *Drosophila athabasca*. *Genetics* 46: 1053-60
78. L'Héritier, P. 1970. *Drosophila* viruses and their role as evolutionary factors. *Evol. Biol.* 4:185-209
79. Matheson, A.C., Arnold, J.T.A. 1973. Resistance to carbon dioxide, an anoxic stress in *Drosophila melanogaster. Experentia.* In press
80. Matheson, A.C., Parsons, P.A. 1973. The genetics of resistance to long-term exposure to CO_2 in *Drosophila melanogaster*; an environmental stress leading to anoxia. *Theor. Appl. Genet.* In press
81. Mather, K. 1942. The balance of polygenic combinations. *J. Genet.* 43: 309-36
82. Mather, K. 1943. Polygenic inheritance and natural selection. *Biol. Rev.* 18:32-64
83. Brooks, M.A. 1965. The effects of repeated anesthesia on the biology of *Blattella germanica* (Linnaeus). *Entomol. Exp. Appl.* 8:39-48
84. Edwards, L.J., Patton, R.L. 1965. Effects of carbon dioxide anaesthesia on the house cricket *Acheta domesticus* (Orthoptera: Gryllidae). *Ann. Entomol. Soc. Am.* 58:828-32
85. Gilmour, D. 1961. *The Biochemistry of Insects*. New York & London: Academic Press
86. Prosser, C.L., Brown, F.A. 1961. *Comparative Animal Physiology*. 2nd. Ed. Philadelphia: W.B. Saunders
87. Zeuthen, E. 1953. Oxygen uptake as related to body size in organisms. *Quart. Rev. Biol.* 28:1-12

88. Parsons, P.A., MacBean, I.T., Lee, B.T.O. 1968. Polymorphism in natural populations for genes controlling radioresistance in *Drosophila. Genetics* 61:211-18
89. Westerman, J.M., Parsons, P.A. 1972. Radioresistance and longevity of inbred lines of *Drosophila melanogaster. Int. J. Radiat. Biol.* 21:145-52
90. Westerman, J.M., Parsons, P.A. 1972. Variations in genetic architecture at different doses of γ- radiation as measured by longevity in *Drosophila melanogaster. Can. J. Genet. Cytol.* In press
91. Parsons, P.A. 1966. The genotypic control of longevity in *Drosophila melanogaster* under two environmental regimes. *Aust. J. Biol. Sci.* 19:587-91
92. Breese, E.L., Mather, K. 1960. The organisation of polygenic activity within a chromosome in *Drosophila.* II. Viability. *Heredity* 14:375-99
93. Mather, K. 1966. Variability and selection. *Proc. Roy. Soc. Lond. Ser. B.* 164:328-40
94. Whitten, M.J. 1970. Genetics of pests and their management. In *Concepts of Pest Management.* Ed. R.L. Rabb, F.E. Guthrie, 119-35. Raleigh: N.C. State University
95. Deery, B.J., Parsons, P.A. 1972. Ether resistance in *Drosophila melanogaster. Theor. Appl. Genet.* 42:208-14
96. Rasmusson, B. 1955. A nucleo-cytoplasmic anomaly in *Drosophila melanogaster* causing increased sensitivity to anaesthetics. *Hereditas* 41:147-208
97. Ogaki, M., Nakashima-Tanaka, E., Murakami, S. 1967. Inheritance of ether resistance in *Drosophila melanogaster. Japan J. Genet.* 42:387-94
98. Georghiou, G.P. 1965. Genetic studies of insecticide resistance. *Advan. Pest Control Res.* 6:171-230
99. Crow, J.F. 1960. Genetics of insecticide resistance: General considerations. *Misc. Publ. Entomol. Soc. Am.* 2(1):69-74
100. King, J.C., Sømme, L. 1958. Chromosomal analysis of the genetic factors for resistance to DDT in two resistant lines of *Drosophila melanogaster. Genetics* 43:577-93
101. Oppenoorth, F.J. 1965. Biochemical genetics of insecticide resistance. *Ann. Rev. Entomol.* 10:185-206
102. Grigolo, A., Oppenoorth, F.J. 1966. The importance of DDT-dehydrochlorinase for the effect of the resistance gene *kdr* in the housefly *Musca domestica* L. *Genetica* 37:159-70
103. Tsukamoto, M., Ogaki, M. 1954. Gene analysis of resistance to DDT and BHC. *Botyu-Kagaku* 19:25-32
104. Tsukamoto, M. 1955. Mode of inheritance of resistance to nicotine sulfate in *Drosophila melanogaster. Botyu-Kagaku* 20:73-81
105. Ogita, Z. 1958. The genetical relation between resistance to insecticides in general and that to phenylthiourea (PTU) and phenylurea (PU) in *Drosophila melanogaster. Botyu-Kagaku* 23:188-205
106. Ogita, Z. 1961. Genetical and biochemical studies on negatively correlated cross-resistance in *Drosophila melanogaster.* I. An attempt to reduce and increase insecticide-resistance in *D. melanogaster* by selection pressure. *Botyu-Kagaku* 26:7-18
107. Crow, J.F. 1954. Analysis of a DDT-resistant strain of *Drosophila. J. Econ. Entomol.* 47:393-98
108. Oshima, C. 1954. Genetical studies on DDT-resistance in populations of *Drosophila melanogaster. Botyu-Kagaku* 19:93-100
109. Merrell, D. J. 1965. Lethal frequency and allelism in DDT-resistant populations and their controls. *Am. Natur.* 99:411-17
110. Kroman, R.A., Parsons, P.A. 1960. Genetic basis of two melanin inhibitors in *Drosophila melanogaster. Nature* 186:411-12
111. Deery, B.J., Parsons, P.A. 1972. Variations in the resistance of natural populations of *Drosophila* to phenyl-thiocarbamide (PTC). *Egyptian J. Genet. Cytol.* 1:13-17
112. Parsons, P.A. 1963. A widespread biochemical polymorphism in *Drosophila melanogaster. Am. Natur.* 97:375-82
113. Duke, E.J., Glassman, E. 1968. Drug effects in *Drosophila*: Streptomycin sensitive strains and fluorouracil resistant strains. *Nature* 220:588-89
114. Herskowitz, I.H. 1951. A list of chemical substances studied for effects on *Drosophila*, with a bibliography. *Am. Natur.* 85:181-99
115. Trout, W.E., Hanson, G.P. 1971. The effect of Los Angeles smog on the longevity of normal and hyperkinetic *Drosophila melanogaster. Genetics* 68: s69
116. Miyoshi, Y. 1961. On the resistibility of *Drosophila* to sodium chloride. I. Strain difference and heritability in *D. melanogaster. Genetics* 46:935-45

117. Waddington, C.H. 1959. Canalization of development and genetic assimilation of acquired characters. *Nature* 183:1654-55
118. Sang, J.H. 1956. The quantitative nutritional requirements of *Drosophila melanogaster*. *J. Exp. Biol.* 33:45-72
119. Robertson, F.W. 1963. The ecological genetics of growth in *Drosophila*. 6. The genetic correlation between the duration of the larval period and body size in relation to larval diet. *Genet. Res.* 4:74-92
120. Heed, W.B., Kircher, H.W. 1965. Unique sterol in the ecology and nutrition of *Drosophila pachea*. *Science* 149:758-61
121. McKenzie, J.A., Parsons, P.A. 1972. Alcohol tolerance: an ecological parameter in the relative success of *Drosophila melanogaster* and *Drosophila simulans*. *Oecologia* 10:372–88
122. Barker, J.S.F. 1971. Ecological differences and competitive interaction between *Drosophila melanogaster* and *Drosophila simulans* in small laboratory populations. *Oecologia* 8:139-56
123. Soliman, M.H. 1971. Selection of site of oviposition by *Drosophila melanogaster* and *D. simulans*. *Am. Midl. Natur.* 86:487-93
124. Moore, J.A. 1952. Competition between *Drosophila melanogaster* and *Drosophila simulans*. I. Population cage experiments. *Evolution* 6:407-20
125. Tantawy, A.O., Soliman, M.H. 1967. Studies on natural populations of *Drosophila*. VI. Competition between *Drosophila melanogaster* and *Drosophila simulans*. *Evolution* 21:34-40
126. Gibson, J.B., Miklovich, R. 1971. Modes of variation in alcohol dehydrogenase in *Drosophila melanogaster*. *Experientia* 27:99-100
127. Ward, R.D., Hebert, P.D.N. 1972. Variability of alcohol dehydrogenase activity in a natural population of *Drosophila melanogaster*. *Nature New Biol.* 236:243-44
128. Jacobson, K.B., Murphy, J.B., Knopp, J.A., Ortiz, J.R. 1972. Multiple forms of *Drosophila* alcohol dehydrogenase. III. Conversion of one form to another by nicotinamide adenine dinucleotide or acetone. *Arch. Biochem. Biophys.* 149:22-35
129. Rendel, J. M. 1967. *Canalisation and gene control*. London and New York: Academic Press
130. Hosgood, S.M.W., Parsons, P.A. 1971. Genetic heterogeneity among the founders of laboratory populations of *Drosophila*. IV. Scutellar chaetae in different environments. *Genetica* 42:42-52
131. Mayr. E. 1963. *Animal Species and Evolution*. Cambridge: Harvard University Press
132. Koopman, K.F. 1950. Natural selection for reproductive isolation between *Drosophila pseudoobscura* and *Drosophila persimilis*. *Evolution* 4: 135-48
133. Carson, H.L., Hardy, D.E., Speith, H.T., Stone, W.S. 1970. The evolutionary biology of the Hawaiian Drosophilidae. In *Essays in Evolution and Genetics in honor of Theodosius Dobzhansky*. Eds. M.K. Hecht, W.C. Steere, 437-543. New York: Appleton-Century-Crofts
134. Fisher, R.A. 1930. *The Genetical Theory of Natural Selection*. Oxford: Clarendon Press

IN VITRO SYNTHESIS OF PROTEIN IN MICROBIAL SYSTEMS

3056

Geoffrey Zubay

Department of Biological Sciences, Columbia University, New York City

Introduction

The first useful systems for cell-free protein synthesis were developed in the Zamecnik laboratory, where it was shown that peptide synthesis takes place on the ribosome and requires ATP, GTP, and tRNA (1). Building on this, Nirenberg and his collaborators operationally separated the mRNA from the ribosome by showing that polyuridylic acid added to ribosomes stimulates polyphenylalanine synthesis (2). This discovery was a first step towards the solution of the genetic coding mechanism for proteins. Many other laboratories have contributed to understanding the biochemistry of protein synthesis. In spite of all the progress that has been made several aspects of the process are perplexing even in the bacterium *Escherichia coli*. The biochemistry of protein synthesis has recently been reviewed (3). This review is devoted to studies involving synthesis of complete proteins in cell-free systems.

In the same year that Niremberg began his synthetic messenger studies aimed at elucidating the genetic code, the first complete protein, the coat protein of coliphage f2, was made in a cell-free extract of *E. coli* (4). Since then a number of viral RNAs have been used as messengers to prime the synthesis of complete proteins. By contrast, bacterial messengers have resisted translation until recently, because of the difficulty in isolating bacterial messengers from whole cells and partly because bacterial messengers are probably more prone to translation as they are being synthesized. To simulate whole cell conditions more closely, our laboratory began studies aimed at cell-free synthesis starting from the DNA. In 1967 DeVries and I (5) demonstrated synthesis of a portion of the β-galactosidase enzyme in a DNA-stimulated messenger-directed system. Subsequently a number of laboratories have studied the DNA-dependent RNA-directed synthesis of both viral and bacterial proteins in *E. coli* extracts.

I shall review the field of cell-free protein synthesis for both the RNA- and DNA-dependent systems with particular emphasis on the latter and the uses to which these systems can be put. This is a young field and it is more important to point the way than to review past accomplishments.

Studies with Purely Translational Systems

The first complete protein to be made in a cell-free system was that of the f2 bacteriophage. Prior to this it had been possible to demonstrate only the synthesis of peptide fragments in cell-free systems. The requirements for complete protein synthesis are more demanding than for partial protein synthesis. The system must be presented with an intact messenger that can be faithfully translated at least once before it is degraded. This could be a serious problem in *E. coli* extracts, as nucleases are normally present. The early success with f2 messenger resulted from its availability in pure form, from secondary structure features which make it relatively immune to nucleases and readily susceptible to absorption at the initiation sites by ribosomes, and finally from the ease of detecting the coat protein it encodes. In fact the f2 messenger encodes three proteins but the activity of the coat protein cistron is about 40 times higher than the next most active cistron. Nathans et al (4) rigorously demonstrated de novo synthesis of the coat protein. For this purpose ^{14}C-lysine and ^{14}C-arginine of the same specific activity were used during synthesis. After synthesis the de novo synthesized coat protein was purified and subjected to trypsin digestion followed by chromatography of the resulting peptides. Trypsin attacks the carbonyl groups of all basic amino acids so that all peptides should be labeled except for that originating from the carboxy-terminal end of the protein. This was found to be the case.

The f2 virus is just one of a family of physically and serologically related single stranded RNA coliphages including Rl7, MS2, and Ml2, about which a great deal is known. Coliphage QB is slightly larger and serologically distinct but will be discussed because a great deal of translation work has been done with it also. The subject of viral RNA translation discussed below has recently been treated in considerably greater depth (6); that article should be consulted for more information and references. We begin this section by a description of the RNA coliphage and its in vivo behavior.

Conditional mutants of the RNA coliphages have been used to identify three phage cistrons by complementation tests and by direct analysis of phage proteins (7–9). These are: (*a*) the coat protein, which is the major protein component of the virus particle (129 amino-acids); (*b*) the maturation or A protein, a minor component of the virus particle (approximately 350 amino acids); and (*c*) the RNA replicase or synthetase, which is not present in the mature phage particle (a minimum of about 450 amino acids). Most coat protein mutations (nonpolar mutations) lead to excessive formation of synthetase, because of continued production late in the infectious cycle, and to a less marked increase in maturation protein synthesis. These mutants make no phage particles. Some mutations of the coat protein cistron (polar mutations) cause the same phenotype as synthetase mutations; i.e., the mutants produce little or no synthetase. Hence little phage RNA or protein of any type is made. They inhibit growth of the host cell without causing lysis. Maturation protein mutants produce defective, nonadsorbing, ribonuclease-sensitive particles. When amber maturation protein mutants are used, the synthesis of coat protein and synthetase are essentially the same as after infection with wild-type phage. RNA synthetase mutants produce no virus particles, RNA, or enzymatically detectable synthetase and have no effect on the growth of host bacteria.

Several lines of evidence indicate that the order of genes from the 5′ end of the viral RNA is: maturation protein, coat protein, and synthetase (10). Use of specific RNA fragments as messenger for in vitro protein synthesis tentatively showed that the synthetase cistron is within the 3′ 60% fragment of the RNA molecule, and the coat gene is nearer the 5′ end. Subsequent identification of the synthetase ribosomal binding sequence within the 3′ 60% fragment of the RNA confirmed the location of the synthetase cistron. By the same technique, the initiation sites for maturation and coat proteins were shown to be within the 5′ 40% RNA fragment, whereas oligonucleotides from the interior of the coat protein gene were recovered from the 3′ fragment, placing the coat gene near the center of the RNA molecule, with the maturation cistron to the left and the synthetase gene to the right.

When an *E. coli* cell is infected with a wild-type phage the first phage protein to appear, 5–15 minutes after infection, is the RNA synthetase. Its rate of synthesis increases for the first 25 minutes and then rapidly diminishes. Maturation protein synthesis begins later than that of synthetase, increases in rate, and continues later in the infectious cycle at a reduced rate. Coat protein synthesis begins later than that of synthetase. Coat protein formation then rapidly increases in rate and continues to be made at near maximal rate throughout most of the phage growth cycle. The approximate molar ratios of RNA synthetase-maturation protein-coat protein made during the developmental cycle are 1:2:20. Clearly, in the development of the RNA coliphages, special mechanisms determine the amount of each phage protein made and its time of synthesis. Understanding the regulation of protein synthesis for the RNA phage is complicated by the fact that during the phage replication cycle there are at least three potentially different forms of viral messenger RNA; (*a*) the parental RNA molecule, (plus strand) (*b*) nascent plus strands still partially associated with the replicative intermediate, and (*c*) free single-stranded progeny molecules. The complement of the plus strand (minus strand) does not appear to function as messenger and there is no evidence for single-stranded fragments of phage RNA in infected cells.

The possible mechanisms underlying differential translation of the polygenic phage RNA have been investigated in an in vitro protein-synthesizing system programmed with phage RNA. Under appropriate conditions complete molecules of all three phage proteins can be synthesized by *E. coli* extracts programmed with NA isolated from the phage. The relative amounts of protein made in vitro are in the same order as that observed in vivo, with the coat protein accounting for the vast majority and the maturation protein the least. Thus in cell-free extracts as in whole cells there is differential translation of the three phage genes.

In vitro studies have shown that the *RNA conformation* is important in the differential translation of the 3 phage genes, at least in vitro. With native RNA the coat protein is the predominant product (11). Also under conditions in which polypeptide chain elongation is prevented only one ribosome can attach to the native RNA and the attachment site is at the beginning of the coat cistron (12). In contrast, when the viral RNA conformation is partially disrupted by mild formaldehyde treatment, there is a dramatic increase in initiation of maturation protein and synthetase, assayed by initial dipeptide formation (13) and by identification of the protected ribosomal binding sites (14).

Another line of investigation indicates that the secondary structure of the phage RNA limits initiation of the synthetase cistron. Amber mutations very early in the coat protein gene prevent synthetase production whereas amber mutations occurring later in the coat protein gene do not inhibit synthetase formation. This polar effect of early coat protein mutations on synthetase production has been observed both in vivo and in vitro (6). Recently, determination of the nucleotide sequence of an extensive stretch of the coat protein and adjacent synthetase cistrons has revealed a possible region of hydrogen bonding, 21 nucleotides long, between the synthetase initiation site and codons 24–32 of the coat cistron, which could account for the closed state of the synthetase initiation site (15).

Initiation of the maturation protein is not affected by conformational changes that occur during translation of the coat gene. However, conformation of the mRNA is an important factor in determining the activity of the maturation protein cistron. Thus the RNA replicative intermediate (RI) consists of an intact minus-strand template with one or more nascent single-stranded plus chains extending from it. In vitro, replicative intermediate initiates about five times as much maturation protein, relative to the total protein initiated, as does single-stranded RNA isolated from phage particles (16). Presumably, the conformation of the 5′ end of the single-stranded RNA chains in the RI is different from that of completed RNA molecules, allowing access to the maturation protein initiation site on the nascent chains. After the nascent RNA chains on the RI attain a sufficient length they probably fold so that no more maturation protein can be made.

RNA conformation may not be the only factor in determining the differential translation of the 3 phage genes. In vitro natural messengers require an initiation factor called IF3 for the function of a stable complex with the ribosome. There are some indications that multiple species of IF3 with cistron-selective activities are present in *E. coli*. The first evidence concerns the altered initiation of Rl7 or MS2 proteins by initiation factors from T4-infected cells. Analysis of the oligonucleotide sequences protected by bound ribosomes from T4-infected or uninfected cells, with initiation factors from T4-infected cells, has revealed that appreciable binding occurs only at the beginning of the maturation protein cistron (17). The observed failure to stimulate coat protein initiation could explain why, in cell-free extracts, initiation factors from T4-infected cells fail to promote MS2- or Rl7-directed protein synthesis, although the extracts are active with T4 or T5 mRNA. The deficiency in the T4 initiation factor preparations has been specifically identified as IF3. IF3 activity from uninfected *E. coli* has been fractionated into components, which show messenger or cistron specificity. Assaying IF3 activity with late T4 mRNA or with MS2 RNA, indicated that the ratio of activities with the two messengers varied markedly with different factor fractions. Moreover, various fractions of IF3 differed in their ability to translate MS2 coat or noncoat proteins.

In addition to possible cistron-specific IF3 fractions, an inhibitory protein (called factor i) has been isolated from *E. coli* ribosomes. It has been shown to interact with IF3 and specifically reduce initiation at the coat protein initiation site of MS2 RNA (18). The i protein also binds to MS2 RNA, hence its precise mode of action is not clear at present. It should be emphasized that no mutants for either IF3 or i factors exist. This makes in vivo correlates relating to their messenger specificity most difficult. Artifacts are common in vitro, and without mutants which would be expected to show parallel alteration of a property in vivo, all evidence from in vitro experiments must be interpreted with great caution.

In addition to the regulator role played by RNA conformation and possibly by ribosomes and initiation factors, another mechanism for translational control at the initiation step has been discovered with the RNA phages. It involves the specific binding of proteins to the messenger. Two instances have been demonstrated in vitro: (*a*) inhibition of synthetase initiation by phage coat protein and (*b*) inhibition of coat protein initiation by the RNA synthetase enzyme complex. When coat protein is incubated with phage RNA and the RNA is then used as a messenger in vitro, the formation of synthetase is specifically inhibited. This inhibition occurs at the initiation step as shown by failure to form the initial dipeptide of the synthetase. The effect of coat protein on translation of phage RNA is correlated with formation of a protein-RNA complex (19). These in vitro observations are consistent with the in vivo evidence cited earlier that phage coat protein mutants show an enhanced synthetase formation. Cessation of synthetase formation in the normal in vivo situation could then be explained by the newly formed coat protein reaching a sufficiently high concentration to form a stable complex with the phage RNA.

The other phage protein shown to have translational effects in vitro is RNA synthetase. Purified QB synthetase (the complex of one virus-specified polypeptide and three host-specified polypeptides) inhibits the formation of an initiation complex between ribosomes, QB RNA, and fMet-$tRNA_f$ (20). This effect of the RNA synthetase complex has been attributed to one of the host polypeptides present in the tetrameric enzyme rather than to the phage-specified subunit. The host component appears to be identical to the i factor discussed above which is present in uninfected *E. coli* and inhibits coat protein initiation in vitro. The finding that the i factor associated with the QB synthetase inhibits binding of ribosomes to the coat protein initiation site provides a possible mechanism for efficient conversion of the input RNA molecule from a messenger to a template for replication.

Only one phage protein, RNA synthetase, is detectable in infected cells labeled between 5 and 15 min after infection. This would appear to be inconsistent with the observation noted above that translation of at least part of the coat protein gene is essential for the translation of the synthetase gene. The apparent inconsistency can be resolved if in vivo only partial translation of the coat protein mutants occurs early in infection, i.e., sufficient translation to unblock the synthetase initiation site for ribosome absorption. No evidence for this hold-up in coat protein translation can be found from in vitro studies. Although the mechanism for the hold-up is unclear, failure to make whole molecules of coat protein during the first few minutes of infection ensures the unrestricted translation of the synthetase gene. Clearly here as elsewhere, in vitro protein synthesis studies will play an important role in clarifying the control mechanism.

The work done with single-stranded phages illustrates many of the unique advantages of cell-free translation systems: different forms of messenger can be studied in isolation: the effects of various proteins on translation of the individual cistrons can be examined. Whenever possible, mutant RNAs have been compared with the normal both in vivo and in vitro to insure against drawing conclusions from purely in vitro artifacts. These studies should serve as a model to those who wish to use in vitro methods with other systems.

Most cell-free protein synthesis with DNA-derived messengers has been done with T-phage messengers (much has been done with animal cell messengers but we are discussing only microorganisms here). T-phage infected cells provide the source for the RNA although in some cases messenger made in vitro has been used. Since the pioneering work of Salser, Gesteland & Bolle (21) a number of T3, T4, T5 and T7 proteins have been synthesized from partially purified phage messengers made in whole cells or cell-free systems (22–27). For example, from T4 messengers the lysozyme, α-glucosyl transferase, and deoxynucleotide kinase have been synthesized; from T7 messengers, the lysozyme, RNA polymerase, and ligase have been synthesized. These in vitro synthesized proteins have been detected by direct enzyme assay. Many more proteins have been detected by their characteristic mobilities on polyacrylamide gels. It is likely that most phage proteins can be synthesized in such systems. The problem resides in having a sufficiently sensitive method to detect their synthesis. The potential value of phage messenger-directed protein synthesis systems is twofold: (*a*) it provides an alternate and perhaps more discriminating means of characterizing messenger than hybridizations, (*b*) it provides a means for studying in isolation the effects of a number of variables on translation.

Very little has been done with bacterial messenger-directed systems. Recently Reiness and I (unpublished data) have synthesized a small fragment of the *E. coli* enzyme β-galactosidase in a messenger-directed system using an extremely sensitive assay. Part of the reason for low yields may be that the messenger isolated from whole cells is damaged. The extent of possible damage has not been assessed. Another explanation may be the inability of complete messenger to absorb to ribosomes because of the secondary structure near the initiation site. These difficulties may be overcome in future by use of different conditions for synthesis or different messengers. However, translation of isolated bacterial messengers may be generally difficult since normally in vivo transcription and translation are closely coupled. In spite of this difficulty with β-galactosidase synthesis it appears that some success has been made in synthesizing the alkaline phosphatase of *E. coli* starting from messenger-containing extracts (28). It is noteworthy that in cell-free systems constructed from eucaryotic extracts the eucaryotic messengers can be readily translated. Since eucaryotic messengers are not normally translated until their synthesis is complete, the greater ease of translation of such messengers is in line with their in vivo behavior. Significant translation of bacterial messengers in vitro did not occur until the development of coupled systems involving simultaneous transcription and translation.

Coupled Systems Involving Transcription and Translation

There are two ways in which DNA can be used to stimulate the synthesis of polypeptide chains or proteins. The first uses single-stranded DNA and the DNA serves as messenger directly (29,30). Usually DNA-stimulated systems of this type require drugs like neomycin. Such systems say something about the requirements for translation but probably are of no significance to what occurs in vivo. The second type of DNA-stimulated system uses double helix DNA and a messenger RNA intermediate in a more or less coupled transcriptional-translational system.

As early as 1961 Matthaei & Nirenberg (31) and others observed DNAse-sensitive peptide synthesis in *E. coli* extracts, suggesting that DNA was stimulating synthesis. Between 1962 and 1964 it was shown that cell-free extracts of *E. coli* could be used with T-bacteriophage DNA to make RNA, which in turn stimulates a high level of peptide synthesis (32,33). The polypeptides made in these cell-free systems were never properly characterized and although the conditions in use at that time stimulated peptide synthesis with T-phage DNA, they were curiously impotent with *E. coli* DNA or λ-phage DNA. At about this time our laboratory became interested in the problem of DNA-stimulated protein synthesis in cell-free systems. This interest stemmed from the notion that such systems should be most useful for studying gene regulation. Because the system we developed is most effective in terms of rate of synthesis and capacity for meaningful synthesis, it will be described here in some detail.

As we were most interested in bacterial genes, our first task was to develop a system that would work with bacterial DNA. This was accomplished by varying the components and their concentrations to achieve a maximum of *E. coli* DNA-stimulated peptide synthesis. The incorporation of ^{14}C-leucine into hot trichloracetic acid insoluble substance served as a measure of the amount of synthesis at each stage. Conditions in which 500 μgms of *E. coli* DNA could stimulate about 4.5 mμmoles of leucine incorporation per ml of synthetic mixture (leucine accounts for about 7% of the total amino acid incorporation) were obtained (34).

Utilizing this system, we investigated the possibility of synthesizing specific peptides of the *E. coli* enzyme β-galactosidase (β-gal). To improve detection of specific products of the β-gal gene, λ*dlac* DNA was substituted for E. coli DNA. λ*dlac* DNA has a molecular weight of about 30×10^6 and contains one β-gal gene per *lac* operon per chromosome. As the *E. coli* chromosome has a molecular weight of 2×10^9, and λ chromosome has a molecular weight of 3×10^7, this substitution introduced DNA about 100-fold enriched in the β-gal gene. The λ*dlac* DNA stimulated gross peptide synthesis in the system about as well as *E. coli* DNA.

The enzyme β-galactosidase is a very large tetrameric molecule containing monomers of 1147 amino acid residues. It seemed most unlikely that total enzyme synthesis could be achieved with the relatively crude conditions in use, so the more modest task of synthesizing fractional portions of the enzyme was undertaken. To study the synthesis of a portion of the β-galactosidase molecule by assaying for β-galactosidase enzymatic activity, we used a complementation system developed by Ullman, Jacob & Monod (35). They had found that mutant proteins of β-galactosidase with a defect in the operator-proximal portion of β-galactosidase (corresponding to about the first fifth of the β-gal gene), called the α region, could be complemented by mutant proteins that had their defects elsewhere. The newly synthesized α fragment of the enzyme could be detected by the enzyme activity resulting from addition of protein complement. For convenience, the cell-free extract or S-30 used in the synthetic incubation mixture was prepared from a strain carrying this complementing protein so that complementation could take place during the synthesis step (36). The strain used to prepare the S-30 contains absolutely no β-galactosidase activity, so that any activity detected after the synthesis reaction must result from the synthesis of part or all of the α peptide. In a typical assay, all the components of the cell-free system except the S-30 are mixed together and warmed to the incubation temperature of 37°C. The S-30 is then added and the incubation is continued with constant shaking for a period of one hour. After one hour, an aliquot is removed for enzymatic assay. The assay measures the enzymatic hydrolysis of O-nitrophenyl-galactoside, which yields a yellow color with a maximum at 420 nm. Assays are run for sufficient time to allow accurate determination of the enzymatic activity. In practice, it has been found that assays may be run as long as four days, enabling detection of extremely low amounts of enzyme activity. In some cases parallel assays for gross peptide synthesis were made using ^{14}C-leucine incorporation into hot acid insoluble precipitate. It was possible to detect a small amount of enzymatic activity (5–10 times background) with the system initially developed solely by the leucine assay described above. Further improvements in the system were made with the aid of the enzymatic assay. The superiority of this assay to any assay measuring gross peptide synthesis cannot be overstated. To illustrate this, gross peptide synthesis was measured in parallel with the enzyme activity. Changes in composition of the cell-free system or in mode of preparation of DNA or cell-free extract, which significantly improved the amount of enzyme activity, frequently led to little or no increase in leucine incorporation or even to a decrease in leucine incorporation.

To find optimal concentrations for the components in the system, the concentration of each component was varied singly by a small standard amount, ± 10%, until an optimum was found. All components were tested in series in this way and then the process was repeated for the same ingredients until no further improvement was made. This is a very tedious procedure but it works. Finding new components was less systematic than finding the optimum concentrations, and there is no reason to believe that even under the best conditions found to date that all the desirable components are present. As the cell-free system was improved, it became possible to detect β-gal enzyme activity without the use of the complementing protein (37). For these studies, strains that have a genetic deletion of the entire *lac* region were used to prepare the S-30s. Further improvements in the system were made by using thc gross amount of β-gal synthesized as the ultimate criterion. The conditions developed to optimize β-gal synthesis are effective for the synthesis of a number of other bacterial and viral proteins. Other conditions for cell-free synthesis are in use, which work well with some viral DNAs but poorly if at all with bacterial DNAs. Evidently the transcription and translation of bacterial genes is a more sensitive process and can occur only under a relatively narrow range of conditions. An up-to-date description of the methods for preparation of materials and incubation conditions according to the recipes used in our laboratory has been set forth below.

METHODOLOGY A complete summary of all the procedures used from growth of cells to analysis of enzyme is given.

GROWTH OF CELLS The bacterial cells used in preparation of S-30s are grown at 28°C to mid-log phase in a New Brunswick microferm fermentor in the following medium: per liter distilled water; KH_2PO_4 (anhydrous), 5.6g; K_2HPO_4 (anhydrous) 28.9g; yeast extract, 10g; thiamine, 10–15mg; and 40 ml of 25% glucose added after autoclaving. Fermentors containing 10 liters of medium are inoculated with 1 liter of culture grown overnight on the same medium. About 4 hr after inoculation, cells are collected without otherwise interrupting the growth process in the fermentors. Cells from the fermentor are chilled to 1°C by passage through a copper coil immersed in a water bath and collected in a Lourdes continuous flow centrifuge at a rate of 100 ml/min. The yield of cells is about 10g of wet paste per liter of medium. The cells are removed from the rotor, flattened into pancakes about 1/8 inch thick and frozen at −90°C in a Revco ultradeep-freeze. The cells are stored in this manner overnight.

PREPARATION OF S-30 EXTRACT The S-30 extract is prepared with minor modifications by the method of Nirenberg (1963). Fifty gms of frozen cells are allowed to soften at 4°C for 30 min and homogenized in a Waring blender with 500 ml of buffer I (0.01 M tris-acetate, pH 8.2, 0.014 M magnesium acetate, 0.06 M potassium chloride, and 0.006 M 2-mercaptoethanol). The suspension is centrifuged for 30 min at 10,000 rpm in a large Serval rotor. The sediment of cells is resuspended in 200 ml of the same buffer and recentrifuged. The final sediment is resuspended in 65ml of buffer II (buffer I containing 0.001 M dithiothreitol in place of 2-mercaptoethanol). The cell suspension is lysed in an Aminco pressure cell using pressures of 4000–8000 psi. Immediately after lysis, 1 μmole of dithiothreitol per ml is added to the lysate. No deoxyribonuclease is added to lysate. After two 30-min centrifugations at 30,000 × g, the supernatant is mixed with 8.0 ml of a solution containing: 6 mmoles tris-Ac, pH

8.2, 0.06 mmoles dithiothreitol, 0.17 mmoles $Mg(OAc)_2$, 0.6 μmoles of the twenty amino acids, 0.048 mmoles ATP, 0.54 mmoles Na_3 PEP, 0.16mg PEP Kinase. The mixture is incubated in a light-protected vessel at 37°C for 80 min and dialyzed for 18 hr at 4°C against buffer III with one change (buffer II containing 0.06 M potassium acetate in place of the potassium chloride). The S-30 extract is rapidly frozen in 2-3ml portions and stored at −90°C. Before use, the S-30 is thawed at 4°C and used immediately.

DNA PREPARATION This varies with the source but most of the DNAs employed come from E. coli strains which are doubly lysogenic for say λ and λ*dlac*. The viral strains used carry two additional mutations in the viral genomes that are most helpful in obtaining large yields of virus with a minimum of toil: a temperature-sensitive repressor for virus multiplication and a mutation inhibiting cell-lysis so that the viruses multiply when induced without lysing the cells in which they multiply. The lysogenic cells are grown in medium containing per liter of distilled water: 10 grams tryptone, 5 grams yeast extract, 5 grams NaCl, and 100 ml 1M tris HCl, pH 7.3. Thirty liters of growth medium are inoculated with a 3-liter culture of the lysogenic E. coli grown with aeration overnight at 33°C until a concentration of about 2×10^9 bacterial/ml is reached (about 3 hr). The temperature of the culture is then raised to 42°C for 7 min (it takes about 15 min to reach 42°C in the fermentor) and then lowered to 37°C and aeration is continued about 3 hr. The cells at 37°C are pumped through a copper tube packed in ice water to reduce the temperature to 1°C. Subsequent steps in the isolation are carried out at 2°C. After passing through the cooling system, the liquid is transferred to a Lourdes continuous-flow system and centrifuged at 10,000 rpm in a CFR-2 rotor with a flow rate of 160 ml/min. The wet paste (80-120gms) is stored at 3°C overnight. The solid is homogenized in 775 ml of buffer containing per liter: 10 ml of 1M tris HCl, pH 7.3; 2.0 ml of 1M $MgSO_4$; 1.0 ml of 1M $CaCl_2$; and 4gms of NaCl. This is warmed to 36°C and 140 ml of chloroform is added. The mixture is stirred for 5 min. After about 2 min it becomes extremely viscous due to lysis of the cells. After 5 min pancreatic DNAse (about 100 μgm) is added dropwise until the viscosity has decreased to that of about water. Stirring is continued for 5 min and the mixture is chilled to 5°C. The solution minus the chloroform is centrifuged for 30 min in a Sorvall GSA rotor at 10,000 rpm. The supernatant is very carefully decanted and centrifuged for 2.7 hr in a Sorvall SS-34 rotor at 16,000 rpm. The pellets for this centrifugation containing the viruses are resuspended in 130 ml of the original buffer. To this solution, 98 grams of CsCl is added with stirring. The material is centrifuged for one hr in a Sorvall SS-34 rotor at 13,000 rpm. A viscous layer at the top and some precipitate at the bottom are removed. The opalescent virus-containing solution is centrifuged for an additional 23 hr at 27,000 rpm in a Spinco number 30 preparative rotor to give two virus bands. The lower one, containing the λ*dlac* DNA, is removed. The pooled λ*dlac* virus in a volume of about 15 ml is dialyzed overnight against a solution 0.1 M in NaCl and 0.1 M in sodium phosphate buffer at pH 7.1. The virus-containing solution is adjusted with the above buffer to obtain an OD at 260 nm of 13. 0.2 ml of 25% sodium lauryl sulfate and an equal volume (15 ml) of 88% redistilled phenol are added with shaking. The two-phase system is shaken for 3 min in a 125 ml erlenmeyer flask. Minimum agitation to assure mixing of the phases is used. After shaking, the phases are separated by centrifugation for 3 min at 20,000 × g. The top phase containing the DNA, is easily decanted because of its high viscosity. The phenol extraction is repeated twice (after each extraction the volume of the DNA solution is adjusted to the original 15 ml with the saline-phosphate buffer) and the final DNA-containing solution is dialyzed against 0.02 M versene, pH 8.0 for 16 hr followed by dialysis against 0.01 M tris-acetate, pH 8.0 for 4 hr. The final solution containing about 500γ DNA/ml is stored in a stoppered test tube at 4°C over chloroform.

INCUBATION PROCEDURE The incubation mixture contains per ml: 44 μmoles of Tris-acetate (pH 8.2); 1.37 μmol of dithiothreitol; 55 μmol of KOAc; 27 μmol of NH_4OAc; 14.7 μmoles of $Mg(OAc)_2$; 7.4 μmoles of $Ca(OAc)_2$; 0.22 μmoles each of 20 amino acids; 2.2 μmoles of ATP; 0.55 μmoles of GTP, CTP, and UTP; 21 μmoles of trisodium phosphoenolpyruvic acid; 0.5 μmoles of cAMP; 100 μg of *E. coli* tRNA; 27 μg pyridoxine HCl; 27 μg of TPN; 27 μg of FAD; 11 μg of p-aminobenzoic acid; 27 μg of folinic acid; 16 mg polyethylene glycol 6000. The amounts given for the calcium and magnesium salts are only approximate. In practice the ratio is kept constant and the total divalent cation concentration is optimized for each S-30. The above ingredients are incubated for 3 min at 37°C with the DNA (usually 50 μgm), with shaking, before 6.5 mg of S-30 extract protein is added. Incubations with shaking are allowed to continue for a period of 1–2 hr at 37°C. The shaking should be as fast as possible without producing bubbles or splashing. During the incubation, a viscous pellet is formed in the bottom of the tube or vessel. At the termination of the incubation, the precipitate is gently resuspended and aliquots for assay are removed.

ASSAY PROCEDURE Assays on complete incubation mixtures and controls were carried out at the end of the synthetic period by mixing 0.1 ml of incubation mixture with 1.6 ml of O-nitrophenyl-β-galactoside (ONPG) solution (0.55mg of ONPG per ml of buffer containing 0.1 M sodium phosphate, pH 7.3, and 0.14 M β-mercaptoethanol). The assay tubes were incubated for a length of time sufficient to develop significant yellow color. To terminate the assay, 1 drop of acetic acid was added to each tube to precipitate the protein. The tubes were quickly stirred, chilled on ice, and then centrifuged in the cold for 15 min at 2,000 × g. The supernatant liquid was transferred to a tube containing an equal volume of 1 M sodium carbonate. The optical density was determined at 420 nm by reading against a distilled water blank. The complete system has an OD_{420} at zero time of assay of about 0.035. Under optimum conditions the assays will give an OD_{420} of about 1.0 in 3 min.

The time dependence of enzyme synthesis in the cell-free system has been studied by removing aliquots from the incubation mixture at the desired times and placing them in the assay buffer for enzyme quantitation. RNA and protein synthesis cannot continue in the assay buffer. Significant activity does not appear until 8 min after starting synthesis at 36°C, and this activity increases linearly for about 2 hr at 36°C or 3 hrs at 29°C. The amount of β-gal activity produced in 3 hr corresponds to about 1 μgm of enzyme per ml of synthetic mixture. Parallel experiments where either rifampicin or actinomycin are added during synthesis allow one to calculate a minimum time for the synthesis of β-gal messenger of 3 min (38). This is about 1/3 the in vivo rate and indicates a mRNA growth rate of 22 nucleotides per second. The translation rate is also about 1/3 the in vivo rate. In spite of this rapid rate it can be calculated that the average λ*dlac* DNA molecule (50 μgms of λdlac DNA is present per ml of synthetic mixture) accounts for the synthesis of only 4 complete β-gal polypeptide chains in 3 hr. Considering the rapid rates of synthesis it would seem that it should be possible to make more β-gal per genome over a 3-hr period. Obviously not all of the DNA molecules are transcribing and translating at the same rate. Either many of the DNA chains are damaged or only a few are working at any one time as if the rate of initiation was very slow and limiting. We suspect that the latter is the case. Another factor that decreases the apparent efficiency of the system is that only a fraction of the polypeptide chains are completed. Thus gross peptide synthesis estimated by ^{14}C-leucine incorporation into hot trichloracetic acid gives a value of 12 mμmole leucine incorporated per ml of synthetic mixture. It can be calculated that only about 7% of this leucine is needed to make the observed β-gal enzyme. About 80% of the incorporated leucine is in peptides that cosediment with the ribosome fraction, suggesting that this portion contains incomplete polypeptide chains still attached to the ribosomes. The remainder is incorporated into other proteins of the *lac* operon and the λ phage as indicated by autoradiography of electrophoresed polyacrylamide gels.

Some of the Applications of the Coupled System

With the coupled system it is now possible to synthesize most bacterial and viral proteins in readily detectable amounts. The main question for the future is not can a particular protein be made in the cell-free situation, but rather, what can be learned from studying its synthesis. One of the most useful applications of the coupled system has been in the detection of factors involved in the regulation of transcription and translation. Once factors have been detected and studied in the coupled system they can be purified, if necessary, for further characterization using the coupled system as an assay to guide the purification. Factors are detected by their ability to stimulate or inhibit protein synthesis. Small molecule factors are amenable to direct study, as they can be added to or deleted from the cell-free system at will. Macromolecular factors, present in cell-free extracts prepared from wild-type bacterial strains can be studied through a technique in which the cell-free extract (S-30) is prepared from mutant strains defective in the factor (s), and the system is reconstituted with fractionated portions of a cell-free extract from a normal strain. Recently we have found another method for studying protein factors. Some protein factors do not survive the normal procedure for making S-30 extracts. These factors can be supplied to the S-30 from more delicately prepared protein-containing extracts. Some applications for studying factors by the three aforementioned procedures are described below.

A. Studies of Factors Affecting the Regulation of the *Lac* Operon (38)

One of the most intensively studied and best understood inducible gene clusters is the *lac* operon, which is involved in the breakdown of lactose to its component monosaccharides, galactose and glucose. The *lac* operon consists of three structural genes, which code for β-galactosidase, lactose permease, and galactoside transacetylase as well as a promoter and an operator. The first protein hydrolyzes lactose to its constituent monosaccharides, the second concentrates lactose from outside the cell, and the third catalyzes the acetylation of β-galactosides. The structural genes for these three proteins are adjacent to one another and the controlling elements are located at one end of the gene cluster. In most of the coupled system studies of the *lac* operon the level of β-gal synthesis was used as an index of gene activity. This assumption seems justifiable in retrospect but it does not eliminate the possibility of purely translational control mechanisms.

Small molecules of two classes exert a profound effect on the enzyme yield as a result of what appears to be a gene-regulating function (39). The first class includes inducers of the *lac* operon, of which isopropylthiogalactopyranoside (IPTG) is the most potent known. In systems, isogenic except for the *lac* operon repressor, IPTG stimulates only that system containing the repressor. If repressor is absent, maximum activity is obtained without IPTG. The interaction of IPTG with the repressor is believed to result in the release of the latter from the operator. The amount of IPTG required to induce β-gal synthesis depends upon the amount of repressor present. As predicted by the mass action law, more IPTG is required at higher concentrations of repressor. Quantitative studies have led to the proposal that the binding of two inducer molecules per repressor molecule is required for maximum derepression. The second class of small molecules regulating enzyme yield are those associated with catabolite repression of which cyclic 3′5′ AMP is a prime example. Cyclic AMP is usually added to the coupled system, as its omission results in much lower enzyme activities. The synthesis of galactoside transacetylase has also been shown to depend upon the presence of cyclic AMP in vitro.

Historically the *lac* repressor was first isolated by Gilbert & Müller-Hill (40) using the binding between the repressor and ^{14}C-IPTG as the assay during purification. However, the DNA-directed system provides a superior assay for repressor because of its greater specificity and greater sensitivity. This can be appreciated if one considers that the sensitivity of the inducer binding assay used by Gilbert & Müller-Hill is limited by the affinity constant of the inducer for repressor, which is about 10^6 liters mole^{-1}, whereas the assay provided by the coupled system is limited by the affinity constant of the repressor for the DNA, which is about 10^{10} liters mole^{-1}. This makes the latter assay more sensitive by a factor of $10^{10}/10^6$ or 10^4. The coupled system assay is also superior because it allows for a fuller characterization of the repressor. Thus in this assay the repressor is characterized by its ability to inhibit β-gal synthesis and the reversal of this effect by inducer. In the inducer binding assay one measures only the ability of repressor to bind inducer. Because for most regulated genes one may expect the binding of gene regulating protein to DNA to be much higher than to small molecule inducer or corepressor, as a general rule the coupled system provides a superior probe for isolation of gene regulating proteins.

The coupled system was used to isolate the cyclic AMP receptor protein called CAP (41). For this purpose an S-30 is prepared from a CAP mutant that fails to synthesize appreciable β-gal unless a fraction from wild-type cells containing CAP is added. The stimulation of β-gal synthesis by CAP is completely dependent upon the presence of cyclic AMP. Using this assay CAP can be monitored in crude extracts and extracts purified to varying degrees. It has been claimed that CAP could be isolated using a binding assay of CAP to labeled cAMP (42). Because of the low affinity constant between CAP and cyclic AMP and the binding of cyclic AMP to other components of the cell this latter approach is a questionable one. In fact the latter approach was greatly aided by prior knowledge that CAP could be extensively enriched on phosphocellulose (43).

In the coupled system one can quantitatively vary the amounts of IPTG inducer, cyclic AMP, CAP, and *lac* operon repressor. This has made possible a number of studies on the gene activity as a function of the concentration of various regulatory components. Such studies have helped in determining the mechanism of turning the operon on and off under conditions most closely resembling the in vivo state (39). To demonstrate that cyclic AMP and CAP were sufficient for activating the RNA polymerase for *lac* transcription it was necessary to turn to the use of simpler systems in which transcription could be studied in isolation (44).

B. On the *Ara* Operon The coupled system has also been used to study the *ara* operon. This, like the *lac* operon, is an inducible cluster of structural genes and regulatory elements concerned with a particular catabolic function, i.e., the 3-step conversion of L-arabinose to D-xylulose 5-phosphate. An unlinked permease gene (E) is involved in the active transport of L-arabinose from the external medium. The *ara* operon and the E gene are both induced when L-arabinose is present in the growth medium. Because of this, it seems likely that some if not all of the control factors from the two genetic sites are the same. The model for control of the *ara* operon, postulated by Englesberg and his coworkers (45) contains both a *positive* and a *negative* control site. According to this proposal, the *araC* gene encodes a specific regulator protein. In the absence of L-arabinose, this protein acts as a repressor, P_1, binding to the *o* locus. Thus, the *o* locus is a point of negative control like the *o* locus in the *lac* operon. P_1 is displaced from *o* by L-arabinose. L-arabinose stimulates the conversion of P_1 to an alternate conformation P_2. P_2 has a high affinity for the *I* locus on the DNA. The binding of P_2 to *I* leads to a high level of gene expression for the *ara* operon; the *I* locus is called a site for positive gene control. The *ara* operon, like the *lac* operon, requires cyclic AMP and the CAP protein for gene expression. The promoter element is presumed to reside somewhere in the *I* region, as is the site for binding of CAP.

A DNA-directed cell-free system that synthesizes L-ribulokinase coded by the *ara* operon has been developed. L-arabinose and cyclic AMP are required for the expression of this operon (46–48). The *ara*C gene product is also required and can be supplied either by *de novo* synthesis in the cell-free system or added back from extracts of whole cells (48).

According to the model described above, the *ara*C gene product is required for activating the operon, so it was anticipated that active S-30 extracts would have to be made from strains carrying the *ara*C gene. Therefore it came as a great surprise when it was found that equally effective S-30s could be prepared from strains with or without the *ara*C gene. It was eventually found that *ara*C protein does not survive S-30 preparation and that ribulokinase is made in a DNA-directed cell-free system only after the *ara*C protein itself has been synthesized (48). Therefore the λ*ara* DNA used must have the *ara*C gene on it (the *ara*C gene is normally located immediately adjacent to the control elements of the *ara* operon) for ribulokinase synthesis to occur. When λ*ara* DNA lacking the *ara*C gene is used, no ribulokinase is made. Active *ara*C protein can be supplied from protein extracts but great care is required. Thus we have found that prior to use such extracts must be kept cold, and must contain arabinose and toluene sulfonyl fluoride, a serine protease inhibitor. The *ara*C protein has been purified by affinity chromatography and this protein specifically binds to λ*ara* DNA(49). However, *ara*C protein made in this way must be damaged, because it does not stimulate ribulokinase synthesis in the coupled system. For this reason attempts are being made to find an alternate method for purifying this extremely delicate protein using the stimulatory effect on ribulokinase synthesis to assay the protein at different stages during purification.

The loss of the *ara*C protein activity in the preparation of S-30 extracts has led us to wonder if other gene-regulating proteins might be missing or inactive in the S-30s. This has resulted in a new approach to finding regulatory proteins, which will be described later.

C. On the *Gal* Operon The *gal* operon is another inducible cluster of genes that has been studied in the DNA-directed cell-free system (50). Like *lac* and *ara*, *gal* expression requires a specific small molecule inducer, fucose, as well as cyclic AMP. A particularly interesting observation has been made in the *gal* operon relating to the phenemon of polarity. Polarity is the greatly reduced production of an operator distal protein as the result of a translational block (such as an amber mutation) in an operator proximal gene. There is no direct proof of the cause of polarity, but the most popular theory is that the ribosome cannot traverse the messenger beyond the translational block and it is difficult for unattached ribosomes to absorb at an intracistronic initiation site on the messenger. For some reason polarity is not observed for the *gal* operon in the cell-free system. Recently we have found that this is also true in the *lac* operon (Urm & Zubay, unpublished). Thus one obtains the normal level of distal cistron protein product in the polar situation described above. Finding the reasons for lack of polarity in vitro may lead to an understanding of the necessary factors involved. Perhaps the protein or proteins required to produce polarity have been inactivated in the S-30s. If so, a similar approach could be used to producing active protein as was used to obtain active *ara*C protein containing extracts.

D. On Repressible Gene Systems: The *trp* and *arg* operons Two repressible operons, *trp* and *arg*, have been studied in the coupled system. These operons encode enzymes essential for the synthesis of the amino acids tryptophan and arginine; they have control elements adjacent to structural genes in much the same way as the inducible operon systems. Whereas all 5 structural genes associated with tryptophan metabolism exist in one cluster, the 9 structural genes associated with arginine metabolism are broken up into 6 separate regions each with its own promoter-operator control elements. Also associated with each amino acid system at an unlinked location is a specific regulator gene R which is believed to encode a protein repressor molecule. In vivo it is known that both of these operons are repressed at high levels of amino acid and derepressed at low levels of amino acid. The proposed mechanism for repressing the operon involves a complex formed between the operator site on the DNA and the repressor in combination with either the amino acid or a derivative thereof.

Studies of the *trp* operon in the coupled system have been made with both λ*trp* DNA (51) and λ*trp-lac* DNA (52). The λ*trp-lac* DNA carries a fusion of the *trp* and *lac* operons so that the synthesis of β-gal is under control of the normal *trp* operon control elements. The use of λ*trp-lac* DNA has been very convenient because of the simplicity and sensitivity of the β-gal enzyme assay. When using λ*trp-lac* DNA, synthesis in extracts of *trp* R^- cells is progressively reduced by increased additions of extract from *trp* R^+ cells. No *trp* R^+ product repression is seen when β-gal synthesis is programmed by normal λ*dlac* DNA. This highly sensitive and specific assay has facilitated quantitation and partial purification of the *trp* repressor. The same basic approach is being used to assay for and purify the *arg* repressor (53). The partially purified *trp* repressor has been studied further in a transcriptional system containing, in addition to the partially purified repressor, λ*trp* DNA, RNA polymerase, and the salts and substrates required for transcription. In such a system it has been possible to show that the *trp* repressor specifically inhibits up to 90% of the transcription from the *trp* operon with an absolute requirement for tryptophan (54). In the purified transcriptional system there is little chance of making appreciable quantities of any tryptophan derivative. Therefore the observations made under these conditions provide strong support for the notion that tryptophan itself functions as the corepressor of the *trp* operon. The coupled system could not be used to make this observation, as both amino acid and tryptophanyl tRNA are invariably present. On the other hand the purified transcriptional system could not be used as an assay for *trp* repressor, particularly on crude extracts, because the *trp* R^+ effects are obscured by the nonspecific effects of other proteins. This study demonstrates the relative merits of the coupled system and the simpler transcriptional system at different stages in an investigation.

E. Detection of Relatively Unstable Protein Factors Involved in Regulation Whereas S-30 extracts seem to contain all the proteins necessary for transcription and translation, there is evidence that some regulatory proteins, such as the *ara*C protein (discussed above) are missing as a result of inactivation. Such deficiencies in the S-30 could be exploited in cases where mutants are lacking. Consider the situation in which an S-30 lacks a particular regulatory protein which can be supplied from protein extracts prepared otherwise, as in the case of the *ara*C protein (48). In such a situation an in vitro complementation assay could be developed just as in the case where normal and mutant extracts are available. Thus far two new protein factors have been found in this way: a possible termination factor for bacterial operons and a possible stimulation factor for anabolic operons. The evidence for such factors is described below.

The evidence for a bacterial termination factor came to light in a comparison of the cyclic AMP effects on the stimulation of β-gal synthesis when using λ*dlac* DNA and λ*plac* DNA (55). When λ*dlac* DNA is used in the absence of cyclic AMP only 5% of normal β-gal synthesis occurs. In contrast when λ*plac* DNA is used in the absence of cyclic AMP only a 50% reduction of β-gal synthesis occurs. These differences cannot be due to the *lac* control elements, as these are known to be the same in both DNAs. Both DNAs also contain at least part of the *i* gene which is situated immediately adjacent to the *lac* control elements; this gene is believed to end with a normal termination signal. To account for the high level of β-gal synthesis with λ*plac* DNA in the absence of cyclic AMP, we have hypothesized that the *lac* operon in this DNA is located near an initiation site on the virus genome which "reads through" the *lac* operon due to a deficiency of a normal bacterial termination factor. To test this possibility a nonpreincubated protein extract was added back to the S-30 and λ*plac* DNA was used to stimulate β-gal synthesis. Under these conditions most of the β-gal synthesis in the absence of cyclic AMP was eliminated without any major change in the cyclic AMP stimulated β-gal synthesis. Apparently the nonpreincubated extract contains the factor necessary to prevent "read through". This inhibitory effect is being used as an assay to guide the purification of a protein from the crude extract; the protein has an apparent molecular weight of about 200,000 determined by flow rate on Sephadex. This is comparable in size to the ρ factor (M.W.=200,000) isolated by Roberts (56), which is believed to be involved in the termination of some transcripts of λ bacteriophage.

Another protein factor that appears to be deficient in S-30s shows a stimulation effect on the *trp* operon. This factor stimulates (a maximum of 4-fold stimulation has been obtained) β-gal synthesis when λ*trp-lac* DNA is used but not when λ*dlac* DNA or λ*plac* DNA are used (57). Only the first DNA is responsive to normal *trp* operon control elements. This stimulation factor is being purified using the coupled system as a monitor. Our working hypothesis is that this protein is part of a positive control system for stimulating the *trp* and possibly other anabolic operons just like CAP functions as part of a positive control system for catabolic operons. As yet there is no evidence to indicate whether the stimulation factor affects transcription or translation.

F. Miscellaneous Activities Found in the DNA-Directed Systems In developing the coupled system for protein synthesis it was our hope to simulate intracellular conditions. In this event the cell-free system should be useful for studying many biochemical reactions other than transcription and translation. Thus far very few other types of reactions have been studied. The synthesis of the transfer RNA for tyrosine, $tRNA^{tyr}$ (58), guanosine tetraphosphate (ppGpp), and guanosine pentaphosphate (pppGpp) (59) have all been carried out under nearly the same conditions as used for transcription and translation.

The guanine nucleotides can be synthesized in the coupled system or in more highly purified systems. Studies in partially purified systems (59) show that synthesis occurs on the ribosome. The substrates are ATP and GTP or GDP for the synthesis of pppGpp or pppGpp respectively. Further studies with metabolic inhibitors both in vivo and in vitro suggest that the formation of the phosphorylated guanine nucleotides occurs when the ribosomal site for aminoacyl tRNA is unoccupied, e.g., when amino acids are missing or an aminoacyl tRNA synthetase is defective.

Using DNA from $\phi 80\mathrm{su}^{+}_{\mathrm{III}}$ tRNA it has been possible to synthesize $\mathrm{su}^{+}_{\mathrm{III}}$ tyrosyl tRNA. This tRNA is assayed by its ability to suppress amber mutations in the gene coding for β-galactoside; therefore, it must be capable of accepting and transferring amino acids. A sequence analysis shows that the in vitro synthesized transfer RNA is identical to the in vivo product except for the absence of some methylated nucleotides (60).

G. Studies of a Possible Master Control for RNA Synthesis Involving pppGpp. and ppGpp. In vivo and in vitro studies have shown that the guanine nucleotides, pppGpp and ppGpp, are synthesized on the ribosome in response to amino acid deprivation (59,61). Correlated with the rapid rise in pppGpp and ppGpp upon amino acid deprivation there is a severe inhibition of RNA synthesis. A genetic locus called *rel* is involved in both of these processes, as mutation to the rel^{-} state eliminates both the rapid rise in guanine nucleotides and the parallel inhibition of RNA synthesis. These correlated genetic and biochemical results have led to the hypothesis that one or both of the guanine nucleotides are the causal agents that lead to a slowing down of RNA synthesis.

The effect of the two guanine nucleotides over a range of concentrations was tested on the activities of the four operons described above, *lac*, *ara*, *trp*, and *arg*, and the genome for $\mathrm{su}^{+}_{\mathrm{III}}$ tRNA in the cell-free systems (62). The two guanine nucleotides behave in similar manner in all systems tested. The addition of 30μM ppGpp does not noticeably inhibit tRNA synthesis. Even 300μM ppGpp causes only a 25% inhibition of tRNA synthesis. The synthesis of β-gal directed by λ*dlac* DNA is stimulated 2- to 3-fold at optimal concentrations of ppGpp (100μM). The effects of a variety of guanine nucleotides ppGpp, pppGpp, pGp, pppGp, ppG, pppG (for pppG there is already 250μM present in the coupled system) showed that only the derivatives with a pyrophosphate group on the 2′(3′) position stimulate. Above 100μM the stimulation decreases suggesting that ppGpp interaction is complex. Normally cyclic AMP is present in the coupled system for β-gal synthesis; its omission results in a severe reduction (95%) of enzyme synthesized. In the absence of cyclic AMP most of the β-gal synthesis is believed to result from abnormal initiation sites (39). Under these conditions β-gal synthesis is not stimulated by ppGpp. These observations show the importance of the initiation site for transcription in eliciting the ppGpp effect. The importance of the initiation site is also indicated from studies of the *ara*, *trp* and *arg* operons.

The *ara* and *trp* operons also show stimulation. In striking contrast the *arg* operon activity (measured by N-α-acetyl-L-ornithinase production) is markedly inhibited by ppGpp so that above 100μM this operon is about 10% as active as the other three operons.

The effect of ppGpp on the synthesis of coat protein of MS2 viral RNA has also been examined. Up to 300μM ppGpp there is no effect on the level of protein synthesized from this RNA. This result shows that ppGpp does not affect most steps involved in translation, at least over the range of concentrations examined. This strengthens the notion that transcription is the primary process affected by the guanine nucleotides. We have asked ourselves what possible protein factors could be shared in common for transcription from the *ara, lac, trp,* and *arg* operons and conclude that the RNA polymerase itself is the only known protein used in common by these diverse gene types. Therefore we propose that the guanine nucleotides affect transcription by interacting with the RNA polymerase; an idea which is supported by Cashel's observations (63) that ppGpp binds to RNA polymerase and that at saturating levels this binding results in the inhibition of about 60% of the RNA synthesis that begins with a guanine residue. From all this we conclude that ppGpp or pppGpp influence transcription in a positive, negative, or negligible way according to the manner in which the (p)ppGpp-polymerase complex interacts with the promoter. In this way the guanine nucleotides produced in response to a retardation of protein synthesis could modulate the transcription from all genes. Conclusive proof of this mechanism awaits the isolation of mutant polymerases that do not respond to the guanine nucleotides in vitro and in vivo.

V. Studies of Viruses by Coupled Systems

Numerous T-phage proteins have been made in coupled systems, including T4 α- and β-glucosyl transferases, lysozyme, deoxycytidylate deaminase, and deoxynucleotide kinase, T7 lysozyme, ligase, RNA polymerase, and endonuclease, and T3 s-adenosyl methionine cleaving enzyme, lysozyme, and RNA polymerase (22-27, 64-77). The DNAs of two Bacillus subtilis bacteriophages SP82 and SP5C have also been used for synthesis of deoxycytidylate deaminase (78). The potential of the coupled system for studying various aspects of the life cycle of the phages is very great and many studies in this area are presently underway. With T-phages one can separate in vitro synthesis into two steps, transcription and translation, so that each step can be studied in isolation.

Some in vitro synthesis studies have been made of proteins from the single-stranded DNA virus ϕX174 (79-81). The double stranded replicative intermediate has been used for this purpose. This complex is believed to be active in transcription in vivo as well as in vitro.

Observations on the regulation of λ phage protein synthesis have been made in the coupled system (82). Poor conditions for synthesis were used but in spite of this the results were interesting. Ninety percent of the polypeptide product synthesized are from the early right and left promoters, p_R and p_L, parallel to what happens in vivo. This synthesis is specifically repressed by the addition of purified λ repressor.

VI. Outlook

It is now possible in a cell-free system to synthesize a wide variety of bacterial and phage proteins starting from the genome or in some cases from the mRNA. The absence of the cell wall permits additions of factors that may influence synthesis. This technique promises to be a tremendous aid in elucidating transcription and translation processes. Similar developments for plant and animal cells should be possible and are badly needed.

Literature Cited

1. Hoagland, M.B., Stephenson, M.L., Scott, J.F., Hecht, L.I., Zamecnik, P.C. 1958 *J. Biol. Chem.* 231:241–57
2. Nirenberg, M., Matthaei, J.H. 1961. *Proc. Nat. Acad. Sci. USA* 47: 1588–1602.
3. Haselkorn, R., Rothman-Denes, L.B., 1973. *Ann. Rev. Biochem.* 42: 397–438
4. Nathans, D., Notani, G., Schwartz, J.H., Zinder, N.D. 1962. *Proc. Nat. Acad. Sci. USA* 48:1424–31
5. DeVries, J.K., Zubay, G. 1967. *Proc. Nat. Acad. Sci. USA* 57:1010–13
6. Kozak, M., Nathans, D. 1972. *Bacteriol. Rev.* 36:109–34
7. Jussin, G.N. 1966. *J. Mol. Biol.* 21: 435–53
8. Tooze, J., Weber, K. 1967. *J. Mol. Biol.* 28:311–30
9. Valentine, R.C., Engelhardt, D.L., Zinder, N.D. 1964 *Virology* 23:159–63
10. Jeppensen, P.G.N., Steitz, J.A., Gesteland, R.F., Spahr, P.F. 1970 *Nature* 226:230–37
11. Lodish, H.F. 1968. *Nature* 220:345–49
12. Takanami, M., Yan, Y., Jukes, T.H. 1965. *J. Mol. Biol.* 12:761–73
13. Lodish, H.F. 1971. *J. Mol. Biol.* 56: 627–32
14. Berissi, H., Groner, Y., Revel, M. 1971. *Nature New Biol.* 234:44–47
15. Min Jou, W., Haegeman, G., Ysebaert, M. Fiers, W. 1972. *Nature* 237:82–88
16. Robertson, H.D., Lodish, H.F. 1970. *Proc. Nat. Acad. Sci. USA* 67:710–16
17. Steitz, J.A., Dube, S.K.. Rudland, P.S. 1970. *Nature* 226:824–27
18. Groner, Y., Pollack, Y., Berissi, H., Revel, M. 1972. *Nature New Biol.* 239:16–18
19. Sugiyama, T., Nakada, D. 1970. *J. Mol. Biol.* 48:349–55
20. Kolakofsky, D., Weissmann, C. 1971. *Nature New Biol.* 231:42–46
21. Salser, W., Gesteland, R.F., Bolle, A. 1967. *Nature* 215:588–91
22. Scherzinger, E., Herrlich, P., Schweiger, M., Schuster, H. 1972. *Eur. J. Biochem.* 25:341–48
23. Schweiger, M., Herrlich, P., Millette, R.L. 1971 *J. Biol. Chem.* 246:6707–12
24. Millette, R.L., Trotter, C.D., Herrlich, P., Schweiger, M. 1970. *Cold Spring Harbor Symp. Quant. Biol.* 35:135–42
25. Black, L.W., Gold, L.M. 1971. *J. Mol. Biol.* 60:365–88
26. Wilhelm, J.M., Haselkorn, R. 1971. *Virology* 43:198–208
27. Wilhelm, J.M, Haselkorn, R. 1971. *Virology* 43:209–13
28. Dohan, F.C., Rubman, R.H., Torriani, A. 1971. *J. Mol. Biol.* 58:469–77
29. Salas, J., Bollum, F.J. 1968. *J. Biol. Chem.* 243:1012–15
30. Bretscher, M.S. 1968. *Nature* 220: 1088–91
31. Matthaei, J.H., Nirenberg, M. 1961. *Proc. Nat. Acad. Sci. USA* 47:1580–87
32. Wood, W.B., Berg, P. 1962. *Proc. Nat. Acad. Sci. USA* 48:94–104
33. Byrne, R., Levin, J.G., Bladen, H.A., Nirenberg, M.W. 1964. *Proc. Nat. Acad. Sci. USA* 52:140–48

34. Lederman, M., Zubay, G. 1967. *Biochim. Biophys. Acta* 149:253–58
35. Ullman, A., Jacob, F., Monod, J. 1967. *J. Mol. Biol.* 24:339–43
36. Zubay, G., Lederman, M., DeVries, J.K. 1967. *Proc. Nat. Acad. Sci. USA* 58:1669–75
37. Lederman, M., Zubay, G. 1968. *Biochem. Biophys. Res. Comm.* 32:710–14
38. Zubay, G., Chambers, D.A. 1971. In *Metabolic Regulation*, vol. 5. 297–347 Ed. H. Vogel. Academic
39. Zubay, G., Chambers, D.A., Cheong, L.C. 1970. in *The Lac Operon*. Ed. D. Zipser, J.R. Beckwith. 375–91 Cold Spring Harbor Lab. Quant. Biol.
40. Gilbert, W., Müller-Hill, B. 1966. *Proc. Nat. Acad. Sci. USA* 56:1891–98
41. Zubay, G., Schwartz, D., Beckwith, J. 1970. *Proc. Nat. Acad. Sci. USA* 66:104–10
42. Emmer, M., DeCrombrugghe, B., Pastan, I., Perlman, R. 1970. *Proc. Nat. Acad. Sci. USA* 66:480–87
43. Zubay, G. 1969. In *The role of adenyl cyclase and cyclic 3′5′-AMP in biological systems*. Ed. T.W. Rall, M. Rodbell, P. Condliffe, Nat. Inst. Health. 231–35
44. Eron, L., Block, R. 1971. *Proc. Nat. Acad. Sci. USA* 68:1828–32
45. Englesberg, E., Squires, C., Meronk, F. Jr. 1969. *Proc. Nat. Acad. Sci. USA* 62:1100–07
46. Zubay, G., Gielow, L., Englesberg, E. 1971 *Nature New Biol.* 233:164–65
47. Greenblatt, J., Schleif, R. 1971. *Nature New Biol.* 233:166–69
48. Yang, H., Zubay, G. 1973. *Mol. Gen. Genet.* 122:131–36
49. Wilcox, G., Clemetson, K.J., Santi, D.V., Englesberg, E. 1971. *Proc. Nat. Acad. Sci. USA* 68:2145–58
50. Wetekam, W., Staack, K., Ehring, R. 1972. *Mol. Gen. Genet.* 116:258–76
51. Pouwels, P.H., Van Rotterdam, J. 1972. *Proc. Nat. Acad. Sci. USA* 69:1786–90
52. Zubay, G., Morse, D.E., Schrenk, W.J., Miller, J.H.M. 1972. *Proc. Nat. Acad. Sci. USA* 69:1100–03
53. Urm, E., Kelker, N., Yang, H., Zubay, G., Maas, W. 1973. *Mol. Gen. Genet.* 121:1–7
54. Rose, J., Squires, C., Yanofsky, C., Yang, H., Zubay, G. 1973. *Nature New Biol.* In press
55. Yang, H., Zubay, G. 1973. *Proc. Nat. Acad. Sci. USA* In press
56. Roberts, J.W., 1969. *Nature* 224: 1168–74
57. Yang, H., Zubay, G. 1973. *Proc. Nat. Acad. Sci. USA* In press
58. Zubay, G., Cheong, L., Gefter, M. 1971. *Proc. Nat. Acad. Sci. USA* 68:2195–97
59. Haseltine, W.A., Block, R., Gilbert, W., Weber, K. 1972. *Nature New Biol.* 238:381–84
60. Manley, J., Reiness, G., Zubay, G., Gefter, M. L. 1973. *Arch. Biochem. Biophys.* In press
61. Cashel, M., Gallant, J. 1969. *Nature* 221:838–42
62. Yang, H., Zubay, G., Urm, E., Reiness, G., Cashel, M. 1973. *Proc. Nat. Acad. Sci. USA* In press
63. Cashel, M. 1970. *Cold Spring Harbor Symp. Quant. Biol.* 35:407–13
64. Schweiger, M., Gold, L.M. 1969. *Proc. Nat. Acad. Sci. USA* 63:1351–58
65. Gold, L.M., Schweiger, M. 1969. *J. Biol. Chem.* 244:5100–04.
66. Gold, L.M., Schweiger, M. 1970. *J. Biol. Chem.* 245:2255–58
67. Young, E.T. 1970. *J. Mol. Biol.* 51: 591–604
68. Gold, L.M., Schweiger, M. 1969. *Proc. Nat. Acad. Sci. USA* 62:892–98
69. Schweiger, M., Gold, L.M. 1969. *Cold Spring Harbor Symp. Quant. Biol.* 34:763–70
70. Brody, E.N., Gold, L. M., Black, L.W. 1971. *J. Mol. Biol.* 60:389–93
71. Natale, P.J., Buchanan, J.M. 1972. *Proc. Nat. Acad. Sci. USA* 69:2513–17
72. Herrlich, P., Schweiger, M. 1971. *Mol. Gen. Genet.* 112:152–60
73. Herrlich, P., Schweiger, M. 1971. *Mol. Gen. Genet.* 110:31–35
74. Herrlich, P., Schweiger, M. 1970. *J. Virol.* 6:750–53
75. Herrlich, P., Scherzinger, E., Schweiger, M. 1971. *Mol. Gen. Genet.* 114: 31–34
76. Summers, W.C., Jakes, K. 1971. *Biochem. Biophys. Res. Comm.* 45:315–20
77. Wilhelm, J.M., Johnson, G., Haselkorn, R., Geiduschek, E.P. 1972. *Biochem. Biophys. Res. Comm.* 46:1970–77
78. Schweiger, M., Gold, L.M. 1970. *J. Biol. Chem.* 245:5022–25
79. Bryan, R.N., Suguira, M., Hayashi, M. 1969. *Proc. Nat. Acad. Sci USA* 62: 483–89
80. Gelfand, D.H., Hayashi, M. 1969. *Proc. Nat. Acad. Sci. USA* 63:135–37
81. Gelfand, D.H., Hayashi, M. 1970. *Proc. Nat. Acad. Sci. USA* 67:13–17
82. Gesteland, R.F., Kahn, C. 1972. *Nature New Biol.* 240:3–6

3057

CONTROL OF GENE EXPRESSION IN BACTERIOPHAGE LAMBDA

Ira Herskowitz

Department of Biology, University of Oregon, Eugene, Oregon

I. INTRODUCTION

The purpose of this review is threefold: (*a*) to provide some of the background work and ideas on phage λ to enable the nonspecialist to follow current studies on λ; (*b*) to present a usable overview of λ—what is probably true, what may be true, and what remains to be determined or discovered; (*c*) to present general features of λ as an organism under molecular control—the features that make λ interesting and amenable to study.

I feel that control of gene expression and cellular development in higher organisms will not simply be a rehash of λ control circuits. In fact different temperate phages such as P2, P1, and P22 have evolved mechanisms of gene control with varying differences from λ (41,44). I do, however, feel that the development of ideas concerning gene control in phages such as λ, and the rigor with which they have been put to test, make λ an intellectual model to emulate in studies of control of gene expression.

This is a highly personalized account of λ, with accent on genetic aspects. As a complement, I suggest reading other papers with more emphasis on biochemical aspects (43,44,105,173). I have attempted to give credit to the appropriate workers who helped me formulate a view of λ. In many cases, these ideas developed in a number of laboratories in parallel, with the full picture being a fusion of different experiments. Apologies are offered in advance for leaving people out. For a more complete bibliography on λ, see *The Bacteriophage Lambda* (77). In addition, I have tried to cite the most recent work on different topics.

THE INFECTION PROCESS When λ infects *E. coli*, one of two things happens: (*a*) within an hour about one hundred phage are released and the cell dies (the *lytic* response); (*b*) the infected cell survives, and often has acquired new heritable properties (the *temperate* response). These cells are immune to λ infection—that is, λ does not grow in such hosts; and, under appropriate conditions, these cells can produce a full burst of phage particles, with resultant cell death. Such bacteria are called lysogens, since they can give rise to cell lysis. Such cells are said to carry a λ prophage.

In both the lytic and temperate responses, λ genes are expressed in an orderly sequence. The purpose of this review is to discuss the manner in which this sequence is brought about.

II. IDENTIFYING λ GENES

λ genes can be divided into "essential genes", whose products are required for plaque formation on ordinary *E. coli* strains, and the "nonessential genes", those not required for plaque formation (21,107).

ESSENTIAL GENES The fundamental problem in studying mutants defective in essential genes is how to keep them alive. This problem has been circumvented in two ways: (*a*) by isolating mutants that are defective under some conditions but functional enough under other conditions to make plaques and allow propagation (conditional lethal mutants). (*b*) by isolating mutants, defective under all conditions, that are propagated as prophages (absolute defective mutants). In practice, two kinds of conditional lethal mutants are used (21,54): (*a*) *Temperature-dependent mutants*: temperature-sensitive mutants (*ts*) grow at low temperature, 30°C or so, but not at high temperature, 40°–42°C (21,54); cold-sensitive mutants (*cs*) grow at high temperatures but not at low temperatures (31). Such mutations usually result in the replacement of a single amino acid (187). (*b*) *Suppressor-sensitive mutants*: these grow on certain bacterial mutants carrying a suppressor (su^+), but not on wildtype bacteria, (su^-). Suppressor-sensitive (*sus*)[1] mutations result in generation of messenger RNA codons UAG (*amber* codon), UAA (*ochre* codon) or UGA (59,162). Because wildtype (su^-) bacteria have no tRNAs able to recognize these codons, the growing polypeptide chain is prematurely terminated (162). Certain strains of *E. coli* (su^+) carry mutations of tRNAs that result in the recognition of these nonsense codons as sense, with efficiencies ranging from approximately 20–70% for *amber* suppressors (59,156). As all genes coding for proteins should be able to give rise to suppressor-sensitive or temperature-sensitive mutations, all essential genes in λ are potentially identifiable by isolating such mutants.

ESSENTIAL SITES Of crucial importance in studies with λ has been the ability to study mutations that are not conditionally lethal—that is, which are *absolute defective* (51,88,108). In this case the λ mutants are maintained as prophages,

[1] Although "*sus*" is a more general term than "*amber*", "*ochre*" or UGA, most λ *sus* mutants studied are *amber* mutants.

inserted into bacterial DNA, and are replicated passively by their host. Examples are mutations of sites on λ DNA which are necessary for transcription.

NONESSENTIAL GENES Nonessential genes have been identified by isolation of deletion or substitution mutations (21, 107,121), and by mutations that lead to altered plaque morphology.

MUTATIONS AND GENES—COMPLEMENTATION TESTS Conditional lethal mutant hunts have produced hundreds of different λ*sus*, λ*ts*, and λ*cs* mutants (16,20,31,61,63,120,175). The mutations have been grouped into genes by complementation or "*cis-trans*" tests based on principles set forth by Benzer and Lewis (6,102). That is, two mutations lie in different functional units, defined here as genes, if phage carrying one of the mutations can help phage carrying the other mutation to grow. (These mutants "complement" each other.) Mutations are in the same gene if phage are not produced when bacteria are infected with both types. Complementation tests between λ*sus* mutants are especially easy to interpret since *sus* mutations result in the *loss* of a particular protein, hence the loss of function. In general, supplying the missing protein is sufficient to restore growth. This is to be contrasted with the case where the mutant protein *interferes* with the action of the wildtype protein. In this case efficient complementation is observed only where wildtype protein is supplied, *and* the mutant protein is somehow removed or further inactivated. In general, all mutations in one complementation group complement representative mutants defective in other complementation groups.[2]

By these rules, λ conditional lethal mutants have been grouped into 24 complementation groups (called *A* through *Z*, leaving out *x* and *y*)—24 genes essential for plaque formation on wildtype *E. coli*. Four other essential genes (*cro* and three head genes) were not identified in these conditional lethal mutant hunts (50,79,110).[3]

ARRANGEMENT OF GENES ON λ DNA λ phage particles contain double-stranded DNA (molecular weight 32×10^6 daltons) with single-stranded "cohesive ends" 12 nucleotides long (34,188). The genetic map of λ derived from recombination frequencies is shown in Figure 1. The arrangement of λ genes has also been determined by two other techniques, which show that the genetic map is colinear with λ DNA as found in the phage particle. Maps of λ have been constructed by heteroduplex mapping, in which artificial DNA heteroduplexes are made between two DNAs from different genetically-characterized phages and examined in the electron microscope. By measuring the positions of homology and nonhomology, physical position of various substitution mutations or deletions can be determined (35,55,182). A second mapping technique is to fragment λ

---

[2] Exceptions to this behavior are described in reference 120 and Section III.

[3] The *cro* gene is considered an essential gene since λ wildtype with a *cro*⁻ mutation is unable to grow lytically (Section III).

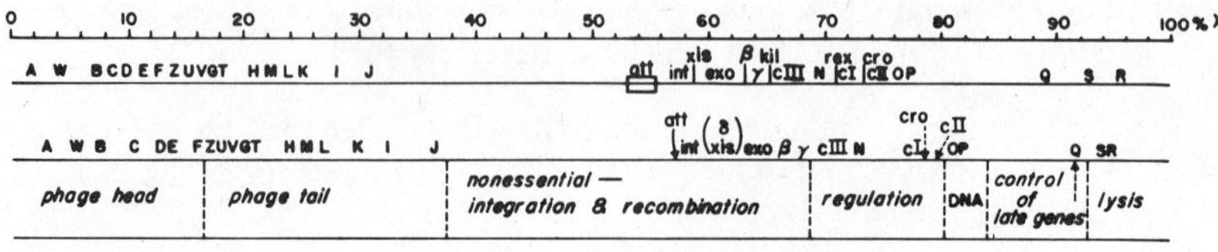

Figure 1 *Physical and genetic maps of* λ. *Line 1* (from top): Percentage length of the λ DNA molecule. 1% corresponds to approximately 465 base pairs. *Line 2*: Genetic map compiled from recombinaton frequencies (21). *Line 3*: Physical map based on electron microscopy of DNA heteroduplexes. The order of genes *delta* and *xis* is not known (3,190). Adapted from Davidson & Szybalski (34).

DNA and determine how close to one of the λ cohesive ends a particular gene is (48). This technique involves infecting bacteria with fragmented λ DNA and coinfecting with λ mutants defective in different genes. Uptake of λ DNA in such experiments requires a λ cohesive end, hence one can determine the minimum size DNA fragment that can contribute a particular gene marker to the coinfecting phage (48).

Are there more essential genes than those identified by presently existing mutations? The existence of λ deletion mutations in different regions puts limits on where any new genes might be found (Figure 2). There are no essential genes in the middle third of the λ map since λ carrying deletions from very close to gene *J* through *int* are viable. In addition substitution mutations deleting λ genes *int* through the left hand endpoint of gene *N* are viable (*bio*256) (149). Between genes *P* and *Q* is a stretch of DNA about 9% of λ length. Phages with deletions of 5–6% in this region (*nin* mutation) are viable (30,55). Between gene *R* and the right hand cohesive end is a stretch of DNA 5% in length, at least 3% of which can be deleted without affecting viability (55).

One can wonder whether λ mutant hunts have systematically missed λ *sus* mutants in genes that require a very low level of protein to give normal growth

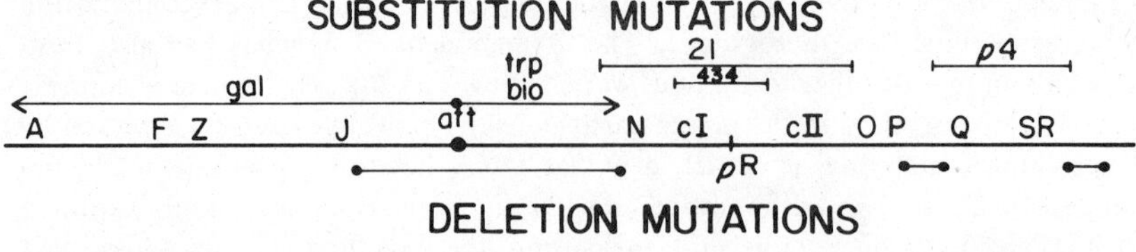

Figure 2 *Substitution and deletion mutations*. Lines indicate regions of λ DNA deleted or replaced by various mutations. Arrows diverging from *att* indicate that *gal, trp,* and *bio* substitution mutations may replace λ DNA beginning at *att* and ending at various positions in the direction of the arrow (56,93,107,120,149). *21* and *434* represent the regions of nonhomology between λ and hybrids λ*imm*^{21} and λ*imm*^{434}. *p4* is described in (55,79). Deletion mutations are summarized in (55,121).

(see for example 101). Even su^- strains have a low level of nonsense suppression (which can be viewed as mistranslation of nonsense codons for sense), on the order of 1% or less; hence it is possible that certain λ*am* mutants in undiscovered genes grow even on these strains. Although such an argument can never be ruled out, it is worth noting that *E. coli* su^- strains do differ on the level of this weak suppression (64), and that some of the λ *amber* mutant hunts were performed with an su^- strain which has particularly low weak suppression (Campbell's strain 594) (120).

In summary, λ DNA is 32×10^6 daltons, enough information to code for approximately 50 proteins of molecular weight 33,000 daltons (76). Approximately 50 have been identified thus far: 28 essential genes by mutations (*A*→*J*, *N*, *cro*, *O*, *P*, *Q*, *S*, *R* and 3 head genes); 11 nonessential genes identified by mutations (*int*, *xis*, *δ*, *exo*, *β*, *γ*, *kil*, cIII, *rex*, cI, cIII); approximately 11 nonessential genes identified by protein bands on acrylamide gels (10 in region between *J* and *att*; and one between cIII –*N*) (76). Not many λ essential genes are yet to be identified. There may be a couple in the left arm. None can be in the central region of the λ map. In the right arm (between *N* and the end), it would be surprising and most interesting if any new genes are to be found—although there is room for up to five genes.

So far we have constructed a map which describes how λ genes are arranged on λ DNA. These genes—so far defined only formally—are arranged in a highly nonrandom fashion. With analysis of what the genes do, and what is required for their transcription, this map reduces to a much simpler visual form (Fig. 1). The map proves to be a useful way to visualize aspects of λ development; it is hoped that once the reader knows the λ map, what genes *N*, *Q*, and cI do, and λ transcription units, the basic processes of λ development can be *reconstructed* rather than *remembered*.

III. INFECTION: THE LYTIC RESPONSE

THE LYTIC GROWTH CYCLE Lytic infection by λ results in release of about 100 new phage per infected cell. The minimum set of events in the growth cycle, therefore, is DNA replication, phage particle synthesis, and cell lysis. One expects λ DNA to carry information for cell lysis and phage particle proteins since it is unlikely that *E. coli* has such proteins. λ genes involved in the steps listed above can be identified by characterizing the block in growth of the various mutants described above (summarized in Table 1). Inferring the normal function of particular genes from behavior of mutants is simplified since λ *amber* mutants lack the particular function rather than modify it.

AWBCDEFZUVGTHMLKIJ: Ten genes are required for formation of normal λ heads: AWBCDEF and three newly discovered genes defined by defective mutations (110,181). *E* codes for the major protein of the phage head (109). *F* is responsible for specificity of head to tail joining (23). *A* is responsible for producing λ cohesive ends (179). Eleven genes are required for formation of normal λ tails: ZUVGTHMLKIJ. *V* codes for the major protein of the tail; *J*

TABLE 1 DNA synthesis, lysis, and particle morphogenesis by λ mutants

	AWBCDEF	ZUVGTHMLKIJ	N	O P	Q	S R
DNA synthesis	+	+	reduced	-	+	+
Lysis	+	+	-	reduced	reduced	-
Particle synthesis						
Heads	-	+	-	reduced	reduced	+
Tails	+	-	-	reduced	reduced	+

determines the specificity of phage absorption and is probably the gene coding for λ tail fiber (17). The role of the other genes in particle morphogenesis is under active investigation (110). Deletion of all these genes does not affect the rate or amount of phage DNA synthesis, or cell lysis (42,150).

SR: Two genes, *S* and *R*, are required for cell lysis. *R* codes for an endopeptidase or "endolysin" (originally thought to be a lysozyme) which breaks bacterial cell walls (22,168). Artificial lysis of λR^-–infected cells yields a normal yield of active phage, indicating that DNA synthesis and particle morphogenesis are normal. *S* gene product is thought to be involved in degradation of the cell membrane (1): S^- mutants produce endolysin but do not lyse and do not shut off bacterial energy production. Artificial lysis of λS^- -infected cells after two or three hours liberates active phage with an exceptionally large yield—as high as 1000 phage per cell—presumably due to failure to shut off bacterial energy production (1,63,73,136).

The genes described thus far affect only a single process of the three necessary for lytic growth (lysis, DNA synthesis, particle formation). Genes *N*, *O*, *P*, and *Q* appear to be involved in more than one of these processes, since mutations in these genes may affect not only DNA synthesis, but cell lysis and particle formation as well (15,39,90).

OP: λ *O* and *P* products are required for DNA replication. λO^- and λP^- mutants replicate less than 1% of one round under nonpermissive conditions, as measured by their failure to shift buoyant density when grown in density labeled medium (176). Reduced production of lysis ability and phage particles can be overcome by increasing the number of copies of replication-defective phages (144,171). Further discussion of the relationship between DNA replication and expression of genes *SRA-J* is presented later.

N and Q: λQ^- mutants produce reduced levels (10-20%) of lysis activity and the particle proteins, as assayed by acrylamide gels or by production of tail fiber antigen (39,76,110). The rate and extent of DNA synthesis by Q^- mutants is similar to that of λ^+ (51,152,161). Q protein thus is required for production of about 23 late proteins.

λN^- mutants are deficient in all three processes: no measurable lysis activity or particle proteins are produced, and DNA replication is reduced to about 5%

the rate of λ wildtype (15,33,39,81,176). The *N* gene, therefore, is required for *all* processes essential for λ lytic growth.

So far we have considered what the λ genes do without considering the sequence in which they act. We next consider the temporal plan of λ growth.

SEQUENTIAL GENE EXPRESSION The general scheme of λ development in infection is similar to that of other phages, in particular the double-stranded DNA phages, T4, T5, T7, P2, and P22. After injection of DNA there is a period of early protein synthesis, followed by DNA synthesis, and a period of late protein synthesis (39,41,47). Such a pattern has been derived from measurement of protein and DNA synthesis as a function of time, and by separating these three phases by use of mutants or protein synthesis inhibitors.

In λ, the periods of early protein synthesis and late protein synthesis can be defined by behavior and properties of λN^- and λQ^- mutants.

(*a*) Blocking early protein synthesis by mutation (in gene *N*) or by chloramphenicol or by λ repressor prevents DNA replication and late protein synthesis (39,176). Therefore, early protein synthesis is necessary for DNA and late protein synthesis.

(*b*) λ requires two gene products, *O* and *P*, for DNA synthesis. Blocking replication by mutation in either *O* or *P* does not affect expression of other early genes, for example λ exonuclease (*exo*) (51,112,134). However, reduced amounts of late proteins are produced.

(*c*) λQ^- produce reduced amounts of proteins made late after infection (33,39,76), but are normal with respect to early protein and DNA synthesis.

In summary, early protein synthesis (to produce *N*, *O*, *P*, and *Q* gene products) is absolutely required for DNA replication and late protein synthesis. DNA synthesis is not essential for early protein synthesis (for example, exonuclease), but stimulates or amplifies late protein synthesis. As might be expected, late protein synthesis is not required for early protein synthesis or DNA replication.[4]

TRANSCRIPTIONAL CONTROL OF λ GENE EXPRESSION: KINETICS AND DIRECTION OF TRANSCRIPTION In parallel with genetic analysis of λ control circuits have been in vivo biochemical studies of λ messenger RNA synthesis. The relationship between λ messenger RNA species and the λ map has been determined by hybridization of λ RNA with DNA of known genetic constitution (26,99,165). λ DNA can be broken into double-stranded pieces by shearing (48,151) and into separate DNA strands by equilibrium density gradient centrifugation (37,86). In addition, for essentially all regions of the λ DNA there exist substitution mutations that replace λ DNA with DNA from either *E. coli* (*gal* and *bio* transducing phages) or from the other lambdoid phages 434, 21, 80 (summarized in Fig. 2).

After infection by λ wildtype, messenger RNA is produced from both "left" and "right" DNA strands and is confined almost exclusively to the right arm

[4] Production of mature λ DNA (linear duplex molecules with single-stranded cohesive ends) does, however, require late gene expression (150, 179, 188).

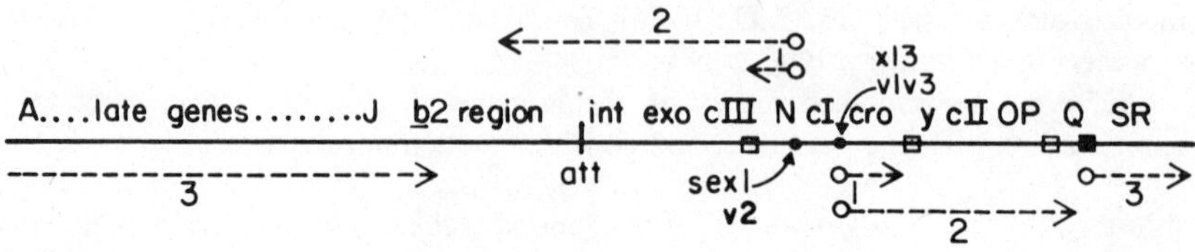

Figure 3 Origin and direction of λ transcription. Arrows indicate direction of transcription and not necessarily individual mRNA species. Arrows pointing leftwards indicate mRNA is transcribed from the *l* DNA strand; rightwards, from *r* strand. Genes cI and *rex* are transcribed from the *l* strand (75,98,159,166). Open boxes indicate proposed mRNA termination sites. Closed symbols indicate promoter sites: circles, *p*L and *p*R; square, late gene promoter. Other symbols are described in the text. The map is not drawn to physical or genetic scale. 1, 2, 3 correspond to Set I, II, III transcription as described in the text.

(early message). After approximately 10 minutes, early transcription is reduced, and transcription from the left arm (from the *r*-strand) predominates (26,99,152,166). Figure 3 summarizes the direction in which λ genes are transcribed.

SEQUENTIAL TRANSCRIPTION λ genes can be divided into three groups on the basis of what functions are required for their transcription:

I. Genes that require no phage products for transcription (*N* and *cro*).
II. Genes that require λ N protein (*int*--cIII, and cII--*Q*).
III. Genes that require λ Q protein (*SRA-J*). (Transcription of cI is considered in Sections IV and V.)

SET I TRANSCRIPTION. Transcription of genes *N* and *cro* occurs immediately after phage infection (74,99,169). This transcription requires *E. coli* RNA polymerase [inhibited by rifampicin (167)] but no phage proteins [not blocked by chloramphenicol (94,99)]. In vitro transcription of λ DNA by RNA polymerase, sigma factor, and termination factor rho produces primarily two messages, those for genes *N* and *cro* [Fig. 5; (7,139)]. Because the *N* and *cro* messages hybridize to different strands of λ DNA, these genes are transcribed in opposite directions, *N* message being transcribed leftward (hence from the *l* DNA strand) and *cro* message rightward (from the *r* DNA strand) (99,169). What sites on λ DNA control the level of transcription of these genes? One can define a *promoter* as a DNA site to which RNA polymerase binds prior to initiation of mRNA synthesis (8,21,105,142). The promoter for *N* gene transcription is defined as *p*L and for *cro p*R (Fig. 5). Location of these two promoters on λ DNA comes from two different kinds of mutations: (*a*) Mutations that lower or abolish transcription of genes *N* or *cro*. The *sex* mutation (Fig. 3; isolation described in Section V) reduces transcription of the *N* gene to 10% the level of λ^+ (56,139). Mutations in the *x*-region (Figs. 3 and 5) reduce transcription of the *cro* gene to 5% wildtype level as measured both in vivo and in vitro (6,75,169). Are these mutations in *p*L and *p*R? It is not presently known whether these mutations affect binding of RNA polymerase to λ DNA. (See 8 for discussion.) (*b*) Promoters can be determined by finding out what DNA must be deleted in order to block transcription of a certain gene. Deletion mutations (obtained in prophages), which remove genes *O*, *P*, and *cro*, and which end within the *c*I gene still allow transcription of *N* (deletion 134, Fig. 4; 125). The *N* promoter therefore lies to the left of such deletions. In similar fashion, deletions removing gene *N*, which end within the *c*I gene, still allow expression of the *cro* gene (93).

The early control region has been elegantly subdivided by use of λ hybrid phages, in particular, λ imm^{434} (51,92) (Fig. 2). By a series of crosses between λ and the related temperate phage 434, a hybrid phage was constructed which carries all λ genes except for the region indicated in Fig. 5. Because this phage has immunity specificity (Section IV) different from λ, this region may be called the "immunity region". Recent work has shown that the *N* mRNA from λ and λ imm^{434} begins with the same sequence (7), indicating that the message RNA *initiation point* lies to the left of the 434 region (as drawn in Fig. 5). Strikingly, *sex* mutations are *within* the 434 region (8,65), on the order of 200 nucleotides from the mRNA initiation site (8). RNA polymerase, therefore, may bind at *p*L and subsequently "migrate" leftwards before initiating RNA synthesis (8).

Termination. In the presence of chloramphenicol or using a λN^- mutant, *N* and *cro* gene transcription terminates at some point to the left of gene *N* and to

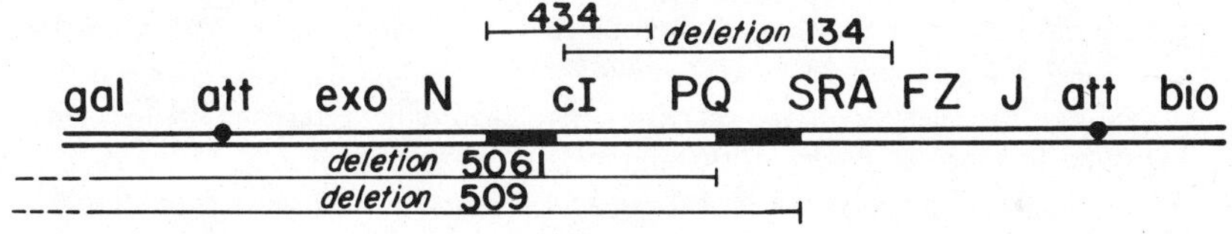

Figure 4 Prophage map showing late gene promoter and site of action of cro product. Lines indicate regions of the prophage deleted. 434 indicates region of nonhomology with λimm^{434}. Filled in segments indicate the region between the endpoints of deletions. Map not drawn to physical or genetic scale. References are described in the text.

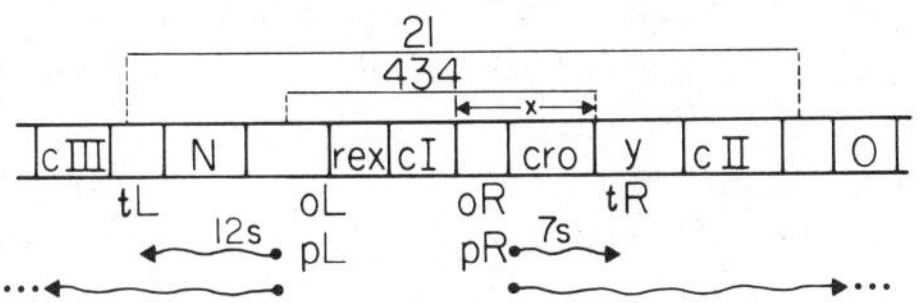

Figure 5 Fine structure of the λ immunity region. 21 and 434 indicate regions of nonhomology between λ and λimm^{21} or λimm^{434}. The *x* and *y* regions are defined with respect to the endpoints of nonhomology with λimm^{434} and beginning of gene *c*II as drawn (51). Arrows indicate direction of transcription. 7*s* and 12*s* refer to specific mRNA species. *o*, operator sites; *p*, promoter sites; *t*, termination sites. *rex* is responsible for exclusion of T4*r*II$^-$ mutants (71,85). Map not drawn to physical or genetic scale. Dotted lines indicate transcription continues in that direction.

the right of the *cro* gene (99,139) (Fig. 5). Production of discrete 12*s* (*N*) and 7*s* (*cro*) message species in vitro requires the action of rho factor, which promotes RNA chain termination (36,139). Sites on λ DNA responsible for this termination have been called *t*L1 (for site near *N*) and *t*R1 (near *cro*) (Fig. 5). Placement of these sites on the λ map is described below.

SET II: N-DEPENDENT TRANSCRIPTION. The *N* gene is required for efficient expression of two blocks of genes—cIII through *int* (from the *l*-strand) and cII through *Q* (from the *r*-strand) (Fig. 3). The deficiency of λN^- mutants in expression of these various genes has been determined in a number of ways.

cIII through int: λN^- mutants produce less than 10% the wildtype level of *int* and *xis* proteins as assayed on acrylamide gels (3). They are likewise defective in processes mediated by *int* and *xis* proteins, namely, in establishing lysogeny, mediating site-specific recombination, and in prophage excision (65,148) (see Section V). λ exonuclease (*exo* gene product) and β protein (β gene product) are reduced to less than 5% (65,133,134).

Evidence that N product is required for expression of genes γ and *kil* is indirect: (*a*) γ product is required for growth on *rec*A^- bacteria. N^- mutants with other mutations (*nin*, described below) allowing growth on rec^+ bacteria do not grow on *rec*A^- hosts (178,189). (*b*) The *kil* gene product participates in killing the bacterial host after prophage induction (69) (see Section V). An N^- mutation blocks the *kil*-dependent pathway responsible for cell death. The *N* gene is also required for production of *trp* enzymes in transducing phages with *trp* genes substituted in place of genes *int* through cIII [λ*trp*, Fig. 2 (56)].

What sites regulate expression of these genes? Are there separate promoters for each gene or a few for groups of genes? Two lines of evidence indicate the (originally) surprising result that the promoter for transcription for genes cIII –*int* is the same as that of the *N* gene itself.

(*a*) λ*sex* mutants have reduced levels of mRNA for the cIII-*int* region (139). In addition, the *sex* mutants are deficient in excision, exonuclease, and in production of *trp* enzymes in λ*trp* fusion strains (56,65,126). Since the *sex* mutation lowers transcription of the *N* gene, and *N* is required for transcription of genes cIII through *int*, one might argue that reduced expression of these genes is due to limitation in *N*. This is not the case, because supplying full levels of N protein with another phage does not restore wildtype levels of exonuclease and *trp* enzymes (56,126;see also 65,148).

(*b*) λ repressor binds near the *N* promoter at operator site *o*L to prevent transcription of gene *N* (130; Section V). In a lysogen, therefore, no exonuclease is produced, as N product is required for exonuclease synthesis. One can ask whether supplying N product can induce exonuclease synthesis from a phage with λ repressor stuck to its operators. In this experiment N protein is supplied to a repressed prophage by infecting the cell with a phage insensitive to λ repressor (e.g., λ with the operators from phage 434, λimm^{434}), which in addition carries a deletion of its exonuclease gene. Under these conditions the prophage *exo* gene is expressed *only if prophage repressor is inactivated* (106). Consequently transcription of *exo* requires a site blocked by λ repressor, presumably *p*L.

cII-O-P-Q: λN^- mutants are deficient in replication and in production of lysis activity and particle proteins. Because genes *O* and *P* are required for λ DNA replication (15,52,176), one might account for the slow replication by λN^- by saying that expression of genes *O* and *P* is reduced (147). A number of arguments suggest that this is the case. Production of *O* and *P* gene product by λN^- phages can be assayed by seeing whether they can stimulate growth of O^- or P^- mutants. One might mixedly infect cells with λN^- and λO^- phages and see whether the O^- phage grows. It does—but this experiment tells one nothing about whether N^- mutants underproduce O protein, as the O^- phage produces normal *N*. For such an experiment, one needs O^- and P^- phages that cannot supply N product to the O^+ phage. The hybrid phage λimm^{21} carries all genes from λ (most importantly genes *O*, *P*, and *Q*) except for the region that includes *N* through *c*II (Fig. 5). In this region are genes analogous to those of λ but from phage 21 (104,183). Growth of $\lambda imm^{21} O^-$ or P^- phages is not restored to wildtype levels by coinfection with λN^- (13,57,81). A normal yield of $\lambda imm^{21} O^-$ or P^- phage is obtained if the λ helper is N^+. By this complementation assay, therefore, λN^- mutants are deficient in production of O and P products. In the presence of N product, the rate of mRNA synthesis in the *c*II-*O* region (inclusive) is increased fourfold (74). Further suggestion that N product is required for replication through stimulation of *O* and *P* gene expression comes from the existence of a mutation that allows expression of genes *O* and *P* in the absence of *N*. The *c*17 mutation[5] is located in the region between genes *cro* and *c*II (the *y* region; Fig. 5) (14,122), and allows *O* and *P* transcription and DNA replication in the absence of N product (115,119,122,154), presumably because it lies to the right of a termination signal *t*R.

N^- mutants are totally deficient in production of the same set of proteins that Q^- mutants underproduce (*S*, *R*, *A*→*J*) (33,39,81). We see in the next section that Q protein is sufficient for expression of these genes. Production of Q product by λN^- phages has been measured in complementation experiments similar to those above: λN^- phages are unable to help growth of λimm^{21} phages deficient in *Q* (32,81). Consequently, N product is required for efficient expression of the *Q* gene. Although the *c*17 mutation bypasses the requirement of N product for *O* and *P* transcription, $\lambda N^- c17$ mutants are unable to transcribe the *Q* gene (18,82). In other words, transcription of the *OP* region is *not* sufficient to guarantee transcription of gene *Q*, presumably because of a termination site between genes *P* and *Q*.

Under two conditions, transcription of the *OP* region (rightwards) can lead to *Q* transcription: (*a*) if N product is supplied by a coinfecting phage (λimm^{434} helps but λimm^{21} does not) (32,81); (*b*) if the $N^- c17$ phage carries a mutation in the region between *P* and *Q* ["N-bypass" or "N-independent" mutations (18,82,164)]. These mutations appear to be modifications or deletions of termination signals rather than the creation of new promoters, as expression of the *Q* gene is still blocked by λ repressor (18).

---

[5] The origin of *c*17 is a bit of a mystery. The most likely guess is that *c*17 is the creation or insertion of a new promoter in a region where such a mutation can be tolerated (138).

Where does transcription of genes *c*II-*O*-*P*-*Q* begin? Just as transcription of *c*III-*int* requires *p*L, transcription of *c*II-*O*-*P*-*Q* requires *p*R, the promoter for the *cro* gene. Mutations (e.g., *x*13, Figs. 3 and 5), located between genes *cro* and *c*I, have been obtained (Section V) which abolish transcription of genes *cro*-*c*II-*O*-*P*-*Q* as assayed both in vivo and in vitro (99,139,169). In addition, deletions that remove *int* through *c*I and end within the *cro* gene abolish expression of genes *O*, *P*, and *Q* (81). As in the case of λ exonuclease and other genes in the *c*III-*int* transcription unit, λ repressor directly blocks expression of genes in the *c*II-*Q* group even when N product is provided by another phage (32,81). Consequently, transcription of genes *c*II-*O*-*P*-*Q* requires a site, presumably *p*R, which is blocked by λ repressor.

Mechanism of N action: Transcription of genes *c*III-*int* and *c*II-*Q* requires prior transcription of genes *N* and *cro*, respectively. In the absence of N product, transcription distal from promoters *p*L and *p*R is terminated between *c*III and *N* and between *cro* and *c*II. Furthermore, transcription of genes *O* and *P* terminates before the *Q* gene. In the presence of N, the genes distal to these termination sites are transcribed. We can say in a formal sense that in the *c*III-*N*, *cro*-*c*II, and *P*-*Q* regions are sites of N action (Fig. 3). These sites are where the presence of N is observed and not necessarily where N itself acts. For example, one might argue that N stimulates transcription of *c*III-*int* and *c*II-*Q* by increasing the number of transcription initiations at *p*R and *p*L, such that the absolute amount of transcription of adjacent regions is increased. This is apparently not true, as the rate of transcription of genes *N* and *cro* is the same in the presence and absence of N (74).

A number of models for N action have been proposed (99,140): (*a*) *antitermination*: In the absence of N, transcription of genes *N* and *cro* terminates. In vitro these terminations are rho dependent. N product might counter messenger RNA termination, for example, by antagonizing rho factor (139). (*b*) *cascade*: Initiation of *c*III-*int* and *c*II-*Q* transcription may require a site that is activated by prior transcription (97). (*c*) *drift*: In the absence of N, RNA polymerase binds to *p*L and migrates to the *N* message initiation site (8). In the presence of N, it may migrate to secondary initiation sites near genes *c*III and *c*II.

It has recently been shown by RNA-DNA hybridization-competition experiments that the *N*-dependent message for the *c*III-*int* region (obtained in vivo) carries part or all of the *N* gene message (128). *N*-dependent transcription from the *l*-strand, and presumably from the *r*-strand as well, therefore, appears to be a continuation of prior transcription. How N product promotes this apparent antitermination of mRNA synthesis is not known. N protein has recently been purified by assaying for factors that stimulate *N*-dependent protein synthesis from λ DNA (38,67,68). Studies of its action are awaited.

SET III: *Q-DEPENDENT TRANSCRIPTION*: The λ *Q* gene is required for full level expression of all λ late genes. λQ^- mutants make reduced levels of lysis activity, phage structural proteins, and late messenger RNA (33,39,76, 81,116,152). The messenger RNA under Q control (late mRNA) is all transcribed from the *r* strand (from left to right as the map is conventionally drawn) (Fig. 3).

The presence of Q product is *sufficient* to turn on late genes of a wildtype λ phage: (*a*) Late gene expression does not require DNA synthesis. The amount of lysozyme or late mRNA produced per gene copy by nonreplicating phages (λO^- or λP^-) is at least as great as in λ^+ infections (81,152). The fact that replication leads to higher late protein synthesis presumably reflects amplification of the number of copies of late genes (171). (*b*) Expression of late genes does not directly require N protein (described below).

What sites (promoters, terminators, etc.) control expression of the late genes, and where does the Q product act? We noted earlier that supplying N product to repressed prophages does not lead to expression of genes cIII-*int* and cII-*Q*. In contrast, late gene expression can be induced from repressed prophages. For example, $\lambda imm^{434} R^-$ (endolysin-deficient) phages can induce endolysin synthesis from a repressed λ prophage (170). Transcription of the prophage *R* gene, therefore, is not directly blocked by λ repressor. The other prophage late genes are also not directly repressed. For example, $\lambda imm^{434} A^-$ phages can grow in a repressed λ lysogen by "turning on" the prophage *A* gene (171).

This induction of late gene expression is dependent upon the *Q* gene and is probably independent of *N*. λQ^- mutants are unable to induce endolysin synthesis from prophages (32). In addition, hybrid phages with the *QSR* region from other λ-like phages (*p*4, Fig. 2; 79) are unable to turn on late gene expression from a prophage. In contrast, *N*-deficient phages with an additional mutation allowing expression of the *Q* gene can still induce expression of the prophage *R* gene (82). Because these experiments were measurements of phage growth and not direct measurements of protein or mRNA, one cannot rule out the possibility that N has some direct stimulatory action on late gene expression (see also 131).

λ late genes can also be induced from defective prophages with extensive deletions of the early genes (including *p*R and *p*L; e.g., deletion 5061, Fig. 4; 80). Induction of expression of representative late genes *R*, *F*, and *K* from a set of prophage deletion strains revealed that between genes *Q* and *S* is a site essential for such induction. Specifically, prophages with deletions that end in the *Q* gene (deletion 5061, Fig. 4) still have full capacity to express genes *R*, *F*, and *K* when supplied with Q product; whereas in deletions that end within gene *S* (deletion 509), expression of genes *R*, *F*, and *K* is practically abolished even though the structural genes *R*, *F*, and *K* are intact (80). Whatever is deleted between genes *Q* and *S* cannot be supplied by the superinfecting phage—hence, the missing function acts only on its own DNA and is by definition a site (or a cis-acting protein). Consistent with the existence of such an essential site regulating late gene expression is the observation that integration of phage Mu into the late genes of a λ prophage inactivates gene expression to the right of the insertion (177). As expression of genes in the left arm of λ DNA requires a site in the right arm, the ends of λ DNA must be joined for efficient late gene expression.

Is this essential site (the region defined by endpoints of deletions) the site of action of Q protein? The evidence is strictly circumstantial: namely, *Q* is required for transcription of late genes, and the *QS* site (the "late gene promoter") is required for late gene expression. Thus far, no mutations have been found in the *QS* region that make Q dispensable for late gene expression. There is, however, a correlation between the source of the *Q* gene and the source of the late gene promoter. That is, phage hybrids with the *QSR* region from other phages (for example, from φ80 or λ*p*4) cannot induce expression of gene *F* from prophages with the *QSR* region from λ. Similarly, λ cannot induce late gene expression from these phages. On the other hand, phages with the *QSR* region of one type can induce late gene expression *in trans* from prophages with the same *QSR* region (79,143).

A different type of evidence indicates that λ late genes share a common site of messenger RNA initiation. A hybrid phage has been constructed in which the arabinose operon has replaced genes in the head region of a λ-φ80 hybrid in such a manner that the *ara* genes *BAD* are transcribed in the same direction as λ late genes. This defective phage has the *QS* region from λ, and the *ara* genes are activated by λ Q product (144). Upon induction of this prophage, λ endolysin is made first and is followed by arabinose isomerase 6 minutes later. If rifampicin is added after endolysin synthesis has begun but before the appearance of isomerase, isomerase is still produced. Whenever rifampicin is added, the relative amounts of isomerase and endolysin finally formed remain constant. The results are consistent with the proposal that endolysin and arabinose isomerase share a *common* rifampicin-sensitive, hence messenger RNA initiation, site. This site presumably is in the *QS* region.

Because late gene transcription does not require prior *r*-strand transcription of early genes, Q product may act by stimulating transcription initiation between genes *Q* and *S*. The molecular basis of *Q*-stimulated transcription is unknown. Q protein might, for example, modify or supplement *E. coli* RNA polymerase to allow recognition of the late gene promoter (81,113).

In what form is late mRNA? λ late gene mRNA may be transcribed as a single species containing all 23 or so gene transcripts, as several species each containing a few gene transcripts, or as separate transcripts for each gene. The last possibility is unlikely, because there are a number of examples of translational polarity among late genes, that is, cases in which *amber* mutations in one gene reduce production of proteins determined by genes distal to it (distal on the genetic map and presumably in this case on the mRNA). Murialdo & Siminovitch (110) have summarized regions of nonsense polarity in the late genes of λ determined from phage growth measurements and effects of *amber* mutations on late protein synthesis:[6] A *WBC* D *EF* ZU *VGTHM LKIJ*. Existence of polarity indicates that at least 4 blocs of late genes are transcribed into polycistronic mRNA. Measurement of late mRNA species has given different results. In one case 5 species with sizes 2–48% of the *A-J* region are found (116). Another report describes a broad range of mRNA species including a major class that might be large enough to carry the entire *A-J* region message (60). A third report describes a single late mRNA species at least 4×10^6 molecular weight (27). Because 15% of the labeled RNA in this peak hybridizes to the right arm of λ DNA, this

---

[6] Regions are italicized. Polarity is from left to right. That is, W^-*am* are deficient in production of *B* gene product.

species may contain *S* and *R* transcripts as well as those for *A* through *J*. Direct demonstration that this RNA peak contains a single RNA species with sequences for genes *S* and *J* has not yet been reported.

The above experiments are consistent with the following representative models for transcription of the λ late genes (80): (*a*) RNA polymerase binds between genes *Q* and *S* at the late gene promoter. It transcribes genes *SR* and *A* through *J* into a single messenger RNA molecule, which might be subsequently cleaved before translation (60). (*b*) RNA polymerase has a single binding site, but as it passes along the late genes, it is able to terminate mRNA chains and reinitiate new chains without being released from the DNA. In any case, for the λ late message which is polycistronic, there are dramatic differences in the amounts of translation of adjacent transcripts. In the polar region *VGT*, 15 times more V protein is made than G, and 27 times more T protein than G (110). These differences may reflect differences in accessibility of the mRNA to ribosome binding (160).

Minor pathway of late gene expression: The major pathway of late gene transcription described above is *Q* gene dependent and is not blocked by λ repressor. A second pathway of late gene expression is revealed by observing that Q^- mutants produce significant though reduced amounts of late proteins and late messenger RNA (approximately 10–20%) (32,39,81). The pattern of late proteins produced by λQ^- phages—that is, the relative amounts of different late proteins assayed on acrylamide gels—is similar to that produced by λQ^+ (76). This low level expression of late genes requires an active *N* gene and can be blocked by λ repressor, indicating that it is initiated from the early rightward promoter (*p*R). Consistent with this view is the existence of λ mutants derived from λQ^- phages, which were selected for their ability to grow without *Q* gene (Q-independent mutants, *qin*) (79,81). These phages carry mutations that are located in the *x*-*y*-*c*II region and increase the expression of genes *O*, *P*, and *Q*. Taken together, these results suggest that this *Q*-independent pathway of late gene expression is due to early *r*-strand transcription that continues into the late genes (32). Whether there exists a mRNA termination signal at the end of the *Q* gene is not known.

SUMMARY OF N AND Q ACTION AND CONTROLLING SITES Transcription of genes *N* and *cro* (Set I, Fig. 3) requires the promoter sites, *p*L and *p*R, respectively. Full level transcription of flanking genes, *c*III-*int* and *c*II-*Q* (Set II, Fig. 3), requires *N* gene product and also requires *p*L and *p*R. In the absence of *N* gene product, three sites (between *N* and *c*III, *cro* and *c*II, and *P*-*Q*; Figs. 3 and 5) lead to termination of prior transcription. *N* gene product, therefore, may act by counteracting this termination. Late gene transcription (*S*, *R*, and *A* through *J*; Set III, Fig. 3) occurs by two modes: (*a*) a high-level, *Q* gene-dependent synthesis, which requires a promoter between genes *Q* and *S*, and (*b*) a low-level, *Q*-independent synthesis, which is probably a continuation of early *r*-strand transcription into the late genes.

THE CRO GENE—SHUT OFF OF EARLY TRANSCRIPTION AND REQUIREMENT FOR LYTIC GROWTH Indication that λ genes were shut off during the lytic growth cycle came first from studies of kinetics of λ exonuclease production. After induction of λ lysogens (or infection with λ), the net amount of λ exonuclease increases for several minutes and then stops. λ carrying mutations in the *x* region which abolish early rightward transcription (Fig. 3) fail to shut off this synthesis (51,132). These x^- mutants fail to make a product responsible for shut off of exonuclease, since exonuclease synthesis is shut off in cells with both λx^- and λ^+ (132). The gene responsible for this shut off (originally called *tof*, "turn-off" or *cro*, "control of repressor and other things") lies within the region of nonhomology with λimm^{434}, as this hybrid phage does not shut off exonuclease synthesis by λx^- (124). Isolation of mutants deficient in the shut off function (which I shall refer to as "*cro*") is described below.

Are other genes besides exonuclease shut off by *cro?* There is good evidence that all early *l*-strand transcription is shut off by *cro*, and some evidence that early *r*-strand transcription is likewise inhibited. Location of a site of action of *cro* product near *p*L indicates that genes *N* through *int* are coordinately regulated by *cro* product. λ^+ cannot shut off exonuclease synthesis produced by λimm^{434} mutants deficient in exonuclease shut off (analogous to λx^- mutants) (124). As λ^+ can shut off exonuclease synthesis from prophage strains with deletions removing part of the *c*I gene and extending rightwards (125), the site of action of the turn off function is within the region of nonhomology with λimm^{434} and to the left of at least part of the *c*I gene (Fig. 4), the region that includes the left operator and promoter. λ carrying a mutation in the left operator (*v*2) is slightly deficient in *cro*-mediated shut off of exonuclease synthesis: λ*v*2 produces about twice as much exonuclease as λ^+ (155). This observation suggests that the site of action of the turn off function may be the same as one of the sites of action of λ repressor. Studies of mRNA synthesis of the *N* through *int* region support the view that this region is coordinately regulated by *cro* product: early *l*-strand messenger RNA is shut off about ten minutes after infection (98,100) and is hyperproduced in x^- mutants (98,100).

Cro product may also inhibit early *r*-strand transcription. Early *r*-strand transcription (from *y* through *O*, inclusive) appears to be reduced between two- and tenfold by ten minutes after infection (26). Whether cro^- mutants continue transcription of this region has not been reported. Defective prophages can be constructed that produce *cro* product and little else ($\lambda N^- c\mathrm{I}_{ts} O^-$ at high temperature; Section V). After infection of such bacteria, the time of appearance of λ^+ is slightly delayed with respect to the time of appearance of λimm^{434} or of λ^+ after infection of cro^- or nonlysogenic cells (72). Growth of λ mutants that carry deletions removing *o*L and ending within the *c*I gene (*N*-defective phages with a *nin* mutation) is also retarded. Since λ*vir* (with mutations at both of its operators) is not retarded, the site of action of the inhibitory factor, presumably *cro* product, may be the right hand operator.

How does *cro* shut off early transcription, for example, transcription of *exo*? Because *cro* product acts near *p*L, it appears to block *exo* transcription indirectly through reducing N synthesis. It also appears to reduce early transcription even in the presence of N protein supplied by another phage (46). An attractive hypothesis is that *cro* product, which behaves formally as a repressor, acts like the λ repressor, *c*I gene product (46,72). The *cro* product has not yet been purified or identified, so a direct test of its binding to λ DNA is not yet possible.

CRO AND LYTIC GROWTH *Cro*⁻ mutants have been isolated in three different ways: (*a*) Selection of mutants able to recover immunity (see Section V). (*b*) Selection of mutants with increased expression of genes in the *c*III through *int* region. In this case the selection exploited the observation that either *red* function or *gamma* function is required for λ growth on *rec*A⁻ bacteria. *N* is required for full expression of the *c*III-*int* transcription unit. Mutants derived from λ*N*⁻*nin*, which form plaques on *rec*A⁻ hosts carry mutations in the *cro* gene. These mutations are located in the *x* region; as expected, their inability to shut off exonuclease synthesis is recessive to *cro*⁺ (29). (*c*) In a screen of temperature-sensitive mutants (79), λ mutants were selected which were able to grow with limiting *Q* product (λQ*am* able to grow in an *E. coli* strain with an inefficient *amber* suppressor). Among the plaque-forming mutants were two with temperature-sensitive growth defects due to mutations in the *x* region. The improved growth of these *Q*-deficient phages might be due to inefficient shut off of transcription of genes *N* and *Q*. Thus far, none of ten or so *cro* mutants are *amber* mutants. No protein product has been identified, so it is not yet clear whether the functional form of the *cro* product is protein or perhaps RNA (see 62).

All of these mutants have been obtained from phages carrying a temperature-sensitive repressor (19,50,56,79). In general, these mutants are unable to grow at 30–32°C or at 42°C, but form plaques at 38–40° (79). The mutation responsible for defective growth is located within the *x* region, and as expected, growth can be restored by λ⁺ but not by λ*imm*⁴³⁴. In the two cases analyzed, these mutants are deficient in early protein shut off (50,56). We shall see below that infection with λ wildtype leads to a burst of repressor synthesis, which is shut off by *cro* product (Section IV). A sufficient explanation for the inability of λ*c*I$_{ts}$ *cro*⁻ phages to grow at low temperature—where repressor is active—is that continued production of repressor blocks lytic growth. Conditions under which the *cro*⁻ mutants grow are conditions in which the level of repression is lowered—growth at intermediate temperature (38–40°C) where the temperature-sensitive repressor is partially inactive (163), or by mutations (*c*II, *c*III, *cy*; see Section IV) which reduce repressor synthesis (45,137). However, inactivation of repressor is not sufficient to allow phage growth, as λ*c*I$_{ts}$ *cro*⁻ phages are unable to grow at 42°C (56,79). As expected, λ*imm*⁴³⁴, with a *cro* product different from that of λ, is unable to help λ*cro*⁻ mutants to grow at 42°C. Unexpectedly, it was found that growth of the λ*imm*⁴³⁴ phage (which itself is *c*I⁻) in the presence of λ*cro*⁻ was reduced 1000-fold (49,79). Because of the inability of *cro*⁻ mutants to shut off *N* through *int* transcription, it has been proposed that hyperproduction of some λ protein blocks λ growth, perhaps by killing the host (49). Attempts to identify the λ genes responsible for this killing have so far been unsuccessful. To account for why these λ*cro*⁻ are able to grow at 38–40°, one is obliged to say that at this temperature repression is sufficient to block hyperproduction but not sufficient to block growth. Clearly more work needs to be done.

In summary, the *cro* gene product acts like a second repressor (46,72). It shuts off early leftward transcription and probably also affects early rightward transcription. Shut off of early transcription is indirectly required for lytic growth to turn off repressor synthesis, and possibly to prevent overproduction of λ proteins which inhibit λ growth.

IV. INFECTION: THE TEMPERATE RESPONSE

In the lytic response, phage are produced and the host cell is killed. In the temperate response, phage are not produced, and the infected cell survives. Some of these surviving cells become lysogens; that is, they are immune to infection by λ and are able to produce phage under appropriate stimuli. To clear up confusion before it arises, let me state that the "temperate", or nonlytic, response does not necessarily lead to formation of a lysogen. Hence it is inaccurate and potentially confusing to refer to the nonlytic response as the lysogenic response. The reason for this distinction is that formation of a stable lysogen (the lysogenic response) requires not only that the host survive phage infection, but also that the phage λ DNA be stably associated with the bacterial DNA.

We saw above that lytic growth yields more phage; hence the minimum set of events for this response is DNA duplication, particle morphogenesis, and cell lysis. Let us try to define the minimum set of events necessary for λ to form a lysogen, or more specifically, to *establish* and *maintain* the lysogenic state.

In a λ lysogen, phage DNA is covalently inserted into the bacterial DNA (reviewed in 65,145,146). This insertion occurs by a site-specific recombination between the λ attachment site (*att*λ) located between genes *J* and *int* and the bacterial attachment site between genes *gal* and *bio*, and results in a prophage map shown in Fig. 4. The prophage DNA is stably maintained in this location because λ proteins necessary for the reverse process ("excision") are not produced (see below).

The λ *c*I gene is solely responsible for maintaining lysogeny. Its central role can be demonstrated by properties of mutants which produce a thermo-labile *c*I protein (103,163).[7] Lysogens made at low temperature are stable when grown at low temperature, and are immune to λ infection. When cultures are shifted to high temperature, immunity to λ infection is lost and lytic growth of the prophage is initiated—each cell produces a full burst of phage within an hour or so (84,103,163).

In summary, establishment of lysogeny requires both *integration* and *repression.* We next consider control of genes required for these processes.

---

[7] Irradiation of lysogens with UV light also abolishes repression (discussed in 24,141). For experimental simplicity, it is often convenient to use λ*c*I-*ts* prophages and induce by shifting lysogenic cultures to high temperature.

INTEGRATION Insertion of λ DNA into *E. coli* DNA occurs by the following scheme: shortly after injection of λ DNA, the cohesive ends join to form a circular molecule. The phage DNA then experiences a recombinational event between phage and bacterial *att* sites, resulting in linear insertion into the bacterial DNA. Formation of λ circles does not require expression of any λ genes. Circularization takes place in the presence of chloramphenicol and upon infection of a λ lysogen (176). DNA synthesis is not required for DNA insertion. Although λO^- and P^- form lysogens with reduced efficiency (1% of infected cells become lysogens), this defect is probably due to lack of ability to establish repression: supplying λ repressor to these phages restores high frequency lysogenization (95). The only gene that appears directly required for insertion is *int*. λ int^- mutants are completely unable to establish lysogeny, although repression is normal. Control of expression of the *int* gene has been discussed above in detail (Section III).

REPRESSION Because the presence of the *c*I product blocks expression of λ genes, and its absence or inactivation leads to expression of λ genes, the *c*I product can be considered as a repressor (89). Its sites of action, operators, were identified by studies of phage mutants able to grow in λ lysogens. λ*vir* carries three mutations flanking the *c*I gene, which allow expression of essential λ genes in the presence of the *c*I product. *v*2 (Figs. 3 and 5; 83,130), to the left, allows expression of gene N. *v*l and *v*3, to the right, allow expression of *cro*, *O*, and *P* (130). The hybrid phage λimm^{434} (see Fig. 5) is also able to grow in the presence of the *c*I product. In this case phage 434 DNA replaces the *c*I gene and the wildtype alleles of *v*2, *v*l, and *v*3 with analogous sites from phage 434 (92,127). Purification of the *c*I product and studies of its action in vitro support and extend the genetic contentions. The *c*I gene codes for a protein that binds to λ DNA specifically at the operator regions. It does not bind to λimm^{434}, and binds with reduced affinity to λ DNA with operator mutations (*o*L, *v*2; *o*R, *v*l*v*3, and *v*3*v*S; 25,118,129,130). Purified repressor is sufficient to block λ transcription in vitro. Using λ DNA as template, *E. coli* RNA polymerase (with sigma and rho factor) produces primarily a 12*s* *l*-strand message (*N* message) and a 7*s* *r*-strand message (*cro* message) (Fig. 5). Addition of λ repressor reduces correctly-initiated λ transcription in general and production of these messages in particular (25,184–186). If the λ DNA carries operator mutations, transcription is not blocked by addition of repressor. Whether the repressor blocks binding or movement of RNA polymerase is unclear at present (see 25,105,184 for further discussion).

REPRESSOR SYNTHESIS—THE ESTABLISHMENT MODE Thus far we have described the requirements for the lysogenic response and some aspects of the λ repressor. We now return to the infection process to see the manner in which lysogeny is established. Remarkably, synthesis of λ repressor appears to be regulated in two different manners, an "establishment mode" of synthesis, described next, and a "maintenance mode" of synthesis (Section V).

Production of the λ repressor after infection is under both positive and negative controls. That is, *c*II and *c*III genes promote high level repressor synthesis, and *cro* is responsible for shutting off its synthesis. The story of control of repressor synthesis begins with the isolation of mutants that form clear plaques rather than the characteristic turbid plaques. The mutations leading to clear plaque phenotype can be put into four groups on the basis of mapping (see Fig. 5), complementation tests, and effect on the frequency of lysogenization—*c*I, *c*II, *c*III, and clear mutations mapping in the *y* region, called *cy*. *c*I$^-$ mutants are unable to lysogenize (less than 10^{-6} infected cells are lysogens); *c*II$^-$ or *cy*$^-$ form lysogens at a frequency of 10^{-4}, and *c*III$^-$ mutants at a frequency of about 10^{-2} (14,91). The lysogens formed by *c*II$^-$, *cy*$^-$, and *c*III$^-$ mutants are completely stable, indicating that the defect in lysogenization by these phages is in the *establishment* of lysogeny, rather than in its *maintenance*. *c*I is required both for establishment and maintenance of lysogeny.

λ*c*I$^-$, *c*II$^-$, and *c*III$^-$ mutants complement each other for lysogenization. Infection of cells with clear mutants from two different groups, for example, λ*c*II$^-$ and λ*c*I$^-$, restores efficient lysogenization (14,91). λ*cy*$^-$ mutants complement *c*II$^-$ and *c*III$^-$ mutants for lysogenization, but not λ*c*I$^-$ mutants (2,14). In other words, λ*cy*$^-$ mutants give rise to lysogens if the coinfecting phage is *c*I$^+$. A formal description of this complementation behavior is that *cy*$^-$ acts on the *c*I gene adjacent to itself. Specifically it was suggested that *cy* is the promoter for *c*I gene transcription (50) (although numerous other possibilities, not elaborated here, remained).

Recent direct measurements of repressor synthesis and *c*I gene transcription provide a simple explanation for the requirement of *c*II, *c*III, and *cy* in the establishment of lysogeny (45,137). Repressor is assayed by its ability to bind λ DNA to nitrocellulose filters, and by an immunological assay in which unlabeled repressor is used to compete with radioactively-labeled repressor for anti-repressor antibodies (45,137). Repressor synthesis begins five minutes after infection, continues at a constant rate for ten minutes, and is then shut off. The amount of repressor produced per infected cell is between 50 and 100 times the amount produced by a lysogen. *c*II$^-$, *c*III$^-$, and *cy*$^-$ mutants are defective in both the rate of repressor synthesis and the final amount of repressor produced. λ*c*II$^-$ mutants produce 1%, λ*c*III$^-$ 15%, and λ*cy*$^-$ 3% as much repressor as λ wildtype upon infection (137). Measurements of *c*I message indicate the difference in rate is probably due to lowered transcription of *c*I by these mutants (159). *c*II, *c*III, and *cy* apparently control lysogenization by stimulating transcription of the *c*I gene.

THE *c*I PROMOTER AND OVERLAPPING TRANSCRIPTION Repressor synthesis by λ*cy*$^-$ is not restored by coinfection with λ*c*I$^-$ (45,137). *cy*$^-$ mutations can therefore be considered mutations in the promoter for *c*I transcription, as they are physically distinct from the *c*I gene (Fig. 5) and affect only the *c*I gene adjacent to the *cy*$^-$ mutation. If the *c*I promoter lies in the *y* region, one can imagine that RNA polymerase binds in the *y* region, transcribes the *x* region from the *l*-strand (its antisense strand), and then transcribes *c*I (Fig. 6). Recent experiments have shown that *c*I transcription is associated with transcription of the *x* region from the *l*-strand (159). This message (anti-*cro* message) has been assayed by isolating double-stranded RNA formed by hybridization between message produced upon infection and *r*-strand *cro* message (see Fig. 5). λ wildtype produces *l*-strand *cro* mRNA with the same time course as λ repressor message and protein. Mutants defective in *c*II, *c*III, or in *cy* produce about the same relative amount of *l*-strand *cro* RNA as they produce repressor or repressor message in comparison with λ wildtype (159). It is not known whether *c*I message is covalently linked with the *l*-strand *cro* RNA. However, these studies indicate that transcription of *c*I, probably initiated within the *y* region, passes across the *x* region before proceeding into *c*I. This promoter for *c*I transcription after infection has been called *pre* ("promoter for repressor establishment"; 137).

IMPLICATIONS OF OVERLAPPING TRANSCRIPTION The observation that the *x* region is transcribed in both directions (Fig. 6) has prompted much speculation concerning whether transcription in one direction interferes with expression of genes transcribed from the other DNA strand. For example, transcription in one direction might interfere with messenger RNA initiation or chain growth from the other strand, or possibly form double-stranded RNA and interfere with translation. One case where transcriptional interference has been invoked is described next (see also 10,96): Phages defective in *c*II, *c*III, or *cy* produce late proteins (*R* and *J* protein) about five minutes earlier after infection than phages with functional *c*II, *c*III, and *cy* (28,112). This early appearance is not due to absence of repressor since *c*I$^-$*c*II$^-$ phages produce late proteins earlier than *c*I$^-$*c*II$^+$. Because late gene expression requires transcription from *p*R (to produce *Q* message), and *c*II, *c*III, and *cy* are required for transcription of *c*I, the early appearance of late proteins may reflect earlier than usual transcription of the *Q* gene due to absence of *c*I gene transcription. That is, *c*II and *c*III may act directly on the *y* region to stimulate *c*I transcription and thereby inhibit *r*-strand transcription. Examples in which *r*-strand transcription may interfere with *c*I transcription are discussed below.

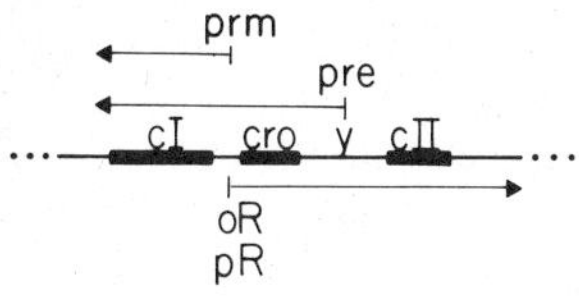

Figure 6 *cI promoters and overlapping transcription. pre*, promoter for establishment of repression; *prm*, promoter for maintenance of repression (137). Arrows indicate direction of transcription. *o*R, right operator; *p*R, right promoter.

ACTION OF *c*II AND *c*III How do *c*II and *c*III stimulate transcription of *c*I? The answer is not known, but sample suggestions are offered: (*a*) *Direct model.* The common requirement of *c*II, *c*III, and *cy* for repressor synthesis suggests that the *cy* region may be the site of action of the *c*II and *c*III proteins, for example, to stimulate binding of RNA polymerase at the *cy* promoter (44,50,159). No direct support for or against this model has been reported. (*b*) *Indirect models.* *c*II and *c*III gene products may inhibit processes antagonistic to *c*I transcription. (*i*) We see in the next section that the *cro* product shuts off repressor synthesis after infection. At early times after infection, one might imagine that *c*II and *c*III stimulate *c*I transcription by antagonizing the action of *cro*. This is almost certainly *not* true, since cro^- mutants still require *c*II and *c*III to produce repressor (137). (*ii*) Rightward transcription from *p*R might interfere with transcription of *c*I. *c*II and *c*III proteins might directly inhibit *r*-strand transcription and thereby stimulate *c*I transcription.

Before leaving *c*II and *c*III, let me raise an interesting apparent paradox (which shall be returned to): Upon infection repressor synthesis requires *c*II and *c*III proteins. How is repressor synthesis maintained in a lysogen, where *c*II and *c*III genes are repressed?

SHUT OFF OF REPRESSOR SYNTHESIS AFTER INFECTION Shut off of repressor synthesis is mediated by the *cro* gene—cro^- mutants continue to produce repressor for at least 15 or 20 minutes longer than cro^+ phages (137). We saw above that *cro* product is responsible for shut off of transcription of both genes *N* and *c*III and probably early rightward transcription as well. *Cro* product, therefore, may cause shut off of repressor synthesis by reducing synthesis of *c*III and *c*II both directly and by reducing synthesis of *N* product. Is the effect of *cro* on *c*II and *c*III sufficient to explain shut off of repressor synthesis? Under the following conditions it is: When λ infects a cell full of *cro* (a lysogen carrying $N^- v1v3O^-$; see below), repressor is not produced (46). Coinfection with a phage insensitive to λ *cro* product (λimm^{434}) leads to production of λ repressor (46). The helping effected by λimm^{434} requires that this phage carry $c\text{II}^+$ and $c\text{III}^+$ genes.[8] In this case stimulation of repressor synthesis by *c*II and *c*III gene products takes precedence over inhibition of repressor synthesis by *cro* product. In infection of wildtype bacteria, however, shut off of repressor synthesis by *cro* product must take place *after* *c*II and *c*III products have stimulated repressor synthesis. Perhaps *c*II and *c*III products are unstable, and a sufficient level of *cro* product builds up to prevent further *c*II and *c*III production. We see below that, in certain circumstances, *cro* product inhibits repressor synthesis independently of its action in *c*II and *c*III. The shut off of the establishment mode of repressor synthesis by *cro* product, therefore, may be a sum of its action on *c*II, *c*III, and on *c*I itself.

---

[8] Note that the inability of $\lambda imm^{434} c\text{II}^-$ to help λ repressor synthesis in the presence of *cro* product suggests the *cro* product prevents the infecting λ phage from producing its own *c*II product. This observation, therefore, demonstrates action of *cro* product on *r*-strand gene expression.

V. CONTROL OF PROPHAGE GENE EXPRESSION: MAINTENANCE OF THE PROPHAGE STATE

Successful lysogeny requires that the prophage DNA be stably integrated into the host DNA, and that the host survive. The maintenance of lysogeny is mediated by the λ repressor: It prevents killing of the host and excision of the prophage DNA. We shall consider first the manner in which the repressor keeps λ genes silent to prevent excision and killing of the host. Identification of the genes responsible for killing of the host after induction and of sites on λ DNA necessary for expression of these genes has stimulated and facilitated many of the facets of λ gene expression already described. The three main promoters for λ lytic growth (*p*R, *p*L, and the late gene promoter) were identified by mutations isolated from prophages. In addition, physiological analysis of defective prophages has revealed a remarkable "switch" mechanism in the control of λ repressor synthesis.

DIRECT AND INDIRECT REPRESSION The repressor is responsible for keeping essentially all λ genes silent in the prophage state. In lysogens at least 90% of the λ-specific mRNA produced is message for the *c*I gene and its neighbor *rex* (74,75,169). The repressor blocks transcription of all other genes in both a direct and an indirect manner by binding to the two operator sites *o*L and *o*R. In doing so it directly prevents transcription of gene *N*, which we saw above is required for full level transcription of genes *c*III-*int* and *c*II-*O*-*P*-*Q*. In fact, λ repressor exerts a double block on these two gene clusters, because even in the presence of N protein (e.g., supplied by another phage) these genes are not transcribed. Repressor therefore ensures that *int* and *xis* genes are not transcribed and the prophage DNA remains inserted into bacterial DNA. Extensive investigation has been carried out concerning the λ genes responsible for killing of the host upon induction of the prophage. This work is summarized in a later section for its own interest and because many of the mutations referred to in this review as well as numerous other λ standbys have been uncovered in these studies.

Transcription of the lysis genes *S* and *R* and the other late genes *A* through *J* is prevented in two ways by λ repressor: the low-level transcription dependent upon N is blocked directly by repressor, and the full-level expression dependent upon *Q* is indirectly blocked, as repressor blocks transcription of gene *Q*. We saw earlier that supplying Q protein to a prophage is sufficient to activate the late genes of the prophage (172,173).

CONTROL OF CELL DEATH Studies of lysogens carrying defective prophages indicate that there are two independent pathways which lead to cell death upon induction of the prophage. Lysogens carrying prophages with a mutation in an essential gene die after induction (survival approximately 10^{-6} or less) even though phage are not produced (12,51,153). More specifically, lysogens carrying *N*-deficient, *Q*-deficient, or any late gene-deficient prophage die upon induction. Lysogens carrying replication-deficient prophages (O^-, P^-, or x^- mutations) survive at a much higher frequency ($10^{-1} - 10^{-2}$) and usually no longer carry a prophage (153). Preventing phage excision with an int^- or xis^- mutation reduces host survival to 10^{-6} or so (66).

The phage genes responsible for killing of the host have been identified by isolating bacterial mutants that survive prophage induction. Induction of replication-deficient prophages (O^-, P^-, or x^-) leads to bacterial survivors with two types of phage mutations (ignoring cured survivors that can be selected against by locking in the prophage with an int^- mutation): (*a*) mutants with a defective *N* gene, and (*b*) *sex* mutants, which exhibit reduced expression of all early *l*-strand transcription (Fig. 3). The reduced expression of these genes leads to inability of λsex^- prophages to excise from bacterial DNA (hence the name, *s*ans *ex*cision) (65). Subsequent analysis (69) indicates that *N* product itself is not responsible for cell death; rather *N* regulates expression of the *kil* gene (and possibly others) which lead to cell death.

Induction of N^- prophages leads to two major classes of bacterial survivors: (*a*) mutants with defective *O* or *P* gene,[9] and (*b*) x^- mutants that have severely reduced early *r*-strand transcription (Fig. 3). Other prophage mutations that allow lysogens carrying λN^- prophages to survive phage induction include (*a*) insertion mutations in the *y* region or in the *c*II gene which lower transcription of distal genes (notably *O* and *P*) (12), and (*b*) mutations in the origin of λ DNA replication (near gene *O*) (42,135). Killing appears therefore to be associated with λ DNA synthesis, although the molecular event is unclear.

In summary, it is possible to keep a λ prophage stably maintained within its host by putting in mutations that block excision of the prophage and killing of the host. The actual nature of the events that lead to cell death are not known and are under investigation (58,69,87).

CONTROL OF REPRESSOR IN THE PROPHAGE STATE It is clear that synthesis of λ repressor in a lysogen is regulated differently from synthesis upon infection. Synthesis of repressor after infection (establishment mode of synthesis) requires genes *c*II and *c*III, and a promoter in the *y* region (Fig. 6). In the prophage state *c*II and *c*III are repressed (9,123,169), indicating that they are not needed to maintain repressor synthesis. We noted earlier that lysogens formed by *c*II$^-$, *c*III$^-$ and cy^- mutants are stable, though they are formed with low efficiency. Repressor synthesis in such lysogens or in lysogens with deletions of *c*II and/or *c*III is the same as in wildtype lysogens (137). Because no message from the *x* region is observed in a lysogen (75,169), *c*I transcription must be initiated at a point between genes *cro* and *c*I. The site for RNA polymerase binding (the promoter for maintenance of *c*I synthesis *prm*) is presumably in the same region (Fig. 6). In the establishment mode, repressor synthesis occurs at a rate five to ten times faster than in the maintenance mode. The increased rate is *not* due to increased copies of the *c*I gene due to replication, as nonreplicating phages (λO^-) produce repressor at four to eight times the rate of prophages (137).

---

9 Although N product is required for full level of *O* and *P* expression, λN^- mutants do produce a low level of O and P (12,147).

Synthesis of repressor in the absence of *c*II and *c*III proteins is under control both by repressor itself and by the *cro* gene product. Experiments described next indicate (*a*) that even in the presence of functional repressor, *cro* product reduces repressor synthesis; (*b*) regardless of the presence of *cro*, inactivation of repressor greatly reduces repressor synthesis.

INHIBITION BY CRO: A MOLECULAR SWITCH Lysogens carrying $\lambda N^-c\mathrm{I}_{ts}O^-$ or $\lambda N^-c\mathrm{I}_{ts}P^-$ prophages are not immune to λ infection at high temperature due to inactivation of the repressor. *A priori* one expects that when such a culture is returned to temperatures where repressor is active, immunity to λ infection should return. Surprisingly, most cells in the culture do not recover immunity at 30°C if they have grown at 42°C for at least two or three generations (50,114,117,157). A small fraction of the cells (0.1%), however, can be found which have recovered immunity. These immune cells, when grown at 42°C, are again deficient in recovery of immunity at 30°C, and likewise give rise to approximately 0.1% immune cells (114,157). Lysogens carrying a $\lambda N^-c\mathrm{I}_{ts}O^-$ prophage (or genetic equivalent) can therefore exist in two interconvertible states at 30°C—an immune state and a nonimmune state.

Lysogens carrying $\lambda N^-c\mathrm{I}_{ts}x^-$ prophages, in contrast, do not exist in two states. After growth at 42°C, all cells in the culture recover immunity at 30°C (19,50,117). Lysogens carrying both $\lambda N^-c\mathrm{I}_{ts}O^-$ (or P^-) and $\lambda N^-c\mathrm{I}_{ts}x^-$ prophages do not recover immunity at 30°; that is, the double lysogen behaves like the O^- single lysogen rather than the x^- single lysogen. Since x^- mutations block transcription of the regions *x*, *y* and genes *c*II through *Q*, a simple explanation is that the x^+ prophage produces something that blocks repressor synthesis. This explanation is supported by the isolation of derivatives of $\lambda N^-c\mathrm{I}_{ts}O^-$ which do recover immunity at 30°C after prolonged growth at 42°C (19,50). These prophages have acquired mutations called *cro* or *Ai* (anti-immunity), which are recessive to $\lambda N^-c\mathrm{I}_{ts}O^-$ for recovery of immunity and are located in the *x* region of the λ map. (Other properties of cro^- mutants have been described in Sections III and IV.)

The existence of an anti-repressor was also revealed by a different observation: lysogens carrying λN^-O^- prophages with *o*R mutations *vlv*3 are not immune (111,155). (Note that this situation is analogous to $\lambda N^-c\mathrm{I}_{ts}O^-$ lysogens at high temperature.) Moving the λN^-vlv30^- prophage into an immune λN^-O^- lysogen on an F′*gal* episome leads to a strain carrying both prophages which is *not immune* (155). Although it has not been shown explicitly, the factor acting in *trans* to cause loss of immunity is presumably *cro*.

How does *cro* inhibit repressor synthesis of the prophage and recovery of immunity after repressor inactivation? Lysogens carrying prophages with deletions extending from within *cro* rightwards recover immunity after inactivation of repressor. Supplying *cro* product to these strains blocks recovery of immunity, indicating that the site of *cro* action lies probably between *cro* and *c*I (11). We saw above that *cro* may act in this region to inhibit rightward transcription. How binding in this region might interfere with *c*I transcription remains a matter of conjecture. We see in the next section that repressor may stimulate its own synthesis by binding to *o*R. *Cro* product might inhibit repressor synthesis directly by blocking transcription or translation of the *c*I gene, or indirectly by blocking repressor binding.

In summary, *cro* and *c*I are involved in a bistable switch. In the immune state, λ repressor is produced and *cro* product is not—repressor blocks *cro* gene transcription. In the nonimmune state, *cro* product is produced, and λ repressor is not. The mechanism by which *cro* product interferes with repressor synthesis is not clear. Further details of this switch are described in (19,44,46,157,159).

REPRESSOR STIMULATION $\lambda N^- c\mathrm{I}_{ts} x^-$ prophages produce normal amounts of repressor at low temperature (137). Again one expects repressor synthesis to continue at high temperature. However, when $\lambda N^- c\mathrm{I}_{ts} x^-$ lysogens are grown at 42°C, no new repressor is synthesized (137). (The repressor used in these experiments was stable even if produced at high temperature.) Likewise, the rate of *c*I gene transcription is reduced upon induction of this strain (158). Are any λ functions, expressed after inactivation of the repressor, responsible for shut off of repressor synthesis? $\lambda N^- x^-$ prophages are defective in transcriptions of *cro* as well as in other rightward transcription, and have reduced levels of *c*III-*int* transcription.

Inactivation of repressor, therefore, probably does not lead to production of any inhibitory products. Rather, it has been proposed that repressor is necessary for its own synthesis (98,158). Two other observations are consistent with this hypothesis: (*a*) Restoration of active repressor by shifting $\lambda N^- c\mathrm{I}_{ts} x^-$ prophages to low temperature after pregrowth at high temperature stimulates transcription of the *c*I gene (158). (*b*) The right hand operator site *o*R is required for the maintenance mode of repressor synthesis. Infection of a repressed $\lambda N^- c\mathrm{I}_{ts} x^-$ lysogen with $\lambda c\mathrm{I}^+ v13c\mathrm{II}^-$ leads to less than 10% the amount of repressor produced in comparison with $\lambda c\mathrm{I}^+ c\mathrm{II}^-$ carrying wildtype *o*R (137). (In these experiments the prophage repressor and superinfecting phage repressor can be distinguished, because only the wildtype repressor can bind efficiently to λ DNA.)

How might repressor stimulate its own synthesis? Two suggestions are offered: (*a*) Repressor might stimulate binding of RNA polymerase to the maintenance *c*I promoter *prm* (Fig. 6). For example, repressor binding at *o*R might facilitate recognition of *prm* by RNA polymerase. (*b*) Repressor might stimulate transcription of *c*I indirectly, by blocking rightward transcription. That is, rightward transcription from *p*R might block *c*I transcription, say by preventing RNA polymerase binding at *prm*. Transcription of the *x* region *per se* does not block *c*I transcription, as the x^- mutants studied still require repressor for *c*I transcription. It is not, however, known whether the x^- mutants used block RNA polymerase *binding* at *p*R.

SUMMARY OF THE MAINTENANCE MODE OF *c*I SYNTHESIS In the prophage state, repressor synthesis is independent of genes *c*II and *c*III, and the *c*I promoter in the *y* region. In the maintenance mode of repressor synthesis the *c*I gene apparently is transcribed from a second promoter (*prm*) located between *c*I and *cro*. In addition, in this mode, repressor synthesis is stimulated by repressor and is inhibited by *cro* product.

VI. LYSIS VS. LYSOGENY

Why has λ evolved the elaborate web of control circuits described above? Since λ has two predominant life styles (44,78)—autonomous replication in the lytic response, or passive replication in the lysogenic response—these control circuits presumably maximize the attainment of these states and mediate the choice between them. These responses can be characterized as follows (Fig. 7):

THE LYTIC RESPONSE *N* gene product stimulates transcription of genes *O* and *P* (DNA replication) and *Q* (late gene regulator). Q product stimulates late gene transcription (genes for cell lysis and particle morphogenesis). In addition, DNA replication amplifies late gene expression by increasing the number of gene copies. *Cro* product ultimately shuts off transcription of genes no longer needed for lytic growth (genes *c*III-*int* and *c*II-*O*-*P*-*Q*). The successful lytic infection yields a full burst of active phage.

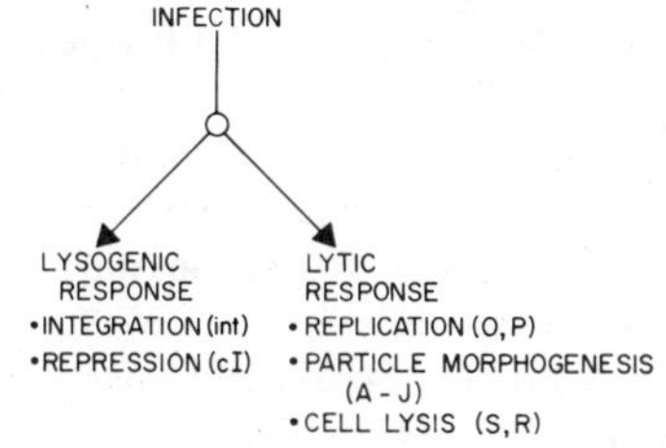

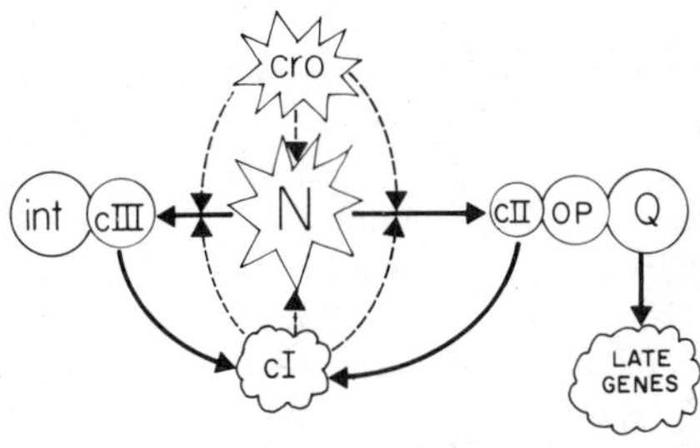

Figure 7 *Summary of lytic vs. lysogenic response.* Lower panel: Arrows indicate stimulatory or inhibitory action of one gene product on synthesis or action of another. Dotted lines, inhibitory; solid lines, stimulatory. The sequence of gene expression (*N*, *cro*): (*int*, *c*III), (*c*II, *O*, *P*, *Q*): (*c*I), (late genes) is represented pictorially. The reader is invited to draw a dotted line from *c*I to *cro*.

THE LYSOGENIC RESPONSE N product stimulates transcription of *int* (integration) and *c*II and *c*III, which in turn stimulate *c*I transcription (repression). The successful lysogenic response yields a stable lysogen.

Both pathways exhibit certain delaying or phasing periods, which lead to ordered transcription of genes and subsequently the ordered action of the gene products. These timing devices are primarily mediated through the regulatory genes *c*I, *c*II, *c*III, *cro*, *N*, and *Q*. *c*I and *cro* behave as negative regulators—their presence blocks λ gene expression. *c*II, *c*III, *N*, and *Q* can be considered as positive regulators, as their presence is required for λ gene expression. Only for the *c*I gene product has the molecular mechanism of action been demonstrated (129). Because the λ repressor may in addition have a positive regulatory role (in promoting its own synthesis), the terms "positive" and "negative" regulator must be considered operational definitions rather than necessarily implying molecular mechanisms. Further analysis of these regulatory gene products is eagerly awaited.

As an example of phasing, the addition of a late gene positive regulator (*Q*) is one mechanism by which genes for cell lysis are transcribed late in the lytic growth cycle, that is, after DNA synthesis. It has also been suggested that the time required to transcribe the region between genes *P* and *Q* contributes to this delay (180). In the lysogenic response the burst of repressor is synthesized after a burst of *int* protein. Establishment of repression without insertion (abortive lysogeny) would be considered unsuccessful lysogeny. Another kind of phasing device is manifested by *cro*. Even though the *cro* gene is transcribed first (at the same time as *N*), its action to shut off early transcription does not occur immediately. This delay may be because translation of *cro* mRNA is delayed or because a certain concentration of *cro* product must be built up. One cannot say that λ pathways have been established in *the* only possible way. For example, the morphogenetic proteins do not necessarily *have* to be produced after DNA duplication. All one can fairly say is that late genes *are* expressed after DNA duplication, and that the net amount of transcription is amplified by DNA duplication.

As defined above, the lytic and lysogenic responses are mutually exclusive—one leads to cell death, the other to cell survival. A striking feature in the choice between the lytic and lysogenic response is that λ apparently embarks a considerable distance on both before it is committed to one or the other. Both responses require transcription of coordinately regulated groups of genes *cro*-*c*II-*O*-*P*-*Q* and *N*-*c*III-*int*. In the course of λ infection, therefore, a cell may contain proteins that stimulate both the lytic response (*cro*, *O*, *P*, and *Q* gene products) and the lysogenic response (*int*, *c*II, and *c*III gene products). Presumably the concentration of these different proteins determines the choice of response. In particular, the relative levels of repressor and *cro* product may be a primary decision making element. Infection of a cell in which *cro* product is already present always leads to lytic growth (19,46,114); infection in the presence of repressor always leads to a temperate response. Are *cro* and *c*I the main switch mechanism in normal infection? Specifically, perhaps the ratio of *cro* product and *c*I product determines which response is chosen. For this to be true *c*I action must take precedence over all lytic processes (e.g., DNA replication) that are in the works. Likewise, *cro* must take precedence over lysogenic processes (i.e., establishment of repression and integration). To a first approximation this appears to be true. At the time when repressor is produced (after *int* transcription and before cell lysis) it blocks processes favorable to lytic growth but not lysogeny. It prevents continued transcription of *N*, *O*, *P*, and *Q*. In addition it blocks DNA replication directly even in the presence of O and P protein, since λ DNA replication appears to require transcription per se near the origin of replication [near genes *c*II and *O* (42,174)]. Although the repressor does not block the action of Q product directly, a low level of late gene transcription is not lethal to the host (69). *Cro* product counters the tour down the lysogenic pathway by inhibiting *c*I synthesis through its shut off of *c*II and *c*III proteins and also by blocking the maintenance mode of repressor synthesis. An appealing feature of the hypothesis that the *cro*/*c*I product ratio determines the life style is that slight alteration of this ratio should be reinforced since *cro* product shuts off *c*I synthesis, and *c*I product shuts off *cro* synthesis. In other words, a period of uncommitted phage growth would rapidly and irreversibly be directed towards lysis or lysogeny with only a slight bias in the *cro*/*c*I product ratio. Assigning *the* decision-making role to this ratio, however, is hazardous and probably a gross oversimplification, as the extent of DNA duplication, among other things, is likely to affect the level of *cro* and *c*I product (96).

What host factors control the lysis-lysogeny decision? It is not possible at this point to make a simple statement describing the host role in the lysis-lysogeny decision (or the decision to withdraw from the prophage state). With a more complete understanding of the mechanism of action of the λ regulatory gene products, it might be possible to define more precisely the crucial phage decision-making elements. Bacterial mutants that affect the various decisions (4,5,24,70) should also contribute to our understanding of the question that one is still left with, "What determines.....?"

Acknowledgments

I would like to thank Penny Toothman for extensive criticism and clarifying discussion; Jim Hicks, Frank Stahl, and Jeff Strathern for discussion and suggestions; Kathleen Drasky and Loretta Schantz for preparation of the manuscript; Don Shaiffer for figures; and Ethan Signer for introducing me to λ.

Literature Cited

1. Adhya, S., Sen, A., Mitra, S. 1971. The Role of Gene *S*. See Ref 77: 743–47
2. Astrachan, L., Miller, J.F. 1972. Regulation of λ *rex* expression after infection of *Escherichia coli* K by lambda bacteriophage. *J. Virol.* 9:510–18
3. Ausubel, F., Voynow, P., Signer, E., Mistry, J. 1971. Purification of nonessential λ proteins. See Ref. 77: 395–406
4. Belfort, M., Wulff, D.L. 1971. A mutant of *Escherichia coli* that is lysogenized with high frequency. See Ref. 77: 739–42
5. Belfort, M., Wulff, D.L. 1973. A genetic and biochemical investigation of the *Escherichia coli* mutant *hfl*-1 which is lysogenized at high frequency by bacteriophage lambda. Unpublished
6. Benzer, S. 1957. The elementary units of heredity. In *The Chemical Basis of Heredity* W.D. McElroy, B. Glass, eds., pp. 70–93. Johns Hopkins Press, Baltimore, Maryland
7. Blattner, F.R., Dahlberg, J.E. 1972. RNA synthesis startpoints in bacteriophage λ: Are the promoter and operator transcribed? *Nature New Biol.* 237:227–32
8. Blattner, F.R., Dahlberg, J.E., Boettinger, J.K., Fiandt, M., Szybalski, W. 1972. Distance from a promoter mutation to an RNA synthesis startpoint on bacteriophage λ DNA. *Nature New Biol.* 237:232–36
9. Bode, V.C., Kaiser, A.D. 1965. Repression of the c_{II} and c_{III} cistrons of phage lambda in a lysogenic bacterium. *Virology* 25:111–21
10. Bøvre, K., Szybalski, W. 1969. Patterns of convergent and overlapping transcription within the b2 region of coliphage λ. *Virology* 38:614–26
11. Brachet, P. Personal communication.
12. Brachet, P., Eisen, H., Rambach, A. 1970. Mutations of coliphage λ affecting the expression of replicative functions *O* and *P*. *Mol. Gen. Genet.* 108:266–76
13. Brachet, P., Green, B. 1970. Functional analysis of early defective mutants of coliphage λ. *Virology* 40:792–99
14. Brachet, P., Thomas, R. 1969. Mapping and functional analysis of *y* and cII mutants. *Mutation Res.* 7:257–60
15. Brooks, K. 1965. Studies in the physiological genetics of some suppressor-sensitive mutants of bacteriophage λ. *Virology* 26:489–99
16. Brown, A., Arber, W. 1964. Temperature-sensitive mutants of coliphage lambda. *Virology* 24:237–39
17. Buchwald, M., Siminovitch, L. 1969. Production of serum-blocking material by mutants of the left arm of the λ chromosome. *Virology* 38:1–7
18. Butler, B., Echols, H. 1970. Regulation of bacteriophage λ development by gene *N*; Properties of a mutation that bypasses *N* control of late protein synthesis. *Virology* 40:212–22
19. Calef, E., Avitabile, A., del Giudice, L., Marchelli, C., Menna, T., Neubauer, Z., Soller, A. 1971. The genetics of the anti-immune phenotype of defective lambda lysogens. See Ref. 77: 609–20
20. Campbell, A. 1961. Sensitive mutants of bacteriophage λ. *Virology* 14:22–32
21. Campbell, A. 1971. Genetic Structure. See Ref. 77: 13–44
22. Campbell, A., Del Campillo-Campbell, A. 1963. Mutant of lambda bacteriophage producing a thermolabile endolysin. *J. Bacteriol.* 85:1202–07
23. Casjens, S. 1971. The morphogenesis of the phage lambda head: the step controlled by gene *F*. See Ref. 77: 725–32
24. Castellazzi, M., George, J., Buttin, G. 1972. Prophage induction and cell division in *E. coli*. II. Linked (*rec*A, *zab*) and unlinked (*lex*) suppressors of *tif*-1-mediated induction and filamentation. *Mol. Gen. Genet.* 119:153–74
25. Chadwick, P., Pirrotta, V., Steinberg, R., Hopkins, N., Ptashne, M. 1970. The λ and 434 phage repressors. *Cold Spring Harbor Symp. Quant. Biol.* 35:283–94
26. Champoux, J.J. 1970. The sequence and orientation of transcription in bacteriophage λ. *Cold Spring Harbor Symp. Quant. Biol.* 35:319–23
27. Chowdhury, D.M., Guha, A. 1973. Characterization of the mRNA transcribed from the head and tail genes of phage lambda chromosome. *Nature New Biol.* 241:196–98
28. Court, D. Personal communication.
29. Court, D., Campbell, A. 1972. Gene regulation in *N* mutants of bacteriophage λ. *J. Virol.* 9:938–45
30. Court, D., Sato, K. 1969. Studies of novel transducing variants of lambda: Dispensability of genes N and Q. *Virology* 39:348–52
31. Cox, J.H., Strack, H.B. 1971. Cold-sensitive mutants of bacteriophage λ. *Genetics* 67:5–17

32. Dambly, C., Couturier, M. 1971. A minor Q-independent pathway for the expression of the late genes in bacteriophage λ. *Mol. Gen. Genet.* 113:244–50
33. Dambly, C., Couturier, M., Thomas, R. 1968. Control of development in temperature bacteriophages. II. Control of lysozyme synthesis. *J. Mol. Biol.* 32:67–81
34. Davidson, N., Szybalski, W. 1971. Physical and chemical characteristics of lambda DNA. See Ref. 77: 45–82
35. Davis, R.W., Davidson, N. 1968. Electron-microscopic visualization of deletion mutations. *Proc. Nat. Acad. Sci. USA 60:243–50*
36. DeCrombrugghe, B., Adhya, S., Gottesman, M., Pastan, I. 1973. Effect of rho on transcription of bacterial operons. *Nature New Biol.* 241:260–64
37. Doerfler, W., Hogness, D.S. 1968. The strands of DNA from lambda and related bacteriophages: isolation and characterization. *J. Mol. Biol.* 33:635–59
38. Dottin, R.P., Pearson, M.L. 1973. Regulation by phage lambda *N* gene protein of anthranilate synthetase synthesis *in vitro*. *Proc. Nat. Acad. Sci. USA* 70:1078–82
39. Dove, W.F. 1966. Action of the lambda chromosome. I. Control of functions late in bacteriophage development. *J. Mol. Biol.* 19:187–201
40. Dove, W.F. 1968. The genetics of the lambdoid phages. *Ann. Rev. Genet.* 2:305–40
41. Dove, W.F. 1971. Biological inferences. See Ref. 77: 297–312
42. Dove, W.F., Inokuchi, H., Stevens, W.F. 1971. Replication control in phage lambda. See Ref. 77: 747–72
43. Echols, H. 1971. Regulation of lytic development. See Ref. 77: 247–70
44. Echols, H. 1972. Developmental pathways for the temperate phage: lysis vs. lysogeny. *Ann. Rev. Genet.* 6:157–90
45. Echols, H., Green, L. 1971. Establishment and maintenance of repression by bacteriophage λ: the role of *c*I, *c*II and *c*III proteins, *Proc. Nat. Acad. Sci. USA* 68:2190–94
46. Echols, H., Green, L., Oppenheim, A.B., Oppenheim, A., Honigman, A. 1973. The role of the *cro* gene in bacteriophage λ development. *J. Mol. Biol.*, submitted
47. Echols, H., Joyner, J.A. 1968. The temperate phage. In *The Molecular Basis of Virology*, Fraenkel-Conrat, H., Ed., p. 532. Reinhold, New York
48. Egan, J.B., Hogness, D.S. 1972. The topography of lambda DNA: isolation of ordered fragments and the physical mapping of point mutations. *J. Mol. Biol.* 71:363–82
49. Eisen, H. Personal communication.
50. Eisen, H., Brachet, P., Pereira da Silva, L., Jacob, F. 1970. Regulation of repressor expression in λ. *Proc. Nat. Acad. Sci. USA* 66:855–62
51. Eisen, H.A., Fuerst, C.R., Siminovitch, L., Thomas, R., Lambert, L., Pereira da Silva, L., Jacob, F. 1966. Genetics and physiology of defective lysogeny in K12(λ): Studies of early mutants. *Virology* 30:224–41
52. Eisen, H.A., Pereira da Silva, L., Jacob, F. 1968. The regulation and mechanism of DNA synthesis in bacteriophage λ. *Cold Spring Harbor Symp. Quant. Biol.* 33:755–64
53. Eisen, H.A., Siminovitch, L., Mohide, P.T. 1968. Excision of lambda prophage: effects on host survival. *Virology* 34:97–103
54. Epstein, R.H., Bolle, A., Steinberg, C.M., Kellenberger, E., Boy de la Tour, E., Chevalley, R., Edgar, R.S., Susman, M., Denhardt, G.H., Lielausis, A. 1963. Physiological studies of conditional lethal mutants of bacteriophage T4D. *Cold Spring Harbor Symp. Quant. Biol.* 28:375–94
55. Fiandt, M., Hradecna, Z., Lozeron, H.A., Szybalski, W. 1971. Electron micrographic mapping of deletions, insertions, inversions, and homologies in the DNAs of coliphages lambda and phi 80. See Ref. 77: 329–54
56. Franklin, N. 1971. The *N* operon of λ: extent and regulation as observed in fusions to the tryptophan operon of *E. coli*. See Ref. 77: 621–38
57. Friedman, D.I., Jolly, C.T., Mural, R.J. 1973. Interference with the expression of the *N* gene function of phage λ in a mutant of *E. coli*. *Virology* 51:216–26
58. Freifelder, D., Kirschner, I., Goldstein, R., Baran, N. 1973. Physical study of prophage excision and curing of λ prophage from lysogenic *Escherichia coli*. *J. Mol. Biol.* 74:703–20
59. Garen, A. 1968. Sense and nonsense in the genetic code. *Science* 160:149–59
60. Gariglio, P., Green, M.H. 1973. Characterization of polycistronic late lambda messenger RNA. *Virology*. In press

61. Georgopoulos, C.P., Herskowitz, I. 1971. *Escherichia coli* mutants blocked in lambda DNA synthesis. See Ref. 77: 553–64
62. Gesteland, R.F., Kahn, C. 1972. Synthesis of bacteriophage λ proteins *in vitro*. *Nature New Biol.* 240:3–6
63. Goldberg, A.R., Howe, M. 1969. New mutations in the S cistron of bacteriophage lambda affecting host cell lysis. *Virology* 38:200–02
64. Gorini, L., Jacoby, G.A., Breckenridge, L. 1966. Ribosomal ambiguity. *Cold Spring Harbor Symp. Quant. Biol.* 31:657–64
65. Gottesman, M.E., Weisberg, R.A. 1971. Prophage insertion and excision. See Ref. 77: 113–38
66. Gottesman, M.E., Yarmolinsky, M.B. 1968. The integration and excision of the bacteriophage genome. *Cold Spring Harbor Symp. Quant. Biol.* 33:735–47
67. Greenblatt, J. 1972. Positive control of endolysin synthesis *in vitro* by the gene N protein of phage λ. *Proc. Nat. Acad. Soc. USA* 69:3606–10
68. Greenblatt, J. 1973. Regulation of the expression of the *N* gene of bacteriophage lambda. *Proc. Nat. Acad. Soc. USA* 70:421–24
69. Greer, H., Signer, E.R. Personal communication
70. Grodzicker, T., Arditti, R., Eisen, H. 1972. Establishment of repression by lambdoid phage in catabolite activator protein and adenylate cyclase mutants of *Escherichia coli*. *Proc. Nat. Acad. Sci.* 69:366–70
71. Gussin, G.N., Peterson, V. 1972. Isolation and properties of *rex*⁻ mutants of bacteriophage lambda. *J. Virol.* 10:760–65
72. Hampacherová, M., Koutecká, E., Neubauer, Z. 1973. Two repressors in bacteriophage lambda. *Mol. Gen. Genet.* 120:133–37
73. Harris, A.W., Mount, D.W.A., Fuerst C.R., Siminovitch, L. 1967. Mutations in bacteriophage lambda affecting host cell lysis. *Virology* 32:533–69
74. Heinemann, S.F., Spiegelman, W.G. 1970. Role of the gene N product in phage lambda. *Cold Spring Harbor Symp. Quant. Biol.* 35:315–18
75. Heinemann, S.F., Spiegelman, W.G. 1970. Control of transcription of the repressor gene in bacteriophage lambda. *Proc. Nat. Acad. Soc. USA* 67:1122–29
76. Hendrix, R. 1971. Identification of proteins coded in phage lambda. See Ref. 77:355–70
77. Hershey, A.D. 1971. Editor, *The Bacteriophage Lambda*. Cold Spring Harbor Laboratory, Cold Spring Harbor, New York 792 pp.
78. Hershey, A.D., Dove, W. 1971. Introduction to Lambda. See Ref 77: 3–13
79. Herskowitz, I. 1971. *Control of late genes, early genes, the cI gene and replication in bacteriophage λ*. Ph.D. thesis, Massachusetts Institute of Technology, Cambridge, Mass.
80. Herskowitz, I., Signer, E.R. 1970. A site essential for expression of all late genes in bacteriophage λ. *J. Mol. Biol.* 47:545–56
81. Herskowitz, I., Signer, E.R. 1970. Control of transcription from the *r* strand of bacteriophage lambda. *Cold Spring Harbor Symp. Quant. Biol.* 35:355–68
82. Hopkins, N. 1970. Bypassing a positive regulator: isolation of a λ mutant that does not require N product to grow. *Virology* 40:223–29
83. Hopkins, N., Ptashne, M. 1971. Genetics of virulence. See Ref. 77: 571–74
84. Horiuchi, T., Inokuchi, H. 1967. Temperature-sensitive regulation system of prophage λ induction. *J. Mol. Biol.* 23:217–30
85. Howard, B.D. 1967. Phage lambda mutants deficient in *r*II exclusion. *Science* 158:1588–89
86. Hradecna, Z., Szybalski, W. 1967. Fractionation of the complementary strands of coliphage λ DNA based on the asymmetric distribution of poly I, G-binding sites. *Virology* 32:633–43
87. Inokuchi, H., Dove, W.F., Freifelder, D. 1973. Physical studies of RNA involvement in bacteriophage λ DNA replication and prophage excision. *J. Mol. Biol.* 74:721–28
88. Jacob, F., Fuerst, C.R., Wollman, E.L. 1957. Recherches sur les bactéries lysogènes défectives. II. Les types physiologiques liés aux mutations du prophage. *Ann. Inst. Pasteur* 93:724
89. Jacob, F., Monod, J. 1961. Genetic regulatory mechanisms in the synthesis of proteins. *J. Mol. Biol.* 3:318–56
90. Joyner, A., Isaacs, L.N., Echols, H., Sly, W.S. 1966. DNA replication and messenger RNA production after induction of wildtype λ bacteriophage and λ mutants. *J. Mol. Biol.* 19:174–86
91. Kaiser, A.D. 1957. Mutations in a temperate bacteriophage affecting its ability to lysogenize *Escherichia coli*. *Virology* 3:42–61

92. Kaiser, A.D., Jacob, F. 1957. Recombination between related temperate bacteriophages and the genetic control of immunity and prophage localization. *Virology* 4:509–21
93. Kayajanian, G. 1968. Studies on the genetics of biotin-transducing, defective variants of bacteriophage λ. *Virology* 36:30–41
94. Konrad, M.W. 1968. Dependence of "early" λ bacteriophage RNA synthesis on bacteriophage-directed protein synthesis. *Proc. Nat. Acad. Sci. USA* 59:171–78
95. Kourilsky, P. 1971. Lysogenization by bacteriophage lambda and the regulation of lambda repressor synthesis. *Virology* 45:853–57
96. Kourilsky, P. 1973. Lysogenization by bacteriophage lambda: I. Multiple infection and the lysogenic response. *Mol. Gen. Genet.* 122:183–95
97. Kourilsky, P., Bourguignon, M.-F., Bouquet, M., Gros, F. 1970. Early transcription controls after induction of prophage λ. *Cold Spring Harbor Symp. Quant. Biol.* 35:305–14
98. Kourilsky, P., Bourguignon, M.-F., Gros, F. 1971. Kinetics of viral transcription after induction of prophage. See Ref. 77: 647–66
99. Kourilsky, P., Marcaud, L., Sheldrick, P., Luzzati, D., Gros, F. 1968. Studies on the messenger RNA of bacteriophage λ. I. Various species synthesized early after induction of the prophage. *Proc. Nat. Acad. Sci. USA* 61:1013–20
100. Kumar, S., Calef, E., Szybalski, W. 1970. Regulation of the transcription of *Escherichia coli* phage λ by its early genes *N* and *tof*. *Cold Spring Harbor Symp. Quant. Biol.* 35:331–39
101. Lew, K.K., Roth, J.R. 1970. Isolation of UGA and UAG nonsense mutants of bacteriophage P22. *Virology* 40:1059–62
102. Lewis, E.B. 1951. Pseudoallelism and gene evolution. *Cold Spring Harbor Symp. Quant. Biol.* 16:159–74
103. Lieb, M. 1966. Studies of heat-inducible lambda bacteriophage. I. Order of genetic sites and properties of mutant prophages. *J. Mol. Biol.* 16:149–63
104. Liedke-Kulke, M., Kaiser, A.D. 1967. Genetic control of prophage insertion specificity in bacteriophages λ and 21. *Virology* 32:465–74
105. Losick, R. 1972. *In vitro* transcription. *Ann. Rev. Biochem.* 41:409–46
106. Luzzati, D. 1970. Regulation of λ exonuclease synthesis: role of the N gene product and λ repressor. *J. Mol. Biol.* 49:515–19
107. Manly, K.F., Signer, E.R., Radding, C.M. 1969. Nonessential functions of bacteriophage λ. *Virology* 37:177–88
108. Mount, D.W.A., Harris, A.W., Fuerst, C.R., Siminovitch, L. 1968. Mutations in bacteriophage lambda affecting particle morphogenesis. *Virology* 35:134–39
109. Murialdo, H., Siminovitch, L. 1971. The morphogenesis of bacteriophage lambda. III. Identification of genes specifying morphogenetic proteins. See Ref. 77: 711–24
110. Murialdo, H., Siminovitch, L. 1972. The morphogenesis of bacteriophage lambda. IV. Identification of gene products and control of the expression of the morphogenetic information. *Virology* 48:785–823
111. Murotsu, T., Horiuchi, T. 1971. Prophage induction by the product of x region of superinfected λ operator-mutants. *Japan. J. Genet.* 46:365–68
112. McMacken, R., Mantei, N., Butler, B., Joyner, A., Echols, H. 1970. Effect of mutations in the *c*II and *c*III genes of bacteriophage λ on macromolecular synthesis in infected cells. *J. Mol. Biol.* 49:639–56
113. Naono, S., Tokuyama, K. 1970. On the mechanism of λ DNA transcription in vitro. *Cold Spring Harbor Symp. Quant. Biol.* 35:375–81
114. Neubauer, Z., Calef, E. 1970. Immunity phase-shift in defective lysogens: non-mutational hereditary change of early regulation of λ prophage. *J. Mol. Biol.* 51:1–14
115. Nijkamp, H. J. J., Szybalski, W., Ohashi, M., Dove, W.F. 1971. Gene expression by constitutive mutants of λ. *Mol. Gen. Genet.* 114:80–88
116. Oda, K., Sakakibara, Y., Tomizawa, J. 1969. Regulation of transcription of the lambda bacteriophage genome. *Virology* 39:901–18
117. Oppenheim, A.B., Slonim, Z. 1971. Conversion of λ defective lysogens from the non-immune state to the immune state. *Mol. Gen. Genet.* 112:255–62
118. Ordal, G.W. 1971. Supervirulent mutants and the structure of operator and promoter. See Ref. 77: 565–70
119. Packman, S., Sly, W.S. 1968. Constitutive λ DNA replication by λc_{17}, a regulatory mutant related to virulence. *Virology* 34:778–89
120. Parkinson, J.S. 1968. Genetics of the left arm of the chromosome of bacteriophage lambda. *Genetics* 59:311–25

121. Parkinson, J.S., Huskey, R.J. 1971. Deletion mutants of bacteriophage lambda. I. Isolation and initial characterization. *J. Mol. Biol.* 56:369–84
122. Pereira da Silva, L.H., Jacob, F. 1968. Étude génétique d'une mutation modifiant la sensibilité a l'immunité chez le bactériophage lambda. *Ann. Inst. Pasteur* 115:145–58
123. Pereira da Silva, L.H., Jacob, F. 1968. Induction of CII and O functions in early defective lambda prophages. *Virology* 33:618–24
124. Pero, J. 1970. Location of the phage λ gene responsible for turning off λ-exonuclease synthesis. *Virology* 40: 65–71
125. Pero, J. 1971. Deletion mapping of the site of action of the *tof* gene product of phage lambda. See Ref. 77: 599–608
126. Pero, J. Personal communication
127. Pirrotta, V., Ptashne, M. 1969. Isolation of the 434 phage repressor. *Nature* 222:541–44
128. Portier, M.-M., Marcaud, L., Cohen, A., Gros, F. 1972. Mechanism of transcription of the N operon of bacteriophage lambda. *Mol. Gen. Genet.* 117:72–81
129. Ptashne, M. 1967. Specific binding of the λ phage repressor to λ DNA. *Nature* 214:232–34
130. Ptashne, M., Hopkins, N. 1968. The operators controlled by the λ phage repressor. *Proc. Nat. Acad. Sci. USA* 60:1282–87
131. Rabovsky, D., Konrad, M. 1970. Gene expression in bacteriophage λ. I. The kinetics of the requirement for N gene product. *Virology* 40:10–17
132. Radding, C.M. 1964. Nuclease activity in defective lysogens of phage λ. II. A hyperactive mutant. *Proc. Nat. Acad. Sci. USA* 52:965–72
133. Radding, C.M., Echols, H. 1968. The role of the *N* gene product of phage λ in the synthesis of two phage-specified proteins. *Proc. Nat. Acad. Sci. USA* 60:707–11
134. Radding, C.M., Shreffler, D.C. 1966. Regulation of λ exonuclease. II. Joint regulation of exonuclease and a new λ antigen. *J. Mol. Biol.* 18:251–61
135. Rambach, A., Brachet, P. 1971. Sélection de mutants du bactériophage λ incapables de se répliquer. *C.R. Acad. Sci. Paris* 272:149–52
136. Reader, R.W., Siminovitch, L. 1971. Lysis defective mutants of bacteriophage lambda: Genetics and physiology of S cistron mutants. *Virology* 43:607–22
137. Reichardt, L., Kaiser, A. 1971. Control of λ repressor synthesis. *Proc. Nat. Acad. Sci. USA* 68:2185–89
138. Roberts, J. 1969. Promoter mutation *in vitro. Nature* 224:480–82
139. Roberts, J.W. 1969. Termination factor for RNA synthesis. *Nature* 224:1168–74
140. Roberts, J. 1970. The ρ factor: Termination and anti-termination in λ. *Cold Spring Harbor Symp. Quant. Biol.* 35:121–27
141. Rosner, J.L., Kass, L.R., Yarmolinsky, M.B. 1968. Parallel behavior of F and Pl in causing indirect induction of lysogenic bacteria. *Cold Spring Harbor Symp. Quant. Biol.* 33:785–89
142. Scaife, J., Beckwith, J.R. 1966. Mutational alteration of the maximal level of *lac* operon expression. *Cold Spring Harbor Symp. Quant. Biol.* 21:403–08
143. Schleif, R. 1972. The specificity of lambdoid phage late gene induction (lambdoid phage late gene specificity). *Virology* 50:610–12
144. Schleif, R., Greenblatt, J. 1970. Transcription in the lambda-ara phage. *Cold Spring Harbor Symp. Quant. Biol.* 35:369–73
145. Sharp, P.A., Hsu, M.-T., Davidson, N. 1972. Note on the structure of prophage λ. *J. Mol. Biol.* 71:499–501
146. Signer, E.R. 1968. Lysogeny: The integration problem. *Ann. Rev. Microbiol.* 22:451–88
147. Signer, E.R. 1969. Plasmid formation: A new mode of lysogeny by bacteriophage λ. *Nature* 223:158–60
148. Signer, E.R. 1970. On the control of lysogeny in phage λ. *Virology* 40: 624–33
149. Signer, E.R., Manly, K.F., Brunstetter, M. 1969. Deletion mapping of the *cIII-N* region of bacteriophage λ. *Virology* 39:137–40
150. Skalka, A. 1971. Origin of DNA concatemers during growth. See Ref. 77: 535–48
151. Skalka, A., Burgi, E., Hershey, A.D. 1968. Segmental distribution of nucleotides in the DNA of bacteriophage lambda. *J. Mol. Biol.* 34:1–16
152. Skalka, A., Butler, B., Echols, H. 1967. Genetic control of transcription during development of phage λ. *Proc. Nat. Acad. Sci. USA* 58:576–83
153. Sly, W.S., Eisen, H.A., Siminovitch, L. 1968. Host survival following infection with or induction of bacterio-

phage lambda mutants. *Virology* 34:112–27

154. Sly, W.S., Rabideau, K. 1969. The mechanism of λc17cI virulence. *J. Mol. Biol.* 42:385–400
155. Sly, W., Rabideau, K., Kolber, A. 1971. The mechanism of lambda virulence. II. Regulatory mutations in classical virulence. See Ref. 77: 575–88
156. Soll, L., Berg, P. 1969. Recessive lethals: A new class of nonsense suppressors in *Escherichia coli. Proc. Nat. Acad. Sci. USA* 63:392–99
157. Spiegelman, W.G. 1971. Two states of expression of genes *cI*, *rex*, and *N* in lambda. *Virology* 43:16–33
158. Spiegelman, W.G., Heinemann, S.F., Brachet, P., Pereira da Silva, L., Eisen, H. 1970. Regulation of the synthesis of phage lambda repressor. *Cold Spring Harbor Symp. Quant. Biol.* 35:325–30
159. Spiegelman, W.G., Reichardt, L.F., Yaniv, M., Heinemann, S.F., Kaiser, A.D., Eisen, H. 1972. Bidirectional transcription and the regulation of phage λ repressor synthesis. *Proc. Nat. Acad. Sci. USA* 69:3156–60
160. Steitz, J.A. 1969. Polypeptide chain initiation: nucleotide sequences of the three ribosomal binding sites in bacteriophage R17 RNA. *Nature* 224: 957–64
161. Stevens, W.F., Adhya, S., Szybalski, W. 1971. Origin and bidirectional orientation of DNA replication in coliphage lambda. See Ref. 77: 515–34
162. Stretton, A.O.W., Kaplan, S., Brenner, S. 1966. Nonsense codons. *Cold Spring Harbor Symp. Quant. Biol.* 31:173–79
163. Sussman, R., Jacob, F. 1962. Sur un système de répression thermosensible chez le bactériophage λ d'*Escherichia coli. C.R. Acad. Sci. Paris* 254: 1517–19
164. Szpirer, J. 1972. Le controle du développement des bactériophages tempérés. IV. Action spécifique du produit N au niveau d'une barriér de transcription. *Mol. Gen. Genet.* 114: 297–304
165. Szybalski, W., Bøvre, K., Fiandt, M., Guha, A., Hradecna, Z., Kumar, S., Lozeron, H.A., Maher, V.M. Sr., Nijkamp, H.J.J., Summers, W.C., Taylor, K. 1969. Transcriptional controls in developing bacteriophages. *J. Cell. Physiol.* 74: suppl. 1, 33–70
166. Szybalski, W., Bøvre, K., Fiandt, M., Hayes, S., Hradecna, Z., Kumar, S., Lozeron, H.A., Nijkamp, H.J.J., Stevens, W.F. 1970. Transcriptional units and their controls in *Escherichia coli* phage λ: operons and scriptons. *Cold Spring Harbor Symp. Quant. Biol.* 35:341–54
167. Takeda, Y., Oyama, Y., Nakajima, K., Yura, T. 1969. Role of host RNA-polymerase for λ phage development. *Biochem. Biophys. Res. Commun.* 36:533–38
168. Taylor, A. 1971. Endopeptidase activity of phage λ-endolysin. *Nature New Biol.* 234:144–45
169. Taylor, K., Hradecna, Z., Szybalski, W. 1967. Asymmetric distribution of the transcribing regions on the complementary strands of coliphage λ DNA *Proc. Nat. Acad. Sci. USA* 57:1618–25
170. Thomas, R. 1966. Control of development in temperate bacteriophages. I. Induction of prophage genes following heteroimmune superinfection. *J. Mol. Biol.* 22:79–95
171. Thomas, R. 1970. Control of development in temperate bacteriophages. III. Which prophage genes are and which are not *trans*-activable in the presence of immunity? *J. Mol. Biol.* 49:393–404
172. Thomas, R. 1971. Control circuits. See Ref. 77: 211–20
173. Thomas, R. 1971. Regulation of gene expression in bacteriophage λ. *Curr. Top. Microbiol.* 56:13–42
174. Thomas, R., Bertani, L. 1964. On the control of the replication of temperate bacteriophages superinfecting immune hosts. *Virology* 24:241–53
175. Thomas, R., Leurs, C., Dambly, C., Parmentier, D., Lambert, L., Brachet, P., Lefebvre, N., Mousset, S., Porcheret, J., Szpirer, J., Wauters, D. 1967. Isolation and characterization of new *sus* (amber) mutants or bacteriophage λ. *Mutation Res.* 4:735–41
176. Tomizawa, J., Ogawa, T. 1968. Replication of phage lambda DNA. *Cold Spring Harbor Symp. Quant. Biol.* 33:533–51
177. Toussaint, A. 1969. Insertion of phage Mu.1 within prophage λ: a new approach for studying the control of the late functions in bacteriophage λ. *Mol. Gen. Genet.* 106:89–92
178. Unger, R.C., Clark, A.J. 1972. Interaction of the recombination pathways of phage λ and its host *E. coli* K12: Effects on exonuclease V activity. *J. Mol. Biol.* 70:539–48

179. Wang, J.C., Kaiser, A.D. 1973. Evidence that the cohensive ends of mature λ DNA are generated by the gene *A* product. *Nature New Biol.* 241:16–17
180. Watson, J.D. Personal communication
181. Weigle, J. 1966. Assembly of phage lambda *in vitro*. *Proc. Nat. Acad. Sci. USA* 55:1462–66
182. Westmoreland, B.C., Szybalski, W., Ris, H. 1969. Mapping of deletions and substitutions in heteroduplex DNA molecules of bacteriophage lambda by electron microscopy. *Science* 163:1343–45
183. Wilgus, G.S., Friedman, D.I. 1973. *N*-like mutants of λimm^{21}. Manuscript in preparation
184. Wu, A.M., Ghosh, S., Echols, H. 1972. Repression by the *c*I protein of phage λ: Interaction with RNA polymerase. *J. Mol. Biol.* 67:423–32
185. Wu, A.M., Ghosh, S., Echols, H., Spiegelman, W.G. 1972. Repression by the *c*I protein of phage λ: *in vitro* inhibition of RNA synthesis. *J. Mol. Biol.* 67:407–22
186. Wu, A.M., Ghosh, S., Willard, M., Davison, J., Echols, H. 1971. Negative regulation by lambda: repression of lambda RNA synthesis in vitro and host enzyme synthesis in vivo. See Ref. 77: 589–98
187. Yanofsky, C., Helinski, D.R., Maling, B.D. 1961. The effects of mutation on the composition and properties of the A protein of *E. coli* tryptophan synthetase. *Cold Spring Harbor Symp. Quant. Biol.* 26:11–24
188. Yarmolinsky, M.B. 1971. Making and joining DNA ends. See Ref. 77: 97–112
189. Zissler, J., Signer, E., Schaefer, F. 1971. The role of recombination in growth of bacteriophage lambda. I. The gamma gene. See Ref. 77: 455–68
190. Zissler, J., Signer, E., Schaefer, F. 1971. The role of recombination in growth of bacteriophage lambda. II. Inhibition of growth by prophage P2. See Ref. 77: 469–75

T4 AND THE ROLLING CIRCLE MODEL OF REPLICATION 3058

A. H. Doermann

Department of Genetics, University of Washington, Seattle, Washington

Introduction

The genetic map of T4 is circular (1), even though the DNA molecules of the infectious phage are linear. Map circularity results from the fact that, among the particles of a T4 population, the ends of the DNA molecules are randomly located in the map, rather than at a unique genetic site (2)[1]. Two genetic markers located near opposite ends of the genome in one individual phage particle would thus be closely linked in the large majority of the other members of the population. The genomes of T4 are furthermore known to be diploid for the terminal 1–2% of their length, carrying homologous genes at both the right and left ends of the DNA molecule (3). This circularly permuted terminal redundancy appears to play an important role in genetic recombination.

The non-encapsidated T4 DNA, vegetative-state DNA, within the infected bacterium consists in large part of concatemers (4) which are composed of two or more phage genomes in a linear continuum. It is thought that phage capsids (or their precursors) are loaded with DNA by filling them to capacity from such concatemers ("headful hypothesis") (5). At least two ways have been suggested for originating such concatemers within the cell. One would consist of making many copies of the parental genomes which were injected into the host bacterium and forming concatemers by recombination among these daughter molecules. In the case of single infection such recombinations could occur in joining the right and left redundant ends of different copies, yielding a concatemer almost twice as long as the original parental DNA. Following multiple infection, which would generally provide parental molecules with ends in different genetic locations, concatemer formation could be achieved by recombination of the end of one

[1] T4 and T2 are so closely related that data from T2 investigations are treated as though derived from T4, except where precise genetic homologies or specific genetic markers are involved.

molecule with the middle of another. Another method of making concatemers would be for the parental molecule(s) to circularize by recombination between the homologous regions of its redundant head and tail. A nick at some initiation site in one strand of the circle would then permit a replication fork to proceed unidirectionally around the circle again and again. This is the so-called "rolling circle model" (RCM).

It was originally planned to cover here recombination in bacteriophage T4, a subject that was reviewed in 1970 by Gisela Mosig (6) in an opus that cited 170 references. In attempting to organize the advances made since that time, it became increasingly clear that the genetic data could not be convincingly interpreted without first deciding whether concatemers are formed by recombination or by an RCM. This is because the two models would predict widely different frequencies of molecular ends in the vegetative phage population which is multiplying and recombining within the host cell. Since regions near the ends of DNA molecules have exceptionally high recombination frequencies (7–10), the interpretation of formal recombination data will strongly depend on the frequencies and clonality of molecular ends. Such interpretations can become particularly complicated in a case where molecular ends may occur in any map location. The result is that this review became focused on deciding between the alternative routes of concatemer formation.

Most, but not all, of the experiments that permit any conclusions on the molecular nature of the replicating structure have been carried out on the very early stages of replication. The principal support for the RCM comes from a series of papers by Mosig and her collaborators (11–13). In the light of past and present evidence, however, it seems likely that the main supporting experimental evidence put forward was interpreted incorrectly, and should be reconsidered. To this author, it seems justified now to conclude that at least the initial copies of parental genomes do not result from unidirectional replication from a single initiation point as was suggested by Mosig and collaborators. It is much more probable that they are formed from bidirectional synthesis starting from multiple sites of initiation that have specific genetic locations. The experimental substance for the original formulation of the RCM will first be reviewed. Thereafter, doubts about the validity will be discussed.

The Experimental Foundation for the Rolling Circle Model

(A) The model according to Mosig and Werner The RCM was first seriously proposed for T4 by Gilbert & Dressler (14) at the Cold Spring Harbor Symposium of 1968 where no less than four other speakers found reason to use it in explaining their results (15–18). Two of the five papers (14, 15) included the suggestion that initial replications might be completed from noncircularized templates in order to account for polarized segregation of genetic markers located near the ends of DNA molecules (8,19,20). More recently, however, Mosig & Werner (11) concluded that less than one complete round of replication can occur in the absence of circularization of the parental DNA molecule. From these data they formulated a frequently cited version of the RCM. Mosig & Werner carried out experiments with "small" phages isolated from a population of T4 in which they exist at low frequency. Small phages of the type used contain only two thirds of a normal T4 genome, the deleted third being a continuous, randomly located segment (21, 22). Such particles, whose DNA was labeled with ^{32}P were used to infect bromouracil (BU)-labeled bacteria at low multiplicity of infection (MOI) (0.06-0.09) in a BU-containing medium. DNA was extracted from the infected cells and analyzed by density equilibrium centrifugation and by sedimentation through neutral and alkaline sucrose gradients. The following observations were made: (*a*) About one third of the injected incomplete chromosomes did not replicate to a detectable extent, while the remaining two thirds did replicate to varied extents, judging by the shift of the radioactive label toward a density characteristic of hybrid DNA, and by a slight increase in its sedimentation velocity as compared to the injected chromosomes. (*b*) The density distribution of the unsheared DNA suggested that all degrees of replication of the injected incomplete chromosomes are equally likely. (*c*) In a second experiment using ^{3}H-labeled BU, it was found that cells infected with normal genomes quickly initiate new rounds of replication in agreement with Werner's (18, 23) earlier results. Incomplete chromosomes, however, made the diagnostic fully dense DNA at only 4% of the normal rate, an amount that could be attributed to the minor fraction of cells infected by two complementing small phages or by a contaminating normal phage. They proposed the following model for T4 DNA replication: (*a*) Replication starts at a single, genetically fixed initiation site, proceeding in one direction. (*b*) Reinitiation at that site occurs only if the template molecule has formed a circle by recombination in its terminal redundancy, an event impossible in the incomplete chromosome. With appropriate corrections all of their data are compatible with this model.

(B) Locating the single initiation site Mosig (12) attempted to discover the map location of the initiation site in a series of single-burst experiments in which cells were infected with a single small particle differing in two genetic markers from a complete helper phage. The rationale of her experiment was based on the conclusion from earlier work (15) that even in the presence of a complete helper phage the genetic markers on incomplete genomes are not replicated extensively until they are incorporated into a complete genome by recombination. Thus, any marker near the initiation site, but in the direction of replication, was expected to have an advantage because it could be copied once without circularization, thus providing two copies of the phage for rescue into a replicating circle. A marker on the opposite side of the origin would not be replicated even once prior to its rescue. Marker recovery frequency should therefore be high near, and on one side of, the initiation site and drop off gradually as the distance between marker and origin increases. Markers on the opposite side should be recovered less frequently, resulting in an abrupt frequency difference in markers near, but on opposite sides of the origin. Mosig (12) did find such a point in the genome where markers on the clockwise side were recovered from small phage genomes about 85% as efficiently as from complete chromosomes. Nearby markers on the counterclockwise side showed an efficiency of only about 50% of normal. Other treatments of the data, for example, selecting bursts for presence or absence of particular markers, also supported the model. She interpreted these results to indicate that a single initiation site is located in the vicinity of genes 42 and 43, and that replication proceeds in a clockwise direction only. Additional data confirming these results were subsequently published by Mosig, et al (10).

Strong support for the idea that an initiation site is located in the vicinity of genes 42 and 43 comes from the transformation experiments of Marsh, Breschkin & Mosig (13). The transformation system employed was developed by Wais & Goldberg (24). It utilizes *Aerobacter aerogenes* spheroplasts as a medium into which recipient whole T4 is introduced via urea-treated phage particles. The donor consists of single-stranded DNA differing from the recipient in one or more genetic markers. The donor for the experiments of Marsh, et al, was extracted from cells infected with $^{32}P\,^{15}N\,^{13}C$ -labeled small phages (MOI from 0.1 to 0.2). The presence of ^{3}H-thymidine during replication in $^{14}N\,^{12}C$-containing media provided differential labeling of the newly-synthesized DNA and the injected parental DNA. They could thereby be readily separated from each other in alkaline density gradients. DNA extracted from normal (complete) phages with similar genetic markers was used, after denaturation, as a quantitative standard for transformation in each experiment. The RCM of replication as proposed by Mosig & Werner (11) would predict first that the extracted parental DNA should, on a molecule for molecule basis, transform any markers in the entire genome with the same efficiency as the control standard DNA. It would predict further that the newly synthesized strands would transform those markers that are both near the origin and located clockwise to it with an efficiency nearly equal to that of the parental DNA, as at that location a copy should have been made on each parental strand. Newly synthesized DNA would, on the other hand, be expected to transform markers at greater distances with decreasing efficiency, and those located in the counterclockwise third of the map not at all. Both the parental and the newly synthesized DNA behaved in accordance with these predictions when one of two extraction procedures was used, namely, a method in which 88% of the input parental ^{32}P label was recovered. An alternative method of extraction recovered only a nonrandom 40% of parental label. Results with the latter showed less striking agreement with expectation, but they do not diminish the force of the agreement of the first results with the model's predictions.

Observations Incompatible With The Rolling Circle Model

(A) CLONE-SIZE DISTRIBUTION OF SPONTANEOUS MUTANTS One observation that appears to be incompatible with the RCM for phage T2, a very close relative of T4, was made even before DNA was recognized as being composed of two complementary strands in a bihelical structure. In 1951 Luria (25) investigated replication of the genetic material of T-even phages by analyzing the clone-size distribution of spontaneous mutants among single bursts of T2. He found that the distribution was not random, but clonal, indicating that once a mutation had occurred, its frequency increased either because a template had been altered or because the mutant itself was replicating. More significantly in the present context, the frequency distribution of various clone sizes implied exponential multiplication of mutants once formed, that is, successive rounds of doubling. The distribution was not compatible with a model in which the mutation occurred in the template, which thereafter produced a linear succession of mutant copies. Luria's data were shown in 1961, by Steinberg & Stahl (26) to fit closely to an exponental multiplication hypothesis also when replication is conceived as taking place as a random-in-time process in a steady-state pool of vegetative phages. This is a more realistic approximation to the actual situation than the original idealized treatment given by Luria. The data fit the theory exceptionally well over the critical range of clone sizes from 1–16 mutants per clone. These are the clones made from mutations in the near-terminal replications. It is difficult to see how exponential replication in the late stages of multiplication can be reconciled with an RCM that is expected to make concatemers, the latter assumed to be the immediate precursor of matured phage genomes. Although they did not elaborate on it, Kozinski, Kozinski & James (27) already in 1967, and more recently Miller, Kozinski & Litwin (28), called attention to this difficulty for any hypothesis of a giant molecule growing at one end and maturing at the other.

(B) Lack of a covalent link between parental and early replica DNA

At the Cold Spring Harbor Symposium in 1968, which first directed serious attention to the RCM for T4, another question was raised by both Kozinski and Denhardt [see discussions following the papers of Frankel (17) and of Werner (18)]. Production of a concatemer from a rolling circle requires formation of a covalent bond linking the nicked parental strand with the initial copy from the other, still circular, parental strand. While Kozinski, Frankel and Werner had each attempted to find such a connection, none had succeeded. Kozinski's group has reiterated its inability to find such a covalent connection in four papers (27, 29–31). To date the only case of an early covalent attachment of newly synthesized DNA to the injected parental molecule was found by Murray & Mathews (32) who observed a 6% covalent addition of new DNA to parental DNA. In their experiment, however, they used a gene-44 amber mutant, *am*N82, to infect a suppressor-negative host. Warner & Hobbs (33) have shown that *am*N82 infecting a restrictive host synthesizes less than 0.4% of the amount of DNA produced following wild-type T4 infection. It is probably also significant that the attached material found by Murray & Mathews hybridized poorly with T4 DNA. In the absence of more precise information on the function of gene 44 in T4 DNA synthesis it is difficult to evaluate the exceptional result. For the present, at least, the covalent attachment of bona fide early replicated T4 DNA, expected from the RCM, seems to lack convincing demonstration.

(C) REPEATED REPLICATION OF FRAGMENTS OF PARENTAL MOLECULES In a series of publications, Kozinski and many collaborators have been engaged in a continuing study concerned with the fate of the parental DNA molecule after its injection into the host cell and with the replication process, particularly during the early stages of phage DNA increase. Several methods that have consistently played a role in these experiments include the following. Parental phage is labeled with a radioactive isotope (usually ^{32}P), which identifies the original strands or a smaller contribution from them. Synthesis of new T4 DNA in the infected cell is carried out in the presence of bromodeoxyuridine, which substitutes for thymine in the daughter strands, causing a large density increase and thus serving to identify the new copies. Analysis of the composition of the DNA extracted from infected cells is carried out by velocity sedimentation through sucrose gradients and by density equilibrium centrifugation in CsCl. Information on the intermolecular distribution of segments of the variously labeled strands is obtained by combining the centrifugation techniques with controlled shearing of the molecules using whole mature T4 molecules labeled with a second radioisotope (usually ^{3}H) as a standard. Repeated use of these methods has resulted in a number of observations that appear to conflict with the predictions of the RCM.

The first of these was that fragments of parental molecules could be replicated not only once, but repeatedly. In these experiments they selectively inhibited the formation of recombination enzymes by carefully timing the addition of chloramphenicol (CAM) to the infected bacterial culture (27). CAM added at 5 minutes or later allowed a large part of the parental label to go to hybrid density, indicating a significant amount of replication. The addition of CAM at 7 minutes allowed more replication, but also resulted in appreciable cutting of the parental molecules. In both cases, however, the parental label was, for the most part, separable from the newly synthesized DNA simply by centrifugation under the denaturing conditions of an alkaline density gradient. The more significant result for present considerations was seen when the density-reversal technique was applied. The procedure was to allow replication to proceed as before until most of the parental label was in molecules of hybrid density. Many of these molecules were shorter than one phage-equivalent length. At that point half of the culture was extracted in order to check the fraction of parental isotope that had moved to hybrid density. The other half of the culture was transferred to a medium in which the density label, bromodeoxyuridine, had been replaced by thymidine. After incubation in that medium, the culture was extracted and the DNA centrifuged in a neutral CsCl gradient. Application of that method showed that the parental ^{32}P, which was largely in the hybrid density location at one time-point, moved back to light density with continued incubation in light density medium. It had exchanged its BU-labeled partner strand for a new strand, indicating that a second round of replication had taken place. A similar result was obtained when the experiment was repeated in a gene-30 amber mutant (ligase deficient) which showed even more clearly the repeated replication of fragments shorter than one phage-equivalent length (34, 35). While a single replication might be allowed, Mosig & Werner's (11) interpretation of their experiments, which forms a substantial part of the experimental foundation of the RCM, does not permit reinitiation on a noncircular template. Exchange of partner strands during a second replication as observed in these experiments would be forbidden.

(D) The delayed appearance of strands longer than one phage equivalent Another finding of Kozinski's group which appears to be at variance with the RCM is related to the time of appearance of single DNA strands longer than one phage equivalent unit and the prevention of their appearance by imposition of conditions that inhibit molecular recombination. Miller, Kozinski & Litwin (28) argue that, if formed by continued replication according to the RCM, long single strands should be detectable in infected bacteria "before two or three phage-equivalent units of DNA per bacterium are synthesized." They could show, however, that less than 2% of parental label appeared in strands significantly longer than one phage equivalent as late as 9 minutes after infection, a time when about 20 phage-equivalents of DNA had been synthesized. Identical results (not shown in the paper) were obtained when nonradioactive parental phage was used to infect bacteria in a medium to which ^{32}P was added two minutes after infection. While data are not given for multiplicities of infection lower than five, a personal communication from Kozinski indicates that similar results were obtained even at multiplicities of approximately one. A multiplicity of parental templates, all starting replication at the same time, could not, therefore, account for the data. They furthermore noted that conditions previously shown to inhibit molecular recombination (27) also inhibited formation of long single strands. Thus, addition of chloramphenicol at five minutes after infection, while allowing synthesis of 50–100 phage equivalents of T4 DNA, almost completely inhibited long single-strand formation. On the other hand, chloramphenicol added at eight minutes after infection or later (in which case molecular recombination does occur) failed to prevent the appearance of a large amount of long single-stranded T4 DNA. These data were interpreted to mean that concatemers result from recombination and are not the immediate product of replication.

(E) The parallel encapsidation of parental and newly synthesized DNA In the same paper discussed in the preceding paragraph, Miller et al (28) reported a further observation. The question was considered whether the radioactive label in one strand of the parental DNA, being circular in structure and continuously required as a template, was thereby prevented from maturing into infectious phage particles. They labeled the parental DNA (MOI = 5) with ^{32}P and the early synthesized DNA with a pulse of ^{3}H-thymidine which was chased after one minute. The culture was lysed at intervals thereafter, and the trichloracetic acid precipitable radioactivity determined with and without deoxyribonuclease treatment. Infectious phages present in the lysates were also scored. The enzyme-resistant DNA was interpreted as being DNA removed from the replicating pool, probably encapsidated in phage heads. The results showed that parental and progeny DNA are encapsidated at the same rate, and in parallel with the appearance of mature, infectious phage particles. They indicate, therefore, that half of the parental label was not selectively retained in the replicating pool and thereby prevented from entering mature phage.

It should be noted, on the other hand, that a study of the kinetics of T4 DNA synthesis led Altman & Lerman (36) to support the idea that part of the injected parental DNA is conserved in the replicating pool. They also used radioisotopes to label the parental DNA (^{32}P) distinguishably from the DNA synthesized (^{3}H) during an early pulse (0 –7 minutes at 37°C) of thymidine to the medium. They found that about 45% of the parental label and about 35% of the incorporated pulse label failed to find their way out of the replicating pool and into mature phages. These results are, however, at variance with the very early results reported by Hershey & Burgi (37) who, following comprehensive accounting of the parental label in all fractions of an experiment with T2, concluded that the 50% losses of label generally observed in parent to progeny transfer resulted from random losses and have "nothing to do with the mechanism of transfer." It seems possible that the synchronizing procedure employed by Altman & Lerman, which involved the use of fluorodeoxyuridine and L-*p*-fluorophenylalanine, as well as anaerobic conditions, may have caused the transfer of some label to an inactive fraction of the intracellular DNA pool which does not participate normally in replication.

(F) MULTIPLE INITIATION SITES The hypothesis that there is only a single initiation site in the T4 genome has been challenged by two kinds of evidence. One study, by Delius, Howe & Kozinski (38) is based on electron microscope (EM) examinations of T4 replicating complexes extracted from host bacteria 5–8 minutes after infection (MOI = 10) at 37°C. The parental phage was marked with BU and ^{32}P, and the infection carried out in medium supplemented with ^{3}H-thymidine and cells devoid of BU. Partially replicated parental molecules were separated from the others by their density, which was intermediate between the fully heavy and the completely hybrid molecules. EM examination provided several important observations. (*a*) Replicative loops of many sizes containing two Y-shaped branch points were frequently seen. The double-stranded nature of the arms of the loops was demonstrated by partial denaturation in the presence of T4 gene-32 protein. (*b*) The branch points of most of the loops contained one single-stranded whisker. When a whisker was visible at both branch points of a loop, they were invariably located in *trans* position. (*c*) In single phage-equivalent lengths, distinct multiple loops were seen frequently. (*d*) Four loops contained secondary loops of which one contained a tertiary loop. (*e*) Secondary initiation at a site must, in this material, have been infrequent, however, since sonication followed by density analysis of the same DNA extracts that had been examined by EM showed that no fully light ^{3}H label could be demonstrated.

The fact that within a 50 μ stretch of DNA multiple loops were frequently found (in one molecule 5 are shown) is incompatible with the idea of a single initiation site or with rolling circle replication as demanded by the model of Mosig & Werner. The observation of secondary loops located at some distance from both branches of the primary loop must furthermore be interpreted either as indicating bidirectional replication of the original loop, or a rather high incidence of potential initiation sites.

Howe, Buckley, Carlson & Kozinski (39) followed up the EM study with another series of experiments that similarly indicates the existence of multiple initiation sites. Here, too, partially replicated parental DNA molecules were isolated with the help of differential density (BU) and radioactive isotope (^{32}P and ^{3}H) labeling. Cells were extracted 4–6 minutes after infection with T4 at 37°C. The distribution of the newly replicated light DNA within the replicating heavy parental molecules was investigated by controlled shearing. It was shown that the whole molecules contained sufficient isotope that they must have replicated a total of 18 μ of DNA, and that this material was distributed in stretches 6–3 μ in length, thus demanding 3–6 well separated areas of replication. This result was obtained following both single and multiple infection. The estimate of the newly replicated segment lengths was confirmed by alkaline sucrose gradients, which showed that all the radioactivity was found in fragments 4–6 μ long and attached to the parental molecules only by hydrogen bonds.

In the same paper another approach was used to characterize further the initiation sites of T4 DNA. The principle on which it is based follows. *E. coli* exonuclease I attacks single-stranded DNA only, degrading it from the 3′ end of the molecule until a double-stranded region is encountered. Thus, single-stranded recipient DNA can be protected by addition of single-stranded donor DNA that can hybridize to it. If, then, the newly synthesized DNA of partially replicated molecules represents a specific genetic region, it will, as a donor, give protection to only a unique region of the recipient. Even the use of saturating concentrations of such a donor will leave much of the recipient unprotected. If replication initiates at random map locations, saturating concentrations of donor should provide nearly complete protection, equivalent to that given by DNA extracted from mature phage. A small number of genetically specific sites would give intermediate values. In the experimental application, newly synthesized T4 DNA was isolated from partially replicated molecules by use of density and radioactive labeling. The result of the test is again compatible only with the hypothesis that DNA replication is initiated at a small number of genetic sites, appreciably greater than one. Quantitative evaluation of the data led to the estimate of 8–10 sites per molecule.

It should be mentioned that, as pointed out by the authors (39), the methods used here are likely to lead to an underestimate rather than an overestimate of the number of initiation sites. That results from the fact that none of the three methods used, EM, controlled shearing density analysis, or exonuclease I protection, will adequately resolve replicative loops which are located too close to each other. Another difficulty might be encountered if some initiation sites are used in preference to others, particularly under the changing conditions in the early stages of infection.

(G) Evidence for repeated replication in the absence of circularization
Another recent investigation yielded results incompatible with the RCM of the Mosig & Werner (11) type. Doermann (40) carried out experiments to provide a direct test of a plausible alternative explanation for the data of Mosig & Werner (11) and of Marsh et al (13). Warner & Hobbs (33) had previously made quantitative tests of the amount of T4 DNA synthesized in cells infected with an amber mutation in one of 21 DNA genes of the phage. Many of the mutations seriously reduced the amount of T4 DNA synthesized. The locations of these on the "physical" map of the T4 genome (22,41) shows that a randomly located 33% deficiency of the chromosome, as was present in Mosig's small phages, would almost invariably delete one or more genes essential to efficient DNA synthesis. Thus, the alternative explanation of the Mosig & Werner (11) and the Marsh et al (13) data would invoke the absence of genes essential for DNA synthesis. In his test Doermann used two petit mutants of T4 (42) as a source for purified preparations of small particles that differed from those employed in the previously described experiments. While the deficiency that characterized Mosig's small particles was about one-third of the genome, the average deficiencies in three samples of particles used in these experiments were 29, 23, and 15% of the genome, respectively. It was anticipated that the smaller the deficiency, the greater would be the chance that a single phage would carry all the genes required for normal DNA synthesis. The experiment differed in another important respect. Replication was measured, not by density change, but by estimating the amount of ^{3}H-thymidine incorporation into T4 DNA, thus not restricting the observation to the first replication of the genome. The ^{3}H -labeled T4 DNA was determined by DNA:DNA hybridization.

The results agreed with expectation: Infection with the shortest petit sample (longest deficiencies) yielded the least DNA per infected cell, 9%, compared to similar infections by complete phages of the same genotype (control). Infections by the longest petits yielded the most DNA (45% of control), while intermediates yielded 26%. Since control burst sizes ranged from 20–51, genomes in every petit sample must have initiated replication repeatedly. The petit particles from mutant stocks, like those used in the studies of Mosig and her collaborators, are terminally deficient (43), and thus have no terminal redundancies that allow circularization by recombination. It should be pointed out, however, that these data should not be overinterpreted in the quantitative sense because it was necessary to equate ^{3}H-thymidine incorporated with DNA synthesized. To do so assumes that the labeled thymidine is similarly distributed in the DNA derived from control on the one hand, and from the experiment on the other. Physiological conditions or differences in the genetic inputs of infecting petit and normal phage particles may distort that similarity. Nevertheless, it seems unlikely that such distortion would be great enough to alter the general conclusion that, given the required T4 genes, T4 DNA is effectively replicated even in the absence of circularization.

Other Related Experiments

A number of other experiments appear to bear less critically on the question whether rolling circles or recombination are primarily responsible for concatemer formation. However, because they have been interpreted as being either in agreement or in disagreement with the RCM, it seems justified to discuss them briefly here. Werner (18,23) originally proposed the "pinwheel" variation of the RCM to account for the data from his ingenious experiments for determining the number and distribution of growing points in vegetative T4 DNA. He interpreted his results as eliminating models of replication that allow the initiation sites to be duplicated. This restricted him to models that require that the first template used be replicated over and over again. Like RCM generally, that is in conflict with the previously discussed data of Luria (25) on the clone-size distribution of spontaneous mutants, which requires that once a mutation occurs, the mutant must be permitted to multiply in an exponential way. While Luria's data reflect best the replications that occur in the later stages of virus multiplication, Werner's data are restricted to the early stages (eclipse period). His results indicate that growing points are clustered on only a fraction of the vegetative DNA. It seems possible that the likelihood of reinitiation on a particular initiation site may be different at early times than in the later stages of the latent period when the milieu in which replication takes place is probably quite different. In his second paper (18) on this subject, Werner also showed that his data are incompatible with a model in which initiation occurs at random sites in the genome. On the other hand, the data fit with the hypothesis that replication is unidirectional from a single genetically specific origin, and would thus be compatible with Mosig & Werner's model. It was not shown, however, that the data would exclude hypotheses which postulate several initiation sites, nor was the effect of bidirectional replication considered. In view of the results discussed in the previous section, particularly the results of Delius et al (38) and Howe et al (39), those conditions now demand serious consideration.

Circular DNA forms have been demonstrated in lysates of T4-infected cells by Bernstein & Bernstein (44). They incubated phage-infected cells for various periods in a medium containing ^{3}H-thymidine of high specific activity. The lysates were spread on Millipore filters and dried. Autoradiographs were then made. The photographs reveal not only linear and branched-linear structures, but also unbranched circles as well as circles with one, two, and four (one case) linear arms attached. The sizes of the circles ranged from 1–21 phage-equivalent lengths, the majority ranging from 2–10. In order to account for these structures they proposed an RCM that is adapted to forming multimeric circles, as well as unit circles.

While these circular structures may be made from such a model, it should be noted here that recombination among replicating linear molecules could generate all the types of structures which Bernstein & Bernstein describe: (*a*) Linear concatemers would be formed by head to tail recombination between two separate molecules. (*b*) Unbranched circles would be formed by head to tail recombination of a continuous molecule, either unit-length or longer. (*c*) Circles with two arms could represent loops formed by replication from an internal initiation site that had not yet reached either end of the template. (*d*) One-branched linear molecules would be formed similarly, but where one template end has been reached, the other has not. (*e*) A one-armed circle could result from recombination between one arm of a linear branched molecule with the linear segment still undergoing replication. As recombination is prerequisite to forming any circular structure from the infecting linear molecules, and even the multimeric circles depend on recombination for their existence, it is suggested that seeing these circles provides no critical evidence that requires a circular molecule to accomplish repeated replication. Of course, the one-armed circle formed as suggested would indeed be an example of rolling circle replication. It does not, however, provide a reason to consider circle formation as essential for replication.

Another argument used against the RCM comes from the experiments made by Kozinski (30) to decide whether one of the two strands of the parental DNA molecule plays a quantitatively more important role in replication than the other. From the RCM one might predict that after circularization a particular one of the two strands would be nicked at the initiation site, and, after the first replication, would generally be available for encapsidation. The strand retaining its circularity would, however, remain engaged with the rolling circle replication and therefore be much less likely to be matured. Together with the labeling system characteristic of his investigations, Kozinski made use of the method of Guha & Szybalski (45), which enabled him to separate the L and R strands of T4 DNA extracted from mature phages. The amounts of R and L strands in his experimental samples were then determined by specific hybridization to the separated strands. Investigating a wide variety of conditions that covered both the early and late stages of the latent period, Kozinski was unable to find any difference in the degree to which parental L and R strands appeared in mature progeny phage. This consideration is here regarded as less compelling than some others because the possibility remains that, even with a single initiation site, a mechanism may exist to randomize the decision that selects the strand to be preserved in circular form.

Data on the comparative frequency with which genetic markers are rescued from radiation-inactivated phages during co-infection with helper particles (cross reactivation) have independently given rise to the hypothesis that there are four sites in the genome at which replication can be initiated. The following observation was first made by Womack (46) in analysing the rescue of genetic markers from ultraviolet-inactivated T4. When the frequency of rescue of various markers is plotted against their position in the genetic map, the markers in four regions show significantly higher rescue frequency than those in the intervening segments. A similar result was noted by Levy (47) in experiments with T4 inactivated by decay of ^{32}P incorporated in its genome. Campbell (48,49) has reported the same result in his investigation of the nature of damages induced by X-rays. To explain his results and those of Womack and of Levy, he proposed that replication is arrested or retarded when it encounters a radiation damage. A marker situated between an initiation site and the nearest damage would be effectively copied, while one beyond the damage would not. Rescue of the former, available in multiple copies, would thus be more probable than rescue of the latter, and a peak of rescue efficiency in the genetic map would, on that hypothesis, signal an initiation site. This is not, of course, the only possible interpretation of the data of Womack, Levy, and Campbell. The previously discussed findings of Delius et al (38) and of Howe et al (39), which suggest three or more specific initiation sites do, however, lend the hypothesis additional credibility.

Discussion and Summary

In summary, the weight of present evidence forces abandonment of the RCM as envisioned by Mosig & Werner. There appears to be compelling evidence in favor of dismissing the main supporting arguments. The experiments that were interpreted to mean that reinitiation is forbidden on a linear template (11) have apparently been misconstrued. If the small phages of Mosig & Werner had contained all the T4 genes required for DNA synthesis, they evidently would have replicated their DNA repeatedly, as shown by the experiments of Kozinski's group (27, 34, 35) and those of Doermann (40). Furthermore, the evidence for a single initiation site now seems less convincing than the data presented by Delius et al (38) and by Howe et al (39), which both appear to demand multiple initiation sites in the T4 DNA molecule. Other arguments against the RCM, which have not been refuted by the model's supporters, include its incompatibility with the data on the clone-size distribution of spontaneous mutants (25), as well as the repeated failure to demonstrate a covalent connection between parental and early-synthesized T4 DNA (27, 29–31). Furthermore, single strands of T4 DNA longer than one phage-equivalent are not found in the intracellular DNA pool until many genome copies have accumulated, and their appearance seems to be dependent on conditions that are conducive to molecular recombination (28). Finally, existing evidence, although not unanimous, suggests that parental DNA is moved from the replicating pool to mature phage with the same probability as newly synthesized DNA, indicating that a parental circle is not preferentially retained as a replicating template.

It is not yet clear why Mosig & Werner (11) and Marsh et al (13) obtained the results they did. It seems possible, however, as suggested by Howe et al (39) and by Campbell (49) that the indication of a single initiation point may have been imposed on the small phage experiments by the supposition that one of the initiation sites is, in fact, located in that region of the T4 genome that contains a cluster of genes required for the synthesis of its DNA. Genes 41–45, each essential to T4 DNA synthesis, are located within a segment comprising only about 6% of the molecular map (22,41), and it is precisely in that segment in which the single initiation point proposed by Mosig (12) and Marsh et al (13) is located. This also coincides with one of the initiation points suggested by Campbell's interpretation of the cross-reactivation data (49). Because the presence of the gene 41–45 region is almost surely required for synthesis of any T4 DNA, it would not be surprising if an initiation site situated there would be utilized significantly more often than any other in small phages deficient for one third of the remainder of the genome.

The very small amount of replication observed by Mosig & Werner could be attributed to the absence of other T4 genes required for DNA synthesis. Gene 1, which codes for the deoxyribonucleotide kinase that is responsible for phosphorylation of three of the four nucleotide monophosphates, is located about a quarter of the total molecular map from gene 43 in the clockwise direction (22, 41). Warner & Hobbs (33) have shown that less than 0.3% of the normal amount of T4 DNA is synthesized in a restrictive host when the infecting phage carries an amber mutation in gene 1. Gene 32, which codes for the DNA unwinding protein (50), is located about a quarter of the map from gene 43, but in the counterclockwise direction. An amber mutation in gene 32 reduces T4 DNA synthesis in a restrictive host to about 2% of normal (33). Furthermore, Huberman, Kornberg & Alberts (51) have shown that gene 32 protein stimulates the polymerase activity of gene 43 protein in vitro. The large majority of Mosig & Werner's small phages would be expected to be deficient for one of these genes and would therefore not be expected to replicate more than a token amount of T4 DNA. The remaining particles, which do have all the genes so far discussed, would almost invariably be deficient for gene 30, which codes for the T4 ligase, and is located about 32% of the molecular map beyond gene 1 in the clockwise direction. Warner & Hobbs (33) showed that an amber mutation in that gene reduced T4 DNA synthesis to 8% of normal. It is suggested that almost none of the small particles of Werner & Mosig contained all of the genes needed for T4 DNA synthesis, and that this is why they produced very little DNA compared to normal synthesis in T4-infected cells.

The observation made by Mosig & Werner (11) and by Marsh et al (13) that replication progressed primarily in the clockwise direction on the T4 map appears, for the present, to be more difficult to explain, but can nevertheless be accounted for by making an ad hoc assumption. For that purpose it is assumed that an endonuclease sensitive segment exists in the genome immediately counterclockwise to the initiation point in the gene 41–45 region. Most small phages carrying genes 41–45 and gene 1, all of which appear to be indispensable for DNA synthesis, would lack gene 30 and have to depend on the less efficient host ligase to repair nicks. An unrepaired nick would, however, probably arrest a DNA growing point, and thus prevent counterclockwise replication in the majority of single infections by Mosig & Werner's small phages. Kozinski (35) suggested that endonuclease sensitive segments of the genome would explain his results, which indicated that the parental DNA of gene 30 amber mutants is degraded, soon after infection, to fragments of about one-fourth the normal length. Campbell (48) pointed out the apparent correlation between these quarter-size degradation fragments and four equally spaced regions of the genome in which recombination is significantly higher than expected from the molecular maps of Mosig (22) and Childs (41). It is interesting that one of these high recombination regions is located between genes 41 and 43, suggesting frequent nicks immediately counterclockwise to the presumptive initiation site. If the hypothesis proves to be correct, it would explain why counterclockwise replication was largely prevented in Mosig & Werner's small phages.

In conclusion, it appears probable that bidirectional replication from a number of genetically fixed starting points is the primary mechanism by which the T4 genome is reproduced. It is thereby implied that ends of DNA molecules are replicated into clones of ends, and that concatemers are formed by recombination. As mentioned in the discussion of the experiments of Bernstein & Bernstein (44), it is nevertheless expected that circular molecules will be formed by recombination, and if recombination occurs in a branched linear molecule, continuous replication on the resulting circular template could persist. It seems clear, however, that rolling circle replication cannot be the prevailing mechanism either for the earliest nor for the near terminal stages of replication in the T4 infected cell.

ACKNOWLEDGMENTS

I wish to acknowledge colleagues, too numerous to mention individually, for many discussions that helped me reach the primary conclusion presented here. In particular I wish to express my gratitude to Bruce Alberts, Harris Bernstein, Thomas Broker, Douglas Campbell, and Andrzej Kozinski who also provided unpublished manuscripts to assist me.

My research has been supported by U.S. Public Health Service Research Grant GM 13280 from the National Institute of General Medical Sciences.

Note added in proof: In discussing Gisela Mosig's interpretation of the work from her group, I have consistently equated circularization of the injected parental DNA with rolling circle replication. Dr. Mosig has, however, called to my attention that the rolling circle model was originally proposed by others (see references 14, 16, 17, 18, and 23) and that her own proposals need not necessarily include rolling circle replication. I regret that I have apparently read into her papers conclusions she did not intend to convey.

Literature Cited

1. Edgar, R.S., Wood, W.B. 1966. Morphogenesis of bacteriophage T4 in extracts of mutant-infected cells. *Proc. Nat. Acad. Sci. USA* 55:498–505
2. Thomas, C.A. Jr., Rubenstein, I. 1964. The arrangements of nucleotide sequences in T2 and T5 bacteriophage DNA molecules. *Biophys. J.* 4:93–106
3. MacHattie, L.A., Ritchie, D.A., Thomas, C.A. Jr., Richardson, C.C. 1967. Terminal repetition in permuted T2 bacteriophage DNA molecules. *J. Mol. Biol.* 23:355–63
4. Frankel, F.R. 1968. Evidence for long DNA strands in the replicating pool after T4 infection. *Proc. Nat. Acad. Sci. USA* 59:131–38
5. Streisinger, G., Emrich, J., Stahl, M.M. 1967. Chromosome structure in phage T4. III. Terminal redundancy and length determination. *Proc. Nat. Acad. Sci. USA* 57:292–95
6. Mosig, G. 1970. Recombination in bacteriophage T4. *Advan. Genet.* 15:1
7. Doermann, A.H. 1964. Recombination in bacteriophage T4 and the problem of high negative interference. *Proc. XI Int. Congr. Genet.* 2:69
8. Doermann, A.H., Parma, D.H. 1967. Recombination in bacteriophage T4. *J. Cell. Physiol.* 70(Suppl. 1):147–64

9. Michalke, W. 1967. Erhöhte Rekombinationshäufigkeit an den Enden des T1-Chromosoms. *Mol. Gen. Genet.* 99:12–33
10. Mosig, G., Ehring, R., Schliewen, W., Bock, S. 1971. The patterns of recombination and segregation in terminal regions of T4 DNA molecules. *Mol. Gen. Genet.* 113:51–91
11. Mosig, G., Werner, R. 1969. On the replication of incomplete chromosomes of phage T4. *Proc. Nat. Acad. Sci. USA* 64:747–54
12. Mosig, G. 1970. A preferred origin and direction of bacteriophage T4 DNA replication. *J. Mol. Biol.* 53:503–14
13. Marsh, R.C., Breschkin, A.M., Mosig, G. 1971. Origin and direction of bacteriophage T4 DNA replication. II. A gradient of marker frequencies in partially replicated T4 DNA as assayed by transformation. *J. Mol. Biol.* 60:213–33
14. Gilbert, W., Dressler, D. 1968. DNA replication: The rolling circle model. *Cold Spring Harbor Symp. Quant. Biol.* 33:473–84
15. Mosig, G., Ehring, R., Duerr, E.O. 1968. Replication and recombination of DNA fragments in bacteriophage T4. *Cold Spring Harbor Symp. Quant. Biol.* 33:361–69
16. Amati, P., Favre, R. 1968. Phage DNA synthesis in bacteria infected with T4 light particles. *Cold Spring Harbor Symp. Quant. Biol.* 33:371–74
17. Frankel, F.R. 1968. DNA replication after T4 infection. *Cold Spring Harbor Symp. Quant. Biol.* 33:485–93
18. Werner, R. 1968. Initiation and propagation of growing points in the DNA of phage T4. *Cold Spring Harbor Symp. Quant. Biol.* 33:501–07
19. Doermann, A.H., Boehner, L. 1963. An experimental analysis of bacteriophage T4 heterozygotes. I mottled plaques from crosses involving six rII loci. *Virology* 21:551–67
20. Womack, F.C. 1963. An analysis of single-burst progeny of bacteria singly infected with a bacteriophage heterozygote. *Virology* 21:232–41
21. Mosig, G. 1966. Distances separating genetic markers in T4 DNA. *Proc. Nat. Acad. Sci. USA* 56:1177–83
22. Mosig, G. 1968. A map of distances along the DNA molecule of phage T4. *Genetics* 59:137–51
23. Werner, R. 1968. Distribution of growing points in DNA of bacteriophage T4. *J. Mol. Biol.* 33:679–92
24. Wais, A.C., Goldberg, E.B. 1969. Growth and transformation of Phage T4 in *Escherichia coli* B/4, *Salmonella*, *Aerobacter*, *Proteus*, and *Serratia*. *Virology* 39:153–61
25. Luria, S.E. 1951. The frequency distribution of spontaneous bacteriophage mutants as evidence for the exponential rate of phage reproduction. *Cold Spring Harbor Symp. Quant. Biol.* 16:463–70
26. Steinberg, C., Stahl, F. 1961. The clone-size distribution of mutants arising from a steady-state pool of vegetative phage. *J. Theoret. Biol.* 1:488–97
27. Kozinski, A.W., Kozinski, P.B., James R. 1967. Molecular recombination in T4 bacteriophage deoxyribonucleic acid. I. Tertiary structure of early replicative and recombining deoxyribonucleic acid. *J. Virol.* 1:758–70
28. Miller, R.C. Jr., Kozinski, A. W., Litwin, S. 1970. Molecular recombination in T4 bacteriophage deoxyribonucleic acid. III. Formation of long single strands during recombination. *J. Virol.* 5:368–80
29. Kozinski, A.W., Kozinski, P.B. 1965. Early intracellular events in the replication of T4 phage DNA. II. Partially replicated DNA. *Proc. Nat. Acad. Sci. USA* 54:634–40
30. Kozinski, A.W. 1969. Unbiased participation of T4 phage DNA strands in replication. *Biochem. Biophys. Res. Commun.* 35:294–99
31. Miller, R.C. Jr., Kozinski, A.W. 1970. Early intracellular events in the replication of bacteriophage T4 deoxyribonucleic acid. V. Further studies on the T4 protein-deoxyribonucleic acid complex. *J. Virol.* 5:490–501
32. Murray, R.E., Mathews, C.K. 1969. Addition of nucleotides to parental DNA early in infection by bacteriophage T4. *J. Mol. Biol.* 44:233–48
33. Warner, H.R., Hobbs, M.D. 1967. Incorporation of uracil-^{14}C. into nucleic acids in *Escherichia coli* infected with bacteriophage T4 and T4 amber mutants. *Virology* 33:376–84
34. Kozinski, A.W., Kozinski, P.B. 1968. Autonomous replication of short DNA fragments in the ligase negative T4 AM H39X. *Biochem. Biophys. Res. Commun.* 33:670–74
35. Kozinski, A.W. 1968. Molecular recombination in the ligase negative T4 amber mutant. *Cold Spring Harbor Symp. Quant. Biol.* 33:375–91

36. Altman, S., Lerman, L.S. 1970. Kinetics and intermediates in the intracellular synthesis of bacteriophage T4 deoxyribonucleic acid. *J. Mol. Biol.* 50:235–61
37. Hershey, A.D., Burgi, E. 1956. Genetic significance of the transfer of nucleic acid from parental to offspring phage. *Cold Spring Harbor Symp. Quant. Biol.* 21:91–101
38. Delius, H., Howe, C., Kozinski, A.W. 1971. Structure of the replicating DNA from bacteriophage T4. *Proc. Nat. Acad. Sci. USA* 68:3049–53
39. Howe, C.C., Buckley, P.J., Carlson, K.M., Kozinski, A.W., 1973. Multiple and specific initiation of T4 DNA replication. *J. Virol.* In press
40. Doermann, A.H. Replication of T4 chromosomes lacking terminal redundancies. Unpublished
41. Childs, J.D. 1971. A map of molecular distances between mutations of bacteriophage T4D. *Genetics* 67: 455–68
42. Doermann, A.H., Eiserling, F.A., Boehner, L. 1973. Genetic control of capsid length in bacteriophage T4. I. Isolation and preliminary description of 4 new mutants. *J. Virol.* In press
43. Parma, D.H. 1969. The structure of genomes of individual petit particles of the bacteriophage T4D mutant E920/96/41. *Genetics* 63:247–61
44. Bernstein, H., Bernstein, C. 1973. Circular and branched circular concatenates as intermediates in phage T4 DNA replication. *J. Mol. Biol.* In press
45. Guha, A., Szybalski, W. 1968. Fractionation of the complementary strands of coliphage T4 DNA based on the asymmetric distribution of the poly U and poly U, G binding sites. *Virology* 34:608–16
46. Womack, F.C. 1965. Cross-reactivation differences in bacteriophage T4D. *Virology* 26:758–61
47. Levy, J.N. 1972. *Biological and physical effects of radiophosphorus decay in bacteriophage T4D.* PhD thesis. Univ. Washington, Seattle. 151 pp.
48. Campbell, D.A. The genetics of X-irradiated bacteriophage T4. I. On the recombinant increase and the problem of the recombinational topography. Unpublished
49. Campbell, D.A. The genetics of X-irradiated bacteriophage T4. II. The genetic structure of radiation-induced damages. Unpublished
50. Alberts, B.M., Frey, L. 1970. T4 bacteriophage gene 32: A structural protein in the replication and recombination of DNA. *Nature* 227:1313–18
51. Huberman, J.A., Kornberg, A., Alberts, B.M. 1971. Stimulation of T4 bacteriophage DNA polymerase by the protein product of T4 gene 32. *J. Mol. Biol.* 62:39–52

ASPECTS OF MOLECULAR EVOLUTION

3059

Walter M. Fitch

Department of Physiological Chemistry, University of Wisconsin, Madison, Wisconsin

This article is a critique of the meaning and importance of some of the recently published work in a field whose development as a special area began with papers by Zuckerkandl & Pauling (1) and Pauling & Zuckerkandl (2) entitled *Chemical Paleogenetics, Molecular Restoration Studies of Extinct Forms of Life* and the paper by Margoliash (2a) on the evolution of cytochrome *c*. The primary emphasis is on conclusions and inferences derivable from amino acid sequences, followed by a section on nucleotide sequences. These are preceded by a brief indication of some of the literature that antedates the origin of the genetic code as it now stands.

This will not treat immunoglobulin evolution, reviewed by Gally & Edelman (3) and Smith, Hood & Fitch (4) or related material on the evolution of enzymes and proteins reviewed by Smith (5) and Arnheim (6). The functional interpretation of these changes has made significant progress [Dickerson (7, 8) and Takano et al (9)]. For the evolution of metabolic pathways, see Brew (10), Vogel (10a), and Lejohn (11).

Abiogenic Evolution

Those interested in evolutionary processes after there was life can hardly be disinterested in the preliminaries. Two of the best general sources of this topic are the review of Lemmon, *Chemical Evolution* (12) and the book by Kenyon & Steinman, *Biochemical Predestination* (13). Other general references include Oparin (14), Keosian (15), Calvin (16) and Margulis (17).

A discussion of the origin of our planet, its primitive atmosphere, its geology and extraterrestrial chemistry preceding organic evolution is contained in references 17–21. A number of papers have also appeared on the origin of various organic monomers (22–33), optical isomery (34–36) and the condensation of monomers (37–52).

Origin of the Genetic Code

The book, *The Genetic Code*, by Woese (53) is a good starting point in this area. Two subsequent papers by Crick (54) and Orgel (55) are excellent presentations of an informed view and lead to very readable expositions of the alternative theories on the code's origin by Woese (56). Woese sets out four. The first is the "lethal mutation" theory, first stated by Sonneborn (57) but also independently by Fitch (58). It contends that the code was selected to minimize lethal or deleterious mutations by maximizing the likelihood that a mutation would either not change the amino acid at all, or would change it as little as possible.

The second is the "vocabulary expansion" theory first espoused by Jukes (59) and later by Crick (60). It suggests that originally there were only perhaps 15 amino acids encoded with complete degeneracy in the third nucleotide position of the codon, followed later by the expansion of the number of amino acids encoded by using the third position purines and the pyrimidines to distinguish between the newly introduced and the older, previously encoded amino acid.

The third is the "ambiguity reduction" theory, due I believe to Dr. R. L. Metzenberg (unpublished) and first clearly enunciated in Fitch (58) who quoted him. It suggests that originally groups of codons coded for groups of amino acids and that selection proceeded to differentiate within the set. Thus U in the middle position of the codon might originally have specified indiscriminately any of the amino acids val, leu, ile, met, or phe. Later, the first nucleotide position was used to refine the choices. This theory seems to permit the gradual accumulation of small but beneficial changes from an originally chemical process to an ultimate biological one.

The fourth is the "direct interaction" theory of Woese (61). It suggests that there is a differential affinity between the various amino acids and the various nucleotides such that the final assignment of amino acids to codons was largely deterministic. The determinism of this theory, as opposed to the more stochastic features of the preceding three, gives this theory its special appeal. See Woese (56) for a more complete discussion of these theories.

More recent papers on this subject have appeared (62–65). Lesk (63) expands on the "ambiguity reduction" by coupling it to an expansion of the number of nucleotides used in coding from an original two (adenine, inosine) through various stages to today's four.

Saxinger & Ponnamperuma (64) immobilized a different amino acid on each of nine different polyvinyl-amine gel columns and studied the selectivity coefficients (K) of five 5′-mononucleotides (A, C, G, U, and I). The amino acids, when ranked in order of increasing K, had essentially the same order (gly, lys, pro, met, arg, his, phe, trp, tyr) for all five monunucleotides. They report other similar evidence and see hope for the "direct interaction" theory in such results where, for example, tyrosine has from 5 to 10 times the K value for nucleotides as does glycine. However, where one amino acid is preferred over another regardless of the nucleotide, selection may have little foothold. How is glycine to become associated with a primordial t-RNA anticodon XCC if all the other amino acids bind to cytidylic acid better than glycine does? Raszka & Mandel (66), using proton magnetic resonance to study amino acid binding to homopoly A, I, and

U found the same order of binding to the polynucleotides as Saxinger & Ponnamperuma for mononucleotides.

Mackay (67) noted that the frequency of amino acid usage is correlated with the number of codons so as to minimize the information required in the coding. The interpretation is unclear, however. The correlation between amino acid frequency and number of codons may be like the correlation between amino acid frequency and expected relative frequency of codons given only the nucleotide composition of the DNA as calculated by King & Jukes (68). King (69) pointed out that the correlation could occur, even in the presence of selection, given a reasonable amount of randomness. For example, if threonine and leucine are both improvements in a position of a functioning protein, the fact that there are 50% more leucine than threonine codons means that the leucine improvement will be found first 50% more often. This could lead in time to the correlations observed and would not imply that evolution found it advantageous to minimize the information required for encoding proteins.

The Transition to the Living State

The problem in this context is not "How does one draw a line to divide a continuum into living and nonliving?" but rather "How can we provide a model consistent with thermodynamics and general scientific laws that spans the gap between stochastic organic chemistry and an intergrated hierarchy of well regulated processes characteristic of living cells?" This area has much activity (70–75). The first reference, *Mathematical Challenges to Neodarwinism* seems not to be saying that our theories are inadequate (as we know), but rather to be misinterpreting what we have said so as to conclude that life is impossible. Waddington (71) on the other hand, explores much of the ground with a constructive attitude. Eigen (73) provides a thermodynamic, information-theoretic basis for the development of continuously more complex, regulated, self-reproducing processes. He has stated so much so well that this work will be the starting point for much useful future work. Theodoris & Stark (75) suggest that polynucleotides preceded polypeptides because it is more important to accumulate information than it is to reproduce.

Detecting Similarity

Investigators frequently compare two proteins to see how similar they are. If the similarity appears significant, it is usually inferred that the two proteins had a common ancestral gene. Two proteins possessing a common ancestor are properly called homologous and the proteins are divergently related. Significant similarity could also be produced by convergent evolution, in which case the proteins are analogous rather than homologous. The usage of homology to mean similarity is confusing and should be avoided. Only a few sets of molecules, fungal and metazoan cytochromes (76), and alpha and beta hemoglobins (77), have been proven homologous and none analogous. Nevertheless where significant similarity is observed, the presumption of homology is strong.

COMPOSITION TESTS So many interesting proteins are known only by composition that a good method for discovering significant similarity from composition would be useful. The limitations of these methods, illustrations of which are given below, inhere in the data and no method can be expected to improve greatly upon them. The earliest method is due to Metzger et al (78). Their difference index is the sum of the differences between the mole fraction of each amino acid in the two proteins. This is multiplied by 50 to get a range that goes from 0→100, zero indicating complete identity of composition, 100 indicating no amino acids in common. For 630 comparisons, their index had a mean of 25 ± 8 (standard deviation). Unfortunately, the pairs, ATP creatine transphosphorylase-prothrombin, DNAse-Carbonic anhydrase, and hyaluronidase-leucine aminopeptidase, all have indices less than 10 whereas lysozyme-alpha lactalbumen, which are significantly similar in sequence, has an index of 22.6.

A statistical improvement on the method has been introduced by Marchalonis & Weltman (79) who, in effect, sum the squares of the differences expressed as mole percent. Their index has a mean (820 pairs tested) of 300 ± 120 but the curve is skewed and only 2% of the values are less than 100. The improvement is modest. Alpha and beta human hemoglobins are barely significantly similar (index = 95) whereas shark immunoglobulin has a lower index with each of the proteins bovine pepsin, lobster glyceraldehyde 3-phosphate dehydrogenase, and tobacco mosaic virus coat protein. As the authors suggest, the procedure may be a good screening device, but inferences of homology are hazardous. Harris et al (79a) introduced the deviation function which is the square root of this index.

The procedure of Shapiro (80) is better designed to prove that two cytochromes are dissimilar than that hemoglobin and myoglobin are similar. Moreover, differences can always be made significant. If cow and chicken cytochromes give too low a chi-square value, 2 cows and 2 chickens will double chi-square without any increase in the degrees of freedom. If one objects that the sample size was doubled by examining the same datum twice, then substitute one sheep and turkey for one cow and chicken and get, since their sequences are identical, the same result. The problem is that the samples are not independent.

SEQUENCE TESTS Tests for similarity that depend upon a knowledge of the amino acid sequence have proliferated (81–95). They are sufficiently numerous, and relatively unevaluated, that their listing with a brief description and evaluation is called for. The first systematic procedure was that of Fitch (81), who estimates the minimum number of nucleotide replacements (mutation distance) that must have occurred since any presumptive common ancestor. To minimize the problem of unsuspected partial internal deletions and insertions (commonly called gaps because of their effect on sequence alignments) he examines sequences [called spans by McLachlan (87)] considerably shorter (e.g., 30) than the total lengths of the protein. All possible combinations of sequence pairs, one from each protein, are examined. This involves comparing the ith amino acid from one protein to the jth amino acid of the other, *i-j* being called the register shift. The two amino acids serve as the basis of entry to a table of mutation values. Fitch uses values representing the minimum number of nucleotides that must be changed to convert a codon for one amino acid to that of the other. Thus his mutation values run from 0 to 3. These values are summed over the sequence examined and become the statistic whose probability needs evaluation. Knowing the amino acid composition of the proteins, one can find p_i, the probability the two randomly selected amino acids will have codons different by *i* or more nucleotides. The probability of any given result can be obtained from the multinomial distribution. In Fitch's method, the summed values are a mutation distance and low values suggest possible homology. A subsequent paper (82) improved the procedure and gave a method for calculating the overall probability *P* that the result could occur by chance. It is important to distinguish between the probability that a specific pair of subsequences would have a specified mutation distance if one had made a single random sampling and the probability that the overall result would have occurred by chance. The former fails to take into account the thousands of pairs of subsequences examined in order to find the one that is reported. Other investigators use other values and many arrange them such that high sums suggest homology.

Manwell (83) introduced 3 methods. His first counts identities, thus focusing on the degree of similarity, which should be high in homologous proteins. Thus, by looking only at p_{ident}, (= Fitch's p_0 with $1 - p = p_1 + p_2 + p_3$), Manwell's estimate simplifies to a binomial rather than a multinomial equation. He uses, however, a Poisson distribution. Unfortunately, neither suffices by itself to give an overall *P* value. His second method compares amino acids according to three polarity groupings. Computational procedures are not provided. The third is a variant of mine but without quantitation. The work suffers from several defects: (*a*), the first two methods are less sensitive than Fitch's; (*b*), the last two methods are not quantitated; (*c*), they all depend upon unwarranted assumptions, the worst of which is that the introduction of a few gaps to align the sequences is not critical; and (*d*), alignment of the sequences before the test is tantamount to assuming the similarity one is testing for. It is perhaps not surprising that Manwell found RNAse and lysozyme to be homologous. Those not too critical gaps can be introduced in $129!/(124!\,5!) = 2.75 \times 10^8$ different ways. Haber & Koshland (84) used computer searching procedures to show that the introduced gaps would probably account for the results. My own procedures showed no similarity (unpublished).

Sackin (85) gives three methods. For human alpha hemoglobin vs whale myoglobin, his best method gives a $P > 0.01$ whereas Fitch (82) using human beta rather than alpha hemoglobin got a $P < 10^{-6}$. The difference in sensitivity is even more pronounced for the lysozyme-lactalbumin comparison.

Haber & Koshland (84) introduced two methods. The first, the identity comparison, was in principle the same as Manwell's. The second method divides the amino acids into comparison sets numbering anywhere from 9 to 14. An amino acid may belong to more than one comparison set and does so on the basis of a presumed relatedness of the set members. Thus chemically closer but genetically more distant amino acids may be members of the same set (e.g., phe-trp) while genetically close but chemically distant ones may not (e.g., gly-glu). The procedure is like the identity comparison (mutation values = 0,1) except the question asked is not are the paired amino acids alike but are they members of the same set? Procedures such as this, including weighting schemes such as that proposed by Sneath (96), are potentially more sensitive in discovering structural similarities than other genetic code oriented methods. On the other hand, to the extent that nature may be able to create reasonable approximations of the same structures from genetically unrelated origins, and where one wishes to detect real homology (i.e., common ancestry) rather than analogy (convergence), such methods should be avoided. Haber & Koshland give no overall *P* values for known protein pairs by which to judge their method's relative effectiveness. With a method that emphasizes structure (including alpha helices?) rather than genes, it may not be surprising that they observed other features more prominent than the suggested, partial, internal gene duplication of hemoglobin (97). Their method is, however, better than Manwell's and they cannot detect the RNAse-lysozyme similarity he claimed (83). One peculiar discrepancy between Haber & Koshland (84) and Fitch (82) appears. In a collection of 200 random, independent pairs of protein sequences examined by Fitch, the result most deviant from expectation (cf 82, Fig. 5) was quite moderate compared to two results in only ten pairs examined by Haber & Koshland (84).

Gibbs & McIntyre (86) introduced two methods. The first is a "runs" index that considers the number of consecutive, matching amino acids in two complete sequences. It does not appear to be very sensitive in picking out distantly related proteins and the authors give mostly results for a "diagonals" index. They compute the total number of identities along the complete alignment for each possible register shift (change of register shift ≡ change of diagonal). This method also lacks sensitivity in that the lysozyme-lactalbumin comparison gave an overall $P = 2.5 \times 10^{-4}$. Examining various subsequence lengths and altering other parameters, the Fitch method (unpublished) gets no $P > 10^{-15}$. Lack of sensitivity may explain their failure to observe the two similar portions of hemoglobins that Fitch did (97).

The procedures to this point have been largely unweighted in that two amino acids were alike (or similar in Haber & Koshland's comparison sets) or not. The statistic then was the (sometimes) modified sum of zero-one answers. Fitch's method was weighted according to the genetic code. The simplicity previously obtained by asking zero-one questions was achieved at the cost of sensitivity. In effect, these methods throw away useful information. It is useful despite Juke's assertion (98) that "Because of the degeneracy of the code, base difference comparison is a much less precise method of comparing polypeptide sequences than is amino acid difference." Not all amino acid replacements involving codons one nucleotide different occur with equal frequency. While isoleucine-valine replacements are plentiful, I know of no arginine-cysteine replacements that have been fixed in the course of evolution. McLachlan (87), using a table of amino acid replacement frequencies, designed a weighting scheme such that each amino acid pair had a "value" ranging from 0–9. It thus has the character of

Fitch's mutation value table (99) except that (*a*), nine rather than zero means identity; (*b*), the expanded range of values permits a potentially greater sensitivity; and (*c*), the utilization of observed substitutional frequencies to assign the values should achieve that potential. There is one other difference. McLachlin also uses a weighting scheme on the position value of the amino acid pair in the length of sequence examined, or "span" in her terminology. For example the pairs on the ends of the span are given only 1/3 or 1/5 the weight of the middle pair. No evidence is provided that this is beneficial and it is not clear to me why it should be. The overall benefits of these differences seem to be suggested by the fact that her overall *P* for alpha vs beta hemoglobins is between 10^{-30} and 10^{-38}. While this is very good it should be kept in mind that they were obtained using the entire length *after* inserting gaps to maximize the homology. Fitch (unpublished) gets 10^{-22} for a length examined (span) of only 11 (not 148) amino acids if the register shift is restricted so that not more than 5% of the alpha chain is unpaired. Moreover, that was without gaps. Inserting gaps reduced the probability more than six orders of magnitude.

The procedures given to this point all make a compromise between examining long sequences to maximize sensitivity and short lengths to avoid the disruption of gaps. If significant homology is found, then the gaps can be located (100). It would be better to avoid the compromise in the first place, and Needleman & Wunsch (88) have devised a method that overcomes that problem. It is based upon a matrix whose coordinates are the amino acid positions of the two sequences being examined and whose cells are filled with whatever mutation values one is using for the amino acid pairs so defined. If smaller values are assigned to better matches, one's object is to find, subject to some constraints, a path through the matrix such that the sum of the cells' values is a minimum. The path along a diagonal corresponds to successive amino acid pairs in the sequences. To change diagonals in the path is equivalent to introducing a gap for which a penalty of arbitrary amount is added. The beauty of the method, apart from its ability to examine the entire sequence at once, is the elegant and simple method of finding the optimal path. Its major drawback is the need for numerous Monte Carlo trials for each new pair of proteins examined in order to determine the improbability *P* of the result occurring by chance. Using mutation values equivalent to 0, 1, 2, and 3 with a gap penalty of 3.09, they compared beta hemoglobin and myoglobin and got an overall *P* of $< 10^{-16}$. I know of no other method this sensitive. Despite this sensitivity, they found no evidence of the RNAse-lysozyme similarity claimed by Manwell (83). Dayhoff (89) claims that the method of Needleman & Wunsch can be made still more sensitive by using the mutation values in her figure 10–1, and a gap penalty equal to the value assigned to the pair of amino acids (cysteine-tryptophan) least commonly substituting for each other. Sankoff (90) has criticized the method but seems not to have understood Needleman & Wunsch's constraint that successive cells in the path through the matrix should have coordinates (g,h) and (i,j) such that either $|g - i| = 1$ and/or $|h - j| = 1$.

In a novel information theoretic approach to the same problem, Reichert (91) framed the question as "What is the minimum information required to transform one of two sequences into the other?" The answer is then compared with the results for random sequences of the same composition. Three types of information are specified and summed: (*a*), the location (e.g., the 38th nucleotide); (*b*), the process (e.g., insertion, deletion, replacement); and (*c*), the nucleotide (A,C,G,U). The total information $I \simeq 2N + 4.32G + F(\ell)$ where N is the number of nucleotides replaced, G is the number of gaps and $F(\ell)$ is a function of the various locations. The result seems particularly sensitive to the F term, which has the least meaning in any intuitive biological sense, particularly when the mutational events in time are not ordered in accordance to their position in the sequence. Alternative assumptions could subordinate the effect of $F(\ell)$. The parameters depend upon the relative probability one assumes for various events. If $F(\ell)$ is reduced essentially to a constant, the effective equation is reduced to two terms that are equivalent to the Needleman & Wunsch computation. Indeed, if one were to assume that a deletion/insertion mutation occurred only 0.05517 as often as a nucleotide replacement, the equation would become $I \simeq 2N + 6.18G$ plus a constant which, except for the constant, is essentially twice the Needleman & Wunsch statistic. Reichert's method thus contains the virtues of the Needleman & Wunsch method while providing, at least potentially, a powerful information theoretic base.

Bauer, in a series of publications (92–95 and others) proposes not only that histones are derived from a small ancestral polypeptide but that from that *same* ancestral polypeptide there clearly derived the proteins kallidin, ACTH, glucagon, insulin, RNAse, phospholipase, cytochromes, F_2 phage coat protein, immunoglobulins, ferredoxin, haptoglobin, fibrinopeptides A and B, and chymotrypsinogen. The errors are numerous. A string of 11 consecutive matching nucleotide pairs is asserted to have the probability of occurring by chance of $(0.25)^{11}$. This ignores the fact that thousands of strings have been examined to find the one presented and that in many of those positions the probability of asserting a match is 1.00 (not 0.25) since the fourfold ambiguity of the third nucleotide position of codons always permits a match. Neither does he take into account the effect of gaps which he freely inserts. His so-called pseudo-symmetrical sequences suffer similar problems.

Recombination

HOMOLOGOUS AND NONHOMOLOGOUS RECOMBINATION Biologists have been using the word nonhomologous with two meanings that are conflicting when used at the level of the gene. Homology has been used by biologists since the time of Darwin to denote common ancestry and nonhomology the lack of common ancestry. Geneticists have, since the time of Bridges (108), used nonhomology to indicate a misalignment of chromosomes. The result is that beta and delta hemoglobins are homologous by ancestry but nonhomologous by alignment and hemoglobin Lepore can be the result of either a homologous or a nonhomologous crossing-over, depending on the author's view. I propose some clarifying nomenclature that preserves as much as possible the usage of the past and that has been found useful in phylogenetic studies where homology is an inadequate criterion of suitability for constructing trees. It is vital to draw a distinction between the same homologous locus in two different taxa (e.g. alpha hemoglobin) and different homologous loci in the same taxon (e.g. alpha and beta hemoglobin). The former is called orthologous from the exact correspondence between the ancestry of the gene and the taxa while the latter is called paralogous from the parallel nature of genes' history following the gene duplication event (6, 110, 111). Thus depending upon the gene loci involved at synapsis, crossing-over may be either orthologous, paralogous, or nonhomologous. Even though the genes involved are homologous, the codon alignment need not be, as in the case of partial internal gene duplication. To avoid confusion with the terminology relating to the genes, I suggest conjugate, in its sense of having a common derivation, and non-conjugate to describe the codon alignment. Examples of each kind of recombinant are discussed below.

NONHOMOLOGOUS CROSSING-OVER (*a*) *Gene fusions.* Tryptophan synthetase is composed of two products from contiguous genes in *E. coli* (105) and other bacteria but is composed of a single gene product with two functional regions in fungi (106, 107), suggestive of gene fusion. In the same operon, in *Salmonella typhimurium*, Grieshaber & Bauerle (104) presented evidence that the *trpB* gene product has two catalytic functions, a glutamine amidotransferase and an anthranilate phosphoribosyl transferase activity on different parts of the enzyme. A similar result has been observed for the aspartokinase-homoserine dehydrogenase activities in *E. coli* (103). For these cases, gene-fusion is not the only possible explanation. (*b*) *Gene Duplication.* The earliest suggestion of gene duplication of which I am aware is for the bar locus of *Drosophila* of Bridges (108) in 1936 and, for proteins whose sequence gave evidence of this, for hemoglobins by Ingram (109) in 1961. An alternative explanation of sequence similarities could be convergent evolution. Fitch (110) has provided a method which allows one to distinguish common ancestry from convergence. In effect, it asks whether the ancestral alpha and beta hemoglobin genes were more alike at some past time than they are today. This requires an objective method for reconstructing the ancestral sequences, which has also been provided by Fitch (76) and will be discussed further on. Comparing an alpha chain ancestral to carp and higher vertebrates and a beta chain ancestral to frog and higher vertebrates, Langley & Fitch (77) found that the ancestral sequences were more alike than would be expected, their nucleotide difference being nearly eleven standard deviations below the mean expectation. Convergent proteins would have given values above the mean. The probability of the result being so far below the mean is $< 10^{-15}$ and so we may safely assert that the alpha and beta hemoglobins are the result of gene duplication provided the two genes are, within themselves, orthologous genes.

No other pairs of proteins have been proven to have arisen by gene duplication, although the significant similarity in sequence leads to the presumption of that event for the serine proteases (112, 113), lysozyme-lactalbumen (114), prolactin-growth hormone (115), corticotropin-melanotropin-lipotropin (116, 117). This collection of sequences in an optimal alignment is found in Dayhoff (89). The beta chain of human haptoglobin is similar to the serine proteases (118) as well as prothrombinase (119).

In perhaps one of the most enterprising ventures into gene duplication, Adelson (120) has attempted the correlation of the embryological origin of the protein secreting glands (generally entodermal) and the sequences of their proteins. The evidence for the wide range of relationships proposed is sufficient to warrant further study of this interesting matter. Strydom (120a) has since found a strong relationship between mammalian trypsin inhibitors and two snake toxins.

The initial result of gene duplication is a pair of tandem genes. Tandem alpha hemoglobin genes have been asserted for mouse (121). This has now been asserted for humans as well (122) where Ostertag, Von Ehrenstein & Charache have evidence for *two* alpha variants in addition to the normal chain in a single heterozygote. The evidence, however is adduced from the composition of the mixed peptide and awaits confirmation by sequencing. The observation (122a) that homozygotes for hemoglobin Tongariki make no normal A/A_2 makes two alpha loci most unlikely. Harris, Wilson & Huisman (123) make a similar claim for the Virginia white-tailed deer where three alpha chains were observed in a single animal. Four nonallelic structural genes have been proposed for human gamma hemoglobin (124). The argument depends upon an analog of Dalton's law of definite and multiple combining proportions, which explains how different hemoglobin F variants are present in amounts approximating 5.5, 11, and 22% of the total F. Unfortunately they must also postulate differential activity at the various loci, which introduces sufficient flexibility to explain most results in more than one way. The evidence for two or more loci does appear strong, nevertheless.

CONJUGATE PARALOGOUS CROSSING-OVER Hemoglobin Lepore (Boston) was shown by Baglioni (125) to be identical to the usual delta hemoglobin for the first 87 residues and like beta hemoglobin for the last 32 residues. The obvious conclusion was that Lepore was recombinant. The fact that the amino terminus is like the delta chain does not prove that the delta gene lies to the left of the beta gene. Only the knowledge that a person homozygous for Lepore has no beta or delta hemoglobin at all (126) permits that conclusion.

If Lepore is a conjugate paralogous cross-over between the delta and beta hemoglobin genes that leaves no unaltered beta and delta genes, then there should also be a complementary cross-over set in which unaltered delta and beta genes exist on either side of the mixed gene for which the product contains the beta hemoglobin portion on the amino terminus rather than the carboxyl terminus and vice versa for the delta portion. This has been termed anti-Lepore and two such gene products have now been found (127–129). The beta and delta hemoglobins are present as required, but that can as well be explained by the propositi being heterozygous.

CONJUGATE ORTHOLOGOUS CROSSING-OVER This type of recombination can only be observed when examining amino acid sequences if there are two alleles differing in at least two amino acids. Galizzi (130) crossed two rabbits whose beta hemoglobin genotypes were HSN/HSN and HSN/NNS where H, S, and N stand for the amino acids histidine, serine, and asparagine respectively and the three letter positions are residues 52, 56, and 76 respectively. In a single litter of ten, six had the paternal genotypes HSN/HSN, two had the maternal HSN/NNS and two were apparently different. Data are not given that would tell us whether the variants result from somatic or germ line changes. One variant has the phenotype HSN/HSS. The HSS product could result from a simple cross-over in the maternal line. The other variant has the phenotype NNS/HSS. This variant is the same product as before but since its allele is maternally derived, the previous explanation cannot apply. Galizzi concludes, "Somatic cross-over between the two polymorphic alleles is the most plausible single mechanism for the observed rearrangements." I would suggest that two recombinant events in ten offspring is too improbable to be believed. A simpler explanation, requiring only a single nucleotide replacement, is that one of the male germ cells suffered an A→G replacement in the second position of the asparagine 76 codon at a time when only a few germ cells had yet developed, thus leading to a "mosaic" set of sperm in which ~20% carried the HSS gene. It is unfortunate that the mother was not homozygous NNS and that F_2 generation studies were not performed.

There is an example from *E. coli* of conjugate orthologous crossing-over. Yanofsky (131) reported two mutants of tryptophan synthetase in which the wild type glycine had been replaced by an arginine and a glutamate respectively. When the two mutants were crossed, the wild type was recovered as a recombinant, i.e., GAR(x)CGX → GGX.

NON-CONJUGATE ORTHOLOGOUS CROSSING-OVER The effect of this form of recombination is to lengthen or shorten the gene and its product. Its occurrence in the course of evolution is exemplified by the three gaps in human alpha hemoglobin and the one in beta that are required for their optimal alignment (cf. Figure 1).[1] Its occurrence in contemporary times is illustrated by hemoglobin Gun Hill (102).

There seem to be three distinguishable classes that result in these partial deletions and insertions. One is apparently facilitated by the complementary base pairing of nonconjugate regions. The suggestion is most obvious in hemoglobin Gun Hill (cf. Table 1, I 3) where the deletion of five consecutive

[1] The hemoglobin events not accounted for include the lamprey glutamate between positions 83 and 84, the carp methionine between positions 70 and 71 and the missing alpha-48 residue.

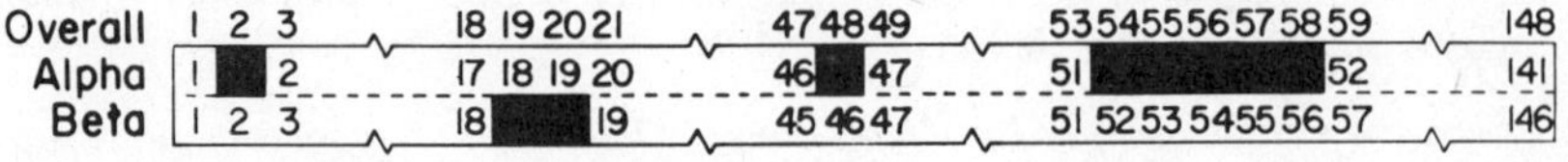

Overall	1 2 3	18 19 20 21	47 48 49	53 54 55 56 57 58 59	148
Alpha	1 2	17 18 19 20	46 47	51 52	141
Beta	1 2 3	18 19	45 46 47	51 52 53 54 55 56 57	146

Figure 1 Hemoglobin Alignment. The common hemoglobins of amniotes contain 141 (alpha) and 146 (beta) amino acids: Any attempt to maximize the correspondence between the two sequences will require the introduction of 4 gaps comprising 9 residues as shown, giving a total of 148 positions overall. So that conjugate positions will have the same number in both chains, and since residues deleted also need a number, all discussion in the text is in terms of the overall number and the figure may be used to obtain the more common numbering for any individual chain except that bovid beta hemoglobins are also missing residue 2.

Table 1 Insertion/Deletion Mechanisms

I. Simple Recombination (codons between base-pairing regions lost)	
1. Human Hb β^{6} deletion	CCX-GAR-GAR pro *glu* glu
2. Human Hb β^{45-47} deletion	UUY-UUY-GAR-UCX-UUY-GGX phe phe *glu* *ser* *phe* gly
3. Human Hb β^{92-96} deletion	GAR-CUX-CAY-UGY-GAY-AAR-CUX-CAY glu leu *his* *cys* *asp* *lys* *leu* his
4. Proinsulin pig 48-49 deletion	GGX-GGX-CCX-GGX-CUR gly gly *pro* *gly* leu
5. Proinsulin cow 53-57 deletion	GGY-CUX-CAR-GCX-CUX-GCX-CUX gly *leu* *gln* *ala* *leu* *ala* leu
6. Proinsulin dog 34-41 deletion	CGX-GAR-GUX-GAR-GAY-CCX-CAR-GUX-GGX-GAR-GUX-GAR arg glu *val* *glu* *asp* *pro* *gln* *val* *gly* *gly* val glu
II. Complex Recombination (codon of first amino acid lost, incompletely base-paired)	
1. Human Hb β^{25} deletion	GUX-GGX-GGX-GCX *val* gly gly ala
2. Human Hb β^{58-61} deletion	GGX-AAY-CCX-AAR-GUX-AAR *gly* *asn* *pro* *lys* val lys
3. Ancestral deletion Hb β^{19-20}	UAY-GCY-AAY-GCH *tyr* *ala* asn ala
4. Ancestral deletion Hb α^{54-58}	GAY-GCR-CUH-AUG-AAY-AAY-GCX-AAR *asp* *ala* *leu* *met* *asn* asn ala lys
III. No apparent base-pairing potential	
1. Ancestral bovidae deletion Hb β^{2}	GUG-CAY-CUX val *his* leu

amino acids in the beta sequence (107–113), leu-his-cys-asp-lys-leu-his) could arise from the mispairing of the two leu-his coding regions. Altogether, six such cases are shown. Moreover, the first three, since they occur in the current population, cannot have their potential base pairing regions arising from faulty reconstruction of ancestral sequences. A cross-over anywhere in the base-paired region could eliminate the codons for the amino acids known to be missing. Not shown because the pairing regions are 59 codons apart, is the creation of alpha-2 haptoglobin by crossing-over between alanine codons 12 and 71 of alpha-1

Legend to Table 1. Codon numbering is such that conjugate codons possess the same number. This is the "overall" number given in Figure 1 which shows where the gaps occur in the hemoglobin sequence along with the alpha and beta numbering. This numbering excludes an extra 8 residues on the amino terminus of the lamprey hemoglobins, a lamprey glutamate between residues 83 and 84, and a carp methionine between residues 70 and 71. For proinsulin, the number corresponds to that of horse, rat, monkey, and man which require no gaps to align them to each other and to the listed sequences. The coding used, given in terms of the messenger, is A = adenine, C = cytosine, G = guanine, U = uracil, R = A or G, Y = C or U, X = any, H = not G. The amino acids encoded are given below the message. The horizontal bars above and below the message show regions that could base-pair with the same complement assuming the ambiguous nucleotides are properly specified. The distance between these regions matches precisely the number of codons lost. Amino acids lost are shown in script-type and their codon numbers correspond to those given in the description to the left. Where the deletion occurred in some ancestral species, the sequence used was that of the ancestor preceding the deletion event. These were determined by the method of Fitch (76). Where alternative ancestral amino acids were possible without requiring additional mutations, an arbitrary choice was made and where that choice would affect the pairing potential it is given below: In I6, valine-34 could also be alanine; valine occurs here in rat and cow. In II3, tyrosine-19 could also be cysteine; tyrosine occurs here in the lamprey, serine in the carp alpha. Alanine-20 is a transition state between the glycine in most alphas and the carp proline. In II4, alanine-55 could be glutamate (one lamprey); alanine is the most common beta residue. Leucine-56 (lamprey) could also be i-leucine or valine (betas). Asparagine-59 (all betas) could also be serine (lamprey and alphas). Lysine-61 (all betas) could also be glutamine (all alphas). In IIIl, histidine-2 (most species) could also be asx (llama). In group I, the specification of the amino acids removed is arbitrary in that in every case the same final sequence will be obtained after the deletion by shifting the group one or more residues so long as the residue admitted to the group on one end matches the residue dropped on the other. References are as follows: I1, Hemoglobin Leiden (241); I2, Hemoglobin Rio de Janeiro (141); I3, Hemoglobin Gun Hill (102); Proinsulin sequences pig (242), cow (243), dog (244); III, Hemoglobin Freiburg (245); II2, Hemoglobin Tochigi (246); Hemoglobin sequences used in ancestor formation are referenced in Langley & Fitch (77).

haptoglobin (132, 133). This recombination produced an insertion of 59 residues and occurred between distinguishable alleles in a heterozygote.

The second class has a reasonable suggestion of base-pairing but a cross-over in the pairing region would miss by one or two nucleotides of removing the codons for the amino acids known to be missing. Perhaps in the region of the cross-over, the fidelity of replication may be reduced. It is curious that in all four cases, it is only the left-most amino acid whose absence cannot be accounted for. The ancestral deletions could fall in this class due to errors in reconstruction. It is only necessary to alter one nucleotide in the ancestral sequence to permit the transfer of II 3 and II 4 to class I. But while a single nucleotide change could do the same for the first two cases, there is no ancestral reconstruction error to fall back upon. These are extant hemoglobins.

The third class shows no potential for base pairing in the region of the eliminated codon and therefore suggests a completely different mechanism. Only one case is shown because the other hemoglobin instances involve ancestors probably so remote from the insertion/deletion event that intervening mutational events could account for the apparent absence of base-pairing regions. In the case shown, the removal of beta-2 is the reverse of the insertion that presumably occurred very early in the evolution of the beta chain, and the presumed ancestral sequence involves the same valine in position 1 and i-leucine in position 3 in either case.

Where the data are best, it is clear that the possible base sequence identity is most frequently present, so that one feels that it is commonly involved in insertion/deletion processes notwithstanding the fact that the potential pairing region is often only three to five nucleotides long.

UNCLASSIFIED CROSSING-OVER For the most part, insertions and deletions occurring at the amino or carboxyl termini of sequences seem devoid of any easy rationale. Among the hemoglobins, frog beta is missing residues 1→6 (134), sheep and goat beta C are missing 1→5 (135, 136), lamprey has 8 additional residues on the amino terminus (137, 138), and, on the carboxyl terminus, myoglobin has an extra 6 residues (139), a human alpha variant (Constant Spring) has an extra 31 (140), and a human beta variant (Tak) has an extra 10 (141). Of these, hemoglobin Constant Spring is the only variant with a simple explanation, viz. the normal terminator codon was UAR (R = purine) which suffered a point mutation changing it to the glutamine codon CAR and the message was read on until a terminator was reached 31 codons farther or until there was no further nucleotide sequence.

Hemoglobin Tak's first additional amino acid is threonine, which is at least two nucleotides different from any terminator. In an attempt to look for recombination, the Tak sequence was examined against the remainder of the beta chain sequence in all possible reading frames (Fitch, unpublished). There is no alignment that has fewer than 5 mismatched nucleotides and that assumed an optimal arrangement of three amino acids whose order is not yet known. Thus it would seem clear that the Tak addition does not arise from within the beta gene.

Myoglobin is paralogous to the hemoglobins and extends beyond position 148 an additional six amino acids. No pair of the additional residues (myoglobin, Tak, Constant Spring) shows any apparent homology in this region.

There are two interesting imbalances in the data examined. Out of 9 instances of nonterminal insertions or deletions discussed above in the alpha and beta hemoglobins, only one occurred in the alpha chain and none of the five known human variants occurs in the alpha chain. This strongly reinforces the conclusions of much other data, to be discussed later, that the alpha chain is less tolerant than beta of changes in its structure.

The other imbalance is that out of 13 nonterminal insertions and deletions discussed, only two are insertions. I cannot claim that the sample is random since I deliberately restricted my examples to haptoglobin (historically first), hemoglobin, and proinsulin (numerous and clear). Nevertheless, there appears to be a bias against observing insertions as opposed to deletions as modifiers of chain lengths.

Three potential cases of partial gene duplication (insertion) deserve special consideration. The first is the similarity between two hemoglobin regions 66 amino acids apart beginning at overall residue 59 (beta 57, alpha 52), extending 24 amino acids, and showing similarity to the terminal 24 amino acids (97). These portions are, basically, the E and H helices. This similarity is particularly pronounced in the beta sequence. We have previously alluded to the fact that others (84, 86) have not found this similarity, or they believed it to be insignificant but were using less sensitive methods. McLachlan (142) has more recently made a comprehensive search for such internal duplications. Using her earlier method (87) but a revised mutation value table, she found no significant internal similarities in hemoglobins, chymotrypsin, carboxypeptidase, pepsin, or histones. She has no difficulty in discovering the E and H helix similarity; it is the significance that is doubted. The relative sensitivities of our two methods are unclear and my original report was at a time when I could not provide overall P values.

There is one test that could assist in deciding the issue. If the similarity we see today is the remnant of an earlier internal duplication, then that vestige should be more pronounced in ancestral sequences. To that end, I have revised (unpublished) my original protein sequence comparison method (81, 82) to handle nucleotide sequences rather than amino acid sequences, with the proviso that register shifts are in multiples of three and that probabilities of a match at each of the three nucleotide positions of a codon must be computed independently. This leads to the trivial change of computing the sum of three binomials rather than computing one tetranomial. I then tested human beta hemoglobin messenger using the fourfold degenerate codons (CGX, CUX, and UCX) for the three amino acids (arg, leu, and ser) with six codons only when there was no other information to assist in the decision. I also used my ancestor program (110) to reconstruct an ancestral alpha-beta hemoglobin message. The results of testing these two messengers for internal similarity gave an overall P value for present-day beta message of 0.04 and for the alpha-beta ancestral message of 0.17. Since this is a larger P value, I conclude that there is little reason to believe that the "suggested" (that was the word in the original title) partial duplication in fact occurred. Of course the homology between hemoglobin and myoglobin means that such a partial duplication would have preceded the gene duplication that led to independent myoglobin and hemoglobin genes. There is thus the remote possibility we still have not reconstructed a gene sequence sufficiently ancestral to detect the original event. It would thus appear that the structural constraints involved in the E & H helices were sufficiently similar and strong that we have one of the few cases of (not quite, yet) demonstrable convergence. In my judgment, the evidence of McLachlan against partial gene duplication is, for all the cases mentioned above, superior to the evidence for the duplication.

The second case is bacterial ferredoxin for which the significant similarity was independently observed by several (97, 143–145). It may be time to shed a modicum of doubt upon this partial gene duplication. If one compares the presumed homologous halves of *Clostridium butyricum* ferredoxin in the first two nucleotides only of the codons, one discovers that they differ by 17 nucleotides and random sequences would be expected to differ by 38.4± 3.2. The observation is therefore 6.75 standard deviations from the mean for a $p \simeq 10^{-11}$. If one adds in two more *Clostridia*, a *Micrococcus*, a *Chromatium*, and a *Desulfovibrio*, the ancestral ferredoxin halves, with all the ambiguity resulting from uncertainties in reconstruction, still are at least 14 nucleotides different while expected to be 27.3±3.6. This is only 4.5 standard deviations from the mean for a $p \simeq 3 \times 10^{-6}$. In other words, the ancestral forms do not seem to be getting more similar. If the certainty of this last duplication is in doubt, then Eck & Dayhoff's conjecture (144) that ferredoxin arose from a primordial tetrapeptide is without evidentiary support. Indeed Fitch (146) earlier pointed out that 29 amino acids chosen at random from a pool with the composition of ferredoxin could generally be converted into ferredoxin in fewer genetic events than it takes to transform Eck & Dayhoff's tetrapeptide into ferredoxin.

In a third case, Engel (147) presents "the first evidence bearing upon the evolution of regulatory sites from ancestral catalytic sites." The enzyme is bovine liver glutamate dehydrogenase. The optimum alignment of the 50 residues asserted to show homology requires 71 nucleotide replacements to account for their coding difference. Although at 1.42 replacements per codon, that is barely better than chance, Engel asserts that the 12 identical residues paired would occur with a $p \simeq 4 \times 10^{-6}$. There are at least five errors in the computation: (*a*), the paired sequences of 50 amino acids were not discovered by the mechanism for which the overall probability P was calculated, viz., nonoverlapping pairs of sequences; (*b*), the probability for a single identity in a pair of amino acids chosen at random assumes all amino acids are equally frequent, which gives a spuriously low value of p_0; (*c*), the overall P should be for 12 or more identities, not 12, since you wish to know the probability of an event as rare as the one observed, not just of a subset of such rare events which necessarily has a lower P; (*d*), there are at least 660 ways (not 352) of pairing two subsequences each 50 long such that no two such pairs at the same register shift overlap; and (*e*), corrected for the error in (*a*), the last number would be still larger. The evidence here is inadequate to the conclusion.

Clupeine Z has been asserted to be a probable nonconjugate orthologous cross-over between YI and YII alleles (146). There are two kinds of evidence. The first is that if one were to assume that all three descended according to the usual bifurcating phylogeny and the minimally necessary mutations entered thereon, the leg descending to clupeine Z would possess no mutations. The second is that all the mutations by which clupeine YI differs from Z lie to the right of all those by which YII differs from Z. The joint probability of these two events was calculated at $< 3 \times 10^{-3}$. The same computation for hemoglobin Lepore (Boston or Baltimore) gives a $P = 3.5 \times 10^{-3}$. Thus the clupeine cross-over would be on firmer ground than the Lepore if one had to depend upon this kind of evidence rather than genetic analysis. Even if this is a cross-over, we cannot be sure YI and YII are allelic. If they belong to separate loci, this is our first case of nonconjugate paralogous crossing-over.

Clupeine illustrates the hazards of trying to account for the evolution from primordial oligopeptides. Black & Dixon (148) suggested a plausible scheme for clupeine stemming from a pentapeptide. To illustrate how easy the game is, Fitch showed how a hexapeptide could get to the same end products with fewer genetic events including the cross-over. Recently Ycas (149), in a study of periodic proteins, has suggested a third scheme.

FRAMESHIFTS Although Magni & Puglisi (150) and Whitfield, Martin & Ames (151) have presented evidence that a large portion of mutants in yeast and *Salmonella* are frameshifts, and although Inouye et al (152) and Berger, Brammer & Yanofsky (153, 154) have conclusively demonstrated their existence in vitro, there is no evidence of their ability to survive the rigors of selection. A method that it was hoped would detect them has been devised (155), and a possible case at the amino terminus of ferredoxin to explain the difference between the plant and bacterial proteins was investigated. The evidence was dubious, overall $P \sim 0.18$ of occurring by chance. No more likely example has come to light. Perhaps a statistical search of this type is futile in that the region must be fairly long to give statistically warrantable results, in which case the resulting sequence is probably so far from optimal that much rapid evolution ensues immediately thereafter, thus destroying the traces of the original event.

If we are willing to look upon compensatory frameshifts of three nucleotides as frameshifts rather than as two nonconjugate orthologous crossings-over, there is an excellent candidate in the fish alpha hemoglobins (156, 157) in which overall residues 64–70 are his-gly-lys-lys-val-ile-met but only properly relate to the other corresponding alpha residues gly-his-gly-lys-lys-val-ala if we assume the removal of the glycine from the amino-terminus and the insertion of the methionine on the carboxyl-terminus of the second sequence.

Phylogenetic Trees

Sokal & Sneath (158) introduced the terms cladogram and phenogram to distinguish between a diagram depicting the geneological relationships of a set of organisms and one depicting a relationship among a set of character states. This has been of heuristic value and since we are all examining phenotypes, our results are by definition phenograms. Nevertheless biologists hope that, by the proper choice of the characters to be examined, the resulting phenogram will be as close to the cladogram as possible. That hope is expressed in our terming these diagrams phylogenies and phylogenetic trees. Compared to most taxonomic studies, the data of the molecular biologist are sparse. Nevertheless, from the very beginning of the use of defined algorithms on molecular sequences (99), the resulting phylogenies have been in very good agreement with classical biological opinion. Other molecularly based methods have been less satisfactory but they all require some estimate of the "distance" separating any pair of taxa.

DISTANCE MEASUREMENTS All distance measurements express a sense of the degree of differentness between the two taxa. All of them, whether directly molecularly based or not, suffer from lacking a one-to-one correspondence between the number of genetic events underlying the differentness of the taxa and the actual measurements made. Worse still, the error is nonlinear in that the more distant two organisms are, the larger is the fraction of all genetic events that remain unseen. The handicap is not the same for all measures however. The biologist measuring polygenic traits is almost completely removed from the question. At the other end of the spectrum, the sequencer of the genes will be able to count the base differences in the DNA and estimate the total number of genetic events. The molecular evolutionist is currently trying to approach that latter happy state of affairs. Among the measurements dependent upon sequence specificity but not requiring direct knowledge of the sequence are DNA hybridization and immunological cross reactivity. DNA hybridization techniques involve two aspects of importance to this discussion. One is the rate at which thermally dissociated DNA reanneals upon cooling. The curve is complex and indicates that some DNA sequences exist as only one (unique) copy per genome while others (frequently in the so-called satellite DNA) are in multiple (repetitive) copies. Either group may be studied but the unique strands are in some ways more useful for determining genetic distance. The other aspect is the temperature at which reannealed DNA melts. If the reannealed DNA was formed from two closely related species, not all the bases will pair because of mutations fixed since their common ancestor. The result is a lowering of the melting temperature of about 1°C for every 1.5% mismatched pairs (159). Perhaps the most useful introduction to this area is the paper by Kohne (160). Some other papers of interest are (159–170). This group of references is biased toward works on repetitive DNA, the last three being speculative regarding their role. The paper by Britten (169) is a good general discussion. On the basis of the observation that satellite DNA is more plentiful in rapidly speciating taxa, Mazrimas & Hatch (170) suggest that repetitive DNA is useful in recombination enhancement. The distance measurement used for phylogenetic studies here has been percent nucleotide change (160).

Immunological methods are of long standing but their quantitative use for phylogenetic studies as currently employed really derives from Sarich's immunological distance (171), which is based on a microcomplement fixation technique and is defined as 100 log *I* where *I* is the factor by which the antiserum concentration must be raised to bind (fix) as much complement as a standard amount of the original antigen. Sarich & Wilson (172) have shown that immunological distance is proportional to the percent of amino acid differences in the sequence. For several proteins it would appear that each 1% amino acid difference increases the immunological distance by about 5 (173.) Since one is measuring a surface phenomenon, denaturation of the antigen is an important consideration (173a, b). Only surface amino acids are antigenic in native proteins and thus, as molecular weight increases, the fraction of different amino acids overall needed to increase the distance 1 unit should increase with native material. One would also expect that the test would be sensitive to the number of antigenic determinants on the site. This seems not to have been a problem however.

Many distance measures depend on a knowledge of the amino acid sequence itself. They are nearly as common as the tests of similarity in the preceding section. Indeed every such test is in principle a measure of distance and most have been suggested for that use. The only fundamental requirement is that the measure be over the entire sequence. In seeking similar sequences we hope to see distances that are significantly less than unrelated sequences would give. To be useful for building trees, all such pairwise distances must be significantly smaller and, additionally, it must be a reasonable inference that the most closely related pairs give the smallest distances.

TREE BUILDING METHODS Once one has a matrix of distances there are a number of ways that one can manipulate the data to produce trees. The most comprehensive yet readable source about these procedures is the new book by Sneath & Sokal (158). Most phylogenies using amino acid sequences are created by a methodology in principle like that of Fitch & Margoliash (99) even if the distance matrix is not composed of minimal mutation distances. The major exceptions are the procedures related to that given in the *Atlas of Protein Sequences* (89) which does not depend upon a distance matrix but rather adds new sequences into a pre-existing tree structure to form a new branch according to whichever nodal sequence is most similar. It is unusual in that it constructs presumptive ancestral sequences as it proceeds to add each new sequence.

There are three major questions associated with tree construction. The first question is, "How suitable are the data?" Sequence-homology is an insufficient criterion. For example, since the alpha-beta hemoglobin gene duplication antedates the divergence of the mammals by hundreds of millions of years, any attempt to construct a phylogeny using the alpha hemoglobin from man and donkey and the beta hemoglobin of chimpanzee and horse leads to a phylogenetic absurdity. Distances must be based on the comparison of orthologous genes or gene products. The comparison cannot be between paralogous genes or gene products except where one wishes to get the historical relationship of the genes (a gene phylogeny) rather than the history of the taxa (a species phylogeny). The potential dangers are real and great. Immunologic studies on the lysozymes of birds show that the duck is closely related to the chicken but very distantly related to the goose. That being biologically unreasonable, the data could be made to fit the expected phylogeny from which it might be concluded that goose lysozyme had undergone extensive and rapid evolution. In setting out to test this possibility Arnheim & Steller (174) discovered that the black swan had two activities, one of which cross-reacted with anti-chicken lysozyme but not anti-goose lysozyme, the other of which cross-reacted in the reverse relation. This suggests that the goose and chicken lysozymes are nonallelic, that is, they are paralogous, not orthologous. Unfortunately, in the absence of a species with both gene products, there is nothing inherent in the sequences of a single gene product that assures one that he is dealing with orthologous rather than paralogous gene products. The less precise the data base, the greater the chance for this type of error.

The second question is, "How do I recognize the best phylogeny?" One can compare the result to the classical phylogenies, but that is suitable only when testing a method with good data or testing data with a good method. The interesting phylogenetic questions do not lie where such an appeal is possible, and if they did, the method would not be needed. There are two kinds of alternatives. One builds the criterion into the algorithm that creates the tree. We may call this an internal criterion. Cluster analysis (175, 176) successively divides the taxa into two groups such that at every division the sum of squares of the between-group distances is maximized and the within-group distance sum is minimized. Such procedures get the best tree but the standard tends to be more removed from the biological reality one is attempting to discover than are other criteria and also, at least with the cluster analysis criterion, there are grave problems in handling large numbers of taxa.

Before examining external criteria we need to consider another problem, the numbers of possible trees. A tree is said to be rooted if the location of the ultimate ancestor is defined. It is the only node that has two rather than one or three branches extending from it because it is the only node without an ancestor. If the ultimate ancestor is unspecified, the tree is unrooted and two adjacent internal nodes are not distinguishable as ancestor-descendant. The number of possible rooted trees for n taxa is the product of n-1 successive odd numbers (177). For unrooted trees, it is n-2 successive odd numbers, i.e., $T_u = \Pi(2i - 1)$ over $(1 \leq i \leq n - 2)$. The number of legs (internodal intervals) is $2n$-2 and $2n$-1 respectively.

For even moderate numbers of taxa, the number of trees is too large to permit an exhaustive search of them to find the best tree, and most algorithms are designed for the third question, "How do I find the best tree while looking at only a small fraction of the total?" Normally one looks for a tree that should be close to the best [e.g., by always joining next the two nearest taxa (or previously joined groups of taxa)] and then testing all possible rearrangements of a limited class (e.g., the interchange of neighboring branches). These types of procedures require criteria exterior to the tree building process, and there are a number of them. One of them is to assign lengths (distances) to each internodal branch by some appropriate method and then sum them between the pairs of taxa and declare that tree best for which these recomputed distances best match the original distances. One method of doing this was given by Fitch & Margoliash (99). Other statistical approaches can be found in Sneath & Sokal (158).

Another criterion is the maximum parsimony principle which states that the best tree is the one that accounts for the available information with the least amount of overall change. Contrary to the allegations of those who oppose its use, it does not assume that nature always follows the most direct path in changing from one state to another. What it does say is that if two taxa have amino acid A and two others amino acid B, it is a more efficient explanation of the data to suggest that the A-containing taxa are more closely related to each other than either is to the B-containing taxa. Given such a relationship, it assumes that an A-B interchange occurred. It is a maximum likelihood explanation. There is no denial than an A-B change might have occurred twice or that the change went via intermediate C, but their presumption is unwarranted. The original effort in this direction is perhaps that of Wagner (178, 179) and such trees have been called Wagner trees. Further development has been by Farris (180). Fitch (76) gave a simple algorithm that, for any given phylogenetic tree and the orthologous nucleotide sequences of the present-day taxa, determines: (*a*), the minimum number of nucleotide replacements required in their descent; (*b*), all possible ancestral forms for each node consistent with that minimum; and (*c*), the expected number of nucleotide replacements along each branch of the tree. One can get the messenger sequence from the amino acid sequence knowing the genetic code, but the method has the limitation that there cannot be more than one ambiguous nucleotide position for which replacements are required in any given codon. Thus a few of the positions containing arginine, leucine, or serine will require care in their translation into messenger codons. More recently Moore, Barnabas & Goodman (181) and Hartigan (181a) have mathematically proved the correctness of my algorithm for unambiguous nucleotides. The former have improved upon the method in an important way in that they can treat arginine, serine, and leucine cases so as to determine absolutely the codon(s) that permit the most parsimonious solution. The method of tree evaluation in the *Atlas of Protein Sequences* (89) is related to parsimony considerations in that the number of amino acid replacements is minimized. It is not clear that the true minimum number of amino acid replacements for a given phylogeny is always obtained by their method. However, the tree that gives the minimum number of amino acid replacements is not necessarily the tree that gives the minimum number of nucleotide replacements.

Many of the techniques of tree building would discover the most parsimonious tree were it not for parallel and back mutations, which cause the distances to be underestimated in various degrees. Parsimony procedures reveal their occurrence.

Estabrook (182) has provided a method whereby one does not need to examine all possible phylogenies to be sure of having the best by some criterion. It is an important step in the right direction but does not sufficiently reduce the magnitude of the search to be useful in many cases.

TREE BUILDING ATTEMPTS The failure to obtain from a given set of data a phylogeny that the outside biologist is happy with, generally lies with the distance measure rather than the tree building process. One of the surest signs that the distance measure is inadequate to the task is when the authors do not present a derived tree and in its place give a general discussion of how the measure can be interpreted in terms of a classical phylogeny. No attempts are known to me that use distances based upon composition. General discussion has occurred for measures such as DNA/haploid cell (168, Fig. 3) and "information" of various types (183, Figs. 3, 4, 6, 7).

Methods based on DNA hybridization and immunological distance tend to be handicapped by not having all possible pairwise distances at hand. Nevertheless Kohne (160) has managed a quantitative analysis of some primate data on DNA hybridization in terms of the accepted phylogeny, and Wilson & Sarich (184) have managed the same for their immunological tests. In the latter paper, Wilson & Sarich argue, from the assumption that amino acid replacements occur at regular rates over time, that the paleontological dating of hominid evolution is wrong. Kohne argues from his data that the assumption of Wilson & Sarich is demonstrably wrong.

Gibbs et al (185) formulated a distance measure on the basis of the frequency with which one amino acid follows another. This is an adaptation of Gibbs & McIntyre's *Runs Index* (86). Using this distance measure, they analyze 24 cytochrome *c* sequences and obtain a phylogeny in which the rattlesnake is located among the primates, wheat among the insects and the kangaroo is associated with the turtle. Compare this with Figure 2. They state, "Amino acids are treated as the same or not the same with no weighted measure ... such as that, based on the genetic code, used by Fitch & Margoliash ... as it is simpler and there is no evidence that the method based on the genetic code gives a better result and not just a different one."

In their other paper (86), they form phylogenetic trees from both their runs index and their diagonal index. In both cases the first eukaryotic split separates out the vertebrate group before the plants, fungi, and insects diverge from each other and the first vertebrate split separates out the rattlesnake-primate group before the fish, frog, birds, and nonprimate mammals diverge from each other. They note, "Both classifications show the close similarity between rattlesnake and primate cytochromes *c*. This similarity was noted by Smith (186) but not found by Fitch & Margoliash." Now Smith noticed, correctly, that the rattlesnake cytochrome *c* was more like the human cytochrome *c* than like any other known cytochrome. Gibbs & McIntyre's procedure must be emphasizing that datum. But it is also true, and more important, that there is not one pair of cytochromes *c* among all the mammals that differs half as much as human and rattlesnake cytochrome *c* differ from each other. Because their results disagree with ours, they "seriously question the value of interpreting classifications of this sort strictly in terms of phylogeny [references to Fitch & Margoliash]" and also "doubt the value of using the minimum mutational distance measure used by Fitch."

Others have been more successful. Using a variation on our method, Goodman's laboratory has examined mammalian evolution using numerous hemoglobin sequences (187–189). Boyer et al (190) examined ten each of primate beta and delta hemoglobins in terms of known phylogeny. Using a variation of the method in the *Atlas of Protein Sequences*, Boulter et al (191) examined higher plant phylogeny using 15 cytochrome *c* sequences. Doolittle's lab has examined 18 artiodactyl (192) and six hominoid (193) fibrinopeptides in terms of the accepted phylogeny and finds them consistent. Prager et al (194) found that the most parsimonious tree for five bird lysozymes had one minor change from traditional relationships. The nonplant cytochromes *c* have been expanded to 29 different sequences from 33 taxa altogether (195) and, including plants and protozoans, to 49 different sequences from 55 taxa altogether (204 and Fitch & Margoliash, in preparation). Smith, Hood & Fitch (4) have examined a large body of immunoglobulin sequence data, constructing gene phylogenies and interpreting them in terms of the various theories of antibody diversity. Strydom (195a) has examined the relationships of 16 snake toxins (both orthologous and paralogous) and Staehelin (195b) those for five calcitonins. Alternative versions for these plus a few others (e.g., lactalbumen, tobacco mosaic virus coat proteins) can be found in the *Atlas of Protein Sequences* (89).

Phylogeny Dependent Information

Once one has a phylogeny, whether derived from sequences or from other biological sources, together with a method of reconstructing ancestral sequences that gives the maximum likelihood estimate of the nature and location (nucleotide, codon, and tree branch) of the nucleotide replacements and therefore the amino acid replacements that occur in the gene's evolutionary history, a large number of interesting questions come under attack.

ARE MUTATIONS FIXED RANDOMLY ALONG THE GENE? Fitch & Margoliash (196) examined the distribution of nucleotide replacements required in the descent of twenty cytochromes *c*, asking if the distribution were Poisson. There were too many positions that had never varied at all and there appeared to be a few that had fixed too many mutations. King & Jukes (68) later denied the correctness of this. Subsequently, Fitch & Markowitz (195) produced statistical results that demonstrated that the data can only be accounted for if there is a class of invariable codons plus at least two additional classes of differing variability. This was followed by Uzzell & Corbin (197) who showed that the data fit a negative binomial distribution, which is tantamount to saying the data fit for codons with a distribution of variability. The same nonrandomness has been shown for the distribution of nucleotide replacements in the evolution of alpha hemoglobin (198) and beta hemoglobin (199). Despite the evidence, Jukes (118) still maintains a faith in the Poisson distribution of fixations. However, he does not count fixations except in a limited sense. He examines all the amino acids present in a given codon site and, using each only once, asks what is the most parsimonious way of interconverting their codons. The number of nucleotide replacements on that tree, regardless of its consistency with classical phylogeny or with other codon positions, is his estimate of the number of observed replacements at that site. Compared to my phylogenetic method, he states that "it seems less presumptious to . . . assume that the distribution of [Jukes'] observed replacements has the same pattern as the distribution of the undoubtedly larger number of actual replacements." Such a method, rejecting most of the data available, must be less sensitive. To get more usable data, Jukes examined not merely the conjugate orthologous sites of a single gene, but pooled for use as single-site data, all the amino acids in the conjugate paralogous sites of myoglobin, alpha, beta, gamma, and delta hemoglobins. Now his most parsimonious tree may not only fail to agree with the phylogeny of the species, but also with the phylogeny of the genes. At any rate his data are reasonably Poisson distributed.

The fraction of the gene that is invariable is one result one gets from the procedure of Markowitz (200) and used in Fitch & Markowitz (195). It depends upon the range of species, increasing as the range of species narrows. We distinguish between invariable and variable-but-unvaried codons (unfortunately, invariant has been used in both senses). As the range of species examined is narrowed, the number of nucleotide replacements observed necessarily decreases and as a consequence the number of variable-but-unvaried codons would be expected to increase, but there is no a priori reason for the number of invariable codons to change. Nevertheless they do increase, indicating that the number of codons at which evolutionary change is acceptable is smaller in a narrower range of taxa than in a broader range. Indeed, by extrapolating to zero range, Fitch & Markowitz are able to estimate the number of covarions (*CO*ncomitantly *VARI* able cod*ONS*) in mammalian cytochrome *c*, the number of codons which in a given species is capable of an evolutionarily acceptable change in the amino acid encoded. The number of covarions in mammalian cytochrome *c* was estimated at ten. Fitch & Margoliash (in preparation) have now obtained a slightly revised overall estimate of 12 covarions using 49 different sequences, with, perhaps, a suggestion of one or two more in mammals and one or two less in plants, although the limits of accuracy do not permit one to conclude that there are indeed fewer covarions in plants than in mammals.

The number of covarions in alpha hemoglobin has been estimated at 50 (198), in beta hemoglobin at 39 (199), in fibrinopeptide A at 18 (201), and in insulin C peptide at 18 (Fitch, unpublished).

The question about the distribution of mutations over codons has been answered in terms of nucleotide replacements fixed in the course of evolution. One can ask the same question about human hemoglobin variants. Vogel (202) has examined the variants for both the alpha and beta chains by two methods. The first is a "runs" test, a run being one or more consecutive residues that have all varied (or where none has varied). Given the number of varied residues, one can calculate the expected number of runs and its standard deviation. Neither cistron showed any nonrandomness by this criterion. Where the data are more plentiful, the second (Poisson) test is more sensitive. Vogel gives the expected and found distributions for each cistron and concludes the distribution is random. But he made no statistical analysis and his own data show that the distribution in the alpha chain is nonrandom. This has been confirmed (203) but there are two biases in the data: (*a*), the variants are not counted according to how many times they have occurred but only whether they have occurred at least once; and (*b*), the electrophoretic method generally used for detecting the presence of a variant could be expected to find all the variants of histidine positions, none of the variants of phenylalanine positions, and intermediate numbers of others, yet the implicit assumption of a Poisson process test is that detectable events are equiprobable at all sites. The meaning of the demonstrated nonrandomness of the distribution of human hemoglobin variants in alpha (but not beta) chains is therefore unclear.

ARE MUTATIONS FIXED RANDOMLY IN THE CODON? If appropriate recognition is made of how many ways an amino acid's codons can change to that of another amino acid by a change of nucleotide in the first, second, or third position of the codon, one can compute the expected distribution of nucleotide replacements according to their position, assuming one nucleotide replacement to be as likely as another. Fitch (204) has done this, examining the nucleotide replacements fixed during the evolution of cytochrome *c* and alpha and beta hemoglobins. The results (Table 2) show a significant nonrandomness for cytochrome *c* with an excess of mutations fixed in the first and a deficit in the second nucleotide position. For alpha hemoglobin, the probability of occurring by chance has increased more than 3 orders of magnitude so that the result is only of borderline significance. Note that the excess mutations are still in the first nucleotide position. The result for beta hemoglobin is not significantly different from random. There is a gross correlation between the degree of randomness and the relative rates of evolution (77) in these three cistrons in that beta hemoglobin is evolving fastest and shows the most randomness while cytochrome *c* is evolving slowest and shows the least randomness. This suggests that as the opportunities for change become more restricted by selective constraints, the opportunities accepted are more easily seen to be nonrandom.

IS THERE NONRANDOMNESS IN THE NUCLEOTIDE REPLACEMENTS? Vogel & Roehrborn (205) examined the then known variants of human hemoglobin and noted the excess of expected transitions compared to expected transversions (transversions are purine-pyrimidine interchanges, others are transitions). Derancourt et al (206) subsequently came to the opposite conclusion, while Fitch (207) was examining both the human hemoglobin variants and the replacements occurring during the evolution of cytochrome *c* and discovering, in both instances, a significant excess over expectation of G→A replacements. Replacements are considered in terms of the messenger RNA. Expectation is calculated on the assumption that all nucleotides are equally likely to accept the next replacement which can be any one of the other three nucleotides with equiprobability. The apparently startling part of the result was that, unless the mutation mechanism were strand-specific, one should have seen a comparable excess of C→U replacements. The essential observations have also been made

Table 2 Evolutionary Nucleotide Replacements by Codon Position

		1	2	3	p
Cytochrome *c*	Expected	263	292	87	
	Found	329	234	79	$< 10^{-6}$
Alpha Hemoglobin	Expected	148	156	43	
	Found	177	138	32	0.005
Beta Hemoglobin	Expected	119	130	38	
	Found	131	123	32	0.3

Probabilities are based on a chi-square for 2 degrees of freedom. From Fitch (204). Further explanation in text.

by Vogel (202), Lehmann & Carrell (208), Vogel, Derancourt & Zuckerkandl (209), and Zuckerkandl, Derancourt & Vogel (210). Vogel (211, 212) believes that the "results—for the hemoglobin variants and for the other proteins—, . . . reflect nonrandomness of the mutation process itself. They render it very unlikely that it [nonrandomness] is caused only by a bias in ascertainment of the mutations (211)." On this we politely disagree in that significant nonrandomness only rejects the null hypothesis and that rejection does not require a choice between the "mutation process" and "bias in ascertainment"; it only means that at least one of the underlying assumptions, upon which the calculations of the expected outcome is predicated, is false. The incorrect assumption, I believe, is that all nucleotides are equally likely to accept the next replacement. We have already seen that neither the codons nor the nucleotide positions within the codon are equally variable so that it would hardly be surprising should that be reflected in data such as are being examined here. Indeed in an extensive examination of 235 and 186 alpha-hemoglobin nucleotide replacements occurring in the first and second nucleotide positions respectively during evolution, Fitch (198) showed that the difference between the number of replacements in the two directions of interchange between any two nucleotides in either position was not detectably different from zero. Thus the composition of the first two nucleotide positions of the alpha hemoglobin gene is in an approximate equilibrium but not at an equimolar composition of the four bases as would eventually be reached if the assumption underlying the earlier calculation of expectations were true. I am not concerned, as Vogel states (212), that we shall run out of G. It is sufficient, in my view, to explain the data if there are phenotypic constraints. I suggest two. If the interchanges of ala-thr, lys-arg, asp-asn, val-ile/met, gly-ser, glu-lys, and gly-asp/glu are more commonly allowable than other sets of amino acid pairs relatable by a specific pair of replacing nucleotides, there will be an apparent excess of G-A interchanges at the messenger level (198). If the secondary structure of the messenger RNA is important, some positions may tolerate replacements only in the stages AU—GU—GC (213). Only transitions would be acceptable in such positions.

RATE OF EVOLUTION To determine the rate of evolution, one needs two data: (*a*), a well dated time-point in the past when two species diverged; and (*b*), the number of mutations fixed in the two lines since that divergence. The molecular biologist generally accepts the first datum from the paleontologist. The second datum requires a count of observed fixations plus an allowance for the unseen fixations, i.e., multiple changes at one site observable only as one difference. Zuckerkandl & Pauling (1) made the first such estimate (without allowance for unseen fixations) and found a unit evolutionary period[2] of 20 for alpha hemoglobin. The need for the allowance was first explicitly recognized by Margoliash & Smith (214) who suggested, in the first of three approaches to be considered here, that the number of events counted (the number of differing amino acid positions between two species) be corrected as if they were the result of a Poisson process over the codons of the gene. They recognized that perhaps some codons were not variable because any alternative amino acid was deleterious. Following the first estimate of the number of invariable codons in cytochrome *c* (215), Fitch & Margoliash (216) calculated the unit evolutionary period[2] to be 26 million years/residue change/variable codon or 21 million years/nucleotide replacement/variable codon. Dickerson (7) has found unit evolutionary periods of 1, 6, and 20 for fibrinopeptides, hemoglobin, and cytochrome *c*. King & Jukes (68) got 1, 5, and 12 respectively.

A second approach is to recognize that, as a result of selection, some amino acids are more likely to be replaced than others and that there are differential probabilities with respect to the replacing amino acid. This makes the relationship between observed distance and total distance (observed plus unseen) dependent upon the initial frequency of amino acids, but Dayhoff (89) has used simulation techniques on proteins of "average" composition to obtain (Figure 9–9, p. 95) an expected relation between observed and total amino acid substitutions. In my judgment, it is the most accurate estimator available because its empirical base allows for the fact that not all positions are equally variable.

In a third approach to the problem, Holmquist (217, 218) has done a thorough analysis of the effect of a number of successive mutations on a codon in terms of the difference from the original encoded amino acid that will be observed in the finally encoded amino acid with strict adherence to stochastic properties and a complete knowledge of the genetic code. Our problem is the reverse, namely to go from the observed to an estimate of the number of successive mutations, but in a later paper (219) he provides (his Table 1, p. 215) the answers to what would otherwise be a complicated process for the would-be user. Holmquist is trying to determine all the nucleotide substitutions, not just those that change the amino acid encoded, so his numbers are larger than comparable figures from Dayhoff's

[2] The term unit evolutionary period when first used (214) was defined as millions of years per residue change in two diverging genes. Later (7, 240) that was changed to millions of years per residue change per 100 amino acids in two diverging genes. This latter is used here and earlier reported values have been changed to this unit in the text. The reciprocal of the unit evolutionary period equals 5 paulings. The pauling is a rate for a single gene whereas the unit evolutionary period is looking at both separating lines.

method (89) or the expected number of observable minimum base differences. While the aim is laudable, it is doubtful that the model does what is claimed for it vis-a-vis the real world. There is no justification for the assumption that nucleotide replacements changing the encoded amino acid are as frequent as those that do not. Indeed, in the one case where information is available [viz. phage coat protein messenger (213)] nine out of ten replacements did not change the encoded amino acid. Holmquist's computation of expected minimum base differences says that if there are 25 to 30 amino acid differences per 100 positions, one should see 2 more base differences than amino acid differences. In an old tree of 24 cytochromes *c*, 12 pairs of sequences differed by 26 to 29 amino acids (average = 27). The minimal number of base differences averaged 11 more than that (smallest value, 8 more), a 550% error in the prediction property of his model for observable minimum base differences.

All three methods permit a transformation of amino acid differences into total differences, observed plus unobserved, that are intended to linearize the data with respect to the true total. There is no evidence that supports Holmquist's view that the slope for his transformed data is closer to one than the others.

A fourth procedure is the ancestral sequence reconstruction discussed above (76). When the number of nucleotide replacements along each branch of the tree connecting two taxa are summed, the result is the minimal phyletic distance and may be considerably greater than the minimal pairwise distance. This is minimal since there is no allowance for unseen mutations. Now the number of amino acid differences between human and tobacco hornworm cytochrome *c* (in 103 positions) is 26 and Holmquist's model estimates that there will have been 14 unseen replacements and a total of 42 mutations fixed (presumably only 28 of which change the encoded amino acid). Analysis of 49 cytochrome *c* sequences shows that there are at least 70. Thus all three approaches seriously underestimate the minimal number of replacements we know has occurred from ancestral reconstruction.

The first three above procedures incorrectly assume that all codon positions of a gene are equally likely to receive the next mutation (save in some degree Dayhoff's empirical method). The solution to this problem lies in discovering the number of variable positions available to the next mutation. Fitch & Markowitz showed it is considerably less than the length of the gene. In fact, it is the number of covarions, and with that the issue would be solved were it not for the fact (220) that the variable codons in one group are not the same as those in another. This creates the formidable problem of allowing for the quantitatively unknown manner in which the set of covarions changes as mutations are fixed. Holmquist & Jukes (221) have suggested the "varion". It appears to be similar to the number of variable residues in the method of Fitch & Markowitz (200) for a given range of species except that it uses only two species and examines the number of amino acids with a minimum of one and two base coding differences (222, 223). This is among the most erratic parameters in the comparison of amino acid sequences. In the set of 24 cytochrome *c* sequences examined above, there were 12 pairs for which there were at least as many multiple-base minimal differences as one-base

Table 3 Evolutionary Rates

Protein	Fixations	Covarions	Rate
Cytochrome *c*	5	12	3.3
Alpha Hemoglobin	22	50	3.4
Beta Hemoglobin	31	39	6.0
Insulin C	8	18	3.4
Fibrinopeptide A	10	18	4.4

Fixations are the observed nucleotide replacements observed since the common ancestor of horse and pig. Covarions as determined by method of Fitch & Markowitz (195). Rate is in amino acid changing fixations/covarion/10^9 years in one line of descent assuming the most recent protoungulate ancester existed 64 million years ago (Van Valen, personal communication).

minimal differences. This would give a Poisson exponent in their method $\simeq \infty$! For pairs of sequences with the same common ancestor and minimal base differences, the Poisson exponent more often than not varied by a factor of more than two. It is a curious method also in that it sets the expected number of one-base minimal differences equal to the observed.

ARE THE RATES CONSISTENT WITH THE EVOLUTIONARY CLOCK AND NEUTRAL MUTATIONS? The interest in rates has had two stimuli in recent times. One is the idea that if the rate of substitution is reasonably uniform over sufficient time, then knowing the rate and number of substitutions, is equivalent to knowing how long ago the common ancestor of two sequences existed. This has the potential value of fixing time points that are not obtainable through paleontological dating. Unfortunately, the most interesting dates are the most distant ones and it is here that the corrections that have been developed so far are greatest and least trustworthy.

Since replacements can occur only in covarions, rates ought to be constant per covarion regardless of the gene examined, if the clock hypothesis is correct. Table 3 shows the data for products of five different genes for which there is sufficient information to estimate the number of covarions and for which both the pig and horse gene products are known. The results are in reasonable agreement, given the method, and this would be true regardless of the accuracy of the paleontological dating since the same date is common to all.

The second stimulus is the problem of the extent to which neutral mutations account for the sequence differences between species. Kimura (224), in an attempt to explain more diversity among proteins than seemed accountable for by selection, suggested that most observed amino acid replacements were not selected for and demonstrated that newly arising alleles for which there was neither advantage nor disadvantage would go to fixation at a rate proportional to their occurrence. This, he assumed, should be reasonably uniform over time. That part of the process of change for which stochastic rather than selective effects determined the outcome was given the name Non-Darwinian Evolution by King & Jukes (68). In the ensuing controversy, one crucial issue has been the extent to which evolutionary rates among proteins have been uniform over time.

Since the first evolutionary tree for cytochrome *c* (99), it has been obvious that something was aberrant about the rattlesnake. There are two difficulties in using the rattlesnake sequence to disprove the stochastic uniformity of evolutionary rate. One is the possibility of sequence errors in that the amount of available material was very small for the procedures at that time and difficulties were encountered. The second difficulty is in showing that it is not simply an observation from the tail of the expected distribution. Jukes & Holmquist (225) have attempted to analyze this case and concluded that the rate of cytochrome *c* change in the rattlesnake was significantly high. However, the one value of significance given would only prove that the rattlesnake is significantly further from birds than the turtle is. That conflicts with biological notions as was pointed out earlier (99).

Ohta & Kimura (226) examined the differences between pairs of orthologous hemoglobin and cytochrome *c* sequences taken such that no pair shared a common line of descent with any other orthologous pair so that within a pair set for any one gene, the pairs were independent. Within two of the sets, beta hemoglobin and cytochrome *c*, the observed variance, s^2, in rates of divergence, compared with the expected variance, σ^2, were significantly larger by an F test. They concluded that variations in rates "are larger than expected from chance."

In a more comprehensive study, Langley & Fitch (77) performed a maximum likelihood estimate of the number of nucleotide replacements that occurred in the descent of 4 vertebrate gene products. The constraints were that the total replacements must equal those observed in their descent and that the process be Poisson and time uniform. There are two requirements for uniformity of evolutionary rates: (*a*), that from any ancestor, the two descending lines have equal expected numbers of replacements when summed over all four genes; (*b*), that within each branch, the relative number of replacements in the four genes is the same as their overall rate of evolution. Since one obtains estimates for the expected number of replacements on each branch for each gene, one can test both requirements. Both tests fail, meaning that there is significant variation beyond that inherent in the stochastic nature of the assumptions. Thus it is concluded that substitutions are not uniform over time. Moreover, the failure to meet the second requirement means that differential generation times or differential numbers of cell divisions cannot account for the results either, so that the substitutions are not uniform on these bases either. Although this does not rule out the occurrence of neutral mutations, it appears that the neutral mutation theory can not account for these results without ad hoc hypotheses of the type that assert that the numbers of neutral mutations available in the gene pool fluctuate considerably from time to time and gene to gene and species to species; but it was primarily because that fluctuation did not appear to exist in the first place that the neutral hypothesis had any appeal.

But while our analytical techniques have sharpened sufficiently to observe the departures from uniformity, they also provide us with the kind of overall

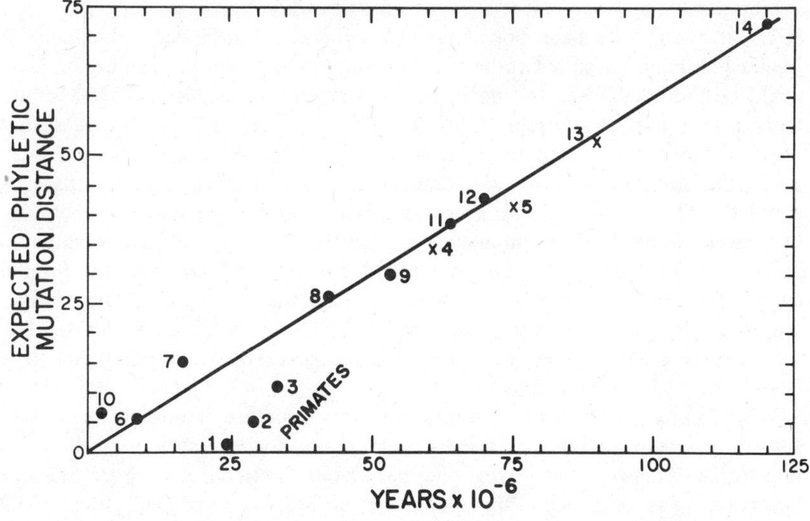

Figure 2 Phyletic Mutation Distance as a Function of Paleontological Time among Mammals. Each point is a node on a phylogenetic tree whose paleontological dating was given by Van Valen without knowledge of the phyletic distances which were determined by Langley & Fitch (61) using the gene products alpha and beta hemoglobin, cytochrome *C* and Fibrinopeptide A. We followed Simpson (247) in joining the rodents and lagomorphs to form the cohort Glires. This made our phylogeny different at three points from Van Valen's view of mammalian evolution so that he provided the dates marked as an X with distinct cautionary remarks and perhaps ambivalence as well.

averages that are currently best suited to an evaluation of the evolutionary clock hypothesis. Figure 2 is a plot (Langley & Fitch, in preparation but based on data from reference 77) of the maximum likelihood number of nucleotide replacements to a given ancestral node versus the paleontological times of mammalian divergence as estimated, without knowing our results, by Dr. Leigh Van Valen. Except for the primates, the agreement is excellent and hopefully not fortuitous.

Nucleic Acids

This is a field in its infancy since not many sequences are yet known except for the t-RNA's. Nevertheless, intrepid souls have ventured into these murky waters and both Dayhoff (89) and Cedergen, Cordeau & Robillard (227) have phylogenies of t-RNA's. I feel that much more data will be required before we can have any confidence in the details of such phylogenies since we have no way of distinguishing the orthologous from the paralogous sequences. To emphasize this point, Squires & Carbon (228) have found an *E. coli* glycine t-RNA that is considerably more similar to one of two *E. coli* valine t-RNA's than the two valine t-RNA's are to each other.

Since amino acid sequences leave ambiguities when translated into messenger RNA, it might have been hoped that the procedues of Tinoco, Uhlenbeck & Levine (229) and Delisi & Crothers (230) for finding secondary structure in RNA could be modified so as to find it in the ambiguous messenger. That assumes there is secondary structure in messenger RNA. Figuero et al (231), suggest that "natural proteins select the amino acid codons of their messenger RNA's in such a way that antiparallel auto-complementarity of six or more successive bases is avoided." The evidence is in a previous paper (232) which I find unconvincing. Moreover, alpha hemoglobin messenger has been shown to have substantial hyperchromicity (232a). To show it is possible, they produced a putative human cytochrome *c* messenger that is asserted to have no autocomplementarity longer than five bases. It contains the sequence UUCACAGUG $\overline{\text{CCA}}$,CACUGUGGA (nucleotides 42→62). The sequence has nine successive auto-complementary bases if the $\overline{\text{CCA}}$ is treated as the end loop.

White, Laux & Dennis (233) found secondary structure in ambiguous messengers of several proteins, but there are so many possible structures that it is impossible to believe that the real ones have been identified. Knowing that Fier's laboratory was sequencing the messenger of MS2 coat protein, Fitch (234) brashly attempted to guess the sequence of the portion not yet published. Unfortunately, the reported amino acid sequence was incorrect in this region and his guess proved wild when the messenger was sequenced (235). Mark & Petruska (235a) have a similar procedure for finding putative messenger structures.

Information of consequence lies available in the sequenced material already at hand. Has, for example, nature reduced the inapproriate out-of-frame reading of messengers by using the ambiguous third position to install out-of-frame terminators or to minimize out-of-frame initiators? For example, Asp-Lys can be encoded GA $\overline{\text{U·AA}}$R or GAC·AAR (R= purine). The former case has the out-of-frame terminator UAA. Fitch (234) examined three viral messengers and found no significant increase over random in the number of such terminators, suggesting that initiation rather than termination is the crucial control point. There was also no significant decrease over random in the number of such out of frame initiators.

How then is initiation controlled? All efforts have left doubt because of the apparent absence of generality and the paucity of known messengers sequenced. There frequently appear to be terminators preceding the initiating AUG but not always in reading frame [e.g., phage λ RNA in region of the *Q* gene if the Lebowitz, Weissman, Radding (236) material is a messenger, as seems likely]. Also initiators seem frequently to be on the end loops of helical stalks but not always [e.g., Qβ maturation gene (237), R17 replicase gene (238)]. There appear to be cases where the regions containing the initiating AUG can exist in two fairly stable forms, one with the AUG at the beginning of an end loop and one with the AUG base paired suggesting (234) that these may be alternatives utilized for regulating translation. Another example now is seen in VA RNA, which is made by cells infected with adenovirus-2. It has been sequenced by Ohe & Weissman (239) who have proposed a secondary structure for it. Although they discuss it as a potential messenger, there is no real evidence to support this. It does have, however, an alternative structure only slightly less stable than the one the authors proposed in which nucleotides 79→95 GUGAUCC AUGCGGUUAC form a stalk with AUG at the beginning of a 4 nucleotide end loop, in which a terminator occurs 2 codons before the AUG and in which no subsequent terminators occur in phase with the AUG.

Note added in proof

Two additional human hemoglobin deletion mutants have been observed (248). Both of them are beta chain variants. Thus within the human population, of the seven insertion/deletion mutants observed, there are seven beta and zero alpha variants further emphasizing the relative importance of the alpha chain. This also alters the nonterminal insertion/deletion ratio among the variants discussed from 1/8 to 1/10, emphasizing the apparent bias towards deletions at the observational level. These two deletions do not fit well into the categories of Table 1 since, in the region of the deletion, not more than two consecutive identical base pairs can be found that are as many nucleotides apart as nucleotides were deleted. Finally a C-terminal frameshift mutation in human alpha hemoglobin has been observed that alters the last three amino acids and extends five more amino acids plus a terminator beyond the normal arginine-142 (249). The frameshift is proven by its consistency with hemoglobin Constant Spring in the region distal to the normal C-terminus.[3]

[3] Appreciation is given for the valuable assistance of Mr. Frank Iltis. This work was supported by NSF grant GB 3227X.

Literature Cited

1. Zuckerkandl, E., Pauling, L. 1962. In *Horizons in Biochemistry*, ed. M. Kasha, B. Pullman, 189–225, New York: Academic
2. Pauling, L., Zuckerkandl, E. 1963. *Acta Chem. Scand.* 17:S9

2a. Margoliash, E. 1963 *Proc. Nat. Acad. Sci. USA* 50:672

3. Gally, J.A., Edelman, G.M. 1972. *Ann. Rev. Genet.* 6:1
4. Smith, G., Hood, L., Fitch, W.M. 1971. *Ann. Rev. Biochem.* 40:969
5. Smith, E.L. 1970. In *The Enzymes*, ed. P. Boyer, 267–339, New York: Academic
6. Arnheim, N. 1973. In *The Antigens*, ed. M. Sela, New York: Academic. In press
7. Dickerson, R.E. 1971. *J. Mol. Evol.* 1:26
8. Dickerson, R.E. 1971. *J. Mol. Biol.* 57:1
9. Takano, T., Kallai, O.B., Swanson, R, Dickerson, R.E. 1973. *J. Biol. Chem.* In press
10. Brew, K. 1970. *Essays Biochem.* 6:93

10a. Vogel, H.J. 1965. In *Evolving Genes and Proteins*, ed. V. Bryson, H. Vogel, 25–40, New York: Academic

11. LeJohn, H.B. 1971. *Nature* 231:164
12. Lemmon, R.E. 1970. *Chem. Rev.* 70:95
13. Kenyon, D.H., Steinman, G. 1969. *Biochemical Predestination*, New York: McGraw-Hill, 301 pp.
14. Oparin, A.I. 1957. *Origin of Life*, New York: Dover, 495 pp.
15. Koesian, J. 1964. *Origin of Life*, New York: Rheinhold, 118 pp.
16. Calvin, M. 1969. *Chemical Evolution*, Oxford: Oxford Univ., 278 pp.

17. Margulis, L. 1970. *Origins of Life*, New York: Gordon & Breach, 376 pp.
18. Wolman, Y., Haverland, W.J., Miller, S.L. 1972. *Proc. Nat. Acad. Sci. USA* 69:809
19. Lawless, J.G., Kvenvolden, K.A., Peterson, E., Ponnamperuma, C., Moore, C. 1971. *Science* 173:626
20. Lawless, J.G., Kvenvolden, K.A., Peterson, E., Ponnamperuma, C., Jarosewich, E. 1972. *Nature* 236:66
21. Sagan, C. 1972. *Nature* 238:77
22. Cairns-Smith, A.G., Ingram, P., Walker, G.L. 1972. *J. Theor. Biol.* 35:601
23. Fox, S.W., Windsor, C.R. 1970. *Science* 170:984
24. Hulett, H.R., Bar-Nun, A., Bar-Nun, N., Bauer, S.H., Sagan, C. 1970. *Science* 170:1000
25. Bar-Nun, A., Bar-Nun, N., Bauer, S.H., Sagan, C. 1970. *Science* 168:470
26. Sagan, C., Khare, B.N. 1971. *Science* 173:417
27. Hulett, H.R., Wolman, Y., Miller, S.L., Ibanez, J., Oro, J., Fox, S.W., Windsor, C.R. 1971. *Science* 174:1038
28. Khare, B.N., Sagan, C. 1971. *Nature* 232:577
29. Wollin, G., Ericson, D.B. 1971. *Nature* 233:615
30. Sanchez, R.A., Orgel, L.E. 1970. *J. Mol. Biol.* 47:531
31. Stephen-Sherwood, E., Oro, J., Kimball, A.P. 1971. *Science* 173:446
32. Friedmann, N., Miller, S.L., Sanchez, R.A. 1971. *Science* 171:1026
33. Dowler, M.J., Fuller, W.D., Orgel, L.E., Sanchez, R.A. 1970. *Science* 169:1320
34. Garay, A.S. 1968. *Nature* 219:338
34a. Frank, F.C., 1953. *Biochem. Biophys. Acta* 11:459
35. Segal, H.L. 1972. *FEBS Lett.* 20:255
36. Mortberg, L. 1971. *Nature* 232:105
37. Paecht-Horowitz, M., Berger, J., Katchalsky, A. 1970. *Nature* 228:636
38. Degens, E.T., Matheja, J., Jackson, T.A. 1970. *Nature* 227:492
39 Chang, S., Flores, J., Ponnamperuma, C. 1969. *Proc. Nat. Acad. Sci. USA* 64:1011
40. Saunders, M.A., Rohlfing, D.L. 1972. *Science* 176:172
41. Fuller, W.D., Sanchez, R.A., Orgel, L.E. 1972. *J. Mol. Biol.* 67:25
42. Tapiero, C.M., Nagyvary, J. 1971. *Nature* 231:42
43. Lohrmann, R., Orgel, L.E. 1968. *Science* 161:64
44. Bishop, M.J., Lohrmann, R., Orgel, L.E. 1972. *Nature* 237:162
45. Swartz, A.W. 1972. *Biochem. Biophys. Acta* 281:477
46. Ibanez, J.D., Kimball, A.P., Oro, J. 1971. *J. Mol. Evol.* 1:112
47. Ibanez, J.D., Kimball, A.P., Oro, J. 1971. *Science* 173:444
48. Sulston, J., Lohrmann, R., Orgel, L.E., Miles, H.T. 1968. *Proc. Nat. Acad. Sci. USA* 59:726
49. Sulston, J., Lohrmann, R., Orgel, L.E., Miles, H.T. 1968. *Proc. Nat. Acad. Sci. USA* 60:409
50. Sulston, J., Lohrmann, R., Orgel, L.E., Schneider-Bernloehr, H., Weimann, B.J., Miles, H.T. 1969. *J. Mol. Biol.* 40:227
51. Degani, C., Halmann, M. 1972. *Nature New Biol.* 235:171
52. Usher, D.A. 1972. *Nature* 235:207
53. Woese, C.R. 1967. *The Genetic Code*, New York: Harper & Row
54. Crick, F.H.C. 1968. *J. Mol. Biol.* 38:367
55. Orgel, L.E. 1968. *J. Mol. Biol.* 38:281
56. Woese, C.R. 1969. *J. Mol. Biol.* 43:235
57. Sonneborn, T.M. 1965. In *Evolving Genes and Proteins*, ed. V. Bryson, H. Vogel, 377–97, New York: Academic
58. Fitch, W.M. 1966. *J. Mol. Biol.* 16:1
59. Jukes, T.H. 1967. *Biochem. Biophys. Res. Comm.* 27:573
60. Crick, F.H.C. 1967. *Nature* 213:119
61. Woese, C. 1965. *Proc. Nat. Acad. Sci. USA* 54:71
62. Thomas, B.R. 1970. *Biochem. Biophys. Res. Comm.* 40:1289
63. Lesk, A.M. 1970. *Biochem. Biophys. Res. Comm.* 38:855
64. Saxinger, C., Ponnamperuma, C. 1971. *J. Mol. Evol.* 1:63
65. Rendell, M.S., Harlos, J.P., Reise, R. 1971. *Biopolymers* 10:2083
66. Raszka, M., Mandel, M. 1972. *J. Mol. Evol.* 2:38
67. MacKay, A.L. 1967. *Nature* 216:159
68. King, J.L., Jukes, T.H. 1969. *Science* 164:788
69. King, J.L. 1971. In *Biochemical Evolution and the Origin of Life*, ed. E. Schoffeniels, 1, Amsterdam: North Holland Pub. Co.
70. Moorhead, P.S., Kaplan, M.M. (eds.). 1967. *Mathematical Challenges to the Neo-Darwinian Interpretation of Evolution*, Philadelphia: Wistar Univ. Press, Monograph , 140 pp.
71. Waddington, C.H. (ed.). 1968. *Towards a Theoretical Biology*, (3 vols.) Chicago: Aldine Pub. Co.
72. Hulett, H.R. 1969. *J. Theor. Biol.* 24:56

73. Eigen, M. 1971. *Naturwissenschaften* 58:465
74. Black, S. 1971. *Biochem. Biophys. Res. Comm.* 43:267
75. Theodoris, G.C., Stark, L. 1971. *J. Theor. Biol.* 31:377
76. Fitch, W.M. 1971. *Systematic Zool.* 20:406
77. Langley, C.H., Fitch, W.M. 1973. In *Genetic Structure of Population*, ed. N.E. Morton, Honolulu: Univ. Press of Hawaii
78. Metzger, H., Shapiro, M.B., Mosimann, J.E., Vinton, J.E. 1968. *Nature* 219:1166
79. Marchalonis, J.J., Weltman, J.K. 1971. *Comp. Biochem. Physiol.* 38b: 609
79a. Harris, C.E., Kobes, R.D., Teller, O.C., Rutter, W.J. 1969. *Biochem.* 8:24–42
80. Shapiro, H.M. 1971. *Biochem. Biophys. Acta.* 236:725
81. Fitch, W.M. 1966. *J. Mol. Biol.* 16:9
82. Fitch, W.M. 1970. *J. Mol. Biol.* 49:1
83. Manwell, C. 1967. *Comp. Biochem. Physiol.* 23:383
84. Haber, J.E., Koshland, D.E. 1970. *J. Mol. Biol.* 50:617
85. Sackin, M.J. 1971. *Biochem. Genet.* 5:287
86. Gibbs, A.J., McIntyre, G.A. 1971. *Eur. J. Biochem.* 16:1
87. McLachlan, A.D. 1971. *J. Mol. Biol.* 61:409
88. Needleman, S.B., Wunsch, C.D. 1970. *J. Mol. Biol.* 48:443
89. Dayhoff, M.O. 1972. *Atlas of Protein Sequences*, Washington, D.C.: Nat. Biomed. Res. Found., 544 pp.
90. Sankoff, D. 1972. *Proc. Nat. Acad. Sci. USA* 69:4
91. Reichert, T.A., Cohen, D.N., Wong, A.K.C. 1973. *J. Theor. Biol.* In press
92. Bauer, K. 1970. *Int. J. Protein Res.* 3:165
93. Bauer, K. 1971. *Int. J. Protein Res.* 3:313
94. Bauer, K. 1971. *Nature* 58:364
95. Bauer, K. 1972. *Biochem. Biophys. Acta.* 278:606
96. Sneath, P.H.A. 1966. *J. Theor. Biol.* 12:157
97. Fitch, W.M. 1966. *J. Mol. Biol.* 16:17
98. Jukes, T.H. 1971. *J. Mol. Evol.* 1:46
99. Fitch, W.M., Margoliash, E. 1966. *Science* 155:279
100. Fitch, W.M. 1969. *Biochem. Gen.* 3:99
101. Bonaventura, J., Riggs, A. 1967. *Science* 158:800
102. Bradley, T.B., Jr., Wohl, R.C., Rieder, R.F. 1967. *Science* 157:1581
103. Veron, M., Falcoz-Kelly, F., Cohen, G.N. 1972. *Eur. J. Biochem.* 28:520
104. Grieshaber, M., Bauerle, R. 1972. *Nature New Biol.* 236:232
105. Yanofsky, C. 1960. *Bact. Rev.* 24:221
106. Bonner, D.M., DeMoss, J.A., Mills, S.E. 1965. In *Evolving Genes and Proteins*, ed. H.J. Vogel, V. Bryson, 305–18, New York: Academic
107. Manney, T.R., Duntze, W., Janosko, N., Salazar, J. 1969. *J. Bact.* 99:590
108. Bridges, C.B. 1936. *Science* 83:210
109. Ingram, V.M. 1961. *Nature* 189:704
110. Fitch, W.M. 1970. *Syst. Zool.* 19:99
111. Fitch, W.M., Margoliash, E. 1970. *Evol. Biol.* 4:67
112. Walsh, K.A., Neurath, H. 1964. *Proc. Nat. Acad. Sci. USA* 52:884
113. Hartley, B.S., Brown, J.R., Kauffman, D.L., Smillie, L.B. 1965. *Nature* 207:1157
114. Brew, K., Vanaman, T.C., Hill, R.L. 1967. *J. Biol. Chem.* 242:3747
115. Li, C.H., Dixon, J.S., Tung-Bin, L., Pankov. Y.A., Schmidt, K.D. 1969. *Nature* 224:695
116. Harris, J.I., Ross, P. 1956. *Nature* 178:90
117. Li, C.H., Barnafi, L., Chretien, M., Chung, D. 1965. *Nature* 208:1093
118. Barnett, D.R., Lee, T.H., Bowman, B.H. 1972. *Biochemistry* 11:1189
119. Titani, K., Hermodson, M.A., Fujikawa, K., Ericsson, L.H., Walsh, K.A., Neurath, H., Davie, E.W. 1972. *Biochemistry* 11:4899
120. Adelson, J.W. 1971. *Nature* 229:321
120a. Strydom, D.J. 1973. *Nature New Biol.* 243:88
121. Hilse, K., Popp, R.A. 1968. *Proc. Nat. Acad. Sci. USA* 61:930
122. Ostertag, W., von Ehrenstein, G., Charache, S. 1972. *Nature New Biol.* 237:90
122a. Abramson, R.K., Rucknagel, D.L., Shreffler, D.C., Saave, J.J. 1970. *Science* 169:194
123. Harris, M.J., Wilson, J.B., Huisman, T.H.J. 1972. *Arch. Biochem. Biophys.* 151:540
124. Huisman, T.H.J., Schroeder, W.A., Banister, W.H., Grech, J.L. 1972. *Biochem. Genet.* 7:131
125. Baglioni, C. 1962. *Proc. Nat. Acad. Sci. USA* 48:1880
126. Fessas, P., Stamatoyannopoulos, G., Karaklis, A. 1962. *Blood* 9:1
127. Yanase, T., Hanada, M., Sieta, M., Ohya, I., Ohta, Y., Imamura, T., Fujimura, T., Kawasaka, K., Yamaoka, K. 1968. *Jap. J. Human Genet.* 13:40

128. Lehmann, H., Charlesworth, D. 1970. *Biochem. J.* 119:43

128a. Badr, F.M., Lorkin, P.A., Lehmann, H. 1973. *Nature New Biol.* 242:107

129. Ohta, Y., Yamaoka, K., Sumida, I., Yanase, T. 1971. *Nature New Biol.* 234:218

130. Galizzi, A. 1971. *Nature New Biol.* 229:142

131. Yanofsky, C. 1963. *Cold Spring Harbor Symp. Quant. Biol.* 28:581

132. Smithies, O., Connell, G.E., Dixon, G.H. 1962. *Nature* 196:232

133. Black, J.A., Dixon, G.H. 1970. *Can. J. Biochem.* 48:133

134. Chauvet, J.P., Acker, R. 1970. *FEBS Lett.* 10:136

135. Boyer, S.H., Hathaway, P., Pascasio, F., Bordley, J., Orton, C., Naughton, M.A. 1967. *J. Biol. Chem.* 242:2211

136. Huisman, T.H.J., Adams, H.R., Dimmock, M.O., Edwards, W.E., Wilson, J.B. 1967. *J. Biol. Chem.* 242:2534

137. Braunitzer, G., Fujiki, H. 1969. *Naturwissenschaften* 56:322

138. Li, S.L., Riggs, A. 1970. *J. Biol. Chem.* 245:6149

139. Edmundson, A.B. 1965. *Nature* 205:883

140. Weatherall, D.J., Clegg, J.B. 1972. In *Synthesis, Structure and Function of Hemoglobin*, ed. H. Martin, L. Nowicki, 237–39, Munich: J.F. Lehmann's Verlag

141. Lehmann, H. 1972. In *Synthesis, Structure and Function of Hemoglobin*, ed. H. Martin, L. Nowicki, 359–79, Munich: J.F. Lehmann's Verlag

142. McLachlan, A.D. 1972. *J. Mol. Biol.* 64:417

143. Tanaka, M., Nakashima, T., Benson, A.M., Mower, H.F., Yasunobu, K.T. 1966. *Biochemistry* 5:1666

144. Eck, R.V., Dayhoff, M.O. 1966. *Science* 152:363

145. Jukes, T.H. 1966. *Molecules and Evolution*, New York: Columbia Univ. Press, 285 pp.

146. Fitch, W.M. 1971. *Nature* 229:245

147. Engel, P.C. 1973. *Nature* 241:119

148. Black, J.A., Dixon, G.H. 1967. *Nature* 216:152

149. Ycas, M. 1972. *J. Mol. Evol.* 2:17

150. Magni, G.E., Puglisi, P.P. 1966. *Cold Spring Harbor Symp. Quant. Biol.* 31:699

151. Whitfield, H.J., Jr., Martin, R.G., Ames, B.N. 1966. *J. Mol. Biol.* 21:335

152. Inouye, M., Akaboshi, E., Tsugita, A., Streisinger, G., Okada, Y. 1967. *J. Mol. Biol.* 30:39

153. Berger, H., Brammar, W.J., Yanofsky, C. 1968. *J. Mol. Biol.* 34:219

154. Berger, H., Brammar, W.J., Yanofsky, C. 1968. *J. Bacteriol.* 96:1672

155. Fitch, W.M. 1970. *J. Mol. Biol.* 49:15

156. Hilse, K., Braunitzer, G. 1968. *Z. Physiol. Chem.* 349:433

157. Powers, D.A., Edmundson, A.B. 1972. *J. Biol. Chem.* 247:6694

158. Sneath, P.H.A., Sokal, R.R. 1973. *Numerical Taxonomy, Principles and Practice of Numerical Classification*, San Francisco: W.H. Freeman & Co. In press

159. Laird, C.D., McConaughy, B.L., McCarthy, B.J. 1970. *Nature* 224:149

160. Kohne, D.E. 1970. *Quart. Rev. Biophys.* 3:327

161. Britten, R.J., Kohne, D.E. 1968. *Science* 161:529

162. Southern, E.M. 1970. *Nature* 227:794

163. Walker, P.M.B. 1971. *Nature* 229:306

164. Southern, E.M. 1971. *Nature* 232:82

165. Sutton, W.D., McCallum, M. 1971. *Nature* 232:83

166. McCarthy, B.J., Farquhar, M.N. 1972. *Brookhaven Symp. Biol.* 23:1

167. Rice, N.R. 1972. *Brookhaven Symp. Biol.* 23:44

168. Britten, R.J., Davidson, E.H. 1969 *Science* 165:349

169. Britten, R.J. 1972. *Brookhaven Symp. Biol.* 23:80

170. Mazrimas, J.A., Hatch, F.T. 1972. *Nature New Biol.* 240:102

171. Sarich, V.M. 1969. *Syst. Zool.* 18:286

172. Sarich, V.M., Wilson, A.C. 1967. *Science* 158:1200

173. Wallace, D.G., Wilson, A.C. 1972. *J. Mol. Evol.* 2:72

173a. Arnon, R., Maron, E. 1971. *J. Mol. Biol.* 61:225

173b. Arnheim, N., Sobel, J., Canfield, R. 1971. *J. Mol. Biol.* 61:237

174. Arnheim, N., Steller, R. 1970. *Arch. Biochem. Biophys.* 141:656

175. Edwards, A.W.F., Cavalli-Sforza, L.L. 1965. *Biometrics* 21:362

176. Cavalli-Sforza, L.L., Edwards, A.W.F. 1967. *Am. J. Human Genet.* 19: 233

177. Fitch, W.M., Margoliash, E. 1968. *Brookhaven Symp. Biol.* 21:217

178. Wagner, W.H., Jr. 1961. In *Recent Advances in Botany*, 841–44, Toronto: Univ. Toronto Press

179. Wagner, W.H., Jr. 1969. In *Systematic Biology*, 67–99, Washington, D.C.: Nat. Acad. Sci.

180. Farris, J.S. 1970. *Syst. Zool.* 19:83

181. Moore, G.W., Barnabas, J., Goodman, M. 1973. *J. Theor. Biol.* 38:459

181a. Hartigan, J.A. 1973. *Biometrics* 29:53
182. Estabrook, G.F. 1971. *J. Theor. Biol.* 21:421
183. Reichert, T.A., Wong, A.K.C. 1971. *J. Mol. Evol.* 1:97
184. Wilson, A.C., Sarich, V.M. 1969. *Proc. Nat. Acad. Sci. USA* 63:1088
185. Gibbs, A.J., Dale, M.B., Kinns, H.R., MacKenzie, H.G. 1971. *Syst. Zool.* 20:417
186. Bahl, O.P., Smith, E. 1965. *J. Biol. Chem.* 240:3585
187. Goodman, M., Barnabas, J., Matsuda, G., Moore, G.W. 1971. *Nature* 233:604
188. Barnabas, J., Goodman, M., Moore, G.W. 1971. *J. Comp. Biochem. Physiol.* 39B:455
189. Barnabas, J., Goodman, M., Moore, G.W. 1972. *J. Mol. Biol.* 69:249
190. Boyer, S.H., Crosby, E.F., Noyes, A.N., Fuller, G.F., Leslie, S.E., Donaldson, L.J., Vrablik, G.R., Schaefer, E.W., Jr., Thurmon, T.F. 1971. *Biochem. Genet.* 5:405
191. Boulter, D., Ramshaw, J.A.M., Thompson, E.W., Richardson, M., Brown, R.H. 1972. *Proc. Roy. Soc. London* B181:441
192. Mross, G.A., Doolittle, R.F. 1967. *Arch. Biochem. Biophys.* 122:674
193. Doolittle, R.F., Wooding, G.L., Liu, Y., Riley, M. 1971. *J. Mol. Evol.* 1:74
194. Prager, E.M., Arnheim, N., Mross, G.A., Wilson, A.C. 1972. *J. Biol. Chem.* 247:2905
195. Fitch, W.M., Markowitz, E. 1970. *Biochem. Genet.* 4:579
195a. Strydom, D.J. 1973. *J. Comp. Biochem. Physiol.* 44B:269–81
195b. Staehelin, M. 1972. *J. Mol. Evol.* 1:258–62 62
196. Fitch, W.M., Margoliash, E. 1967. *Biochem. Genet.* 1:65
197. Uzzell, T., Corbin, K.W. 1971. *Science* 172:1089
198. Fitch, W.M. 1972. *Haematologie und Bluttransfusion*, 10:199
199. Fitch, W.M. 1972. *Brookhaven Symp. Biol.* 23:186
200. Markowitz, E. 1970. *Biochem. Genet.* 4:579
201. Margoliash, E., Fitch, W.M. 1970. *Miami Winter Symposium*, W.J. Whelen, ed., Amsterdam: North-Holland Pub. Co., 33
202. Vogel, F. 1969. *Humangenetik* 8:1
203. Fitch, W.M. 1973. *J. Mol. Evol.* 2:181–86
204. Fitch, W.M. 1973. *J. Mol. Evol.* 2:123–36
205. Vogel, F., Roehrborn, G. 1965. *Humangenetik* 1:635
206. Derancourt, J., Lebor, A., Zuckerkandl, E. 1967. *Bull. Chem. Soc. Biol.* 49:577
207. Fitch, W.M. 1967. *J. Mol. Biol.* 26:499
208. Lehmann, H., Carrell, R.W. 1969. *Brit. Med. Bull.* 25:14
209. Vogel, H., Derancourt, J., Zuckerkandl, E. 1971. *Peptides*, 339–46. Amsterdam: N. Holland Pub. Co.
210. Zuckerkandl, E., Derancourt, J., Vogel, H. 1971. *J. Mol. Biol.* 59:473
211. Vogel, F. 1972. *J. Mol. Evol.* 1:359
212. Vogel, F. 1972. *Humangenetik* 16:71
213. Fitch, W.M. 1972. *Humangenetik* 16:67
214. Margoliash, E., Smith, E.L. 1965. In *Evolving Genes and Proteins*, ed. V. Bryson, H. Vogel, 221–42, New York: Academic
215. Fitch, W.M., Margoliash, E. 1967. *Biochem. Genet.* 1:65
216. Fitch, W.M., Margoliash, E. 1968. *New York Acad. Sci.* 151:359
217. Holmquist, R. 1972. *J. Mol. Evol.* 1:115
218. Holmquist, R. 1972. *J. Mol. Evol.* 1:134
219. Holmquist, R. 1972. *J. Mol. Evol.* 1:211
220. Fitch, W.M. 1971. *Biochem. Genet.* 5:231
221. Holmquist, R., Jukes, T.H. 1972. *J. Mol. Evol.* 2:10
222. Jukes, T.H., Holmquist, R. 1972. *J. Mol. Biol.* 64:163
223. Holmquist, R., Cantor, C.R., Jukes, T.H. 1972. *J. Mol. Biol.* 64:145
224. Kimura, M. 1968. *Nature* 217:624
225. Jukes, T.H., Holmquist, R. 1972. *Science* 177:530
226. Ohta, T., Kimura, M. 1971. *J. Mol. Evol.* 1:18
227. Cedergren, R.J., Cordeau, J.R., Robillard, P. 1972. *J. Theor Biol.* 37:209
228. Squires, C., Carbon, J. 1971. *Nature New Biol.* 233:274
229. Tinoco, I., Jr., Uhlenbeck, O.C., Levine, M.D. 1971. *Nature* 230:362
230. Delisi, C., Crothers, D.M. 1971. *Proc. Nat. Acad. Sci. USA* 68:2682
231. Figueroa, R., Soto, A., Gonzalez, G., Pieber, M., Romero, C., Toha, J.C. 1972. *J. Theor. Biol.* 36:321
232. Friedman, E., Gonzalez, G., Gampel, Z., Pieber, M., Romero, C., Toha, J.C. 1969. *Z. Naturforsh.* 24:419
232a. Gianni, A.M., Giglioni, B., Ottolenghi, S., Comi, P. 1972. *Nature New Biol.* 240:183

233. White, H.B. III, Laux, B.E., Dennis, D. 1972. *Science* 175:1264
234. Fitch, W.M. 1972. *J. Mol. Evol.* 1:185
235. Min Jou, W., Haegeman, G., Ysebaert, M., Fiers, W. 1972. *Nature* 237:82
235a. Mark, A.J., Petruska, J.A. 1972. *J. Mol. Biol.* 72:609
236. Lebowitz, P., Weissman, S.M., Radding, C.M. 1971. *J. Biol. Chem.* 246:5120
237. Billeter, M.A., Dahlberg, J.E., Goodman, H.M., Hindley, J., Weissmann, C. 1969. *Nature* 224:1083
238. Nichols, J.L. 1970. *Nature* 225:147
239. Ohe, K., Weissman, S.M. 1971. *J. Biol. Chem.* 246:6991
240. Nolan, C., Margoliash, E. 1967. *Ann. Rev. Biochem.* 37:727
241. de Jong, W.W.W., Went, L.N., Bernini, L.F. 1968. *Nature* 220:788
242. Chance, R.E., Ellis, R.M., Bromer, W.W. 1968. *Science* 161:165
243. Nolan, C., Margoliash, E., Peterson, J.D., Steiner, D.F. 1971. *J. Biol. Chem.* 246:2780
244. Peterson, J.D., Nehrlich, S., Dyer P.E., Steiner, D.F. 1972. *J. Biol. Chem.* 247:4866
245. Jones, R.T., Brimhall, B., Huisman, T.H.J., Kleihauer, E., Betke, K. 1966. *Science* 154:1024
246. Shibata, S., Miyaji, T., Uede, S., Matsuoka, M., Tuchi, I., Yamada, K., Shiuhai, N. 1970. *Proc. Japan Acad.* 46:440
247. Simpson, G.G. 1945. *Bull. Am. Mus. Nat. Hist.* 85
248. Wajcman, H., Labie, D., Schapira, G. 1973. *Biochem. Biophys. Acta* 295:495
249. Seid-Akhavan, M., Winter, W.P., Abramson, R.K., Rucknagel, D.L. 1972. *Blood* 40:927

3060

DEVELOPMENTAL GENETICS OF *DROSOPHILA* IMAGINAL DISCS

John H. Postlethwait and Howard A. Schneiderman

Department of Biology, University of Oregon, Eugene, Oregon,
and Center for Pathobiology, University of California, Irvine, California

Introduction

The aim of developmental biology is to discover how the genetic instructions for development are implemented by the epigenetic processes of development. In recent years *Drosophila* has been increasingly used to answer many questions concerned with development, and the imaginal discs have received special attention. These small nests of cells are set aside early in development for the formation of the adult integument. The entire surface of the adult fly is formed from imaginal discs so that any mutation that affects the cuticle of an adult fly directly or indirectly alters the development of imaginal discs. Many such mutations have been detected, described, and analyzed(180). This review examines the genetic basis of the development of the imaginal discs of *Drosophila.* Other aspects of *Drosophila* developmental genetics such as chromosome puffing (6), behavioral genetics (16a), genetics of embryogenesis (48, 351), spermatogenesis (131, 156), oogenesis (159), and biochemical genetics (64, 307) have recently been reviewed.

Summary of some special features of imaginal discs The imaginal discs of *Drosophila* possess a number of special features that have made them particularly attractive systems for both geneticists and developmental biologists:

(*a*) Imaginal discs share the many virtues of the insect integument. Each disc is composed of a single layer of cells, it secretes a cuticle replete with identifiable markers such as bristles and sensilla. For the most part, the cuticular contributions of individual cells can be recognized.

(*b*) Surgery and culture are convenient. Like fragments of insect integument (341, 342), imaginal discs can easily be isolated from a larva, surgically altered, sucked into a glass micropipette and transplanted into the body cavity of another host insect where they undergo prolonged and extensive growth. The fact that they are physically separate from each other and distinct from the rest of the developing animal makes them even more manageable than fragments of the integument. When they are transplanted to a larva, they respond to hormonal stimuli and undergo metamorphosis and differentiation along with the host, producing specific adult structures (306). Alternatively, they can be cultured in an adult fly abdomen where the hormonal conditions do not cause metamorphosis of the implant but do permit it to grow (19, 20). Proliferation rate can be controlled by using as a host an adult fly that has been maintained on "sugar food" where the discs are not able to grow or differentiate but remain alive for more than 15 days (68). At any time they can be extracted from the adult and reimplanted into a larva for metamorphosis (115).

(*c*) In imaginal discs two basic developmental processes, determination and differentiation, are widely separated in time and can be separately analyzed (116, 119). Determination is recognized when a particular fate such as "legness" or "wingness" is passed from a cell to its offspring during culture (91). It is a developmental decision, propagated by cell heredity. A cell is considered as not yet determined for a single fate if a clone of cells derived from it exhibits several fates after metamorphosis (200).

(*d*) Imaginal discs also provide a special opportunity to analyze nonheritable decisions made by cells, such as the decision to become a specific part of a disc that gives rise to a particular bristle. Bryant (33) has proposed the term "specification" to identify the process by which these nonheritable commitments are made. As we shall see, the process of specification underlies the formation of spatial patterns within an imaginal disc.

(*e*) Through the use of X-ray induced somatic crossing-over techniques it is easy to produce genetic mosaics in which cells have visible cell autonomous genetic markers. In such mosaics it is possible to follow the fate of the progeny of an individual imaginal disc cell and the interaction of populations of genetically marked cells.

(*f*) Because of *Drosophila*'s well-known genetics, it is possible to induce and select mutants of various sorts, including cell-autonomous mutants which are useful as markers in mosaics, and mutants that interfere with various developmental processes. Analysis of the latter type of mutant may be expected to throw light upon the normal processes of development and these provide the basis for this review.

STRUCTURE AND DEVELOPMENT OF IMAGINAL DISCS Imaginal discs occur to various degree in all orders of insects with complete metamorphosis. In some insects, such as Lepidoptera, only certain parts of the adult originate from imaginal discs, for example wings, antennae, and genitalia. Other parts such as the adult legs arise from the progeny of the same cells that formed the larval legs (158). Among the higher Diptera, including *Drosophila*, the entire adult integument except for the abdomen arises from imaginal discs. There are ten pairs of major imaginal discs and a genital disc. Each is characterized by its location in the larva, by its size and shape, and most importantly by the part of the adult that it forms. The integument of the adult head arises from three pairs of discs; the external parts of the adult thorax are formed by three pairs of leg discs, a pair of dorsal prothoracic discs, wing discs, and haltere discs. The genitalia are formed from the genital disc. Each segment of the abdomen is formed from four small nests of abdominal histoblasts which are continuous with the larval epidermis.

In *Drosophila*, the imaginal discs are first evident histologically in the late embryo (163) or early larva (9) as thickenings of the epidermis. Later they become invaginated to form a sac-like structure, attached to the larval epidermis by a stalk. Unlike other epidermal cells, which secrete a cuticle at each larval molt, the imaginal discs do not secrete a cuticle until metamorphosis.

The imaginal discs have a remarkably simple organization. They are basically hollow folded sacs that consist of a single layered columnar epithelium. During the four days of larval life the epithelium becomes folded as the discs grow by cell division. Their cells remain undifferentiated, however, and resemble the relatively undifferentiated cells of a young embryo. This state of affairs continues until the onset of metamorphosis at which time the number of cells in an individual disc ranges from a few hundred in the smaller discs to many thousands in the larger ones. At metamorphosis the epithelium of the imaginal discs undergoes a spectacular morphogenetic movement known as eversion, which converts the tightly folded epithelium of the disc into an extended adult structure. The shape of the cells changes from tall columnar to squamous and the kinds of junctions between the cells also change (218). Differentiation and cuticle synthesis follow, and each disc secretes a specific part of the cuticle of the adult fly.

In this review we have adopted the following philosophy: if a mutation alters a process, then the process is biologically significant. It is easy when studying an organism to become convinced that a process that we as observers *think* is important will also *be* important to the organism. This is not necessarily the case. However when we study a process that is changed by a mutation, then we can have confidence that the process is not only interesting to the observer, but significant to the organism as well. This review analyzes the ways in which genetic mutations alter the determination and differentiation of imaginal discs. We will examine in turn mutations that alter the following steps in the development of imaginal discs: (*a*) the initiation of developmental commitments in cells; (*b*) the maintenance of those commitments; (*c*) the formation of spatial patterns; and (*d*) the differentiation of cells during metamorphosis.

Genetic Techniques Used in Developmental Studies

MORPHOLOGICAL STUDIES The classical method of developmental genetics is to describe the morphological abnormalities associated with a particular mutation. When this is done carefully, one can trace the (often) pleiotropic abnormalities to a single initial aberrant process. For example, the defect leading to small eyes and duplicated antennae in *eyeless* (*ey*, 4—2.0), can be traced to cell death in particular regions of the imaginal disc (62,63). But pleiotrophy may be due to independent gene action in two different tissues (autonomous pleiotropy) or to gene action in a single tissue, which only secondarily affects other regions (relational pleiotropy). These possibilities have been distinguished for *split* (*spl*, 1-3), which causes both rough eyes and doubled thoracic bristles. An examination of the morphology of *split* tissue in genetic mosaics revealed that the eye and bristle effects can be expressed autonomously (284), thus ruling out relational pleiotropy. Similar experiments should be done with *ey* to ascertain whether antennal duplication is due to autonomous or relational pleiotropy.

PRODUCTION AND ANALYSIS OF GENETIC MOSAICS AND CHIMERAS Several techniques exist that permit the formation of individuals, tissues, or cells composed of more than one genotype. The term mosaic is used when cells that are part of an organism are caused to undergo a change of genotype, whereas the term chimera is used when cells or parts of separate organisms are brought together. Thus, mosaics are formed genetically (using gynandromorphs and somatic crossing-over) and chimeras are formed mechanically (various kinds of transplantations). Genetic mosaics and chimeras allow one to answer questions about cellular autonomy. For example, when cells of one genotype are present in the same animal as cells of a second genotype, does each cell type continue to control its own phenotype, or do the cell types interact?

(*a*) *Gynandromorphs*. (see 126). As the name implies, gynandromorphs are organisms that are part male and part female. Since *Drosophila* males are XY or XO and females are XX, the male is hemizygous for X linked genes. A heterozygous female cell can thus be changed to a hemizygous male cell by the loss of an X chromosome. Any recessive genes on the X retained in the male cells may then be expressed, while their effect is hidden in the heterozygous female cells.

Gynandromorphs can be produced by X irradiation of parents (215), or by the action of some meiotic mutants, such as *claret nondisjunction* (178, 287, 337), *paternal loss* or *mitotic loss inducer* (126) all of which produce X chromosome loss in the progeny. But the method most frequenty employed makes use of the spontaneous elimination of an unstable ring X chromosome (126, 133–135, 213). Analysis of gynandromorphs obtained from *claret* or ring chromosomes permits one to make several generalizations about their developmental properties. Since gynandromorphs in general consist of large contiguous areas of either male or female cells, chromosome elimination must occur rather early in development. If one assumes only one elimination event per animal, then the later the elimination the smaller the fraction of XO tissue in the gynandromorph. If there is no preference for a given body part to be a certain sex, then among a large number of gynandromorphs the frequency of XO cells in a specific body part will reflect the average fraction of XO cells in gynandromorphs. Data obtained by ring chromosome elimination (126, 139, 227, 238), and *claret nondisjunction* (73), indicate that any given structure is male about half the time in gynandromorphs. So mosaicism is initiated in gynandromorphs in an early (perhaps the first) cleavage division.

A second conclusion can be drawn from a fact already mentioned. Since a fly consists of broad regions of male or female tissue with little intermixture, cells must maintain their spatial relationship with adjacent cells, only infrequently migrating away from their original neighbors. A third conclusion follows from the observation that any two cuticular markers on the adult surface may be of different sexes (73). This means that the earliest cleavage planes must be at indeterminate angles, a conclusion which has a cytological precedent (337).

By refining and extending a technique originally devised by Sturtevant (287), Garcia–Bellido & Merriam (73) were able to use gynandromorph data to estimate the relative distances that imaginal disc precursors were apart at the time they became clonally distinct from the rest of the embryo. The procedure depends on the fact noted above that the male–female borderline in gynandromorphs lies in variable directions because the earliest planes of nuclear division are in variable directions. Therefore, the frequency with which a borderline lies between any two imaginal disc precursors depends upon their distance apart at the time they were set aside to produce the discs in question. These relative distances can be deduced from the relative frequencies with which the structures derived from the two discs are of opposite sex in the adult. This method produces a two-dimensional map of the relative locations of imaginal disc precursors in the early embryo at the time the imaginal discs are set aside (73, 126, 139).

A drawback to the usefulness of gynandromorphs is that in general one is limited to using X linked genes that have a dominant allele on the ring chromosome. However, by double meiotic crossing-over it is possible to transfer to the ring X, autosomal fragments translocated to a rod-X chromosome (176).

(*b*) *Somatic crossing-over.* Whereas mosaicism in gynandromorphs is initiated very early in development and only the sex chromosomes are involved, somatic crossing-over can produce mosaicism in any mitotically dividing cells at any developmental stage, and both the sex chromosomes and the autosomes can be involved. Somatic crossing-over occurs between chromatids and not between whole chromosomes (277), and apparently involves multiple chromosome breaks (124, 125, 190). It causes a single heterozygous cell to produce two daughter cells each of which is homozygous for loci distal to the position of the crossover. Single cells in imaginal discs (and elsewhere) become homozygous for recessive marker genes while the cells in the rest of the animal are hetereozygous. For example, if the animal is doubly heterozygous for *y* and *sn* (i.e. *y* +/+*sn*) (*y*(1 = 0.0), *yellow* rather than dark brown bristles; *sn* (1–21.0), *singed* rather than straight bristles), it will normally display a wild-type phenotype. However, if one of its cells undergoes somatic crossing-over, then one of the daughter cells might be homozygous *yellow* (*y* +/*y* +) and the other homozygous *singed* (+*sn*/+*sn*). As each cell divides it will generate a clone of cells, which, after metamorphosis, can be seen as a coherent patch of yellow or singed bristles, surrounded by wild-type bristles.

Somatic crossing-over occurs spontaneously, but its frequency can be increased by temperature (27, 152, 277, 282), *Minute* mutants (277), ring chromosomes (28), or other genetic factors (170, 241, 338, 339), high oxygen pressures (273, 274), and increased maternal age (29, 262). For developmental studies, the most useful inducer of somatic crossing-over is X-irradiation (1, 12, 13, 61, 172, 214, 266), since it can be applied at specific developmental stages. Experiments varying dose rate (124, 125), and employing dose fractionation (190), have shown that the recombinational events occur during or very shortly after irradiation.

For a given dose, X-rays of 100kV appear to be more active than those of 55kV, and a high dose rate produces a higher frequency of somatic crossing-over than does a lower dose rate (124, 125). Although several authors have concluded that somatic crossing-over occurs primarily in the proximal heterochromatin (1, 12, 14, 27, 28, 152, 277, 338), it is clear that euchromatic somatic recombination occurs as well (281). Garcia-Bellido (72) has produced an extremely useful somatic recombination map, which indicates that while X-ray induced somatic crossing-over occurs more frequently near the centromere than might be expected from meiotic map distance and mitotic chromosome length, euchromatic crossing-over also occurs at a substantial rate. Somatic crossing-over experiments involving chromosome rearrangements (192) have resulted in a "hairpin" model of chromosome pairing in *Drosophila* somatic cells in which centromeres arrange themselves together, telomeres come in proximity to each other, and homologous euchromatic regions pair (191).

Clones of cells arising from somatic crossing-over can conveniently be used to study cell lineage, morphogenesis, cell migration, cell division rates and determination, as well as cell interactions and autonomy. For example, clone size in the adult depends on the stage of irradiation, because the number of cell divisions that are to occur before metamorphosis decreases with larval age. Hence the average clone size following irradiation at various stages can be used to estimate the cell numbers and rates of cell division in the discs at these stages. For example, the leg disc begins its developmental career as a group of about twenty cells in the embryo, and grows throughout larval life with an average cell cycle time of about 15 h, until the time of metamorphosis (36). The wing and antennal discs grow somewhat faster, with an average doubling time of about 11h (31, 74, 227). The abdominal tergites represent an unusual situation, since the abdominal histoblasts do not appear to increase in cell number during larval life, but do show rapid growth during the initial stages of metamorphosis (76, 110). A striking result from the somatic crossing-over studies is that in many parts of the adult, clones are markedly nonrandom in shape. The clearest case is that of the leg in which clonally related cells form long narrow strips of tissue often extending over several leg segments (36). Some clones have been found to be over 100 cells long and only a few cells wide. This indicates that there is a predominantly radial alignment of daughter cells in the developing disc, which may be one of the important factors determining its structure. Similar longitudinal clonal patterns are present in wing discs (32, 74). Furthermore, the fact that most clones are continuous and with smooth outlines persuades us that there is little individual cell movement in these systems.

Although at present somatic crossing-over in imaginal discs can only be assayed after metamorphosis by looking at cuticular structures (see 126 for lists of useful mutants), for some purposes it would be useful to be able to visualize clones in unmetamorphosed discs, internal organs, and larval organs. A class of mutants with great potential for these studies is temperature sensitive cell autonomous lethals (5, 243); clones of dead and dying cells could be identified by staining techniques (270). Such mutants could be used to map the cell lineage of internal organs in much the same way that cuticular markers have been used to map the cell lineage of the integument. Another group of mutants with potential as internal markers is that of the null alleles of several enzymes (126, 148, 271, 308), which can be assayed histochemically in mosaics.

The main advantage in using somatic crossing-over to produce somatic mosaics is that we can induce the event at specific stages of development. A disadvantage of using X-rays to induce somatic recombination is that the radiation doses necessary to produce significant amounts of somatic crossing-over, kill large numbers of cells (52, 263, 270), which sometimes causes abnormalities in development (229, 317, 331). Cell death is usually followed by increased cell division (128, 263), which can be detected by an increase in clone size, reflecting a decrease in the number of progenitor cells.

(*c*) *Organ transplantation.* Another method of producing organisms composed of several genotypes is to transplant an organ of one genotype into the body cavity of a host of a second genotype (54). This technique has been described in the introduction and is used extensively. Another approach has been devised by Bhaskaran & Sivasubramanian (18) who transplanted mature imaginal discs just beneath the puparium of prepupal hosts. The discs join up with the integument, evert and complete their metamorphosis. By this means it is possible to transplant discs to various locations in the body and examine the way tissues from different discs join up to form the continuous integument. Transplantation of imaginal discs into adults can be coupled with injection of an ecdysone to cause metamorphosis of the implant (226a). With these transplant methods, it is possible to control precisely which organs will be of which particular genotype, whereas with somatic crossing-over and gynandromorphs, chance must be relied upon to provide the type of mosaic desired, and the precise genotype of every tissue in the mosaic animal cannot always be ascertained.

(*d*) *Cell dissociation and reaggregation.* A tissue composed of several genotypes can be produced by mechanical or enzymatic separation of imaginal disc cells into single cells and cell clumps, followed by reaggregation (66, 199, 216). After transplantation into a larval or adult host, the results of developmental interactions can be observed after metamorphosis. This technique provides a greater intermixture of cells than occurs in the previous methods and several genotypes can be intermingled into one tissue. In addition, cells that in normal development never come in contact with each other can be juxtaposed.

(*e*) *Nuclear transplantation.* The tour de force of chimera methods is the production of a single cell whose cytoplasm is from a cell of one genotype and whose nucleus is of a second genotype. This has recently been accomplished by the transplantation of nuclei (plus a small amount of adhering cytoplasm), from one cleavage stage embryo into a fertilized or unfertilized egg (98, 141–144, 204, 261, 357). The interaction of nucleus and cytoplasm can then be investigated. Although this method is technically more difficult than the previous ones, it can be used to assay nuclear changes in development (see also 111).

(*f*) *Somatic cell fusion.* This method, used with success in studying genetic regulation in mammals (51), is potentially useful for similar studies in imaginal discs. Although the authors are not aware of any success with somatic cell mating with *Drosophila* imaginal discs, cells of a *Drosophila* cell culture line have been fused (15). There are also reports of somatic cell fusion between cultured mosquito cells and human cells (358). We suspect a system could be devised using nonallelic temperature-sensitive cell autonomous lethals (5, 239, 243, 284a) to select for hybrid imaginal disc cells. Such a technique would allow one to ask questions of general significance such as, what will be the fate of a hybrid cell formed from an imaginal leg cell and an imaginal antennal cell?

CONDITIONAL MUTANTS Much of the elucidation of bacterial physiology depended upon the isolation of conditional mutants—genetic variants that grow in one environment but not in a second. This kind of mutant is also useful in the study of imaginal disc development. One type of conditional mutation requires a particular nutritional supplement to a defined medium to survive (56, 327, 328). A more generally useful type of conditional mutation requires a particular temperature to express a mutant phenotype (290). Temperature sensitive mutants have been used extensively to study the development of imaginal discs, and a large number of temperature sensitive pupal lethals displaying numerous defects in their discs have recently been isolated and analyzed (5).

SOME OTHER USEFUL DEVELOPMENTAL TECHNIQUES Several recent embryological techniques that promise to prove useful to developmental geneticists should be mentioned.

(*a*) *Methods to increase permeability. Drosophila* eggs are notoriously impermeable to exogenous substances like radiolabeled compounds and even histological fixatives. These difficulties may be overcome by causing females to oviposit premature eggs (356), or by dechorionation followed by treatment with Triton X-100 (250).

(*b*) *Maintenance of frozen tissues.* Many *Drosophila* tissue culture lines are maintained in living adult hosts, and must be transferred about every two weeks. A liquid nitrogen preservation technique using dimethyl sulfoxide has been devised that permits tissues to be stored for many months (83). Some mutant stocks are tedious to maintain, especially in view of the apparent instability of many temperature sensitive mutants (Wright, personal communication). Recently a technique has been perfected for freezing larval ovaries in a glycerin solution, followed by thawing and transplantation into genetically marked host larvae. The implanted ovaries attach to the donor's oviduct in almost 2/3 of the cases, and about 20% of the offspring arise from the mutant ovary (30).

(*c*) *In vitro cell culture.* Although imaginal disc cells have been refractory to in vitro culture, recently Mandaron (186, 187) developed a medium that permits complete metamorphosis in vitro. And Schneider (252) has developed a method to culture imaginal disc cells obtained from embryos.

The Initiation of Developmental Commitments

The egg of *Drosophila*, like that of most insects, does not undergo cleavage in the usual sense. Instead, after fertilization, the nuclei undergo 8 rapid and synchronous cleavage divisions over the course of about 100 minutes and migrate to the peripheral cytoplasm—the cortex or periplasm. The embryo is now called a preblastoderm. The nuclei in this preblastoderm embryo divide synchronously four more times to form a syncytium composed of about 4000 cells, of which about 3500 are on the surface. At this time, 160 minutes after egg deposition, the plasma membranes form simultaneously over the entire egg surface and 3500 cells are formed all at once (65, 146). The embryo is now called a blastoderm. Gastrulation and the remainder of embryogenesis follows. At 25°C the larva hatches from the egg about 22 hours after fertilization and egg deposition.

A number of experiments have focused on the question of when during development the cells or nuclei become determined to construct imaginal discs. To answer this question it has become necessary to analyze oogenesis itself to discover to what degree the egg cytoplasm laid down by the mother plays a causative role in imaginal disc determination. A good starting point is an examination of sterile mutants. Mutants that cause female sterility are of two types: those that cause abnormalities in the initial development of the reproductive system resulting in no mature eggs, and those that cause abnormalities in egg development within the reproductive system resulting in eggs that cannot support normal development (see 159 for extensive review). This latter group of mutants is called maternal effect lethal mutants.

FEMALE STERILE MUTANTS RESCUABLE BY WILD TYPE SPERM In maternal effect mutants in *Drosophila*, the phenotype of the individual depends not only on its own genotype, but also on the genotype of its mother. Consider the genetics of three sex linked recessives which show this maternal effect: *rudimentary* (*r*, 1–54.5), *fused* (*fu*, 1–59.5), and *deep orange* (*dor*, 1–0.3) (182). If homozygous mutant females are mated to hemizygous mutant males, then the offspring die as embryos (45–47). But if heterozygous females are mated to hemizygous mutant males, then all the offspring survive. It is concluded that the mutant mother fails to provide her eggs with a substance indispensable for normal development. A third mating, that of homozygous mutant females to wild-type males, results in death of hemizygous mutant males, but survival of about one third of the heterozygous females (59). These observations suggest that when the sperm introduces a wild-type allele into an egg from a mutant female, it rescues the egg by supplying the missing required factor.

Offspring of *r* mothers can be spared not only by the introduction of a wild-type nuclear gene into the zygote, but also by injection of wild-type cytoplasm from unfertilized eggs into preblastoderm embryos (204). Many preblastoderm embryos injected with 1.5% of their volume of cytoplasm from unfertilized wild-type eggs were able to complete both embryonic and postembryonic development and emerged as adults, whereas not a single uninjected control egg was able to complete embryonic development. Experiments with *dor* mothers yielded similar results and eggs were enabled to develop to late embryonic stages (80). Therefore, wild-type egg cytoplasm, manufactured under the direction of the maternal genome, contains factors that can repair preblastoderm eggs from *r* or *dor* mothers. Apparently *r* and *dor* are blocked in the production of small molecules—pyrimidines in *r* (198, 286), and pteridines in *dor* (80). Injection of cytoplasm repairs this defect. Recent studies indicate that injection of pyrimidines alone will also repair the defect in *r* (204).

The influence of a wild-type gene can also be investigated in gynandromorphs. While a heterozygous *fu* (or *r*) zygote from a homozygous *fu* (or *r*) female will survive, gynandromorphic eggs consisting of half *fu* (or *r*) and half *fu*/+(or *r*/+) cells will seldom survive (58). In those few gynandromorphs that managed to survive, the anterior of the adult was always wild-type. This result suggests to us that when some embryonic organs in the anteriors have a wild-type phenotype, they sometimes provide enough of the fu^+ and r^+ substance for cells in the rest of the embryo so that the larva can hatch. In contrast, wild-type cells in posterior parts of the embryo are unable to rescue the rest of the embryo possibly because they do not produce enough fu^+ and r^+ substance.

The pattern of lethality in these three genes is somewhat different. Genetically *dor* offspring from *dor* females undergo abnormal cleavage division, partial blastoderm formation, and abnormal gastrulation (45–47), although there is variability dependent on genetic or environmental difference (132). Generalized post-gastrula defects occur with *fu* and *r*. We can conclude then from studies of these mutants that wild-type females put into their eggs substances that promote normal blastoderm formation, gastrulation, and primary organogenesis. Collection and study of more female sterile mutants with a maternal influence will increase our understanding of the nature of factors controlling this process. Recent experiments show that maternally influenced lethals similar to *dor*, *r*, and *fu* represent about 1% of all EMS induced sex linked lethals and about 10% of all female steriles (153). Even temperature-sensitive maternally influenced lethals have been produced (137).

MATERNAL EFFECT AND SEXUAL PHENOTYPE A second type of maternal effect lethal alters the sex ratio. One such mutant reminds one of Portnoy's mother in that sons are stifled due to a maternal influence. Females homozygous for the X chromosomal gene *sonless* (*snl*, 1-56.3) (44), mated to *snl* males produce no sons, but almost the normal number of daughters. Matings of *snl* females to snl^+

males produce a rare son, whose single X chromosome is *snl*$^+$ presumably as the result of nondisjunction (42, 43). So, as in the case of *dor*, *fu*, and *r*, a wild-type zygotic gene rescues the otherwise doomed progeny of a mutant mother. Females homozygous for another gene *abnormal oocyte* (*abo*, 2–38) also produce fewer sons than normal (244).

The female counterpart of *snl* has also been described (236). Genetically *daughterless* (*da*, 2–39) females produce only male offspring (16). Genetically female (XX) offspring of a homozygous *da* mother cannot escape lethality even though they may be disguised as males by the *transformer* gene (*tra*, 3–45) (288), which causes a genotypic female to show a male phenotype, or the *double sex* gene (*dsx*, 3–48.1), which causes both XY and XX individuals to be intersexual (41). In contrast, transvestism by the sons of *snl* females due to *dsx* does provide some survival advantage, and genetically female offspring of a homozygous *snl* mother masquerading as males because of *tra* are less viable than normal XX offspring (43).

Sandler has suggested that *abo* and *da* regulate the activity of X and Y heterochromatic loci, presumably ribosomal RNA (245). From these mutants we can conclude that embryos are sexually dimorphic at an early stage, and that wild-type females provide in the cytoplasm of the egg some factor necessary for the survival of sons, and something else for daughters.

These genes resulting in a maternal influence on sex ratio are not to be confused with the *maternally inherited* sex ratio abnormalities in *Drosophila* (233). The maternally inherited condition which results in male death is passed strictly through the female, with no indication of nuclear inheritance, and can be transferred by injection of a normal female with the contents of a dying male. The infective agent has been shown to be a spirochaete (234).

MATERNAL EFFECT MUTANTS AFFECTING SPECIFIC ORGANS The maternal effect mutants described above permit the conclusion that the normal mother produces substances in the egg necessary for normal embryonic development. In the absence of these substances the embryos showed many abnormalities as early as cleavage. However, there are other maternal effect mutants which have much more specific effects and cause particular organs (*a*) to be missing, (*b*) to be formed abnormally, or (*c*) to be replaced by a different body part. These mutants, discussed below, indicate the degree to which the initiation of cellular determination is controlled by the egg cytoplasm.

(*a*) *gs*. Homozygous females of the mutant *grandchildless* (*gs*) of *Drosophila subobscura* are not visibly different from the wild-type. But, when mated to either *gs* or wild-type males, *gs* females produce both sons and daughters that lack germ cells and are therefore sterile (272). Mutants that may be similar have been recently found in *D. Melanogaster* (310). Larval gonads from the offspring of *gs*/*gs* females transplanted into *gs*$^+$/*gs*$^+$ larval hosts retain their sterility, while reciprocally transplanted gonads are fertile (289). Likewise, a *gs*$^+$/*gs*$^+$ larval ovary implanted into a *gs*/*gs* larva produces fertile progeny (289). Autonomy therefore implicated the gonads themselves in the maternally influenced sterility.

Fielding (60), in comparing the oogenesis of homozygous *gs* females and their normal sibs, reported that while most of a developing embryo from a *gs/gs* female was normal, a specific region of the egg posterior, the pole plasm, was observed to be crumbling and vacuolated. These embryos develop into animals whose rudimentary gonads contain no germ cells, although otherwise they are normal. The normal pole plasm contains basophilic structures called polar granules, which appear to be composed of ribonucleoprotein. As the cleavage nuclei multiply and migrate to the periphery of a normal egg and become cellularized to form a single cell layered blastoderm, these polar granules become incorporated into the pole cells (185). The pole cells can be traced through development and are seen to form the germ cells (185) and perhaps the cuprophilic cells of the gut (232). Besides *gs*, the mutation *fs*(*1*)*N* (1–0.0) (49) in *D. melanogaster*, also affects the pole plasm. In eggs laid by *fs*(*1*)*N* females, the polar granules are few or absent, pole cells generally fail to form, and the embryo dies.

Mutation is not the only way by which *Drosophila* pole plasm can be altered. The pole plasm can be destroyed by ultraviolet irradiation (95, 96, 204), or its position shifted by centrifugation (149). In both cases no germ cells are formed and adults derived from the altered eggs are sterile. These observations show that the polar cytoplasm of the egg is involved in germ cell determination.

The biochemical nature of the factor(s) that cause cells to embark upon a germ cell developmental pathway is being investigated in several laboratories. Mahowald & Gehring (92), have begun studies aimed at this question using *gs* animals. They find that the ultrastructure of the polar granules of eggs from *gs* and gs^+ females appears the same, but that in *gs*, cleavage nuclei do not enter the polar plasm directly from the yolk. Rather, nuclei from the lateral periplasm secondarily invade the pole plasm, incorporate the polar granules, but fail to form pole cells. Okada et al (204) have taken a different approach involving cytoplasmic injection into eggs. By precisely timed UV irradiation of the posterior region of newly laid *D. melanogaster* eggs [a technique first applied by Geigy (95)], they produced sterility in 99% of the resulting adults. Such eggs could be rescued by injection of polar plasm from unirradiated eggs so that 42% of the animals that survived the injection regained fertility. In contrast, injection of cytoplasm from the anterior pole of unirradiated eggs into the irradiated eggs failed to restore fertility. Clearly, cytoplasmic injection into eggs can be used both as a bioassay for germ cell determinants and to identify the biochemical nature of the active components.

(*b*) *amx*. A second mutant type altering initiation of developmental commitments produces eggs that develop into adults abnormal in imaginal disc derivatives. Mutant *almondex* (*amx*, 1–27.7) females are sterile when mated to *amx* males, but produce a few heterozygous daughters when mated to amx^+ males (264). But 80% of the surviving progeny of *amx/amx* females are abnormal, and the irregularities are restricted to the thorax or abdomen. Thoracic aberrations included crippled or absent legs and blisters on the wings, while the abdominal sternites were displaced or altered in size or shape. The genitalia were apparently normal.

Shannon's photographs and descriptions demonstrate abnormalities similar to those obtained after experimental insults applied to developing eggs (96). For example, ultraviolet irradiation of the lateral periplasm of early cleavage eggs produced crippled or missing legs, isolated or rudimentary sternites, tergite fusion, bent wings, and short or missing head bristles (202). As in the offspring of *amx* females, the genitalia were normal, and posterior defects were more frequent than anterior irregularities. Defects similar to these may also be produced by removal of some egg cytoplasm by puncturing (143), or by centrifugation of early cleavage eggs (140). It is tempting to suggest that the abnormalities in adults that were subjected to UV, centrifugation, or puncture are due to alterations of the same underlying factors as are abnormalities in adult that were the progeny of *amx/amx* mothers. This possibility has been strengthened by recent studies of embryos and larvae obtained from *amx/amx* females, which reveal that they also display ectodermal abnormalities (265). Apparently *amx/amx* females produce eggs in which there are defects in specific regions that are involved in causing cells to embark upon particular developmental programs.

(*c*) *bicaudal and tuh*. The third type of mutant, which appears to alter an initial determination, causes one body part to be replaced by a different body part, a phenomenon called homoeosis (11). One of the most interesting mutants of this type is *bicaudal* (38). Females homozygous for a specific second chromosome or heterozygous for this chromosome and a *vg* deficiency, when mated to any male, lay eggs that die as embryos due to head abnormalities. In the most extreme cases, the larval head is replaced by an abdomen in mirror image symmetry to the normal abdomen. The micropyle and apparently the pole cells are not involved in the mirror image copy. Dorsally, at the end of each abdomen, are mirror image posterior spiracles connected by tracheae. Ventrally each larval segment contains several rows of small hooks. The anterior row points anteriorly, but the other rows generally point posteriorly. Thus, hooks indicate polarity and their orientation is reversed in the mirror image duplicates.

Although perfect mirror image patterns occur in low frequency (about 1%) this genotype gives rise to a large number of animals with various kinds of other head abnormalities. In some, a single posterior-most abdominal segment replaces the head, while in others several kinds of mouth hook irregularities occur. These phenotypes, caused by the mother's genotype rather than the individual's genotype, indicate that factors in the egg that normally determine anterior structures are missing or abnormal in the eggs of *bicaudal* females.

In flies, double abdomens can be produced not only genetically, but also by several experimental procedures. In chironomid flies centrifugation of cleavage stage embryos toward one of the poles results in mirror image double abdomens or mirror image double heads (84, 212, 352, 354). Ultraviolet irradiation of the anterior pole can also produce double abdomens in chironomids (353). It can be demonstrated that UV primarily affects the cytoplasm of the egg (a maternally derived component) rather than the nuclei, since irradiation of the anterior 1/8 of the egg *prior* to nuclear immigration yields double abdomens (150). Furthermore, it appears as if the UV altered factor is a nucleic acid, since visible light will decrease the frequency of double abdomens (151) and photoreversibility of UV damage is associated with nucleic acids (see 151 for literature).

Although in eggs from *bicaudal* mothers the determination of the larval anterior is altered to that of the larval posterior, it has not been demonstrated whether the determination of imaginal disc cells is also altered in these embryos. In normal animals, it has been shown by mixing genetically marked anteriors with posteriors or whole embryos either half way through embryogenesis (260), or at the cellular blastoderm stage (39), that the anterior and posterior imaginal cells are developmentally different. To test whether eggs from maternal effect mutants such as *bicaudal* affect imaginal as well as larval determination one could examine the phenotype of an embryo from a mutant mother, cut it in half and transplant it into an adult for maturation, and then back into a larva for metamorphosis (123), to see what adult parts arise from specific embryonic or larval areas.

While it is not clear whether *bicaudal* affects imaginal structures, another maternally influenced syndrome does involve a change of anterior adult parts into posterior adult parts. The mutant stock *tumorous head* (78), causes antennae to be replaced by legs, eyes to be replaced by abdominal tergites, and parts of the proboscis to be replaced by genitalia (223). The *tumorous head* stock contains two mutant genes. One, *tuh-3* (3-58.5) is located in a chromosomal region where many homoeotic mutants map, and is responsible for the homoeotic effect. The second, *tuh-1* (1-64.5), increases the penetrance of *tuh-3* and controls the maternal effect (79, 268).

The penetrance of *tuh* is higher at higher temperatures, and the temperature-sensitive phase is during embryonic life (349). Associated with the tumorous head stock is extensive embryonic and early larval lethality (78). Apparently the eggs laid by the mutant mothers are defective in the factors which cause determination of normal anterior embryonic parts and usually result in embryonic death. Schubiger (unpublished) has recently observed polarity defects in larval offspring of *tuh* mothers, which are reminiscent of those in *bicaudal*. A few embryos, less affected, evidently manage to complete embryonic and larval development, but some of the basic defects in the head remain and the adults derived from such embryos display the homoeotic effect in the head region.

EVIDENCE FROM NUCLEAR TRANSPLANTATION EXPERIMENTS Nuclear transplantation experiments (98, 141–144, 204, 261, 357) have produced evidence that the developmental capacity of cleavage nuclei has not become restricted. For example Okada, Kleinman & Schneiderman (204) have shown that nuclei withdrawn from the anterior region of an embryo at the 256–nuclei stage and injected into a 32–nuclei embryo can give rise to either anterior or posterior structures in the resulting adult flies. Illmensee (143) has conducted similar experiments in which embryos resulting from nuclear transplantation into unfertilized eggs were cultured in adult abdomens and then caused to metamorphose into adult structures by transplantation into host larvae. The experiments revealed that cleavage nuclei (possibly even a single nucleus) from various egg regions, and as late as the syncytial blastoderm stage, can support development of all adult structures normally derived from imaginal discs. These experiments prove that before cellularization the blastoderm nuclei have no developmental restriction. In his most recent experiments Illmensee (144) has injected older nuclei from five different regions of early gastrula embryos (3–1/2 h after egg deposition) into unfertilized eggs and reports that the eggs can initiate development and continue to grow through larval life and in one case to the pupal stage. Thus, even after cellularization, the embryonic nuclei remain multipotent.

Although embryonic *nuclei* remain multipotent long after cellularization, this is not true of the embryonic *cells* which become restricted in their developmental capacities soon after cellularization. Thus Chan & Gehring (39) mixed genetically marked cells from anterior or posterior halves of blastoderm embryos with whole embryos, cultured them and caused them to metamorphose. Their results showed that the anterior embryonic cells from blastoderm stage embryos produce only anterior adult cells and vice versa. These embryological experiments thus support and extend the data from genetic variants indicating the determinative influence of cytoplasm on the embryonic nuclei. They also demonstrate that some determinative events occur soon after blastoderm formation. Illmensee's recent experiments (144) suggest that in the case of nuclei these determinative events may be reversed by transplanting the nuclei from determined embryonic cells into egg cytoplasm. Final proof of this reversibility of determinative events awaits the transplantation of nuclei from determined cells such as those of imaginal discs into unfertilized eggs.

SUMMARY AND CONCLUSIONS Studies of mutants with a maternal effect permit the conclusion that normal mothers deposit factors in the egg which are necessary for the following:

(*a*) normal cleavage, blastoderm formation, gastrulation, and primary organogenesis (*fu*, *dor*, *r*);

(*b*) normal development of individuals with a male phenotype (*snl*);

(*c*) normal development of individuals with a female phenotype (*da*);

(*d*) the commitment of germ cells (*gs*);

(*e*) production of normal thoracic and abdominal derivatives (*amx*);

(*f*) development of larval anterior rather than larval posterior (*bicaudal*); and

(*g*) development of anterior rather than posterior structures in the imago and perhaps the embryo (*tuh*).

Experiments using embryological techniques demonstrate the following:

(*a*) Cytoplasmic factors from eggs laid by wild-type animals can partly repair the damage due to *dor* and can completely repair the damage due to *r*.

(*b*) Damage to polar cytoplasm prior to its interaction with embryonic nuclei prevents germ cells from forming. This effect can be repaired by injection of polar cytoplasm from normal eggs.

(*c*) Thoracic imaginal disc derivatives and abdominal histoblast derivatives as well as embryonic and larval structures can be disturbed by altering the periplasm of the egg before its interaction with nuclei.

(*d*) Polarity of fly embryos can be reversed by destruction of periplasm prior to nuclear migration.

(*e*) Single syncytial blastoderm nuclei are competent to form all imaginal disc derivatives, and single gastrula nuclei are competent to support the development of unfertilized eggs.

(*f*) After cellularization of the blastoderm, some developmental restrictions have been imposed upon cells but restrictions on nuclei can be reversed by removing them from their cells and permitting them to divide in egg cytoplasm.

Together these results show that the initiation of developmental commitments depends upon substances elaborated by the mother and built into the egg during oogenesis. The developmental capacity of nuclei is not restricted until the nuclei invade the periplasm of the egg and apparently become intimately associated with the maternal cytoplasm and surrounded by cell membranes. The biochemical nature of the determining factors and the genetic mechanisms underlying cellular commitments are unknown, but close study of maternal effect mutants ought to shed light on these problems. In fact, using these mutants the contributions of the nurse cells can be distinguished from those of the follicle cells since the follicle cells are mesodermal in origin and the nurse cells are clonal relatives of the oocyte. By using Illmensee's (144) technique of transplanting donor pole cells into host blastoderm embryos, one can produce an ovary whose nurse cells and oocytes are of one genotype, and whose follicle cells are of a second genotype.

There is no strong experimental support for the idea that any substances in the egg are specific imaginal disc determinants although there may exist determinants for embryonic regions such as head or abdomen. It is tempting to embrace the view that the later a structure appears in evolution, the later it appears in ontogeny. Since imaginal discs were a relatively recent evolutionary innovation of insects, it seems unlikely that determination specific for imaginal discs would be initiated at the outset of embryonic development. In the case of the head for example, it seems probable that determination for embryonic "headness" would occur before determination for any of the head structures specifically associated with the imago. The same pattern-forming systems that determine larval structures would also determine imaginal structures. Thus any maternal effect mutant that affects imaginal head structures might be expected to affect larval head structures, for both the larval and the imaginal head are derived from parts of the embryonic head. If this argument is valid, one should not expect to find many maternal effect mutants that affect specific imaginal discs without affecting specific larval structures as well, and vice versa. Under this view, the determinants that exist in specific regions of the unfertilized egg are probably for both larval and imaginal structures, not for one or the other. The recent studies of *amx* (265) and *tuh* embryos (Schubiger, unpublished observations) are consistent with this view: in both mutants, maternal effect abnormalities in imaginal structures are associated with abnormalities in the embryo and the larva as well. The discovery of a maternal effect mutant that acted only on imaginal disc development would disprove this suggestion, but no such mutants have been reported at the time of writing.

Evidence for the existence in the unfertilized egg of specific determinants for both larval and imaginal structures also comes from recent surgical defect experiments. Nöthiger & Strub (202) produced defects in adults after ultraviolet irradiation of 20 min-old eggs. They found no topographical correlation between the site of irradiation and the abnormal region of the imago. However, Illmensee (143, 144) using surgical puncture instead of ultraviolet, found that constant defects were produced in both larvae and imagos by puncturing eggs in specific locations. Puncture defects in unfertilized eggs or 2–5 min-old eggs produced abnormalities in particular regions of the larva and resulted in the absence of specific imaginal discs (143, 144). Evidently defects in the periplasm of the egg result in defects in embryos, which lead to defects in both larvae and adults. Indeed, the experiments discussed in the rest of this review indicate that most aspects of imaginal disc development are under direct, rather than maternal genetic control.

The Maintenance of Developmental Commitments

Having established that cleavage nuclei embark on specific developmental pathways soon after they interact with the maternal periplasm to form the cellular blastoderm, we may ask whether these initial commitments of cells are irreversible. Are there mutants that alter developmental commitments once these are initiated? There are many kinds of mutants that block the development of one or a group of organs at various stages of their development (180, 351). But most of these deal with the proper execution of a selected developmental program rather than a change from one program to another. The most informative mutants with regard to the question of altering developmental commitments are the homoeotic mutants.

HOMOEOTIC MUTANTS (*a*) *Homoeotic phenotypes*. Homoeosis denotes the replacement of one body part by another (11). The alteration usually involves part of a disc and the area affected is uniform in some cases but variable in others. For example, in *Antennapedia*, the antenna may have only small regions changed to mesothoracic leg tissue or may be almost completely leg-like.

There are many mutants that cause a homoeotic change (94), such as eye to wing or antenna to leg (Table 1). Most of these homoeotic mutations are located very near each other on the third chromosome. It is one of the few cases in higher organisms in which genes with seemingly related functions are clustered.

Of these homoeotic mutants, only *tuh* has been shown to have a maternal effect and thus might affect the initiation of determination. The others must act later in development, and several techniques have been employed to ascertain the exact stage in which the gene acts.

TABLE 1. Summary of Homoeotic Mutations and Transdeterminations Affecting Imaginal Discs of *Drosophila melanogaster*

Autotypic structure (normal determination of imaginal disc)	Allotypic structure (changed determination of imaginal disc)	Homoeotic mutant			Temperature-sensitive period	Transdetermination after culture	Phenocopy
		name	symbol	locus			
Proboscis	antenna	proboscipedia (25)	*pb*	3-47.7	3rd instar (315,322)	+ (344)	-
	leg	proboscipedia (25)	*pb*	3-47.7	3rd instar (315,322)	+ (344)	-
Antenna	leg	aristapedia (10)	*ss*^{a}	3-58.5	second quarter (102,109,311,313 of 3rd instar 314,323)	+ (86)	+ (21,85,97,107 247,317,331)
		Antennapedia (57,99,168,169,174,275,276)	*Antp*	3-47		+	+
		aristatarsia (209)	*art*	3		+	+
		Nasobemia (87)	*Ns*	3-48		+	+
		1(4)29 (89,90)	*1(4)29*	4		+	+
		tumorous head (78,196,223)	*tuh-3*	3-58.5	embryo (349)	+	+
	wing	Pointed wing (26)	*Pw*	3-94.1		+ (86)	-
	genitalia	turmorous head (78,196,223)	*tuh-3*	3-58.5	embryo (349)	+ (2,82)	-
Eye	wing	ophthalmoptera (103)	*opt*^{G}	2-68 (70,179)	2nd instar (221)	+ (251)	-
	wing	ophthalmoptera (160)	opht	1-5 (207)	2nd and 3rd instar	+	-
	abdomen	tumorous head (78,196,223)	*tuh*-3	3-58.5	embryo (349)	-	
	antenna*	dachs-dachsous (332)	*d*	2-31.0		+ (257)	+ (53)
Prothoracic leg	antenna	-	-	-		+ (254)	-
	wing	-	-	-		+ (254)	-
Meso- or Metathoracic leg	prothoracic leg	Polycomb (173)	*Pc*	3-47			-
		Extra sex comb (127)	*Scx*	3-47			-
		Multiple sex comb (299,302)	*Msc*	3-48			-
		extra sex comb (127)	*esc*	2-54.9			-
		reduplicated sex combs (355)	*rsc*	1			-
Humerus	wing	Hexaptera (130)	*Hx*	2			+ (317,331)
Wing	haltere	tetraltera (312)	*tet*	3-48.5		- (67,292)	+ (319)
		Haltere mimic (180)	*Hm*	T(2;3)		-	+
		Contrabithorax (175,176,177)	*Cbx*	3-58.8		-	+
	mesothorax*	-	-	-		+ (292)	-
	eye	-	-	-		+ (292)	-
	leg**	podoptera (106)	*pod*	multifactorial		+ (292)	-
Haltere	wing	bithorax (26,175,176,177)	*bx*	3-58.8	late 2nd and early 3rd instar (154)	+ (93)	+ (101,114,129,183)
		post bithorax (175,176,177)	*pbx*	3-58.8		+	+
		Ultrabithorax (175,176,177)	*Ubx*	3-58.8		+	+
		tetraptera (8)	*ttr*	3-51.3		+	+
		halteroptera (209)	*hl*	3-		+	+
	Abdomen	Contrabithoraxoid (177)	*Cbxd*	3-58.8		- (93)	-
1st abdominal segment	metathorax	bithoraxoid (175,176,177) *bxd*	3-58.8			-	
	2nd abdominal segment	Ultraabdominal (177)	*Uab*	3-58.8		-	-
Genitalia	antenna	III-10 (267)	III-10	3-46.6		+ (118,122,193)	-
	leg	III-10 (267)	III-10	3-46.6		+ (118,122,193)	-

* The transformations of wing to thorax and eye to antenna (91,257) probably should not be represented as transdeterminations since in each case one part of a disc changes to another part of the same disc. Regeneration might be a more appropriate term. The reverse transformations of thorax to wing and of antenna to eye (86) are never found, since these structures duplicate.

[illegible] the evidence is not compelling. (106)

(*b*) *Clonal analysis.* One might suppose that homoeotic mutations act on the primary determinative events in the embryo when imaginal discs acquire their specificity. This can be tested by a clonal analysis, which identifies the time at which a decision is made between two alternatives in a homoeotic mutant. If the decendants of a single cell can embrace two separate phenotypes, then the original cell had not been irrevocably committed to either phenotype. Thus, an analysis of clones of cells will provide an estimate of the time at which a decision between the two possibilities occurs. The antennal appendage of $Antp^R$ is actually a mixture of antennal cells and leg cells. Clones in $Antp^R$ antennae including both antennal and leg regions can be found derived from cells genetically marked by somatic crossing-over at the start of the third instar (226, 228). This result shows that antennal imaginal disc cells in $Antp^R$ as late as the beginning of the third instar retain the option to become leg or antenna, and implies that the homoeotic gene can alter determination at this late time. It indicates that this homoeotic mutation does not disturb the initial process of determination of a disc but rather makes the determined state unstable.

(*c*) *Temperature sensitive period.* Several of the homoeotic mutants display a wild-type phenotype if grown at one temperature and a mutant phenotype if cultured at a second temperature. By changing the environmental temperature at different developmental stages, one can ascertain the developmental period during which a wild type gene product is necessary for normal development. The temperature sensitive period for several alleles of aristapedia [ss^{a-F}, ss^{a-B}, ss^{a-40a} in *D. melanogaster*; ss^{A} in *D. hydei*; ss^{a-D} and ss^{a-S} in *D. pseudoobscura*), has been shown to be in the third instar (102, 109, 311, 313, 314, 323).

In the case of six *aristapedia* alleles in three *Drosophila* species, the mutant phenotype is more severe at 17°C than at 29°C, which is a reversal of the situation with most temperature sensitive mutants in *Drosophila* (290). Heat sensitive mutations probably cause proteins to be reversibly thermolabile (138, 242), and occur frequently after treatment with mutagens that cause missense mutations (291). Cold sensitive mutations like *aristapedia*, on the other hand, are infrequent, and are clustered in a few cistrons rather than spread throughout the genome as are heat sensitives (189, 242). In bacteria, cold sensitive mutations appear to alter the regulatory functions of allosteric proteins (203), or ribosomal proteins (112). O'Donovan & Ingraham (203) have suggested that cold sensitive mutants would have lesions affecting repression and induction. It is tempting, therefore, to speculate that the *aristapedia* protein might exert its effect by an interaction of some kind requiring allosteric specificity.

The temperature-sensitive phase of *eyeless-ophthalmoptera* is in the second instar (221). Unlike the situation in other homoeotic mutants, in *ophthalmoptera* the phenotype is more normal at low temperatures. Thus, the eye is normal at 17°C but wing cells replace eye cells at 29°C. The phenotype requires the eye reducing gene *eyeless-2* (ey^2, 4-2.0), in addition to opt^G (2-68.0), the homoeotic gene (70). The temperature effect apparently is due to opt^G, since ey^2 animals have the same phenotype at 17°C and 29°C (221).

Another eye-wing homoeotic stock, *loboid-ophthalmoptera*, also requires an eye reducing gene, *loboid* (*ld*, 3-102), as well as a homoeotic modifier (*opt*, 1-5) (207,

208). In *ld-opt* in which the homoeotic modifier is in the first chromosome, low temperature increases penetrance, while in *ey-opt* in which the homoeotic modifier is in the second chromosome, low temperature decreases penetrance. The temperature sensitive periods for the two eye-wing homoeotic stocks overlap (205, 206, 221). Since two separate homoeotic genes are involved, and since the temperature relations are reversed, it seems possible that *ld-opt* and *ey-opt* interfere with different functions in the genetic regulation of the eye and wing states of determination.

Proboscipedia is especially interesting in that the proboscis develops into an antenna-like appendage at 15°C, while it grows into a leg-like appendage at 29°C (315, 322). The nature of the temperature relations of *pb* indicates that the molecular control mechanisms may be more complicated than those involved in other homoeotic mutants.

We have concluded from the studies of temperature sensitivity of homoeotic mutants that a wild-type gene product (most likely a protein) is necessary during later larval development to maintain the initial developmental commitments. The results however, do not indicate when the gene is transcribed, or translated, but only when the wild-type gene product is necessary. The following experiments demonstrate that transcription of the wild-type allele is necessary during the third instar to maintain the initial developmental commitments and that the organism does not make a stable molecule such as a masked mRNA early in development which maintains initial commitments in later development.

(*d*) *Necessity for the wild-type homoeotic allele.* By the use of somatic crossing-over, one can effectively remove a dominant wild-type allele from a cell at any stage of development after which mitosis occurs, leaving a cell that is homozygous for a recessive mutant allele. If after metamorphosis, the cell or its descendants have a mutant phenotype, this means that a wild-type allele is still necessary to suppress the mutant phenotype after the crossing-over event (75). Lewis (176) used this method to ascertain when the wild-type allele of *bithorax* is necessary. *bx* causes mesothoracic structures to replace metathoracic ones, thus changing the halteres into wings. Although *bx* is on the third chromosome, a translocation of bx^+ to a first chromosome bearing y^+ and sn^+ allowed Lewis to show that if somatic crossing-over occurred even near the end of the third larval instar, then occasionally a *y sn* mesothoracic bristle or hair would appear on the metathorax. Evidently a wild-type *bithorax* allele is necessary even as late as the third instar to insure normal development.

When analogous experiments are performed using *aristapedia* it is found that small clones of ss^a tissue develop autonomously if mosaics are induced by somatic crossing-over in the first larval stage (240). Mosaics induced later show that ss^+ is necessary at least until 20 hours before pupariation to insure the development of an arista (antenna) rather than tarsus (leg), but after this time homozygous ss^a cells develop into arista (224).

These observations show that normal gene function in the third instar is necessary to maintain the developmental commitments initiated in the young embryo. They eliminate the possibility that determination is maintained from the blastoderm stage by a stable molecule such as masked mRNA, a conclusion also reached (188, 239a, 336) from studies on the effects of actinomycin D on disc development. They also reemphasize that the homoeotic alteration in these cases does not take place in the early embryo.

TRANSDETERMINATION Besides these *genetic* alterations which affect the maintenance of the determined state, certain *epigenetic* changes cause cells to alter their developmental program. These changes are called by Hadorn *transdetermination* (116).

(*a*) *Phenocopies.* Treatment of developing wild-type larvae with deleterious agents can produce phenotypes resembling those of various mutants. *Drosophila* larvae exposed to nitrogen mustards (21), X-rays (317, 331); boron compounds (97, 107, 247), or 5-fluorouracil (85), give rise to adults in which parts of the antenna are replaced by parts of legs. These agents are effective if applied in the third instar, which indicates that they destabilize the machinery involved in maintenance of antennal determination, as do the ss^a and *Antp* mutations (discussed above).

The *bx* phenotype can be copied by exposing early embryos to ether or heat shock (101, 114, 129, 183). The most sensitive developmental period is during formation of the cellular blastoderm, which suggests that these harmful agents disturb the initiation of the determination for thoracic segments.

Since phenocopies are often produced only during a specific developmental period and sometimes result in specific irregularities (239a), the period of sensitivity to phenocopy agents has been regarded as an indicator of the time of determination of the structures affected (104). But because of the unspecific nature of phenocopy agents, caution is necessary in interpreting their effects. Another problem in interpreting and in confirming phenocopy data is that different wild-type stocks of *Drosophila* behave differently, but typically, in regard to sensitivity to phenocopy agents and the kind of phenocopy produced. Thus some stocks produce an *aristapedia* phenotype frequently in response to a specific phenocopy agent whereas others produce a phenotype with leg defects (107). The reasons for these differences in response are not fully understood but may be due in part to the existence of subthreshold alleles ("isoalleles") which are able to express themselves when animals are exposed to phenocopy agents. This same phenomenon underlies the process of genetic assimilation in which selection improves the animal's hereditary ability to respond to an environmental stress (335).

Notwithstanding the problems attending the analysis of phenocopies, their very existence tells us that various extrinsic alterations to a tissue during embryonic and larval stages can channel some of the cells in that tissue into new developmental pathways, perhaps by interfering directly or indirectly with the genetic mechanism which maintains determined states during embryonic and larval stages. In many cases the mode of action of phenocopy agents may be quite simple. For example, most phenocopy agents have in common the ability to kill cells, and this might stimulate compensatory regeneration. In those cases where phenocopies are homoeotic it may be that the alteration of determination produced by the phenocopy agent is a response to the increased growth of the tissue during regeneration.

(*b*) *Transdetermination in culture.* Another procedure that leads to alterations in determination similar to those produced by phenocopy agents is in vivo culture. Vogt (320) transplanted imaginal antennal discs from late third instar larvae into early third instar to allow extra time for growth, and found leg structures after metamorphosis. But the importance of this change in determination was not appreciated until Hadorn initiated a remarkable series of experiments almost two decades later (115). Hadorn took advantage of the fact that entire imaginal discs or disc fragments can be cultured in the abdomens of adult hosts where they grow but do not differentiate (20). Such tissue lines can be propagated for many years by serial transplantation of disc fragments in the abdomens of new adult hosts (115, 118) and the karyotypes tend to remain normal (237). The developmental potential of the tissue lines can be assayed at any time by implanting fragments into larvae where they metamorphose with the host and secrete a cuticle. When the cultured discs are initially tested they usually produce more of the original structure (305) i.e. a fragment of leg tissue usually gives rise to more leg tissue. But eventually transdetermination occurs and structures characteristic of other imaginal discs appear (117, 118). Transdetermination has been observed after culture of the male (115, 118) and female (193) genital disc, haltere disc (93), wing disc (67, 292), prothoracic leg disc (254, 259), the eye disc (251), the eye-antennal disc (2, 82), the antennal disc (86), and the labial disc (343, 344, 346). The literature on transdetermination has been well reviewed recently (89, 91, 94, 116, 119, 208, 345). The present discussion will be limited to conclusions drawn from studies of transdetermination and homoeotic mutants without an extensive treatment of the data.

(*a*.) Most of the transdeterminative changes that occur in culture experiments are similar to those that occur in homoeotic mutants (Table 1). For example, *Antp* causes legs to replace antennae, and antennal cells sometimes transdetermine to legs during culture. Of the approximately 14 known transdeterminative changes, only four are not also caused by homoeotic mutants: leg to antenna, leg to wing, wing to eye, wing to leg. This may simply be because the corresponding mutants are yet to be discovered. One might expect leg-to-antennae or leg-to-wing mutants to be pharate adult lethals since they could not eclose. Of the many homoeotic mutants, the following have not yet been found as transdeterminations: wing to haltere (*Cbx,Hm*), prothorax to mesothorax (*Hx*), meso- and metathoracic legs to prothoracic legs (*Msc*), metathorax to first abdominal segment (*Cbxd*), first abdominal segment to second abdominal segment (*Uab*), first abdominal segment to metathorax (*bxd*), and eye to abdomen (*tuh*). The *Hx* and *Msc* transformations have not been observed probably because no one has extensively cultured the dorsal prothoracic disc, apparently due to its fragility (162), nor have long term cultures been initiated with the posterior leg discs. The *Cbxd, Uab*, and *tuh* transformations are probably lacking because of failure of the abdominal histoblasts to grow in adult hosts (Hadorn, personal communication) perhaps because of their special hormonal requirements (110). The other transformation (wing-haltere) has not been reported, despite extensive culture, perhaps due to the very low proliferation rate of haltere cells (93). Evidently there is striking correlation between the changes in determination that occur during in vivo culture, and the changes that occur in homoeotic mutants.

(*b.*) A given disc can transdetermine to any of a number of other discs, but homoeotic mutants usually cause only a single change in determination to occur. For example, the antenna can transdetermine into leg, wing, or eye, but these transmogrifications require separate homoeotic mutant genes, *Antennapedia* (antenna to leg) or *Pointed wing* (antenna to wing). No structures other than leg or antennae are ever observed in *Antp* antennae (228). The eye can transdetermine to wing (251) or genitalia (81, 86), but it requires distinct mutations to cause the same changes, *ophthalmoptera* (eye-wing), *tumorous head* (eye-abdomen) (223). The mouth parts are changed to either antennae or leg by *proboscipedia.* But by altering the culture temperature one change may be obtained to the exclusion of the other. The *tuh* strain results in several changes (antennae-leg; rostralhaut-genitalia, eye-abdomen) but here again there is specificity since only one homoeotic structure arises from a given part of the eye-antennal disc (223). The only possible exception to the rule of specificity in homoeotic mutants is a recently discovered but little described pupal-lethal, which changes genitalia to both leg and antenna (267).

(*c.*) Not all transdetermination steps are readily reversible. For example, haltere transdetermines to wing (93), but wing apparently does not transdetermine to haltere (67, 292), and leg transdetermines to wing more frequently than wing transdetermines to leg (292, 254).

(*d.*) Transdetermination from one disc to another occurs at repeatable frequencies characteristic only of the initial and final type. Thus, the frequency of change from antenna to wing is largely independent of whether the culture was started from antennae [about 80% (86)] or from genitalia that had transdetermined to antenna [about 50% (118)].

(*e.*) As a consequence of *c* and *d* (above), transdetermination occurs in a regular and repeatable sequence. Genitalia transdetermine to leg or antenna; leg transdetermines to antenna or wing; antenna transdetermines to leg, to wing, or to eye; eye and haltere transdetermine to wing; and wing changes mainly to mesothorax, and to a lesser degree to leg, antenna, and eye. Thus, there appears to be a more or less directed flow from genitalia through antenna and leg to wing, and finally to mesothorax. As time in culture progresses, more and more cultures form mesothorax, and then die out since mesothoracic disc tissue grows slowly.

(*f.*) Transdetermination and homoeotic mutations both result in determinative changes in populations of cells rather than single cells. This can be demonstrated by analysis of marked clones of cells showing that a transdetermined area (88), or a homoeotic area (226), can be formed by the progeny of more than one cell. A similar analysis of marked clones in gynandromorphs shows that each imaginal disc is formed by the progeny of more than one cell (36, 200, 227). These results lead to the important conclusion that the processes of determination and changing determination occur in groups of cells, rather than a single cell.

(*g.*) A small number of cells can maintain a determined state when surrounded by cells with another determined state. Nöthiger's faulty mosaics (199) showed that after dissociation and reaggregation even a very small group of genital cells surrounded by genetically distinct wing cells would maintain its determination. Also, patches of homozygous homoeotic cells surrounded by wild-type cells in genetic mosaics differentiate autonomously (176, 240).

(*h.*) Transdetermination seems to be correlated with increased growth; the greater the growth of a tissue during in vivo culture, the greater the incidence of transdetermination (120, 292). Furthermore, transdetermined cells arise from the region of a disc which divides most rapidly (343, 344). The same general type of explanation has been advanced to explain the action of homoeotic genes. According to this view homoeotic genes act primarily by increasing growth rate and only secondarily by promoting a switch in determination during this increased growth (104, 160, 208, 311). However, the evidence supporting this view is not compelling. One of the main arguments in favor of this view is that homoeotic structures are larger than the original structure. This is true in some cases (a leg is larger than an antenna, *Antp*), but in other cases the reverse is true (a haltere is smaller than a wing, *Cbx*), and in some cases the discs are the same size (a pro- and mesothoracic leg, *Msc*). Furthermore, colchicine treatment which increases growth rate (294) decreased the expression of some homoeotic genes (326) but increased the expression of others (211). Of three colchicine doses, which permitted survival of *ophthalmoptera* discs, the higher doses had no effect, whereas only the lowest dose caused an increase in penetrance (211). Therefore, it is difficult to conclude from these data that homoeotic genes act primarily to increase growth rate.

Furthermore, increased growth rate does not insure homoeosis. For example, X-rays kill imaginal disc cells, and cause an excess proliferation of surviving disc cells. As a result, one disc from an animal irradiated in the first instar can give rise to two legs in situ (229). Yet close scrutiny failed to reveal any transdetermined structures arising from these discs. Similarly, in situ surgery results in duplication of preexisting structures but no transdetermination (32). In mutants such as *eyeless*, which cause cell death (63), or *l(1)726ts* (243), excess proliferation occurs in surrounding regions, but again duplication of preexisting structures occurs rather than homoeosis (180). Homoeosis of eye discs occurs in *eyeless* to produce wings in the eyes only if a homoeotic gene is present (103). So homoeotic genes act primarily by changing a disc from one developmental program to another, not primarily by changing growth rate. They appear to stimulate transdetermination in a specific direction.

(*i.*) Although transdetermination requires extensive proliferation, other factors appear to be necessary. Thus Schubiger (255) has shown that the part of a leg disc that regenerates, exhibits extensive transdetermination, whereas the part that duplicates, proliferates well but does not transdetermine.

(*j.*) In both transdetermination and homoeotic mutants only clearly defined alternative cell phenotypes are expressed—intermediates do not occur. A differentiated cell displays either leg features or antennal features, but not both. This indicates that within a single cell only one of the alternative gene sets is expressed.

MODELS TO ACCOUNT FOR TRANSDETERMINATION AND HOMOEOTIC MUTANTS The preceding analysis indicates that homoeotic genes regulate important determinative processes. Further evidence consistent with this conclusion comes from the observations that mutant genes affecting the morphology of thoracic legs but not the antennae (*combgap, dachs, thickoid*) also alter antennal legs formed under the influence of homoeotic genes like *aristapedia* (3, 24, 174, 275, 276, 316, 318, 330, 333), and genes affecting thoracic wings also alter homoeotic wings (207). Conversely mutant genes that normally affect the antenna but not the legs (*aristaless, thread*) do not alter the homoeotic antennal leg of *aristapedia*. Because of the many similarities noted above, between the changes caused by homoeotic mutants and transdetermination, it is reasonable to regard homoeotic mutants and transdetermination as genetic and epigenetic alterations of the same regulatory system.

The overall conclusion from in vivo culture experiments is that the determined state tends to be heritable. Changes in determination occur infrequently, but predictably in specific sequences, and the new determined state is inherited just like the original. This has suggested to Kauffman (155) that the determined state in a disc consists of a selfmaintained stable state composed of several "bistable memory circuits"—entities that can be in one state or another and each with a defined frequency of change. Discs differ from each other in the state assumed by each of the several different circuits. The state of each hypothetical circuit can be chosen to provide for the frequencies of transdeterminations based on transdetermination assymetries (eye to wing is more frequent than the reverse). Then predictions can be made from this model as to the quantitative relationships of various transdetermination steps not employed in the formulation of the model (eye to wing should occur more often than eye to haltere). This kind of model has good predictive value. According to this model, a maternal effect homoeotic mutant would be involved with the initial setting of each circuit in the appropriate state for each disc. The homoeotic mutants whose wild-type allele is necessary during larval life for normal development would alter the stability of one of the two stationary states of a particular circuit. Since different discs share the same circuits, and differ only in the state of each circuit, then destabilizing one of the states of one of the circuits might be expected to alter the stability of the determination of several discs. This could be detected by a change in transdetermination frequencies of several discs from homoeotic mutants. To date, among the homoeotic mutants only *aristapedia* antennal discs have been tested for transdetermination (86), and so little data is yet available on this point. Kauffman's ideas point up the necessity for much more rigorous standardization of culture conditions, injury factors, time and temperature control, as well as frequency scoring techniques in transdetermination studies so that transdetermination frequencies from different discs and different genotypes can be compared on a common basis. That basis probably should be transdetermination events per cell division.

Kauffman's hypothesis of bistable circuits predicts the existence of complementary mutations affecting subsets of discs. These subsets will be composed of discs that share circuits in the same state. Shearn et al (267) have isolated 63 mutants that affect imaginal discs (larval-pupal lethals) and 34 of these affect only some discs. These 34 mutants fall into 13 subsets in terms of the discs affected and most of the subsets form complementary pairs (i.e., some mutants affect only eyes and antennae whereas others affect halteres, wings, and legs etc.) By using these mutants to identify discs that share circuits in the same state, Kauffman has used his model to predict the frequency of transdetermination among discs. Although it is questionable whether all these mutants involve determinative changes, nevertheless Shearn's complementary mutant classes constitute the best available evidence for bistable circuits that is independent of transdetermination frequencies. This intriguing result emphasizes that Kauffman has developed a model with real predictive value and not a mere redescription of the data.

If a series of bistable memory circuits are operating in disc cells, then Kauffman's model can also predict the phenotypic outcome of somatic cell fusion between cells from different imaginal discs. The most stable state of each circuit would probably be maintained in the hybrid, leading not to a domination of one disc type by the other, but rather to a different disc type entirely. The most attractive feature of Kauffman's model is its testability with regard to transdetermination frequencies in homoeotic mutants and the occurrence of mutations that affect particular subsets of discs; a drawback is that it provides no biochemical basis for transdetermination and homoeotic mutants.

A second recent model postulates a specific biochemical mechanism for the action of homoeotic mutants. Kiger (157) has focused on Lewis' (175–177) extensive and elegant data on the *bithorax* complex of homoeotic genes. Although Kiger has retained Lewis' suggestion that there is a gradient of an inducer in the egg, which specifies thoracic and abdominal development, he predicts that the *bx* complex produces a single allosteric protein. Although the model interprets much of the genetic data, and makes some predictions about an additional class of mutants, on the whole it seems not to be easily tested experimentally.

Mutations Affecting Spatial Patterns (*or "Position is Nine Points of the Law"*)

When a mature imaginal disc is cut into fragments that are implanted into larvae and caused to metamorphose, the different fragments metamorphose into different parts of the adult structure. These parts can be identified with precision by particular hairs, bristles, and sensilla, and an accurate anlageplan or fate map of the disc can be constructed (254). It is evident from this result that cells of the disc differ from one another depending on where they are located. They form a spatial pattern within the disc and this pattern is isomorphic with the adult pattern. This spatial pattern can be altered by specific mutations that will be discussed in this section. These mutations are quite different from the homoeotic mutations discussed in the previous section because they do not alter what *disc* a cell belongs to, but rather what *part* of a disc a cell belongs to. As we shall see, the decision to be a specific disc is very different from the decision to be a specific part of a disc. The decision to be antenna or leg can be passed on by a cell to its progeny by cell heredity. It is a "determination" and is an example of a "firm bias" as discussed by Schneiderman (253). Bryant (33) has proposed the term "specification" to identify the nonheritable commitments of imaginal disc cells to be specific parts of a disc. This is an example of a "transitory bias" (253).

Specification is the process by which a cell's position becomes specified in an imaginal disc so that it will produce a particular part of that disc. In addition to its being nonheritable, specification differs in the following fundamental way from determination: the determination of each type of imaginal disc is different but the specification of spatial patterns in all discs appears to be very similar and possibly identical. What a cell differentiates into depends on what disc it belongs to (i.e., its determination) and on what position it occupies within a disc (i.e., its specification).

Five different factors involved in forming spatial patterns in imaginal discs have been shown to be controlled by specific genes. These factors include the following: (*a*) position of structures in the disc as a whole; (*b*) position and polarity of structures within a segment; (*c*) inductive influence between different structures such as bristle-forming cells and bract-forming cells; (*d*) proximity of bristles to each other; (*e*) a cell's competence to respond to the above patterning influences.

Mutations affecting the specification of position in the disc as a whole

Since the major axes of a disc are specified by genetic information, then a mutation might be expected to alter the proximo-distal, medio-lateral, or anterior-posterior axes in an imaginal disc. There are mutants that do this in the legs, antenna, wing, and abdomen.

The mutation *reduplicated* (*rdp*, 1-34.7) causes the legs of a fly to duplicate (136) (unfortunately this strain is lost). The duplication results in two legs or parts of legs that are in mirror image symmetry. A leg may be split at any level; parts distal to the split are mirror image duplicates and each duplicate usually includes both medial and lateral leg structures. Proximal parts not split are often of normal size but may have duplicate structures (judging by sex combs). The mutant *crippled* (*crip*, 2-56) (161), also produces some duplicated legs of this type, and mirror image duplicated antennae occur in several mutant stocks [*erosion* (145); *antennaless* (*ant*, 2-not located) (325); *deformed-recessive-Luers* (Dfd^{r-L}, 3-47.5) (324); *eyeless* (246)]. Modifications similar to the ones discussed above can be produced by irradiation (96, 229), surgery (32, 86, 201, 255, 320, 321), actinomycin (188), borate (107), or colchicine (211). These techniques provide additional data to help understand the developmental processes involved in producing duplicated mutant phenotypes. The case for the leg is best understood.

When a late third instar leg disc is bisected into a medial and a lateral half and metamorphosed immediately, each fragment forms only half a leg—a medial and a lateral half respectively (201, 254). But when these fragments are permitted to grow for a period before metamorphosis, differences in developmental capacity (sometimes denoted "morphogenetic potential") are revealed (255). With increasing culture time the medial fragment regenerates more of the missing half of the leg, and growth is required for this regeneration (256). The lateral half forms a mirror image duplicate of itself.

Similar results are found in adults when freshly hatched first instar larvae are X-irradiated. Two kinds of abnormal legs appear, one of which is duplicated with medial parts missing and lateral parts duplicated (229), following rules similar to those established for *rdp* (136). The other type is basically normal in medial parts but is missing various amounts of lateral pattern elements.

Bryant (32) obtained analogous results after partial bisection of leg discs in situ which divided them into dorsal and ventral sections, and he suggested a model that accounts for these results [see (210) for another model]. The model postulates a gradient of developmental capacity such that, after removal of a section of the gradient, cells at the cut surface can produce only structures lower on the gradient. If medial (or anterior) parts are removed by surgery or X-ray killing [see (197) for review of effects of irradiation on cells], the high part of the gradient can replace the lower parts and thus regeneration occurs. The cells on the cut edge of the lower section can produce only still lower regions, and a mirror image pattern is obtained.

In terms of this model, a mutation like *rdp* might act in one of two ways to produce mirror image appendages. It might remove a section of the gradient by causing localized cell death within the disc, an event readily detectable in whole discs by special staining techniques (270). Cell death occurs in parts of the imaginal discs of a number of mutants and many of these mutants show mirror image duplication of some of the remaining structures. In *eyeless* and *l(1)726ts*, cell death occurs in the eye region of the eye-antennal disc and the antenna frequently shows duplications (63, 243, 246). Similarly, in *vestigial* the presumptive wing blade region of the wing disc undergoes extensive cell death (62) and this is sometimes associated with duplications in the thorax (329).

It is also possible that mutations like *rdp* alter the postulated gradient. A patch of mutant tissue might be expected to behave nonautonomously in a gynandromorph test. When mutants of the *rdp* type are reisolated, a decision between these two alternatives can be made.

Support for this gradient model also comes from data derived from cultured disc fragments. Two possible explanations have been advanced for duplicated mirror image patterns found after culture (344). The first explanation supposed that when each cell in a fragment divided, it would duplicate the cell specific structures for which it was committed and the two daughter cells would then move about and take up corresponding positions in the two patterns (cell-by-cell replication). The second explanation suggested that cells bear a heritable determination only to be a particular disc, and that the positional information within the disc was specified in some way by the fragment as a whole. Experiments involving clones of marked cells in duplicating fragments of genital discs (200, 304), and in antennal or leg discs (225), showed that the progeny of a single cell can make much more of a pattern after culture than it would have in situ, and that some cells divide more rapidly than others. These results rule out the cell-by-cell replication explanation and provide no support for the idea that specified states in the mature imaginal disc are heritable. The fact that some disc fragments regenerate instead of duplicate (255) also argues against this explanation. However, the results permit explanations such as the gradient model discussed above to account for pattern duplication.

Genetically controlled anterior-posterior mirror image duplication also occurs in the wing (77), and in the leg (298) under the influence of *engrailed* (*en*, 2-62.0). Large clones of homozygous *en* cells induced early in development differentiate autonomously both in legs (298) and in wings (77). But when *en* becomes homozygous in wings a few divisions before differentiation, it fails to differentiate autonomously (77). This could be due to a persistence of the effect of the wild-type gene or to an interaction with surrounding cells. Since, however the cell markers used did not permit the identification of *en* cells not showing an *en* phenotype, nonautonomy on the edges of large clones could not be checked, and so these data are difficult to interpret.

Besides these genes affecting the major axes in legs and wings, another mutant stock, *asymmetric* (polygenic) produces occasionally abdominal tergites with two posteriors in mirror image (269). This effect can be mimicked by cauterization (181, 248). The careful experiments of Santamaria and Garcia-Bellido (248) showed mirror image duplication of tergite posteriors occurs when anterior cells were killed, but when posterior cells were killed, duplications of the anteriors were not found (248). Since *asymmetric* and surgery produce identical patterns, this difference in the response of anteriors and posteriors must be biologically meaningful, and reflects a gradient of morphogenetic potential in the tergite similar to that demonstrated in the leg. It is also noteworthy that *asymmetric* flies occasionally have an entire imaginal tergite 180°C reversed (269) and this effect can be copied in larger flies by transplantation and rotation of a prepupal tergite (17). Here again *asymmetric* illustrates gene control of major developmental axes in an imaginal primordium.

Since *en* affects the specification of position of both dorsal and ventral discs of all three thoracic segments (77), the possibility is suggested that the same positional cues are used in all the appendages (347). In *Antp* antennal appendages each location in the antenna produces a very restricted and very predictable part of the leg (228), supporting the idea that phenotypic leg cells can respond appropriately to pattern forming cues in the antennae. In addition, different antennal-leg homoeotic mutants change specific parts of the antenna to leg (69, 90, 102), and these changes are in accord with the leg-antennal changes that occur in the leg antennal phenotypic mosaics in *Antp*.

These results indicate that regarding patterns, the basic difference between cells of one disc and cells of another disc must lie in their response to specification. The differences in their response reflect differences in their states of determination which were established early in embryonic life and propagated by cell heredity during the development of the disc. Further support for this idea comes from the following studies of transdetermination. In transdetermination differently-determined disc tissues in contact with each other seem to be responding to similar positional cues. Thus antennal discs in culture produce transdetermined tarsal structures in direct contact with third antennal segment (86) and leg discs give rise to transdetermined distal antennal parts only in contact with distal leg parts (254, 259). It would be instructive to undertake a careful analysis of transdetermined structures to find out whether it is universally true that only specific parts of one disc arise adjacent to specific parts of other discs as our analysis of the limited data suggests.

A further indication that cells with a different determination can respond appropriately to the same positional cues comes from cell dissociation and reaggregation experiments. Genetically marked antennal cells adjacent to leg cells differentiate autonomously with regard to *disc* of origin, but there is nonautonomy regarding the *part* of the disc formed by adjacent leg and antennal cells (69). For instance, arista and claw often arise side-by-side. We suppose that in these regions of the tissue patterning cues are established for distal disc parts, and these cues transcend disc of origin. The cells then respond to these cues in a manner appropriate to their determination—arista in the case of antennal cells and claw in the case of leg cells.

One kind of alteration in positional information in a disc would lead to changes in the size of a disc and the adult structures it forms. This subject has not been well studied but there are a few intriguing results. For example, in vivo culture of part of the genital disc usually gives rise to many individual anal plates. However, Hadorn (117) found some apparently mutant tissue lines, which he called "anormotypic," that produced anal plates that failed to "arealize" and formed single giant anal plates, each of which had hundreds of bristles. If this anormotypic change causes abnormalities in the positional information in the tissue, and if all discs respond to the same positional information, then transdetermination from anormotypic anal plates of this sort should result in anormotypic structures from other discs. Although no further transdetermination in this line was found, perhaps this kind of change should be looked for in the future.

It is curious that mutants that produce enlarged structures like giant anal plates are practically unknown although there are numerous mutants that produce structures of reduced size. One of the very few mutants known that does produce enlarged structures is *lethal giant discs* [*l(2)gd*, 2-42.7] (37) in which all of the imaginal discs are enlarged. Remarkably, in some cases the enlarged discs duplicate within the larva. In the case of the leg disc, for example, a second leg disc grows gradually from the proximal region of the initial leg disc so that at intermediate stages one disc is complete but the other is smaller. If the discs are caused to metamorphose at this time, the end result is two adult legs in mirror image symmetry, one of which is normal, the other small and incomplete. This contrasts with *rdp* and X-rayed legs which are mirror image duplicates of the same size. The relationship between these two kinds of duplication processes is not yet known but their existence suggests that duplicated structures can arise in development by at least two different mechanisms: removal of distal or medial structures, or budding. Apparently *l(2)gd* induces the proximal region of the disc to regenerate even though distal structures have not been removed. Since both kinds of duplications involve mirror image patterns, it is clear that intercellular communication is involved in both.

Another group of mutations that affect the position of structures in the disc as a whole are mutations that affect segmentation. Several experiments indicate that segmentation of an appendage is not a heritable determination. Thus clones of cells induced late in larval life often encompass parts of two leg segments, for example femur and tibia (36). Also, when the presumptive tarsal segments are removed from a leg disc, the remaining piece regenerates them (255). Thus segmentation appears to be part of the specification of position in the disc as a whole [see (33) for a convincing analysis of this point]. Many mutations affect segmentation patterns. For example, the tarsi of *Drosophila* normally consist of five segments but are reduced to four segments in *four jointed*, to three in the combination ey^D; *four jointed*, and two in *l(2)gd*. The analysis of mosaics involving such mutants may provide clues to the process of specification itself.

The fact that certain parts of discs regenerate readily may help explain why certain mutant defects, such as the absence of all tarsal segments, are rare. In general the loss of distal structures due to cell death would not be evident because the more proximal regions of the disc would undergo compensatory regeneration and respecification of regenerate distal structures. Thus, unless the death of presumptive tarsal cells occurred just prior to metamorphosis, it would not be evident. In the case of the wing, however, although the wing is a distal structure, it appears to include the high point of the gradient of developmental capacity and the presumptive wing blade region of a disc can regenerate the rest of the wing disc including the mesothorax. However, the mesothoracic area of the wing disc cannot regenerate wing and can only reduplicate (86). Hence, in contrast to the lack of mutants deleting the tarsi, there are numerous mutants such as *vestigial* that delete distal parts of wings. The proximal parts that remain are able to duplicate but not regenerate. In *vestigial* (329) and in *scalloped* (147) a duplicated thorax occurs, whereas in *apterous*^{blot} there is a duplication of the anterior parts of the wing (329).

The regenerative capacity of specific parts of discs may also help interpret certain transdeterminations such as eye to antenna and wing to mesothorax. These transformations do not involve a change in determination from one disc to another but instead involve a change from one part of a disc to another part of the same disc and have been designated as region-specific transdeterminations (86). It seems possible that some of these transformations may not be transdeterminations but may represent the regeneration of one part of a disc from another part of the same disc [for example the regeneration of antenna from the eye region of an eye antennal disc (91, 257) and of mesothorax from the wing blade region of a wing disc].

The regenerative capacity of parts of discs also suggests another interpretation for experiments which were said to demonstrate that the decision to be a particular part of a disc was reproduced by a fragment of a disc during proliferation (91). Examples cited were the regeneration of thorax from the proximal part of a wing disc after prolonged proliferation and the regeneration of palps from the anterior fragment of an antennal disc (86). However each of these cases can fit into the model presented above if the structures that duplicated were at the low point of the gradient of developmental capacity which would be expected to duplicate rather than regenerate other structures in the disc. In this regard, we can appreciate why mesothorax is the "sink" in transdetermination experiments. The transdetermination sequence leads to wing disc derivatives. If the distal part of the wing is the high portion of the gradient of developmental capacity, and by chance, distal parts are removed during transfer of the culture from one host to another, then only low gradient regions (thorax) would be left, which can only duplicate.

MUTANTS AFFECTING POSITION AND POLARITY WITHIN A SEGMENT The mutant *eyeless-Dominant* (ey^D, 4-2.0) causes a broadening of the basitarsus of the leg and interruptions of the intersegmental membrane (283). In the male this results in several rather than a single row of stout bristles called sex comb teeth. In mosaics involving ey^+ cells marked with the autonomous bristle color mutant *yellow* in the midst of $ey^D y^+$ tissue, the *yellow* ey^+ cells become incorporated into a mutant spatial pattern as a result of the influence of ey^D adjacent cells. Appropriate controls show that the apparent nonautonomy of cells is due neither to the production of a substance elsewhere in the body (300), nor to somatic crossing over between y^+ and ey^D (301). The conclusion from these studies is that ey^D alters the positional information in the whole basitarsus, thus causing the wild-type cell in a normally nonsex comb region to make sex comb teeth.

Recently pattern formation in ey^D and also in the mutant *shibere* (*shi*, 1-52.2 (217), has been reinvestigated, with special reference to the gaps in the intersegmental membrane which occur in both mutants (220). On the normal fly leg the bristles and hairs point distally and associated with the base of each bristle is an adventitous cuticular projection called a bract formed by a cell adjacent to the bristle-forming cell. The bract is normally on the proximal side of each bristle and thus reflects the polarity of the developing leg (164, 165). Whenever gaps occur in the intersegmental membrane in ey^D and in *shi*, there are localized reversals of bristle and hair polarity. Reversal of polarity associated with intersegmental membrane gaps has also been obtained after X irradiation of first instar larvae (222). As Lawrence (164) and Poodry & Schneiderman (220) noted, bristle orientation in relation to such gaps can be modeled on the basis of a segmentally repeating gradient bounded by the intersegmental membrane. In fact, Bohn (22, 23), has shown that segmentally repeating gradients of morphogenetic potential are found in the legs of cockroaches. So ey^D may act first by altering the segmental gradient itself to provide for a larger sex comb region, and second, by causing gaps in the intersegmental membrane that permit visualization of the interactions of the gradients in two adjacent segments.

A change in bristle direction also occurs in *aristaless* (*al*, 2-0.1), where the posterior scutellars are caused to diverge rather than converge. This effect is nonautonomous in mosaics, a fact that reemphasizes the influence of the surroundings in determining bristle orientation (303).

MUTANTS AFFECTING INDUCTIVE INFLUENCES BETWEEN DIFFERENT CELL TYPES As noted above, the bristles on the distal leg parts are accompanied by bracts on their proximal sides. A bristle organule derives from a single cell, which divides twice. The four daughter cells become the shaft (trichogen cell), the socket (tormogen cell), a sensory nerve, and a sheath cell (171). The mutants *Hairless-two* (H^2, 3 – 69) and *shaven depilate* (sv^{de}, 4 – 2.0) apparently do not affect the spatial arrangement of bristle mother cells, but result in sockets without shafts. In the tarsi of these mutants, sockets without shafts fail to form bracts, while complete organules do produce bracts (297). Evidence from *shi* shows that an isolated trichogen cell will not induce a bract (217). This leads to the conclusion that a complete bristle organule is necessary to promote bract formation. This conclusion is bolstered by studies on genetic mosaics and on animals exposed to metabolic inhibitors. A clone of genetically marked cells on the leg can include a bristle without its bract or vice versa (36), showing that the two are not necessarily clonally related in normal development. Recombinants of dissociated leg cells from two genetically marked leg discs sometimes produce bristle-bract mosaics, isolated tarsal bristles, but never isolated bracts (67, 292). Nitrogen mustard and mitomycin C cause some shafts to be formed with no socket, and these shafts have no bract (293, 295, 296). These data show that a complete bristle organule induces a clonally distinct adjacent epidermal cell to differentiate into a bract. The orientation of the bract depends on the orientation of the bristle organule and in mutants like *spiny legs* (*sple*), in which the bristle orientation is abnormal, the bracts follow the orientation of the bristles (71).

BRISTLE SPACING The bristles on the abdomen are not arranged in as precise a pattern as are bristles elsewhere on the body, yet there is a minimum distance maintained between bristles. Wigglesworth (340) has suggested that in *Rhodnius* bristles are spaced as if they were competing for a bristle-forming morphogen or as if they produced a bristle-forming inhibitor. There are several mutants that alter bristle spacing, and some of these have been studied in genetic mosaics to see whether the mutant varied the hypothetical morphogen or the cellular response [see (280) for review]. We shall mention here three cases. The dominant mutation *Tuft* (*Tft*, 2-53.2) causes tufts of closely spaced bristles on the thorax. Tft^+/Tft^+ regions on a thorax consisting primarily of Tft/Tft^+ cells develop autonomously (4). On the other hand, the dominant *Hairy-wing* (*Hw*, 1-0.0) which also causes extra bristles on the thorax, is slightly nonautonomous (108). Homozygous $y\ Hw^+ sn^+$ cells in a $y^+ Hw\ sn/y\ Hw^+\ sn^+$ background occasionally produce an extra bristle near the $y^+\ Hw\ sn/y^+\ Hw\ sn$ spot. Wild-type cells can evidently respond to chaetogenic factors liberated by *Hw*, and *Hw* must produce too many bristles because it produces a higher level of some bristle-forming factor or a lower level of some bristle-forming inhibitor.

Evidence that bristle promoting factors may be used up or bristle inhibitors formed in the construction of a bristle comes from studies of *achaete* (1-0.0), which normally removes the posterior dorsocentral bristle (pdc) from the thorax. In an *achaete* mosaic, if *ac/ac* tissue occupies the pdc region, then no bristle is formed, whereas if *ac/ac* cells reside at this location, a pdc does form (279), showing that the cells are autonomous. An interesting finding is that if nonbristle-forming *ac/ac* tissue covers the pdc region, then an *ac/ac* cell nearby may make a bristle (278, 279). This shows that chaetogenic factors spread over several cells, and if the cell in the proper position is genetically unable to respond, then a competent cell at an ectopic location may form the bristle. Evidently, in the wild-type thorax the presence of a bristle inhibits an otherwise competent cell nearby from forming a bristle.

MUTATIONS ALTERING A CELL'S COMPETENCE TO RESPOND TO PATTERNING INFLUENCE The four types of mutations discussed above either alter or make apparent four different processes which specify position in imaginal discs. These patterning forces have been called prepatterns (280) or positional information (347), and the process of patterning has been termed specification (33) (see also 220). In some way the mutants described affect the intercellular communication processes by means of which a cell assesses its position in a developing system. Besides the mutants that alter positional information, there are mutants that alter a cell's ability to respond to the information. Stern and his colleagues (280) have exhaustively investigated a large number of pattern variants and have defined a great many mutants of this type. The characteristic of this mutant class is cellular autonomy—even a very small clone of mutant cells will have a mutant phenotype. *Tft* and *ac* discussed above are examples of such mutants.

PATTERN RECONSTRUCTION Investigations of mutant phenotypes and of regeneration in wild-type discs described above have generated a model of pattern formation based upon a gradient of positional information (32, 50, 167, 210, 229). According to this model, the specification of position in a disc is a property of the organ as a whole, and would be lost if the disc was dissociated to single cells. Cells in an imaginal disc are viewed as not rigidly committed to any specific ultimate fate except their determination to belong to a particular disc. They become committed when they occupy a particular position in a disc. This model of pattern formation in imaginal discs differs from a more accepted model which has been outlined in an important and attractive new book (309a) in which it is proposed that the commitment to be a particular *part* of a disc may be reproduced by a fragment of a disc during proliferation (91).

Thus, one model suggests that positional information in a disc is not heritable, whereas the other model suggests that it is heritable. It is instructive to examine the evidence obtained by means of dissociation and reaggregation experiments. In experiments of this kind imaginal discs of two specific genotypes [i.e., *y sn* and *mwh* (3-0.0, several rather than one pair per cell) leg discs] are mixed, treated with trypsin, and disrupted with a microstirrer. The disrupted discs are reaggregated by centrifugation and, usually after a short period of culture in an adult, are implanted into host larvae for metamorphosis. Numerous experiments of this kind reveal that dissociated and reaggregated cells derived from genetically different discs with a similar fate (i.e., *y sn* leg and *mwh* leg) do indeed cooperate to reform integrated patterns after culture and metamorphosis (66, 121, 216, 292, 309). On the other hand discs with separate fates (i.e., *mwh* leg and sn^3; e^{11} wing) do not cooperate to form patterns. As pointed out by Hadorn, Anders & Ursprung (121) and Ursprung & Hadorn (309) there are three possibilities to account for this: (*a*) Similar pattern parts join together by chance; (*b*) the patterns arise from precommitted cell groups due to cell or structure affinities; these precommitted cells move about and choose their definite positions in the pattern; (*c*) the cells randomly aggregate and a new pattern of differentiation is imposed upon the cells by a repetition of the same process of pattern formation that occurs in the normal development of an imaginal disc in situ. The first possibility is unlikely due to the high frequency of reordered patterns (121, 309). Garcia-Bellido has favored the second hypothesis based on histological studies of unmetamorphosed, dissociated, and reaggregated cells. His initial conclusion was that "the arrangement of discrete cells in patterns appears to be a result of the moving about of determined cells which reorganize their previous spatial relationships" (66). More recently he has reaffirmed this idea as follows, "We imagine pattern reconstruction in aggregates of imaginal disc cells as resulting from the specific affinities or recognition properties of cells carrying the characteristics of their previous position in the pattern" (71).

Poodry, Bryant & Schneiderman (216) have sought to distinguish between the alternatives. They have mixed dissociated cells from leg discs of two genotypes in equal quantities with various amounts of dissociated cells from wing discs of a third genotype. They then inquired of how many units (=cells or cell clumps) a leg island is formed, and with how many neighbors does one unit interact. If leg islands surrounded by wing come from a single unit, then no mosaics will be found, whereas if many units form a leg vesicle, then mosaics will frequently be found. Similarly, if a unit is only able to interact with six nearest neighbors, for example, then mosaic leg islands will be found less frequently than if cells can move around and interact with, say, 24 different units. The analysis (216) indicates that fewer than 6 units initiate a leg vesicle, and that units apparently interact only with their nearest 6 or 8 neighbors. These results rule out extensive migration as an important process in pattern reconstruction, and show that a leg island is founded by fewer than a half dozen cells or cell clumps.

Since patterns can be reestablished after reaggregation of dissociated cells, migration is not involved, and a leg island consisting of only a few cells grows to hundreds of cells before differentiation, then cells or their progeny must change their initial commitments for a small part of the disc, and cooperate to form parts in line with reestablished positional cues. This interpretation of pattern reconstuction of dissociated imaginal discs is in conflict with earlier interpretations. As noted by Garcia-Bellido (71) the cell affinity hypothesis predicts specificities that exceed greatly any found in cell dissociation-reaggregation experiments using other organisms (50a). This conflict is resolved under our view of pattern reconstruction. It should be noted that dissociated diploid cells of *Drosophila* do indeed sort out into the same kinds of classes as do vertebrate cells, i.e., mesoderm vs. ectoderm (172a), but we feel it unnecessary to ascribe to *Drosophila* cell recognition properties superior to vertebrates to explain the data for imaginal discs.

We conclude from these experiments that cells certainly retain their determination to form a particular disc after dissociation and reaggregation with other discs, but they do not necessarily retain their positional information i.e., where they formerly were in a disc (216). The conclusion that disc cells are not rigidly committed to a final ultimate fate also finds support in the pattern regeneration experiments described earlier. Those experiments showed that if a disc fragment is metamorphosed, allowing little cell division, then it develops into a restricted part of a disc (200, 255), indicating that positional information is conserved in the absence of cell division. But after growth, positional values evidently change in dividing cells, since after pattern duplication and regeneration descendents of a single cell can form a much greater portion of a disc than the cell would have in situ or in an immediately metamorphosed fragment (200, 225, 304). The importance of cell division is further emphasized by the finding that cell division is necessary for regeneration or duplication (256, 344) and the longer the culture period, the greater the extent of regeneration or duplication (255). Cell division is also required to respecify positional information in *Rhodnius* (167). Indeed Lawrence (166) has suggested that positional information is respecified at every cell division during development. The picture that emerges is that a developing imaginal disc possesses a particular spatial pattern of specified states at all times, but within the disc the particular cell that bears a particular specification may change as the cells divide. However, as Bryant (33) has pointed out, when a cell divides and becomes respecified, there are strict constraints on the possible changes of specification that can occur. A major constraint is its position in the disc, ie., where it lies within the gradient of positional information of that disc.

A SECOND LOOK AT DETERMINATION AND SPECIFICATION In a thorough analysis of the larval development of imaginal discs Nöthiger (200) has outlined the way in which the term "determination" has been used by embryologists. He points out that it is usually used to describe the process by which a specific pathway of development is singled out from among several possible pathways. When "determination" is used in this way it encompasses both heritable and nonheritable cellular commitments. However the evidence presented in this review indicates that some real advantage is gained from separately identifying these two types of commitments. For this reason the term determination has been used in this review to refer solely to heritable cellular commitments. When determination is used in this restricted sense it is easily testable (i.e., can a cell convey its commitment to its progeny?). Furthermore, it can be asserted that determination in imaginal discs does not proceed stepwise because the only heritable cellular commitment in an imaginal disc is the decision to belong to that type of imaginal disc. Other commitments that cells make during the development of a disc do not appear to be heritable and a class of nonheritable commitments have been identified as specifications. The fact that distinct mutations may separately alter determination or specification emphasizes that the two types of commitments are different and should be distinguished.

Lethal Mutations Affecting the Metamorphosis of Imaginal Discs

Metamorphosis is initiated by the secretion of molting hormones, the ecdysones, and the cessation of secretion of juvenile hormones. In response to these extrinsic changes, cells in the imaginal discs cease to divide and the discs evert to form adult structures. At the same time, the abdominal histoblasts, which have been quiescent since they were first set aside in embryonic life, start to divide and form the adult abdomen (76, 110). The disc cells differentiate to form cuticle, hairs, bristles, ommatidia, eye pigments, etc. There is a host of cell specific and autonomous mutants affecting each of these structures [for discussion and reviews see (34, 171, 239)]. We will confine our comments to mutants affecting disc metamorphosis.

As in the case of spatial patterns, there are two types of mutants that fail to undergo metamorphosis—those that lack the humoral agents causing metamorphosis, and those in which the imaginal discs lack the ability to respond to the humoral agents by secreting a cuticle. In both cases the mutant larvae survive but die during the larval-pupal transformation. We shall discuss these as far as they are understood.

MUTANTS WITH ENDOCRINE DEFECTS There are several mutants of *Drosophila* that affect endocrine organs. The best known is *lethal giant larva* (*1*(*2*)*gl*) and its alleles in which larvae fail to metamorphose. However many other tissues of the larva, including all of the imaginal discs, are also affected. The mutant discs behave autonomously and degenerate even when transplanted to wild-type hosts (82, 100). The *1*(*2*)*gl* mutation clearly has manifold effects on many organs besides endocrine glands.

Another group of lethal mutants that may have an endocrine basis has been recently described (267, 285). In these mutants the larva grows well until the time of pupation at which time the discs fail to metamorphose but retain a normal appearance and metamorphose when transplanted to a wild-type host. However defects in many factors in the larva besides endocrine mechanisms (such as nutrition) might prevent metamorphosis in these lethals.

Mutations affecting ecdysone production, release, or transport should be lethal at the end of the first larval instar (or possibly earlier if ecdysone is needed for embryonic development). Mutations interfering with juvenile hormone production and release should be lethal at all embryonic and post-embryonic stages until the time of pupariation. They should also act during adult life to cause female sterility since juvenile hormone is necessary for vitellogenesis. One might also expect mutations that would affect metamorphosis by interfering with the neural mechanisms that normally switch endocrine glands on and off. Thus, a mutant in which juvenile hormone production and release failed to decrease at the end of larval life would probably fail to metamorphose since it is known that experimental application of juvenile hormone analogs to larvae or early pupae suppresses metamorphosis in the abdomen and imaginal discs (7, 35, 40, 184, 231). A simple way to detect such mutants is to produce large numbers of temperature-sensitive lethals and test them in gynandromorphs. Hemizygous tissues with an endocrine mutant genotype would be expected to behave nonautonomously and survive in gynandromorphs in which the endocrine organs were in the heterozygous part of the animal.

Recently Arking (5) has examined more than a hundred temperature-sensitive late lethal mutants and has identified one such mutant that can be caused to pupariate by implantation of wild-type ring glands by injecting with an ecdysone. Postlethwait & Weiser (230) have discovered that the female sterile mutant *apterous-4* has an endocrine defect and can be caused to undergo normal vitellogenesis by application of juvenile hormone analogues.

MUTATIONS AFFECTING THE CAPACITY OF DISCS TO METAMORPHOSE A number of prepupal and late larval lethal mutants have been isolated in which the imaginal discs behave autonomously and fail to metamorphose when tested as hemizygous tissue in gynandromorphs or when transplanted into a wild-type host (5, 267, 285). Some of these mutations affect all discs whereas others affect only some discs. The abnormalities include complete absence of discs, small discs, and normal-looking discs that fail to metamorphose. In most cases the mutant larvae were completely viable, indicating that the imaginal discs are essential only for the development of the adult and not of the larva.

Particularly interesting are those mutants in which the discs look normal but fail to metamorphose. For some reason they have not acquired the competence to metamorphose. Recent experiments have shown that the maturation processes involved in the acquisition by a disc of the competence to respond to hormones and secrete a cuticle may require several days (81, 194, 195, 258), and that not all parts of a disc become responsive to hormones at the same time (81, 258). Competence develops in fragments of larvae containing discs cultured in vitro (252), and in discs cultured in an adult host (123), or a pupa (194). However discs do not acquire the competence to respond to hormones when cultured in the absence of cell division in a starved adult host (195). The molecular basis of the competence to metamorphose is not understood, but the absence of competence in some mutant discs could reflect the absence of membrane-bound or intracellular hormone binding sites, or lesions in the genetic machinery switched on by the hormone. An analysis of mutants, particularly temperature-sensitive mutants whose discs are unable to respond to hormones of metamorphosis, should provide a rational approach to understanding both the development of competence as well as identifying specific processes involved in disc differentiation.

MUTANTS AFFECTING BRAIN DEVELOPMENT Evidence has recently been uncovered which shows that many mutations affecting the development of imaginal discs also affect the development of the insect's brain. This correlation was first observed in *l(2)gl*^{4} where the imaginal optic primordia of the adult brain grow tremendously and invade other parts of the brain during larval life (82). Subsequent analysis of several other larval and pupal mutants with defective discs including *l(2)gd* (37), *l(1)d.lg.–1* (285) and *ey*^{D} (235), revealed that they also had defects in parts of the brain destined to form adult structures (81). In general, mutants that had enlarged or abnormal discs also had enlarged brains apparently due to excessive growth of the imaginal primordia of the adult brain, whereas mutants that had small discs also had small brains (285). Apparently most mutations that cause gross abnormalities in all of the imaginal discs also affect the primordia of the adult brain. The reason for this is not hard to find: the primordia of the adult brain are a special type of imaginal structure, analogous to the imaginal discs of the integument. Thus one might expect many pupal lethals with defective imaginal discs to have defective brains also. Hence, these mutants provide a point of departure not only for analyzing the development of imaginal discs but also for analyzing the development of the insect brain.

MUTATIONS CAUSING NEOPLASMS Another novel feature of the mutants whose discs are unable to metamorphose is the fact that in some cases the abnormal discs are neoplastic. The best studied is *l(2)gl*^{4}. In this mutant the imaginal discs fail to metamorphose when transplanted into larvae. When transplanted into adults the *l(2)gl*^{4} discs grow rapidly into large compact neoplasms that kill their host. These neoplastic cells display many of the morphological features typical of neoplastic cells in general such as reduced intercellular contacts, and changes of the cell and nuclear membranes(82).

The brain of *l(2)gl*^{4} transforms into an even more remarkable neoplasm, a lethal and invasive neuroblastoma that promptly kills hosts that receive transplants. The brains of other mutants with defective discs such as *l(2)gd* also transform into neoplasms. Apparently among the late larval and prepupal lethal mutants that have abnormal imaginal discs are a number that cause neoplastic transformations. As it is easy to induce and select large numbers of these mutants, it should be possible to identify the genetic changes responsible for a particular neoplastic property.

General Conclusions

This review discusses experiments that revise some of our general thinking about *Drosophila* development. Previously fly development had been contrasted to vertebrate development because of its mosaic (i.e., rigid) nature [see for example, (334)]. Recent experiments show that in *Drosophila*:

(*a*) Cells with a different determination (antenna versus leg) can respond to the same positional cues (228). This is analogous to a proximal vertebrate forelimb forming distal forelimb parts when transplanted to the distal part of a hindlimb (348).

(*b*) The imaginal discs have previously unsuspected abilities to regenerate (32, 229, 255), similar to vertebrate limbs (55, 249).

(*c*) Gastrula nuclei transplanted back into unfertilized eggs are able to support all of development (144) just as in the case of vertebrate embryonic nuclei (111).

(*d*) *Drosophila* embryonic cells sort out to the same degree as vertebrate cells (172a), rather than exceeding by far the specificities found in other organisms (71).

(*e*) Specific mutations cause true neoplasms in *Drosophila* which seem to be comparable to vertebrate neoplasms (81, 82).

The review has also pointed out the importance of distinguishing between heritable and nonheritable cellular commitments. The commitment to belong to a particular disc is heritable, is different for each disc, and is denoted as determination. The commitment to be a particular part of a disc is not heritable, may be specified by the same factors in several discs and is called specification. Determination and specification are each altered by distinct mutations, which emphasizes that Nature distinguishes between them. The review also emphasizes that changes in determination or specification require cell division, which calls attention to the importance of cell division in altering the developmental commitments of cells.

These recent results persuade us that *Drosophila* development can be used as an appropriate model for developmental systems in general, and, with the use of genetics, *Drosophila* promises to continue to provide solutions to important developmental questions.

Acknowledgments

The authors would like to thank Professor Peter J. Bryant for his helpful suggestions. The comments of Dr. Masukichi Okada and Professor Richard D. Campbell and Messrs. John Haynie and Craig Roseland are also gratefully acknowledged. We are especially thankful to Mrs. Donna Krueger and the staff of the Center for Pathobiology for their bibliographic assistance.

Unpublished results from the author's laboratories were supported by the following grants: GM 19037 from NIH to J. H. Postlethwait, and GB 29561 from NSF and HD 06082 from NIH to H. A. Schneiderman.

Literature Cited

1. Abbadessa, R., Burdick, A. B. 1963. The effect of X-irradiation on somatic crossing over in *Drosophila melanogaster. Genetics* 48:1345–56
2. Akai, H., Gateff, E., Davis, L. E., Schneiderman, H. A. 1967. Virus-like particles in normal and tumorous tissues of *Drosophila. Science* 157: 810–13
3. Anders, G. 1955. Untersuchen uber das Pleiotrope Manifestationsmuster der Mutante *Lozenge-clawless* (lz^{cl}) von *Drosophila melanogaster. Z. Indukt. Abstamm. Vererbungsl.* 87: 113–86
4. Arnheim, N. 1967. The regional effects of two mutants in *Drosophila* analyzed by means of mosaics. *Genetics* 55:253–63
5. Arking, R. 1973. Unpublished
6. Ashburner, M. 1970. Function and structure of polytene chromosomes during insect development. *Advan. Insect Physiol.* 7:1–95
7. Ashburner, M. 1970. Effects of juvenile hormone on adult differentiation of *Drosophila melanogaster. Nature* 227:187–89
8. Astauroff, B. L. 1929. Studien über die erbliche Veränderung der Halteren bei *Drosophila melanogaster* Schin. *Wilhelm Roux' Arch. Entwicklungsmech. Organismen* 115:427–47
9. Auerbach, C. 1936. The development of the legs, wings and halteres in wild type and some mutant strains of *Drosophila melanogaster. Trans. Roy. Soc. Edinburgh* 58:787–815
10. Balkaschina, E. I. 1929. Ein Fall der Erbhomöosis (die Genovariation "ar-

istopedia") bei *Drosophila melanogaster. Wilhelm Roux' Arch. Entwicklungsmech. Organismen* 115:448–63

11. Bateson, W. 1894. *Materials for the Study of Variation Treated with Especial Regard to Discontinuity in the Origin of Species.* London: MacMillan
12. Becker, H. J. 1956. On X-ray-induced somatic crossing over. *Drosophila Inform. Serv.* 30:101
13. Becker, H. J. 1957. Uber Röntgenmosaikflecken und Defektmutationen am Auge von *Drosophila* und die Entwicklungsphysiologie des Auges. *Z. Indukt. Abstamm. Vererbungsl.* 88: 333–73
14. Becker, H. J. 1969. The influence of heterochromatin, inversion heterozygosity and somatic pairing on X-ray induced mitotic recombination in *Drosophila melanogaster. Mol. Gen. Genet.* 105:203–18
15. Becker, J. L. 1972. Fusions *in vitro* de cellules somatiques en culture de *Drosophila melanogaster* induites par la Concanavaline A. *C. R. Acad. Sci. Ser. D.* 275:2969–72
16. Bell, A. E. 1954. A gene in *Drosophila melanogaster* that produces all male progeny. *Genetics* 39:958

16a. Benzer, S. 1971. From the gene to behavior. *J. Am. Med. Assoc.* 218: 1015–22

17. Bhaskaran, G. 1973. Developmental behaviour of the abdominal histoblasts in the housefly. *Nature* 241:94–97
18. Bhaskaran, G., Sivasubramanian, P. 1969. Metamorphosis of imaginal disks of the housefly: Evagination of transplanted disks. *J. Exp. Zool.* 176:385–96
19. Bodenstein, D. 1938. Untersuchungen zum Metamorphose-problem. I. Kombinierte Schnürungs-und Transplantations-Experimente an *Drosophila. Wilhelm Roux' Arch. Entwicklungsmech. Organismen* 137:474–505
20. Bodenstein, D. 1943. Hormones and tissue competence in the development of *Drosophila. Biol. Bull.* 84:34–58
21. Bodenstein, D., Abdel-Malek, A. 1949. The induction of *aristapedia* by nitrogen mustard in *Drosophila virilis. J. Exp. Zool.* 111:95–114
22. Bohn, H. 1970. Interkalare Regeneration und segmentale Gradienten bei den Extremitäten von *Leucophaea*-Larven (Blattaria). I. Femur und Tibia. *Wilhelm Roux' Arch. Entwicklungsmech. Organismen* 165:303–41
23. Bohn, H. 1970. Interkalare Regeneration und segmentale Gradienten bei den Extremitäten von *Leucophaea*-Larven (Blattaria). II. Coxa und Tarsus. *Develop. Biol.* 23:355–79
24. Braun, W. 1940. Experimental evidence on the production of the mutant "*aristopedia*" by a change of developmental velocities. *Genetics* 25: 143–49
25. Bridges, C. B., Dobzhansky, Th. 1933. The mutant "Proboscipedia" in *Drosophila melanogaster.* A case of hereditary homoeosis. *Wilhelm Roux' Arch. Entwicklungsmech. Organismen* 127: 575–90
26. Bridges, C., Morgan, T. 1923. The third chromosome group of mutant characters of *Drosophila melanogaster. Carnegie Inst. Wash. Publ.* 327:1–251
27. Brosseau, G. E. Jr. 1957. The environmental modification of somatic crossing-over in *Drosophila melanogaster* with special reference to developmental phase. *J. Exp. Zool.* 136:567–93
28. Brown, S., Walen, K., Brosseau, G. 1962. Somatic crossing-over and elimination of ring X chromosomes of *Drosophila melanogaster. Genetics* 47:1573–79
29. Brown, S. W., Welshons, W. J. 1955. Maternal aging and somatic crossing-over of attached X chromosomes. *Proc. Nat. Acad. Sci. USA* 41:209–15
30. Bruschweiler, W., Gehring, W. 1973. A method of freezing living ovaries of *Drosophila melanogaster* larvae and its application to the storage of mutant stocks. *Experientia.* 29:134–35
31. Bryant, P. J. 1970. Cell lineage relationships in the imaginal wing disc of *Drosophila melanogaster. Develop. Biol.* 22:389–411
32. Bryant, P. J. 1971. Regeneration and duplication following operations *in situ* on the imaginal discs of *Drosophila melanogaster. Develop. Biol.* 26: 637–51
33. Bryant, P. J. 1973. Cellular determination and pattern formation in the imaginal discs of *Drosophila. Current Topics in Developmental Biology*, eds. A. A. Moscond, A. Monroy In press
34. Bryant, P. J. 1973. Genetic diseases, Chapt. X. *Diseases of Insects*, ed. G. Cantwell. New York: Marcel Dekker. In press
35. Bryant, P. J., Sang, J. 1968. Lethal derangements of metamorphosis and modifications of gene expression caused by juvenile hormone mimics. *Nature* 220:393–94
36. Bryant, P. J., Schneiderman, H. A. 1969. Cell lineage, growth, and deter-

mination in the imaginal leg discs of *Drosophila melanogaster. Develop. Biol.* 20:263–90

37. Bryant, P. J., Schubiger, G. 1971. Giant and duplicated imaginal discs in a new lethal mutant of *Drosophila melanogaster. Develop. Biol.* 24: 233–63

38. Bull, A. L. 1966. *Bicaudal* a genetic factor which affects the polarity of the embryo in *Drosophila melanogaster. J. Exp. Zool.* 161:221–42

39. Chan, L-N., Gehring, W. 1971. Determination of blastoderm cells in *Drosophila melanogaster. Proc. Nat. Acad. Sci. USA* 68:2217–21

40. Chihara, C., Petri, W., Fristrom, J., King, D. 1972. The assays of ecdysones and juvenile hormones on *Drosophila* imaginal disks *in vitro. J. Insect Physiol.* 18:1115–24

41. Colaianne, J. J., Bell, A. E. 1968. The effect of the *doublesex* (*dsx*) mutant on the action of *daughterless* (*da*) in *D. melanogaster. Drosophila Inform. Serv.* 43:115

42. Colaianne, J. J., Bell, A. E. 1970. Sonless, a sex-ratio anomaly in *Drosophila melanogaster* resulting from a gene-cytoplasm interaction. *Genetics* 65:619–25

43. Colaianne, J. J., Bell, A. E. 1972. The relative influence of sex of progeny on the lethal expression of the *sonless* gene in *Drosophila melanogaster. Genetics* 72:293–96

44. Colaianne, J. J., Bell, A. E. 1972. Linkage relationship between the *rudimentary* and *sonless* loci. *Drosophila Inform. Serv.* 48:20–21

45. Counce, S. J. 1956. Studies on female-sterility genes in *Drosophila melanogaster.* I. The effects of the gene *deep orange* on embryonic development. *Z. Indukt. Abstamm. Vererbungsl.* 87: 443–61

46. Counce, S. J. 1956. Studies of female-sterility genes in *Drosophila melanogaster.* II. The effects of the gene *fused* on embryonic development. *Z. Indukt. Abstamm. Vererbungsl.* 87: 462–81

47. Counce, S. J. 1956. Studies on female-sterility genes in *Drosophila melanogaster.* III. The effects of the gene *rudimentary* on embryonic development. *Z. Indukt. Abstamm. Vererbungsl.* 87:482–92

48. Counce, S. J. 1973. The causal analysis of insect embryogenesis. *Developmental Systems: Insects,* vol. 2 eds. C. H. Waddington, S. Counce-Niklas, 1–156. New York: Academic

49. Counce, S. J., Ede, D. A. 1957. The effect on embryogenesis of a sex-linked female-sterility factor in *Drosophila melanogaster. J. Embryol. Exp. Morphol.* 5:404–21

50. Crick, F. 1970. Diffusion in embryogenesis. *Nature* 225:420–22

50a. Curtis, A. S. G. 1967. *The Cell Surface: Its Molecular Role in Morphogenesis.* London: Logos Press

51. Davidson, R. L. 1971. Regulation of gene expression in somatic cell hybrids: A review. *In Vitro* 6:411–26

52. Ducoff, H. S. 1972. Causes of death in irradiated adult insects. *Biol. Rev.* 47:211–40

53. Enzmann, E. V., Haskins, C. P. 1939. Note of modifications in the morphogenesis of *Drosophila melanogaster* occurring under neutron bombardment. *Am. Natur.* 73:470–72

54. Ephrussi, B., Beadle, G. W. 1936. A technique for transplantation for *Drosophila. Am. Natur.* 70:218–25

55. Faber, J. 1971. Vertebrate limb ontogeny and limb regeneration: Morphogenetic parallels. *Advan. Morphogen.* 9:127–47

56. Falk, D. 1972. Partial characterization of three nutritional mutants in *Drosophila melanogaster. Can. J. Genet.* 14:726

57. Falk, E. 1964. ss^{A}: *spineless-aristapedia, dominant. Drosophila Inform. Serv.* 39:60

58. Fausto-Sterling, A. 1971. On the timing and place of action during embryogenesis of the female-sterile mutants *fused* and *rudimentary Drosophila melanogaster. Develop. Biol.* 26: 452–63

59. Fausto-Sterling, A. 1971. Studies on the sterility phenotype of the pleiotropic mutant *fused* of *Drosophila melanogaster. J. Exp. Zool.* 178:343–50

60. Fielding, C. J. 1967. Developmental genetics of the mutant *grandchildless* of *Drosophila subobscura. J. Embryol. Exp. Morphol.* 17:375–85

61. Friesen, H. 1935. Somatische Deletionen des X-chromosoms von *Drosophila melanogaster. Z. Vererbungslehre* 68:436–42

62. Fristrom, D. 1968. Cellular degeneration in wing development of the mutant vestigial of *Drosophila melanogaster. J. Cell Biol.* 39:488–91

63. Fristrom, D. 1969. Cellular degeneration in the production of some mutant phenotypes in *Drosophila melano-*

gaster. *Mol. Gen. Genet*. 103:363–79
64. Fristrom, J. W. 1972. The biochemistry of imaginal disk development. *Results and Problems in Cell Differentiation, V*, ed. H. Ursprung, R. Nöthiger, 108–54. Germany: Springer. 172 pp.
65. Fullilove, S. L., Jacobson, A. G. 1971. Nuclear elongation and cytokinesis in *Drosophila montana*. *Develop. Biol.* 26:560–77
66. Garcia-Bellido, A. 1966. Pattern reconstruction by dissociated imaginal disc cells of *Drosophila melanogaster*. *Develop. Biol.* 14:278–306
67. Garcia-Bellido, A. 1966. Changes in selective affinities following transdetermination in imaginal discs of *Drosophila melanogaster*. *Exp. Cell Res.* 44:382–92
68. Garcia-Bellido, A. 1967. Histotypic reaggregation of dissociated imaginal disc cells of *Drosophila melanogaster* cultured in vivo. *Wilhelm Roux' Arch. Entwicklungsmech. Organismen* 158: 212–17
69. Garcia-Bellido, A. 1968. Cell affinities in antennal homoeotic mutants of *Drosophila melanogaster*. *Genetics* 59: 487–99
70. Garcia-Bellido, A. 1969. *Opt*^{G}: *ophthalmoptera* of Goldschmidt. *Drosophila Inform. Serv.* 44:52
71. Garcia-Bellido, A. 1972. Pattern formation in imaginal disks. *Results and Problems in Cell Differentiation, V*, eds. H. Ursprung, R. Nöthiger, 59–91. Germany: Springer. 172 pp.
72. Garcia-Bellido, A. 1972. Some parameters of mitotic recombination in *Drosophila melanogaster*. *Mol. Gen. Genet*. 115:54–72
73. Garcia-Bellido, A., Merriam, J. 1969. Cell lineage of the imaginal discs in *Drosophila* gynandromorphs. *J. Exp. Zool.* 170:61–76
74. Garcia-Bellido, A., Merriam, J. R. 1971. Parameters of the wing imaginal disc development of *Drosophila melanogaster*. *Develop. Biol.* 24:61–87
75. Garcia-Bellido, A., Merriam, J. R. 1971. Genetic analysis of cell heredity in imaginal discs of *Drosophila melanogaster*. *Proc. Nat. Acad. Sci. USA* 68:2222–26
76. Garcia-Bellido, A., Merriam, J. R. 1971. Clonal parameters of tergite development in *Drosophila*. *Develop. Biol.* 26:264–76
77. Garcia-Bellido, A., Santamaria, P. 1972. Developmental analysis of the wing disc in the mutant engrailed of *Drosophila melanogaster*. *Genetics* 72:87–104
78. Gardner, E. J. 1970. Tumorous head in *Drosophila*. *Advan. Genet.* 15: 116–46
79. Gardner, E. J., Woolf, C. M. 1949. Maternal effect involved in the inheritance of abnormal growths in the head region of *Drosophila melanogaster*. *Genetics* 34: 573–85
80. Garen, A., Gehring, W. 1972. Repair of the lethal developmental defect in *deep orange* embryos of *Drosophila* by injection of normal egg cytoplasm. *Proc. Nat. Acad. Sci. USA*. 69: 2982–85
81. Gateff, E. A. 1971. *Developmental and histological studies of wild-type and mutant tissues of Drosophila melanogaster*, p. 37. Ph.D. thesis. Univ. California, Irvine. 202 pp.
82. Gateff, E., Schneiderman, H. A. 1969. Neoplasms in mutant and cultured wild-type tissues of *Drosophila*. *Nat. Cancer Inst. Monogr.* 31:365–97
83. Gateff, E., Schneiderman, H. A. 1970. Long term preservation of imaginal disc cell lines at low temperature. *Drosophila Inform. Serv.* 45:68
84. Gauss, U., Sander, K. 1966. Stadienabhängigkeit embryonaler Doppelbildungen von *Chironomus th. thummi*. *Naturwissenschaften* 53:182–83
85. Gehring, W. 1964. "Phenocopies" produced by 5-fluorouracil. *Drosophila Inform. Serv.* 39:102
86. Gehring, W. 1966. Übertragung und Änderung der Determinationsqualitäten in Antennenscheiben-Kulturen von *Drosophila melanogaster*. *J. Embryol. Exp. Morphol.* 15:77–111
87. Gehring, W. 1966. Bildung eines vollstandigen Mittelbeines mit Sternopleura in der Antennen-region bei der Mutante *Nasobemia* (*Ns*) von *Drosophila melanogaster*. *Arch. Julius Klaus-Stift. Vererbungsforsch. Sozialanthropol. Rassenhyg.* 41:44–54
88. Gehring, W. 1967. Clonal analysis of determination dynamics in cultures of imaginal disks in *Drosophila melanogaster*. *Develop. Biol.* 16:438–56
89. Gehring, W. 1970. Problems of cell determination and differentiation in *Drosophila*. *Problems in Biology: RNA in Development*, ed. E. W. Hanly, 231–44. Salt Lake City, Utah: Univ. Utah Press. 454 pp.
90. Gehring, W. 1970. A recessive lethal 1(4)29 with a homoeotic effect in *D. melanogaster*. *Drosophila Inform. Serv.* 45:103

91. Gehring, W. 1972. The stability of the determined state in cultures of imaginal disks in *Drosophila*. *Results and Problems in Cell Differentiation, V*, eds. H. Ursprung, R. Nöthiger, 35–58. Germany: Springer. 172 pp.
92. Gehring, W. 1973. Genetic control of determination in the *Drosophila* embryo. *Symp. Soc. Develop. Biol.*, ed. F. Ruddle. In press
93. Gehring, W., Mindek, G., Hadorn, E. 1968. Auto- und allotypische Differenzierungen aus Blastemen der Halterenscheibe von *Drosophila melanogaster* nach Kultur *in vivo*. *J. Embryol. Exp. Morphol.* 20:307–18
94. Gehring, W., Nöthiger, R. 1973. The imaginal discs of *Drosophila*. *Developmental Systems: Insects*, vol. 2 ed. C. H. Waddington, S. Counce-Niklas. New York: Academic
95. Geigy, R. 1928. Castration de mouches par l'exposition de l'oeuf aux rayons ultra-violets. *C. R. Acad. Sci.* 98:106–08
96. Geigy, R. 1931. Erzeugung rein imaginaler Defekte durch ultraviolette Eibestrahlung bei *Drosophila melanogaster*. *Wilhelm Roux' Arch. Entwicklungsmech. Organismen* 125:406–47
97. Gersh, E. 1946. Chemically induced phenocopies in *Drosophila melanogaster*. *Drosophila Inform. Serv.* 20:86
98. Geyer-Duszynska, I. 1967. Experiments in nuclear transplantation in *Drosophila melanogaster*. Preliminary report. *Rev. Suisse Zool.* 74:614–15
99. Ginter, E. 1969. *Apx*: *Antennapedix*. *Drosophila Inform. Serv.* 44:50–51
100. Gloor, H. 1943. Entwicklungsphysiologische Untersuchungen an den Gonaden der Letalrasse (lgl) von *Drosophila melanogaster*. *Rev. Suisse Zool.* 50:339–93
101. Gloor, H. 1947. Phanokopie-Versuche mit Ather an *Drosophila*. *Rev. Suisse Zool.* 54:637–712
102. Gloor, H., Kobel, H. 1966. *Antennapedia* (ss^{Anp}), eine homoeotische Mutante bei *Drosophila hydei* Sturtevant. *Rev. Suisse Zool.* 73:229–52
103. Goldschmidt, E., Liderman-Klein, A. 1958. Reoccurrence of a forgotten homoeotic mutant in *Drosophila*. *J. Hered.* 49:262–66
104. Goldschmidt, R. 1938. *Physiological Genetics*. New York: McGraw-Hill
105. Goldschmidt, R. 1940. *The Material Basis of Evolution*. New Haven, Conn: Yale Univ.
106. Goldschmidt, R. 1945. The structure of podoptera, a homoeotic mutant of *Drosophila melanogaster*. *J. Morphol.* 77:71–103
107. Goldschmidt, R., Piternick, L. 1957. The genetic background of chemically induced phenocopies in *Drosophila*. *J. Exp. Zool.* 135:127–202
108. Gottlieb, F. J. 1964. Genetic control of pattern determination in *Drosophila*. The action of Hairy-wing. *Genetics* 49:739–60
109. Grigliatti, T., Suzuki, D. T. 1971. Temperature-sensitive mutations in *Drosophila melanogaster*. VIII. The homoeotic mutant, ss^{a40a}. *Proc. Nat. Acad. Sci. USA*. *68:1307–11*
110. Guerra, M., Postlethwait, J. H., Schneiderman, H. A. 1973. The development of the imaginal abdomen of *Drosophila melanogaster*. *Develop. Biol.* 32:361–72
111. Gurdon, J. B. 1964. The transplantation of living cell nuclei. *Advan. Morphol.* 4:1–43
112. Guthrie, C., Nashimoto, H., Nomura, M. 1969. Structure and function of *E. coli* ribosomes. VIII. Cold-sensitive mutants defective in ribosome assembly. *Proc. Nat. Acad. Sci. USA*. 63:384–91
113. Hadorn, E. Personal communication
114. Hadorn, E. 1948. Genetische und entwicklungsphysiologische Probleme der Insektenontogenese. *Folia Biotheor., Leiden*, No. 3, pp. 109–26
115. Hadorn, E. 1963. Differenzierungsleistungen wiederholt fragmentierter Teilstücke männlicher Genitalscheiben von *Drosophila melanogaster* nach Kultur *in vivo*. *Develop. Biol.* 7:617–29
116. Hadorn, E. 1965. Problems of determination and transdetermination. *Brookhaven Symp.* 18:148–61
117. Hadorn, E. 1966. Über ein Anderung musterbestimmender Qualitaten in einer Blastemkultur von *Drosophila melanogaster*. *Rev. Suisse Zool.* 73: 253–65
118. Hadorn, E. 1966. Konstanz, Wechsel und Typus der Determination und Differenzierung in Zellen auf männlichen Genitalanlagen von *Drosophila melanogaster* nach Dauerkultur *in vivo*. *Develop. Biol.* 13:424–509
119. Hadorn, E. 1966. Dynamics of determination. *Symp. Soc. Develop. Biol.* 25:85–104
120. Hadorn, E. 1969. Proliferation and dynamics of cell heredity in blastema cultures of *Drosophila*. *Nat. Cancer Inst. Monogr.* 31:351–64
121. Hadorn, E., Anders, G., Ursprung, H. 1959. Kombinate aus teilweise disso-

ziierten Imaginalscheiben verschiedener Mutanten und Arten von *Drosophila*. *J. Exp. Zool.* 142:159–75
122. Hadorn, E., Garcia-Bellido, A. 1964. Zur Proliferation von *Drosophila*-Zellkulturen im Adultmilieu. *Rev. Suisse Zool.* 71:576–82
123. Hadorn, E., Hürlimann, R., Mindek, G., Schubiger, G., Staub, M. 1968. Entwicklungsleistungen embryonaler Blasteme von *Drosophila* nach Kultur im Adultwirt. *Rev. Suisse Zool.* 75:557–69
124. Haendle, J. 1971. Röntgeninduzierte mitotische Rekombination bei *Drosophila melanogaster*. I. Ihre Abhängigkeit von der Dosis, der Dosisrate und vom Spektrum. *Mol. Gen. Genet.* 113:114–31
125. Haendle, J. 1971. Röntgeninduzierte mitotische Rekombination bei *Drosophila melanogaster*. II. Beweis der Existenz und Charakterisierung zweier von der Art des Spektrums abhangiger Reaktionen. *Mol. Gen. Genet.* 113:132–49
126. Hall, J., Gelbart, W., Kankel, D. 1973. Mosaic systems. *Biology of Drosophila*, ed. E. Novitski, M. Ashburner, New York: Academic
127. Hannah, A., Strömnaes, Ø. 1955. Extra sex comb mutant in *Drosophila melanogaster*. *Drosophila Inform. Serv.* 29:121–23
128. Haynie, J. Unpublished observations
129. Henke, K., Maas, H. 1946. Über sensible Perioden der algemeinen Korpergliederung bei *Drosophila*. *Akad. Wiss. Gottingen Math.-Phys. Nachrichten* 1:3–4
130. Herskowitz, I. 1949. Hexaptera, a homoeotic mutant in *Drosophila melanogaster*. *Genetics* 34:10–25
131. Hess, O., Meyer, G. F. 1968. Genetic activities of the Y chromosome in *Drosophila* during spermatogenesis. *Advan. Genet.* 14:171–223
132. Hildreth, P. E., Lucchesi, J. C. 1967. Fertilization in *Drosophila*. III. A reevaluation of the role of polyspermy in development of the mutant *deep orange*. *Develop. Biol.* 15:536–52
133. Hinton, C. W. 1955. The behaviour of an unstable ring-chromosome of *Drosophila melanogaster*. *Genetics* 40: 951–61
134. Hinton, C. W. 1957. The analysis of rod derivatives of an unstable ring chromosome of *Drosophila melanogaster*. *Genetics* 42:55–65
135. Hinton, C. W. 1959. A cytological study of w^{vc} chromosome instability in cleavage mitosis of *Drosophila melanogaster*. *Genetics* 44:923–31
136. Hoge, M. 1915. The influence of temperature on the development of a mendelian character. *J. Exp. Zool.* 18:241–97
137. Holden, J., Suzuki, D. T. 1973. Temperature sensitive mutations in *Drosophila melanogaster*. XII. The genetic and developmental characteristics of dominant lethals on chromosome 3. *Genetics* 73:445–58
138. Horowitz, N. H., Fling, M. 1953. Genetic determination of tyrosinase thermostability in *Neurospora*. *Genetics* 38:360–74
139. Hotta, Y., Benzer, S. 1972. Mapping of behaviour in *Drosophila* mosaics. *Nature* 240:527–35
140. Howland, R. B. 1941. Structure and development of centrifuged eggs and early embryos of *Drosophila melanogaster*. *Proc. Am. Phil. Soc.* 84:605–16
141. Illmensee, K. 1968. Transplantation of embryonic nuclei into unfertilized eggs of *Drosophila melanogaster*. *Nature* 219:1268–69
142. Illmensee, K. 1970. Imaginal structures after nuclear transplantation in *Drosophila melanogaster*. *Naturwissenschaften* 11:550–51
143. Illmensee, K. 1972. Developmental potencies of nuclei from cleavage, preblastoderm, and syncytial blastoderm transplanted into unfertilized eggs of *Drosophila melanogaster*. *Wilhelm Roux' Arch. Entwicklungsmech. Organismen* 170:267–98
144. Illmensee, K. 1973. The potentialities of transplanted early gastrula nuclei of *Drosophila melanogaster*. Production of their imago descendants by germ-line transplantation. *Wilhelm Roux' Arch. Entwicklungsmech. Organismen* 171:331–43
145. Inouye, I., Takaya, H. 1964. On the occurrence of the hereditary irregularities of antennae in *Drosophila melanogaster*. *Jap. J. Genet.* 38:328–36
146. Jacobson, A. 1967. Early development of *Drosophila montana* and of the lethal hybrid *Drosophila montana* female crossed with *Drosophila texana* male. (Film based on research by J. D. Kinsey at University of Texas, Austin)
147. James, A. A. Unpublished observations
148. Janning, W. 1972. Aldehyde oxidase as a cell marker for internal organs in *Drosophila melanogaster*. *Naturwissenschaften* 59:516–17

149. Jazdowska-Zagrodzinska, B. 1966. Experimental studies on the role of "polar granules" in the segregation of pole cells in *Drosophila melanogaster*. *J. Embryol. Exp. Morphol.* 16:391–99

150. Kalthoff, K. 1971. Position of targets and period of competence for UV-induction of the malformation "double abdomen" in the egg of *Smittia* spec. (Diptera, Chironomidae). *Wilhelm Roux' Arch. Entwicklungsmech. Organismen* 168:63–84

151. Kalthoff, K. 1971. Photoreversion of UV-induction of the malformation "double abdomen" in the egg of *Smittia* spec. (Diptera, Chironomidae). *Develop. Biol.* 25:119–32

152. Kaplan, W. D. 1953. The influence of Minutes upon somatic crossing-over in *Drosophila melanogaster*. *Genetics* 38:630–51

153. Kaplan, W. D., Seecof, R. L., Trout W. E., Pasternack, M. E. 1970. Production and relative frequency of maternally influenced lethals in *Drosophila melanogaster*. *Am. Natur.* 104:261–71

154. Kaufman, T. Personal communication

155. Kauffman, S. 1973. Control circuits for determination and transdetermination. *Science*. In press

156. Kiefer, B. 1973. Genetics of sperm development in *Drosophila*. *Symp. Soc. Develop. Biol.* In press

157. Kiger, J. 1973. The bithorax complex—a model for cell determination in *Drosophila*. *J. Theor. Biol.* In press

158. Kim, C. 1959. The differentiation centre inducing the development from larvae to adult leg in *Pieris brassicae* (Lepidoptera). *J. Embryol. Exp. Morphol.* 7:572–82

159. King, R. 1970. *Ovarian development in Drosophila melanogaster*. New York: Academic. 227 pp.

160. Kobel, H. 1968. Homoeotisch Flügelbildung durch das loboid-Allel "*ophthalmoptera*" (*ld*^{oph}) bei *Drosophila melanogaster*. *Genetica* 39:329–44

161. Komai, T. 1926. Crippled, a new mutant character of *Drosophila melanogaster*, and its inheritance. *Genetics* 11:280–93

162. Lamprecht, J., Remensberger, P. 1966. Pölytane Chromosomen im Bereiche der Prothoracal-dorsal-Imaginalscheibe von *Drosophila melanogaster*. *Experientia* 22:293–95

163. Laugé, G. 1967. Origine et croissance du disque génital de *Drosophila melanogaster* Meig. *C. R. Acad. Sci. Paris* 265:814–17

164. Lawrence, P. A. 1966. Gradients in the insect segment: The orientation of hairs in the milkweed bug, *Oncopeltus fasciatus*. *J. Exp. Biol.* 44:607–20

165. Lawrence, P. 1970. Polarity and patterns in the postembryonic development of insects. *Advan. Insect Physiol.* 1:197–266

166. Lawrence, P. 1971. The organization of the insect segment. *Symp. Soc. Exp. Biol.* 25:379–90

167. Lawrence, P. A., Crick, F. H. C., Munro, M. 1972. A gradient of positional information in an insect, *Rhodnius*. *J. Cell Sci.* 11:815–53

168. LeCalvez, J. 1948. $In(3R)ss^{Ar}$: Mutation "*aristapedia*" Hétérozygote dominant, homozygote léthale chez *Drosophila melanogaster*. (Inversion dans le bras droit du chromosome III). *Bull. Biol. Fr. Belg.* 82:97–113

169. Le Calvez, J. 1948. Observations phénogénétiques sur la mutation *Aristapedia* dominante de *Drosophila melanogaster*. *Arch. Anat. Microsc. Morphol. Exp.* 37:50–72

170. LeClerc, G. 1946. Occurrence of mitotic crossing over without meiotic crossing over. *Science* 103:553–54

171. Lees, A. D., Waddington, C. H. 1942. The development of the bristles in normal and some mutant types of *Drosophila melanogaster*. *Proc. Roy. Soc. Ser. B.* 131:87–110

172. Lefevre, G. 1948. The relative effectiveness of fast neutrons and gamma rays on producing somatic crossing-over in *Drosophila*. *Genetics* 33:112–3

172a. Lessops, R. J. 1965. Culture of dissociated *Drosophila* embryos: aggregated cells differentiate and sort out. *Science* 148:502–03

173. Lewis, E. B. 1947. Polycomb. *Drosophila Inform. Serv.* 21:69

174. Lewis, E. 1956. $Antp^B$: *Antennapedia-Bacon* and $Antp^{Yu}$: *Antennapedia-Yu*. *Drosophila Inform. Serv.* 30:76

175. Lewis, E. B. 1963. Genes and developmental pathways. *Am. Zool.* 3:33–56

176. Lewis, E. 1964. Genetic control and regulation of developmental pathways. *Symp. Soc. Develop. Biol.* 23:231–52

177. Lewis, E. B. 1968. Genes and gene complexes. *Heritage from Mendel*, ed. R. A. Brink, 17–47. Madison, Wisconsin: Univ. Wisconsin Press

178. Lewis, E. B., Gencarella, W. 1952. Claret and nondisjunction in *Drosophila melanogaster* (abstr). *Genetics* 37:600–01

179. Lewis, E. B., Bacher, F. 1969. Opt^B: ophthalmoptera-Bacher. *Drosophila*

Inform. Serv. 44:48

180. Lindsley, D. L., Grell, E. H. 1968. Genetic variations of *Drosophila melanogaster. Carnegie Inst. Wash. Publ.* 627

181. Löbbecke, E. 1958. Über die Entwicklung der imaginalen Epidermus des Abdomens von *Drosophila* ihre Segmentierung und die Determination der Tergite. *Biol. Zentralbl.* 77:209–37

182. Lynch, C. N. 1919. An analysis of certain cases of intra-specific sterility. *Genetics* 4:501–33

183. Maas, A. H. 1948. Über die Auslöskarkeit von Temperature modifikationen wahrend der Embryonalentwicklung von *Drosophila melanogaster* Meigen. *Wilhelm Roux' Arch. Entwicklungsmech. Organismen* 143: 515–72

184. Madhavan, K. 1973. Morphogenetic effects of juvenile hormone and juvenile hormone mimics on adult development of *Drosophila. J. Insect Physiol.* 19:441–53

185. Mahowald, A. P. 1971. Origin and continuity of polar granules. *Results and Problems in Cell Differentiation, II*, ed. J. Reinert, H. Ursprung, 158–69. New York: Springer. 342 pp.

186. Mandaron, P. 1970. Développement *in vitro* des disques imaginaux de la *Drosophile.* Aspects morphologiques et histologiques. *Develop. Biol.* 22:298–320

187. Mandaron, P. 1971. Sur le mécanisme de l-évagination des disques imaginaux chez la *Drosophile. Develop. Biol.* 25:581–605

188. Margulies, L. 1972. Actinomycin D-induced phenocopies in *Drosophila melanogaster* and their relevance to the time of gene action. *Genetica* 43:207–22

189. Mayoh, H., Suzuki, D. T. 1973. Temperature-sensitive mutation in *Drosophila melanogaster*. XVII. The genetic properties of sex-linked recessive cold-sensitive mutants. *Mol. Gen. Genet.* In press

190. Merriam, J. R., Fyffe, W. E. 1972. Somatic crossing over in *Drosophila melanogaster*: I. Dose response curves for X-ray induction and effects of dose fractionation. *Mutat. Res.* 14:309–14

191. Merriam, J. R., Garcia-Bellido, A. 1972. A model for somatic pairing derived from somatic crossing over with third chromosome rearrangements in *Drosophila melanogaster. Mol. Gen. Genet.* 115:302–13

192. Merriam, J. R., Nöthiger, R., Garcia-Bellido, A. 1972. Are dicentric anaphase bridges formed by somatic recombination in X chromosome inversion heterozygotes of *Drosophila melanogaster? Mol. Gen. Genet.* 115: 294–301

193. Mindek, G. 1968. Proliferations- und Transdeterminationsleistungen der weiblichen Genital-Imaginalscheiben von *Drosophila melanogaster* nach Kultur *in vivo. Wilhelm Roux' Arch. Entwicklungsmech. Organismen* 161: 249–80

194. Mindek, G. 1972. Metamorphosis of imaginal discs of *Drosophila melanogaster. Wilhelm Roux" Arch. Entwicklungsmech. Organismen* 169:353–56

195. Mindek, G., Nöthiger, R. 1973. Parameters influencing the acquisition of competence for metamorphosis in imaginal disks of *Drosophila. J. Insect Physiol.* In press

196. Newby, W. 1949. Abnormal growths on the head of *Drosophila melanogaster. J. Morphol.* 85:177–95

197. Newcombe, H. B. 1971. The genetic effects of ionizing radiations. *Advan. Genet.* 16:239–303

198. Nørby, S. 1970. A specific nutritional requirement for pyrimidines in rudimentary mutants of *Drosophila melanogaster. Hereditas* 66:205–14

199. Nöthiger, R. 1964. Differenzierungsleistungen in Kombinaten, hergestellt aus Imaginalscheiben verschiedener Arten, Geschlechter und Körpersegmente von *Drosophila. Wilhelm Roux Arch. Entwicklungsmech. Organismen* 155:269–301

200. Nöthiger, R. 1972. The larval development of imaginal disks. *Results and Problems in Cell Differentiation, V*, ed. H. Ursprung, R. Nöthiger, 1–34. Germany: Springer. 172 pp.

201. Nöthiger, R., Schubiger, G. 1966. Developmental behavior of fragments of symmetrical and asymmetrical imaginal discs of *Drosophila melanogaster* (Diptera). *J. Embryol. Exp. Morphol.* 16:355–68

202. Nöthiger, R., Strub, S. 1972. Imaginal defects after UV-microbeam irradiation of early cleavage stages of *Drosophila melanogaster. Rev. Suisse Zool.* 79:267–79

203. O'Donovan, G. A., Ingraham, J. L. 1965. Cold-sensitive mutants of *Escherichia coli* resulting from increased feedback inhibition. *Proc. Nat. Acad. Sci. USA.* 54:451–57

204. Okada, M., Kleinman, I. A., Schneiderman, H. A. 1973. Unpublished

205. Ouweneel, W. J. 1969. Influence of Environmental factors on the homoeotic effect of loboid ophthalmoptera in *Drosophila melanogaster. Wilhelm Roux' Arch. Entwicklungsmech. Organismen* 164:15–36
206. Ouweneel, W. J. 1969. Morphology and development of *loboid-ophthalmoptera*, a homoeotic mutant of *Drosophila melanogaster. Wilhelm Roux' Arch. Entwicklungsmech. Organismen* 164:1–14
207. Ouweneel, W. J. 1970. Genetic analysis of *loboid-ophthalmoptera*, a homoeotic strain in *Drosophila melanogaster. Genetica* 41:1–20
208. Ouweneel, W. J. 1970. Normal and abnormal determination in the imaginal discs of *Drosophila*, with special reference to the eye discs. *Acta Embryol. Exp.* 2:95–119
209. Ouweneel, W. J. 1970. Report of W. J. Ouweneel. *Drosophila Inform. Serv.* 45:35
210. Ouweneel, W. J. 1972. Determination, regulation, and positional information in insect development. *Acta Biotheor.* 21:115–31
211. Ouweneel, W. J. 1972. Effect of colchicine on the development of homoeotic wing tissue. *Drosophila Inform. Serv.* 48:146
212. Overton, J., Raab, M. 1967. The development and fine structure of centrifuged eggs of *Chironomus thummi. Develop. Biol.* 15:271–87
213. Pasztor, L. 1971. Unstable ring-X chromosomes derived from a tandem metacentric compound in *Drosophila melanogaster. Genetics* 68:245–58
214. Patterson, J. T. 1930. Somatic segregation produced by X-rays in *Drosophila melanogaster. Proc. Nat. Acad. Sci. USA.* 16:109–11
215. Patterson, J. T., Stone, W. 1938. Gynandromorphs in *Drosophila melanogaster. Univ. Tex. Publ.* 3825:1–67
216. Poodry, C. A., Bryant, P. J., Schneiderman, H. A. 1971. The mechanism of pattern reconstruction by dissociated imaginal discs of *Drosophila melanogaster. Develop. Biol.* 26: 464–77
217. Poodry, C. A., Hall, L., Suzuki, D. T. 1973. Developmental properties of *shibire*^{ts1}: A pleiotropic mutation affecting larval and adult locomotion and development. *Develop. Biol.* 32:373–86
218. Poodry, C. A., Schneiderman, H. A. 1970. The ultrastructure of the developing leg of *Drosophila melanogaster. Wilhelm Roux' Arch. Entwicklungsmech. Organismen* 166:1–44
219. Poodry, C. A., Schneiderman, H. A. 1971. Intercellular adhesivity and pupal morphogenesis in *Drosophila melanogaster. Wilhelm Roux' Arch. Entwicklungsmech. Organismen* 168:1–9
220. Poodry, C. A. Schneiderman, H. A. 1973. Pattern formation in *Drosophila melanogaster*. The effects of mutation on polarity and "prepattern' in the developing leg. *Develop. Biol.*
221. Postlethwait, J. H. 1973. Development of the temperature sensitive homoeotic mutant *eyeless-ophthalmoptera* of *Drosophila melanogaster.* Unpublished
222. Postlethwait, J. H. 1973. Pattern formation in imaginal discs of *Drosophila melanogaster* after irradiation of embryos and young larvae. II. Regeneration. In preparation
223. Postlethwait, J. H., Bryant, P. J., Schubiger, G. 1972. The homoeotic effect of "tumorous head" in *Drosophila melanogaster. Develop. Biol.* 29:337–42
224. Postlethwait, J. H., Girton, J. 1973. Development of antennal-leg homoeotic mutants in *Drosophila melanogaster.* In preparation
225. Postlethwait, J. H., Poodry, C. A., Schneiderman, H. A. 1971. Cellular dynamics of pattern duplication in imaginal discs of *Drosophila melanogaster. Develop. Biol.* 26:125–32
226. Postlethwait, J. H., Schneiderman, H. A. 1969. A clonal analysis of determination in *Antennapedia*, a homoeotic mutant of *Drosophila melanogaster. Proc. Nat. Acad. Sci. US.a.* 64:176–83
226a. Postlethwait, J. H., Schneiderman, H. A. 1970. Induction of metamorphosis by ecdysone analogues: *Drosophila* imaginal discs cultured *in vivo. Biol. Bull.* 138:47–55
227. Postlethwait, J. H., Schneiderman, H. A. 1971. A clonal analysis of development in *Drosophila melanogaster*: Morphogenesis, determination, and growth in the wild-type antenna. *Develop. Biol.* 24:477–519
228. Postelthwait, J. H., Schneiderman, H. A. 1971. Pattern formation and determination in the antenna of the homoeotic mutant *Antennapedia* of *Drosophila melanogaster. Develop. Biol.* 25:606–40
229. Postlethwait, J. H., Schneiderman, H. A. 1973. Pattern formation in imaginal discs of *Drosophila melanogaster* after irradiation of embryos and

young larvae. *Develop. Biol.* 32: 345–60
230. Postlethwait, J. H., Weiser, K. 1973. Juvenile hormone induced vitellogenesis in a female sterile mutant of *Drosophila melanogaster* induced by juvenile hormone. *Nature.* In press
231. Postlethwait, J. H., Williams, C. M. 1973. Juvenile hormone and the adult development of *Drosophila.* In preparation
232. Poulson, D. F. 1947. The pole cells of Diptera, their fate and significance. *Proc. Nat. Acad. Sci. USA.* 33:182–84
233. Poulson, D. F. 1968. Nature, stability, and expression of hereditary SR infections in *Drosophila. Proc. Int. Congr. Genet., 11th* 2:91–92
234. Poulson, D. F., Sakaguchi, B. 1961. Nature of "sex ratio" agent in *Drosphila. Science* 133:1489–90
235. Putnam, R., Arking, R. Personal communication
236. Redfield, H. 1926. The maternal inheritance of a sex-limited lethal effect in *Drosophila melanogaster. Genetics* 11:484–502
237. Remensberger, P. 1968. Cytologische und Histologische untersuchungen an Zellstammen von *Drosophila melanogaster* nach Dauerkultur *in vivo. Chromosoma* 23:386–417
238. Ripoll, P. 1972. The embryonic organization of the imaginal wing disc of *Drosophila melanogaster. Wilhelm Roux' Arch. Entwicklungsmech. Organismen* 169:200–15
239. Ripoll, P., Garcia-Bellido, A. 1973. Cell autonomous lethals in *Drosophila melanogaster. Nature New Biol.* 241: 15–16
239a. Rizki, T., Rizki, R., Douthit, H. 1972. Morphogenetic effects of halogenated thymidine analogues on *Drosophila.* I. Quantitative analysis of lesions induced by 5-bromodeoxyuridine and 5-fluorouracil. *Biochem. Genet.* 6: 83–97
240. Roberts, P. 1964. Mosaics involving *aristapedia,* a homoeotic mutant of *Drosophila melanogaster. Genetics* 49:593–98
241. Ronen, A. 1964. Interchromosomal effects on somatic recombination in *Drosophila melanogaster. Genetics* 50: 649–58
242. Rosenbluth, R., Ezell, D., Suzuki, D. T. 1972. Temperature-sensitive mutations in *Drosophila melanogaster.* IX. Dominant cold-sensitive lethals on the autosomes. *Genetics* 70:75–86
243. Russell, M. Personal communication
244. Sandler, L. 1970. The regulation of sex chromosome heterochromatic activity by an autosomal gene in *Drosophila melanogaster. Genetics* 64: 481–93
245. Sandler, L. 1972. On the genetic control of genes located in the sex-chromosome heterochromatin of *Drosophila melanogaster. Genetics* 70:261–74
246. Sang, J. H., Burnet, B. 1963. Environmental modification of the eyeless phenotype in *Drosophila melanogaster. Genetics* 48:1683–99
247. Sang, J., McDonald, J. 1954. Production of phenocopies in *Drosophila* using salts, particularly sodium metaborate. *J. Genet.* 52:392–412
248. Santamaria, P., Garcia-Bellido, A. 1972. Localization and growth of the tergite *Anlage* of *Drosophila. J. Embryol. Exp. Morphol.* 28:397–417
249. Saunders, J. W. 1972. Developmental control of three dimensional polarity in the avian limb. *Ann. N. Y. Acad. Sci.* 193:29–42
250. Sayles, C. D., Procussier, J. D., Browder, L. W. 1973. Radio labelling of *Drosophila* embryos. *Nature New Biol.* 241:215–16
251. Schläpfer, T. 1963. Der Einfluss des adulten Wirtsmilieus auf die Entwicklung von larvalen Aungenantennen-Imaginalscheiben von *Drosophila melanogaster. Wilhelm Roux' Arch. Entwicklungsmech. Organismen* 154: 378–404
252. Schneider, I. 1972. Cell lines derived from late embryonic stages of *Drosophila melanogaster. J. Embryol. Exp. Morphol.* 27:353–65
253. Schneiderman, H. A. 1969. Control systems in insect development. *Biology and the Physical Sciences,* ed. S. Devons, 186–208. New York: Columbia Univ. Press. 379 pp.
254. Schubiger, G. 1968. Anlageplan, Determinationszustand und Transdeterminationsleistungen der männlichen Vorderbeinscheibe von *Drosophila melanogaster. Wilhelm Roux' Arch. Entwicklungsmech. Organismen* 160: 9–40
255. Schubiger G. 1971. Regeneration, duplication and transdetermination in fragments of the leg disc of *Drosophila melanogaster. Develop. Biol.* 26: 277–95
256. Schubiger G. 1973. Regeneration of *Drosophila melanogaster* male leg disc fragments in sugar fed female hosts. *Experientia.* 29:631–32

257. Schubiger, G. 1973. Regeneration in a homoeotic mutant of *Drosophila melanogaster*. Unpublished
258. Schubiger, G. 1973. The competence of leg imaginal discs from young third instar larvae of *Drosophila melanogaster* to differentiate adult structures. Unpublished
259. Schubiger, G., Hadorn, E. 1968. Auto- und allotypische Differenzieraugen aus *in vivo* kultivierten Vorderbeinblastemen von *Drosophila melanogaster*. *Develop. Biol.* 17:584–602
260. Schubiger, G., Schubiger-Staub, M., Hadorn, E. 1969. Mischungsversuche mit Keimteilen von *Drosophila melanogaster* zur Ermittlung des Determinations zustandes imaginaler Blasteme in Embryo. *Wilhelm Roux' Arch. Entwicklungsmech. Organismen* 163: 33–39
261. Schubiger, M., Schneiderman, H. A. 1971. Nuclear transplantation in *Drosophila melanogaster*. *Nature* 230: 185–86
262. Schwartz, D. 1954. Studies on the mechanism of crossing over. *Genetics* 39:692–700
263. Schweizer, P. 1972. Wirkung von Röntgenstrahlen auf die Entwicklung der mannlichen Genitalprimordien von *Drosophila melanogaster* und Untersuchung von Erholungsvorgangen durch Zellklon-Analyse. *Biophysik* 8:158–88
264. Shannon, M. P. 1972. Characterization of the female-sterile mutant *almondex* of *Drosophila melanogaster*. *Genetica* 43:244–56
265. Shannon, M. 1973. The development of eggs produced by the female-sterile mutant *almondex* of *Drosophila melanogaster*. Unpublished
266. Shapiro, N. I. 1941. X-ray and the frequency of somatic mosaics. *Drosophila Inform. Serv.* 15:17
267. Shearn, A., Rice, T., Garen, A., Gehring, W. 1971. Imaginal disc abnormalities in lethal mutants of *Drosophila*. *Proc. Nat. Acad. Sci. USA*. 68:2594–98
268. Siervogel, R. M. 1972. Dosage of the maternal effect gene associated with the tumorous-head abnormality in *Drosophila melanogaster*. *Genetics* 72:377–80
269. Sobels, F. H. 1952. Genetics and morphology of the genotype "asymmetric" with special reference to its "abnormal abdomen" character (*Drosophila melanogaster*). *Genetics* 26: 117–279
270. Spreij, T. E. 1971. Cell death during the development of the imaginal disks of *Calliphora erythrocephala*. *Neth. J. Zool.* 21:221–64
271. Sprey, Th. 1970. Localization of 5'-nucleotidase and its possible significance in some of the imaginal disks of *Calliphora erythrocephala*. *Neth. J. Zool.* 20:419–32
272. Spurway, H. 1948. Genetics and cytology of *Drosophila subobscura*. IV. An extreme example of delay in gene action, causing sterility. *J. Genet.* 49:126–40
273. Stauffer, H. H. 1969. Effect of oxygen on the frequency of X ray induced somatic crossing over in *Drosophila melanogaster*. *Nature* 223:1157–58
274. Stauffer, H. H. 1972. The effect of oxygen on the frequency of somatic recombination in *Drosophila melanogaster*. *Genetics* 72:277–91
275. Stepshin, V. P., Ginter, E. K. 1972. Studies of homoeotic genes Antennopedix and Nasobemia of *Drosophila melanogaster*. I. Some allele relationships between Antennopedix and Nasobemia with Antennapedia, polycomb, extra sex comb. *Genetika* 8:98–104
276. Stepshin, V. P., Ginter, E. K. 1972. Studies on homoeotic genes Antennopedix and Nasobemia of *Drosophila melanogaster*. II. Interaction of genes Antennopedix and Nasobemia with polycomb, multiple sex comb, aristapedia, thread. *Genetika* 9:67–74
277. Stern, C. 1936. Somatic crossing-over and segregation in *Drosophila melanogaster*. *Genetics* 21:625–730
278. Stern, C. 1954. Two or three bristles. *Am. Sci.* 43:213
279. Stern, C. 1954. Genes and developmental patterns. *Proc. Int. Congr. Genet. 9th, 1953*, 355–69
280. Stern, C. 1968. *Genetic Mosaics and Other Essays*. Cambridge, Mass.: Harvard Univ. Press
281. Stern, C. 1969. Somatic recombination within the white locus of *Drosophila melanogaster*. *Genetics* 62: 573–81
282. Stern, C., Rentscheler, V. 1936. The effect of temperature on the frequency of somatic crossing over in *Drosophila melanogaster*. *Proc. Nat. Acad. Sci. USA*. 22:451–53
283. Stern, C., Tokunaga, C. 1967. Nonautonomy in differentiation of pattern-determining genes in *Drosophila*. I. The sexcomb of *eyeless-dominant*. *Proc. Nat. Acad. Sci. USA*. 57:658–64

284. Stern, C., Tokunaga, C. 1968. Autonomous pleiotropy in *Drosophila*. *Proc. Nat. Acad. Sci. USA*. 60:1252–59

284a. Stern, C., Tokunaga, C. 1971. On cell lethals in *Drosophila*. *Proc. Nat. Acad. Sci. USA*. 68:329–31

285. Stewart, M., Murphy, C., Fristrom, J. W. 1972. The recovery and preliminary characterization of X chromosome mutants affecting imaginal discs of *Drosophila melanogaster*. *Develop. Biol*. 27:71–83

286. Strøman, P., Bahn, E., Nørby, S., Sick, K. 1971. Suppression of the phenocopying effect of 6-azauracil by a suppressor mutant of rudimentary in *Drosophila melanogaster*. *2 Europ. Drosophila Res. Conf., April 1–3 (Abstr)*

287. Sturtevant, A. H. 1929. The *claret* mutant type of *Drosophila simulans*: a study of chromosome elimination and of cell-lineage. *Z. Wiss. Zool*. 135:323–56

288. Sturtevant, A. H. 1945. A gene in *Drosophila melanogaster* that transforms females into males. *Genetics* 30:297–303

289. Suley, A. C. 1953. Genetics of *Drosophila subobscura*. VIII. Studies on the mutant grandchildless. *J. Genet*. 51:375–405

290. Suzuki, D. T. 1970. Temperature sensitive mutations in *Drosophila melanogaster*. *Science* 170:695–706

291. Suzuki, D. T., Piternick, L. K., Hayashi, S., Tarasoff, M., Baillie, D., Erasmus, U. 1967. Temperature-sensitive mutations in *Drosophila melanogaster*. I. Relative frequencies among γ -ray and chemically induced sex-lined recessive lethals and semilethals. *Proc. Nat. Acad. Sci. USA*. 57:907–12

292. Tobler, H. 1966. Zellspezifische Determination und Beziehung zwischen Proliferation und Transdetermination in Bein- und Flügelprimordien von *Drosophila melanogaster*. *J. Embryol. Exp. Morphol*. 16:609–33

293. Tobler, H. 1969. Beeinflussung der Borstendifferenzierung und Musterbildung durch Mitomycin bei *Drosophila melanogaster*. *Experientia* 25: 213–14

294. Tobler, H. 1970. Veränderung der Differenzierungs-leistung und der Transdeterminationsfrequenz durch Colchicin in Beinimaginalscheiben von *Drosophila melanogaster*. *Wilhelm Roux' Arch. Entwicklungsmech. Organismen* 165:217–25

295. Tobler, H., Maier, V. 1970. Zur Wirkung von Senfgaslosungen auf die Differenzierung des Borstenorganes und auf die Transdeterminationsfrequenz bei *Drosophila melanogaster*. *Wilhelm Roux' Arch. Entwicklungsmech. Organismen* 164:303–12

296. Tobler, H., Pfluger, M. 1970. Untersuchung zur Wirkung von Mitomycin C auf die Entwicklung der männlichen Vorderbeinscheibe und die Differenzierung des Borstenorgans von *Drosophila melanogaster* nach Transplantation in larvale Wirte. *Wilhelm Roux' Arch. Entwicklungsmech. Organismen* 164:293–302

297. Tobler, H., Rothenbuhler, V., Nöthiger, R. 1973. A study of the differentiation of bracts in *Drosophila melanogaster* using two mutations, H^2 and sv^{de}. *Experientia*. 29:370–71

298. Tokunaga, C. 1961. The differentiation of a secondary sex comb under the influence of the gene engrailed in *Drosophila melanogaster*. *Genetics* 46:157–76

299. Tokunaga, C. 1966. Msc: Multiple sex combs. *Drosophila Inform. Serv*. 41:57

300. Tokunaga, C. 1968. Nonautonomy in differentiation of pattern-determining genes in *Drosophila*. II. Transplantation of eyeless-dominant leg discs. *Develop. Biol*. 18:401–13

301. Tokunaga, C. 1970. The effect on somatic crossing over of an ey^D inserted into chromsome 3. *Drosophila Inform. Serv*. 45:58

302. Tokunaga, C. 1972. Autonomy or nonautonomy of gene effects in mosaics. *Proc. Nat. Acad. Sci. USA*. 69:3283–86

303. Tokunaga, C., Stern, C. 1969. Determination of bristle direction in *Drosophila*. *Develop. Biol*. 20:411–25

304. Ulrich, E. 1971. Cell lineage, Determination und Regulation in der weiblichen Genitalimaginalscheibe von *Drosophila melanogaster*. *Wilhelm Roux' Arch. Entwicklungsmech. Organismen* 167:64–82

305. Ursprung, H. 1962. Einfluss des Wirtsalters auf die Entwicklungsleistung von Sagittalhäften mänlicher Genitalscheiben von *Drosophila melanogaster*. *Develop. Biol*. 4:22–39

306. Ursprung, H. 1967. *In vivo* culture of *Drosophila* imaginal discs. *Methods in Developmental Biology*, ed. F. Wilt, N. Wessells, 485–92. New York: Crowell. 813 pp.

307. Ursprung, H. 1971. Biochemische Genetik von Enzymes in der tierischen Entwicklung. *Naturwissenschaften* 58:383–89
308. Ursprung, H., Conscience-Egli, M., Fox, D. J., Wallimann, T. 1972. Origin of leg musculature during *Drosophila* metamorphosis. *Proc. Nat. Acad. Sci. USA. 69:2812–13*
309. Ursprung, H., Hadorn, E. 1962. Weitere Untersuchungen über Musterbildung in Kombinaten aus teilweise dissoziierten Flügel-Imaginalscheiben von *Drosophila melanogaster. Develop. Biol.* 4:40–66
309a. Ursprung, H., Nöthiger, R. 1972. *Results and Problems in Cell Differentiation, V.* Springer-Verlag. New York. 172 pp
310. Veilleux, B., Romans, P. 1972. Delayed expression of female sterility in two mutants of *Drosophila melanogaster. Can. J. Genet.* 14:740
311. Villee, C. 1942. The phenomenon of homoeosis. *Am. Nat.* 76:494–506
312. Villee, C. 1942. A study of hereditary homoeosis: The mutant tetraltera in *Drosophila melanogaster. Univ. Cal. Publ. Zool.* 49:125–84
313. Villee, C. 1943. Phenogenetic studies of the homoeotic mutants of *Drosophila melanogaster.* I. The effects of temperature on the expression of *aristapedia. J. Exp. Zool.* 93:75–98
314. Villee, C. 1944. Phenogenetic studies of the homoeotic mutants of *Drosophila pseudoobscura.* I. Aristapedia-Dobzhansky and Aristapedia-Spassky. *J. Elisha Mitchell Sci. Soc.* 60:141–57
315. Villee, C. 1944. Phenogenetic studies of the homoeotic mutants of *Drosophila melanogaster.* II. The effect of temperature on the expression of proboscipedia. *J. Exp. Zool.* 96:85–102
316. Villee, C. 1945. Developmental interactions of homoeotic and growth rate genes in *Drosophila melanogaster. J. Morphol.* 77:105–18
317. Villee, C. 1946. Some effects of X-rays on development of *Drosophila. J. Exp. Zool.* 107:261–80
318. Villee, C. 1946. Phenogenetic studies of the homoeotic mutants of *Drosophila melanogaster.* IV. Homoeotic and "growth rate" genes. *Genetics* 31:428–37
319. Villee, C. 1947. A quantitative study of phenocopy production with monochromatic ultraviolet irradiation. *Biol. Bull.* 92:1–9
320. Vogt, M. 1944. Beitrag zur Determination der Imaginalscheiben bei *Drosophila. Naturwissenschaften* 32:39–40
321. Vogt, M. 1946. Zur labilen Determination der Imaginalscheiben von *Drosophila.* I. Verhalten verschiedenaltriger Imaginalanlagen bei operativer Defektsetzung. *Biol. Zentralbl.* 65:223–38
322. Vogt, M. 1946. Zur labilen Determination der Imaginalscheiben von *Drosophila.* IV. Die Umwandlung präsumptiven Rüsselgewebes in Bein- oder Fühlergewebe. *Z. Naturforsch.* 1:469–75
323. Vogt, M. 1946. Zur labilen Determination der Imaginalscheiben von *Drosophila.* II. Die Umwandlung präsumptiven Fühlergewebes in Beingewebe. *Biol. Zentralbl.* 65:238–54
324. Vogt, M. 1947. Zur labilen Determination der Imaginalscheiben von *Drosophila.* III Analyse der Manifestierungsbedingungen sowie der Wirkungsweise der zu Antennen- und Palpusverdoppelungen fuhrenden Genmutation *Deformed-recessive-Luers* (Dfd^{r-L}). *Biol. Zentralbl.* 66:81–105
325. Vogt, M. 1947. Zur labilen Determination der Imaginalscheiben von *Drosophila.* V. Beitrag zur Manifestierung der Mutante *antennaless. Biol. Zentralbl.* 66:388–95
326. Vogt, M. 1947. Beeinflussung der Antennendifferenzierung durch Colchicin bei der *Drosophila* mutante *Aristopedia. Experientia* 3:156–57
327. Vyse, E. R., Nash, D. 1969. Nutritional conditional mutants of *Drosophila melanogaster. Genet. Res.* 13:281–87
328. Vyse, E. R., Sang, J. H. 1971. A purine and pyrimidine requiring mutant of *Drosophila melanogaster. Genet. Res.* 18:117–21
329. Waddington, C. H. 1939. Preliminary notes on the development of the wings in normal and mutant strains of *Drosophila. Proc. Nat. Acad. Sci. USA.* 25:299–307
330. Waddington, C. H. 1939. Genes as evocators in development. *Growth* 1:37–44
331. Waddington, C. H. 1942. Some developmental effects of X-rays in *Drosophila. J. Exp. Biol.* 19:101–17
332. Waddington, C. H. 1943. The development of some "leg genes" in *Drosophila. J. Genet.* 45:29–43
333. Waddington, C. H. 1953. The interactions of some morphogenetic genes in

Drosophila melanogaster. J. Genet. 51:243–58
334. Waddington, C. H. 1956. *Principles of Embryology*. London: George Allen & Unwin Ltd. 510 pp.
335. Waddington, C. H. 1961. Genetic assimulation. *Advan. Genet.* 10:257–93
336. Waddington, C. H., Robertson, E. 1969. Determination, activation and actinomycin D insensitvity in the optic imaginal disk of *Drosophila. Nature* 221:933–35
337. Wald, H. 1936. Cytological studies on abnormal development of eggs of the *claret* mutant type of *Drosophila simulans. Genetics* 21:264–81
338. Walen, K. H. 1964. Somatic crossing over in relationship to heterochromatin in *Drosophila melanogaster. Genetics* 49:905–23
339. Weaver, E. C. 1960. Somatic crossing over and its genetic control in *Drosophila. Genetics* 45-345–57
340. Wigglesworth, V. B. 1937. Wound healing in an insect, *Rhodnius prolixus* (Hemiptera). *J. Exp. Biol.* 14:364–81
341. Wigglesworth, V. B. 1954. *The Physiology of Insect Metamorphosis*. Cambridge, England: Cambridge Univ. Press. 151 pp.
342. Wigglesworth, V. B. 1972. *Principles of Insect Physiology*, 7th Ed. London: Chapman & Hall. 827 pp.
343. Wildermuth, H. 1968. Autoradiographische Untersuchungen zum Vermehrungsmuster der Zellen in proliferierenden Rüsselprimordien von *Drosophila melanogaster. Develop. Biol.* 18:1–13
344. Wildermuth, H. 1968. Differenzierungsleistungen, Mustergliederung und Transdeterminationmechanismen in hetero- und homoplastischen Transplantaten der Rüsselprimordien von *Drosophila. Wilhelm Roux' Arch. Entwicklungsmech. Organismen* 160: 41–75
345. Wildermuth, H. 1970. Determination and transdetermination in cells of the fruitfly. *Sci. Progr. London* 58:329–58
346. Wildermuth, H., Hadorn, E. 1965. Differenzierungsleistungen der Labial-Imaginalscheibe von *Drosophila melanogaster. Rev. Suisse Zool.* 72:686–94
347. Wolpert, L. 1969. Positional information and the spatial pattern of cellular differentiation. *J. Theor. Biol.* 25:1–47
348. Wolpert, L. 1972. Positional information and pattern formation. *Curr. Topics Develop. Biol.* 6:183–224
349. Woolf, C. 1949. The effect of temperature treatments on an early developmental period on tumorous head in *Drosophila melanogaster. Proc. Utah. Acad. Sci. Arts Lett.* 26:139–40
350. Wright, T. R. F. Personal communication
351. Wright, T. R. F. 1970. The genetics of embryogenesis in *Drosophila. Advan. Genet.* 15:261–395
352. Yajima, H. 1960. Studies on embryonic determination of the harlequin fly, *Chironomus dorsalis.* I. Effects of centrifugation and of its combination with constriction and puncturing. *J. Embryol. Exp. Morphol.* 8 (Part 2):198–215
353. Yajima, H. 1964. Studies on embryonic determination of the harlequin fly, *Chironomus dorsalis.* II. Effects of partial irradiation of the egg by ultraviolet light. *J. Embryol. Exp. Morphol.* 12 (Part 1):89–100
354. Yajima, H. 1970. Study of the development of the internal organs of the double malformations of *Chironomus dorsalis* by fixed and sectioned materials. *J. Embryol. Exp. Morphol.* 24:287–303
355. Yanders, A. F. 1957. New mutant. *Drosophila Inform. Serv.* 31:85
356. Yoon, S-B., Fox, A. S. 1965. Permeability of premature eggs from *Drosophila* collected with the 'ovitron.' *Nature* 206:910–13
357. Zalokar, M. 1971. Transplantation of nuclei in *Drosophila melanogaster. Proc. Nat. Acad. Sci. USA.* 68: 1539–41
358. Zepp, H. D., Conover, J. H., Hirschhorn, K., Hodes, H. L. 1971. Human-mosquito somatic cell hybrids induced by ultraviolet-inactivated Sendai virus. *Nature New Biol.* 229:119–21

HUMAN GENETICS[1]

3061

Victor A. McKusick and Gary A. Chase

Division of Medical Genetics, Department of Medicine, Johns Hopkins University School of Medicine, Baltimore, Maryland

This review is concerned mainly with the period January 1, 1970 to January 1, 1973. Because of limitation of space it is more restricted in its coverage than the review of the previous three years, 1967–1970 (158).

The fourth International Congress of Human Genetics convened in Paris in September 1971. The published proceedings (3) constitute a useful record of the state of the field as of that date. The reader is referred to it for review of some areas not covered here, e.g. radiation genetics.

Clinical Genetics, a journal published in Copenhagen and edited by Berg of Oslo, Böök of Uppsala, and Mohr of Copenhagen began publication in 1970. The *Journal of Human Evolution* was first issued in January 1972. A new journal, *Behavior Genetics*, began publication in 1971, and The Behavior Genetics Association was established in the same year.

CYTOGENETICS

New Techniques

After being in the doldrums in the latter 1960s, human cytogenetics got, in the last three years, the new techniques it needed for rapid advances comparable to those of 1955–1965. These methods are of four general types: (*a*) fluorescence staining with quinacrine mustard and related agents (38, 48); (*b*) Giemsa-staining after special treatment of the chromosomes such as pH 9.0, trypsin or urea (26, 68, 79, 89, 94, 145, 212, 222, 229, 238); (*c*) the "reverse" banding method of Dutrillaux & Lejeune (81); and (*d*) special techniques for centromeric or constitutive heterochromatin (19, 29, 51, 112, 228). The four methods are called Q, G, R, and C banding, respectively.

The quinacrine fluorescence method was first described by Caspersson and his colleagues in Stockholm in 1968. Pearson and colleagues at Oxford (190) took up the observation of the Stockholm group (256) that the long arm of the Y has a

[1] Preparation of this review was supported in part by NIH training grants GM00795 and DE00193, and by an NIH Genetics Center grant GM19489.

permanent fluorescent segment and applied it to the study of interphase nuclei, sperm, and meiotic chromosomes.

The centromeric heterochromatin methods and all the Giemsa methods had their beginnings with Pardue & Gall (185) who in the process of developing techniques for *in situ* hybridizing RNA to mouse chromosomes, noted that an area adjacent to the centromere of each chromosome stains more deeply with Giemsa than the rest of the chromosome. The conditions of the hybridizing experiments were designed to produce denaturation of the chromosomal DNA, i.e., to separate the double-stranded DNA intc its constituent strands, followed by renaturation, i.e., restoration of the DNA to its double-stranded state. The areas that stain darkly are those that renatured most rapidly. Theory (40) suggests that the DNA in these regions is more highly repetitive than that elsewhere. Arrighi and Hsu (19) modified the procedures somewhat and applied them to man. They showed that in addition to the constitutive heterochromatin demonstrated at the centromere of each chromosome, chromosomes 1, 9, and 16 have long segments of heterochromatin corresponding to the secondary constriction.. Although all these areas are, it seems, repetitive DNA, that of the secondary constriction of chromosome 9 apparently has a base composition distinct from that in other areas, witness in Table 1 the differences in staining properties according to the technique (these being the main discrepancies between the banding patterns obtained by Q staining and those obtained by G and R staining) [See also (125) and (208)]. Further evidence is provided by the special staining technique of Bobrow et al (30) which stains the secondary constriction region of No. 9 and leaves most centromeric regions unstained. (The method demonstrates this area of No. 9 even in interphase nuclei including those of spermatozoa.)

The various Giemsa methods were derived from those introduced by Pardue & Gall (185) and by Arrighi & Hsu (19) mentioned above. Many workers had noted that when human chromosomes are treated and stained for centromeres, they exhibit a banding pattern similar to that produced by quinacrine fluorescence. Some of the techniques are variations of the denaturing and annealing methods (79). Others (187, 229) depend on exposure of the chromosomes to salt solutions at various concentrations, pH values, temperatures and times, which in the hands of the individual investigators have been found by trial and error to be effective. Yet other approaches have been to treat the chromosomes with urea (25) or with a protease such as pronase (80) or trypsin (213, 238), before staining with Giemsa.

Table 1 Staining Characteristics of Heterochromatin [From Paris Conference (186)]

	Q-banding	G-banding	R-banding	C-banding
1qh	-	+	-	+
9qh	-	-	-	+
16qh	-	+	-	+
distal Yq	brilliant	variable	variable	+

Since the chemical or physical basis is largely unknown, the "banding" procedures are essentially cytoalchemy, not cytochemistry. Weisblum & DeHaseth (241) suggested that quinacrine stains specifically for A-T rich regions in DNA. Reviewing the evidence, Pearson (189) suggested that there is intercalation of the tricyclic acridine nucleus of quinacrine into the DNA helix and that the long sidechain of quinacrine is also important to differential staining, probably because of an attachment outside the helix. Giemsa contains a mixture of dyes: methylene blue, eosin, and azure A and B. Methylene blue and the azures have tricyclic structures. Pearson (189) cited observations that suggest a complexing of the dyes such that the mixture performs the two-phase process of the nucleus and sidechain of quinacrine. Thus the partial correspondence between banding patterns achieved by the two methods may be due to similar molecular mechanisms.

Standardization of Techniques and Nomenclature

In Paris in September, 1971, just before the Fourth International Congress of Human Genetics, a conference was held for standardization of cytological techniques (particularly the new ones discussed above) and of nomenclature. Earlier conferences had been held in Denver (1960), London (1963), and Chicago (1966). The conclusions of the Paris Conference have been published by the National Foundation - March of Dimes (186). Changes in nomenclature from those of the Chicago Conference include the following:

(*a*) Where they mean addition or absence of whole chromosomes, the + or − signs should be placed *before* the appropriate symbol. They should be put *after* a symbol to indicate increase or decrease in length. For examples:

47, XX, +21 Karyotype of female with Down's syndrome

45, XX, −D, −G, +t (DqGq) Karyotype of female with balanced Robertsonian translocation between a D and a G group chromosome

46, XX, 5p− Karyotype of female with cri-du-chat syndrome

(*b*) Unusually long or short secondary constrictions should be indicated with the symbol *h* (for heterochromatin) between the symbol for the arm and the + or − sign; e.g., the Donahue karyotype (78, 218) is symbolized 46, XY, lqh+.

(*c*) Symbols for rearrangements should be placed *before* the designation of chromosome(s) involved and the rearranged chromosome(s) should be placed in parentheses; eg., 46, X, i(Xq) = long-arm isochromosome form of Turner syndrome.

(*d*) The two forms of interphase sex chromatin should be designated *X-chromatin* and *Y-chromatin.*

Figure 1 shows a diagrammatic representation of the human karyotype as studied by the new techniques.

Applications

The new techniques have found application (15a, 165, 189) in the following areas, among many:

(*a*) Each chromosome can be uniquely identified. [Hitherto only 5 chromosomes (nos. 1,2,3,9, and 16) could be positively identified by methods short of autoradiography (6, 14), which is cumbersome.] This property of the new techniques has been put to use in several ways:

(*i*) The Philadelphia chromosome of chronic myeloid leukemia, formerly thought to be the same chromosome (No. 21) as that implicated in mongolism (the Down syndrome), turned out to be the other G group chromosome (45, 182). In nine cases of CML, Rowley (204a) found that deleted 22q material was translocated onto the end of the long arm of one chromosome 9, so that these represented instances of t(22q −; 9q+).

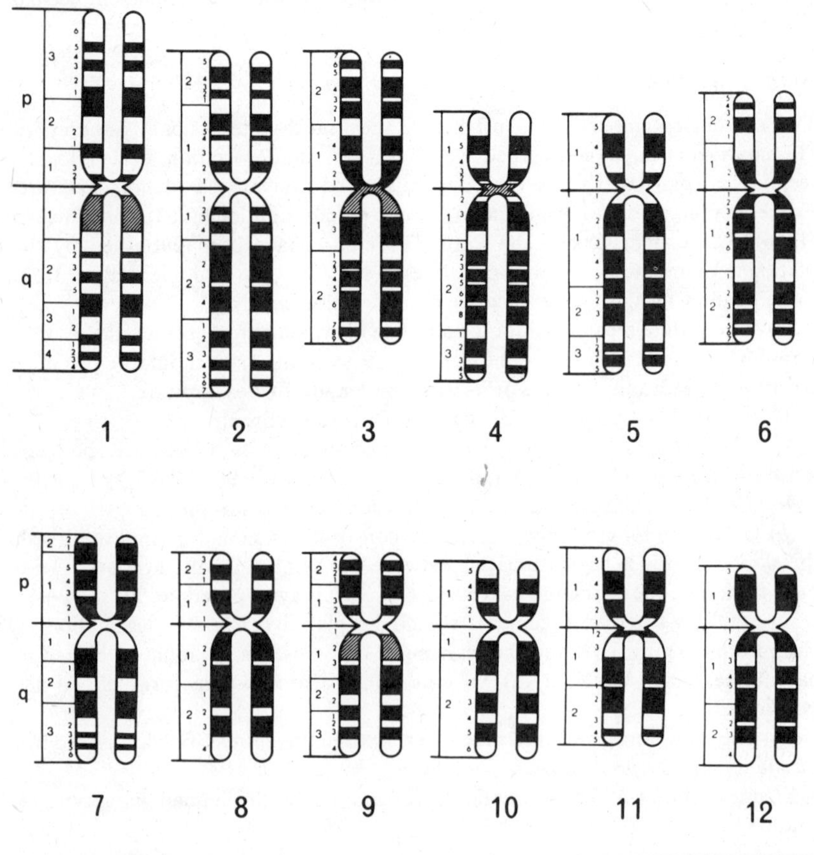

Figure 1 Diagrammatic representation of chromosome bands as observed with the Q-, G-, and R-staining methods; centromere representative of Q-staining method only. (From Paris Conference 1972 (186) with permission of National Foundation-March of Dimes.)

(*ii*) The "mongolism chromosome" turns out to be the smaller of the G group chromosomes (45, 182), i.e., No. 22 *stricto sensu*, not No. 21 as long thought. However, so firmly established is the notion that the Down syndrome is trisomy 21, that the Paris Conference decided that the smaller pair of G group chromosomes should be arbitrarily numbered 21. Thus, the Philadelphia chromosome is No. 22.

(*iii*) The origin of chromosome rearrangements—translocations (18, 39, 46, 59, 95, 209), inversions (90, 173), insertions (107)—can be identified. Niebuhr

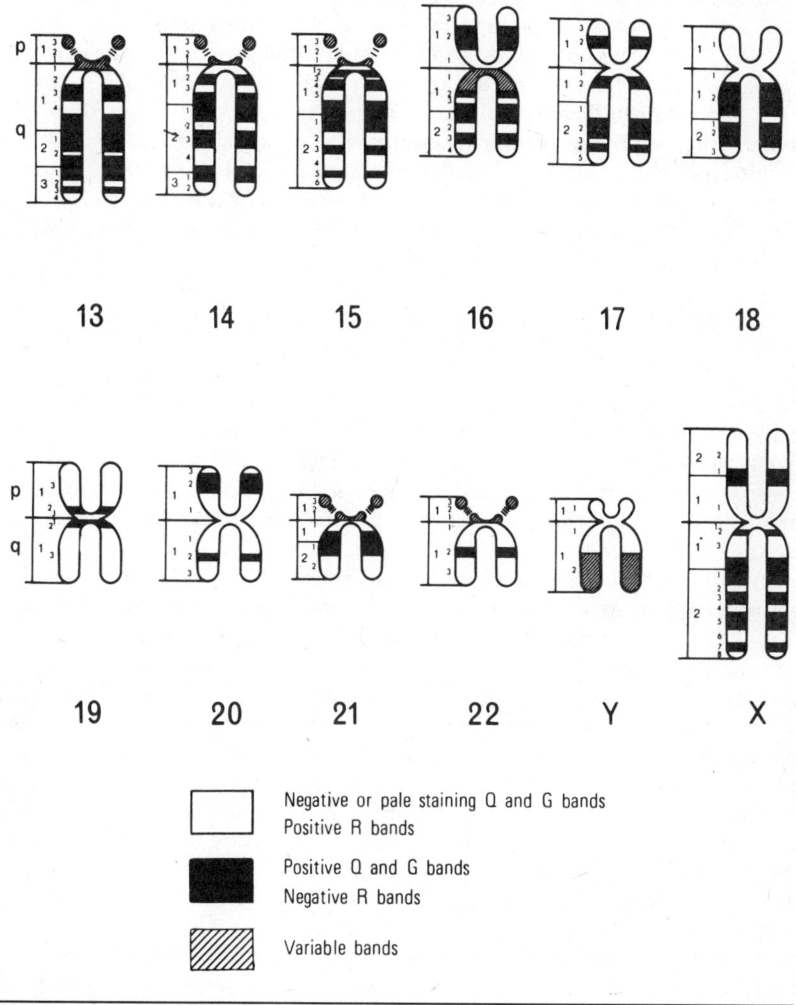

Figure 1 (continued)

(178) used heterochromatin staining to determine whether Robertsonian translocations are dicentric or monocentric. Presumably the frequency of the Down syndrome among the offspring of a 14/21 translocation carrier differs according to whether one or the other or both centromeres are present. Out of 5 cases studied by Niebuhr (178) four were dicentric. The types of translocation were 13/13, 13/14, 15/21, and 21/22. Using quinacrine fluorescence, Caspersson et al (46) identified two 14/21 translocations, one 13/21 translocation, two 13/14 translocations, and one 14/14 translocation among six D/G or D/D cases studied. [See Hecht et al (108a) and Cohen (59) for information on the nature of D/G and D/D Robertsonian translocations as revealed by autoradiography.]

(*iv*) "New" chromosomal syndromes were identified once autosome identification became certain. One such was trisomy 8 (47). Two G-deletion syndromes have been distinguished by quinacrine fluorescence (66, 239). G-deletion syndrome I has a phenotype which Lejeune referred to as "antimonogolism", a "contretype" and results from deletion of chromosome 21. G-deletion syndrome II has a different phenotype and results from deletion of chromosome 22.

(*b*) Normal variations in the chromosome are revealed and can be shown to be inherited; eg.:

(*i*) Fluorescent polymorphism of the centromere of chromosome No. 3 (48).

(*ii*) Polymorphism for fluorescence of giant satellites on D group chromosomes (48).

(*iii*) Unusually short or unusually long segments of constitutive heterochromatin in chromosomes No. 1, 9, or 16 (51, 65). (The "uncoiled" segment of chromosome No. 1 used by Donahue et al (78) as a marker chromosome in the family study that led to assignment of the Duffy blood group locus to that chromosome was studied by Craig-Holmes & Shaw (65) and found to be an unusually long heterochromatic segment, denoted lqh+ in the Paris terminology (186).)

(*c*) Use of polymorphisms to identify the origin of nondisjunction and other chromosomal errors was illustrated by the work of Robinson (202). She found that in all five informative families studied by the quinacrine fluorescence technique the error leading to the Down syndrome must have occurred in the first meiotic division in the mother. Polymorphism in the size and fluorescent intensity in the short arms and satellites of chromosome 21 and in the size of the stalks were used. Four of the five cases had standard trisomy and one had mutant 14/21 translocation. In their one informative case Licnerski & Lindsten (149) concluded that the error was also in the maternal first meiotic division. On the other hand de Grouchy (72) and Juberg & Jones (126) presented families in which transmission of two 2lph− chromosomes demonstrated that nondisjunction occurred in the second meiotic division in the mother.

(*d*) Fluorescence of the long arm of the Y chromosome is evident in interphase nuclei. The so-called F body, or Y chromatin, permits identification of males in the same manner that the Barr body, or X chromatin, permits identification of the normal female. Klinger & Moser (133) described a method for unambiguous identification of both the X and the Y in the same nucleus. In females an F body can be simulated by the fluorescent giant satellite of a D group chromosome (34). In males an F body may be absent because the long arm of the Y chromosome is abbreviated (36, 115). The F body may be absent in about 0.1% of males (189). Sex determining factors appear to be concentrated in the short arm of the Y chromosome, since Yq- persons are completely normal sexually and otherwise (161). The Y chromosome is variable in length, especially in the long arm and particularly in its fluorescent part. In Amish males of the name Beiler who are widely separated genealogically, a short nonfluorescent Y chromosome is a consistent finding and therefore presumably a stable polymorphism (37). Genest & Lejeune (97) found similar stability of a small Y chromosome in a French Canadian kindred and found it also in a French family of the same surname who shared a common ancestor who lived near La Rochelle in France. In a Scottish kindred Jacobs (122, 124) found a pericentric inversion that could be traced back to the 17th century.

Y-bearing spermatozoa show Y chromatin (the F body). Some spermatozoa of XYY males have two F bodies (118). Furthermore, about 2% of the spermatozoa of normal males have two F bodies—an indication of an unexpectedly high frequency of nondisjunction and of prezygotic selection against YY-sperm.

From the application of fluorescence staining to male meiotic chromosomes, Pearson & Bobrow (190) provided definitive proof that it is the short arm of the Y that associates terminally with the short arm of the X chromosome during meiosis.

Fluorescence staining has been used to test the Ferguson-Smith hypothesis (91) that, in cases of XX true hermaphroditism, occult Y material is located on one X as well as the suggestion that Y material has been translocated to an autosome in cases of the XX Klinefelter syndrome. No one has succeeded in demonstrating Y fluorescence on either an X or an autosome in these cases. This result does not disprove the theories, however, since the nonfluorescent parts of the Y are important to sex determination. Bühler et al (44) described changes they interpreted as representing a Y/autosome translocation.

(*e*) The phylogeny of the human karyotype has been studied by the new methods.

(*i*) Comparing the banding pattern of the chromosomes of man and the chimpanzee, de Grouchy et al (73, 233, 234) concluded that chromosomes 3, 6, 7, 8, 10, 11, 13, 14, 16, 19, 20, 21 and 22 are substantially homologous. The others are homologous if a few apparently simple rearrangements are assumed. For example, the chimpanzee has 48 chromosomes and nothing equivalent to No. 2 of man. de Grouchy et al suggest that the human No. 2 represents Robertsonian translocation between two acrocentric chromosomes. Nos. 12 and 17 have a pericentric inversion in the chimpanzee. Secondary constrictions are missing from Nos. 1 and 9 in the chimpanzee. Reciprocal translocation between Nos. 15 and 18 of the chimpanzee restores homology of these chromosomes to those of man.

Studies of DNA (135), of the amino acid sequence of proteins (105), and of immunologic characteristics (252) indicate that man is more closely related to the

African apes than to any other primates. Similarities to the chimpanzee are particularly striking. For example, both the alpha and the beta chains of hemoglobin are identical in chimpanzee and man. Washburn & McCown (240) noted the behavioral similarities of chimpanzee and man. The observation of clinically typical mongolism in a chimpanzee with G trisomy (157) is further evidence of close relationship.

(*ii*) Comings (60) examined the notion that early in phylogeny the human chromosome set underwent tetraploidization. He pointed out that chromosomes 11 and 12 have many morphologic similarities and that LDH A is on chromosome 11 and LDH B is on chromosome 12 (205).

Lyonization

Lyonization of anomalous X chromosomes is the rule. Presumably the reason (as suggested by Sparkes and others) is cell selection; cells in which at lyonization the deficient X chromosome is left as the active one do not survive because of partial X-nullisomy. The rule that the anomalous X is always lyonized does not hold for some X-autosome translocations, probably for a similar reason of cell selection. For instance, in the t (Xp−; 14q+) of Buckton et al (43), the normal X was late-replicating in the balanced females. One female with an unbalanced form of the translocation had the abnormal X chromosome late replicating in all cells. Likewise in the case of the t (Xq−; 14q+) which is referred to as the KOP translocation for the initial letter of the family names of the proband and the two physicians who found it (see p.459), a balanced carrier female with 46 chromosomes had both translocation chromosomes active and the normal X inactive in all cells. Her son, with 47 chromosomes, showed two 14q+ chromosomes (presumably originating by nondisjunction in the second stage of meiosis) and the Klinefelter syndrome. One of the 14q+ chromosomes was late replicating and formed a Barr body. A normal X chromosome was apparently genetically active. Although this person was virtually trisomic for chromosome 14 he showed little or no effects thereof, perhaps because of spreading of inactivation from the X chromosome into the autosomal segment of one of the two 14q+ chromosomes.

Brown & Chandra (42) proposed a mechanism for the random inactivation of the X chromosome in eutherian mammals. The theory is based on the notion that the eutherian situation evolved from an earlier system which has been retained in marsupials such as the kangaroo. In marsupials the paternal X chromosome is consistently inactivated. In accordance with the diagram shown in Figure 2, Brown & Chandra suggest that a locus (SS=sensitive site) on the X chromosome in marsupials synthesizes a single molecule of an "informational entity" (0) which ensures activity of that X chromosome through its effect on a receptor site. In the paternal X chromosome passage through the male produces "imprinting," the effect of which is to render the sensitive site inactive.

In eutherian mammals the sensitive site has been translocated to an autosome but retains the same susceptibility to imprinting during male meiosis. Since the maternally derived sensitive site is no longer on the same chromosome as the

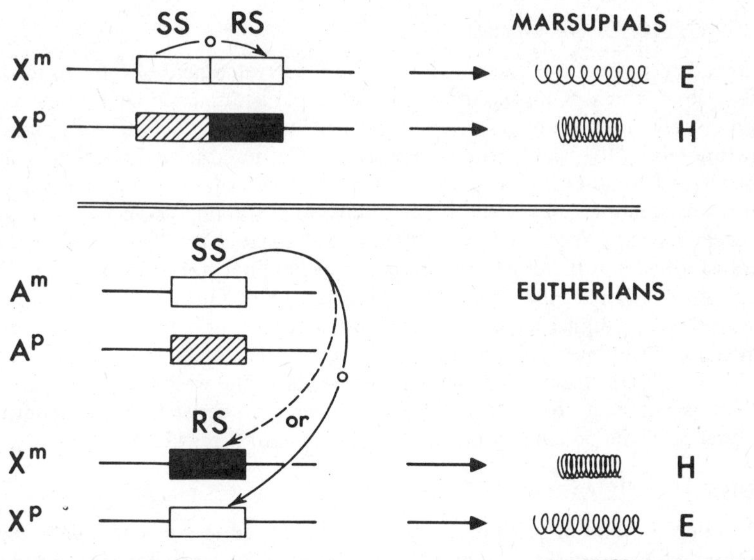

Figure 2 The marsupial (above) and eutherian (below) systems for inactivation of X chromosome, according to theory of Brown & Chandra (42). See text. (Reproduced with permission of Dr. S.W. Brown and of *Proc. Nat. Acad. Sci.*)

receptor site, the single molecule of the informational entity will inactivate either X chromosome at random.

The theory seems consistent with all the facts about lyonization. Observations in triploids, both natural and experimentally produced, are of particular interest. The number of X chromosomes that remain active is the same as the number of maternal haploid sets. From other evidence, most human triploids seem to arise through digyny; as the Brown-Chandra theory would require, most XXY triploids are chromatin-negative and most XXX triploids have a single Barr body. Furthermore, in digynous triploids induced in rabbits by suppression of the second polar body, Bomsel-Helmreich (32) found that two X chromosomes were always active and the G6PD activity, which is thought to be X-linked in the rabbit as in several other mammals, was consistently higher in triploid embryos than in diploid controls.

Other Considerations

Hungerford (119) and Hungerford et al (120) reported studies of the meiotic chromosomes in the male and presented a provisional pachytene map.

Chromosome structure and organization in relation to function, activation, evolution, etc., is a matter of great current interest. Ris & Kubai (200), Thomas (232) and Comings (8e) gave reviews, and Crick (67) and Paul (188) presented general models and theories.

SOMATIC CELL GENETICS

Somatic cell genetics has come of age with the development of cell hybridization, the use of which in chromosome mapping (206) is discussed later. Other genetic processes under study in cultured cells include transduction (62), mutation (216), aging (104), and genetic regulation. The notable usefulness of cultured fibroblasts for the in vitro study of genetic diseases (see p. 445) illustrates one form of somatic cell genetics. The ability to maintain lymphocytes in culture indefinitely has been a boon to somatic cell genetics (1). The lymphocytes are transformed and therefore are referred to as lymphocytoid. They remain diploid and do not display the senescence characteristic of fibroblasts. Their growth in suspension facilitates handling. They continue some differentiated functions, e.g., synthesis of immunoglobulin (28) and can be used for biochemical studies (52) and in cell hybridization. A possible drawback to the wide use of lymphocytoid lines is the fact that the desired characteristics depend on the continuing presence of the potentially oncogenic Epstein-Barr virus (257).

BIOCHEMICAL GENETICS

The third edition of Stanbury, Wyngaarden & Fredrickson's *Metabolic Basis of Inherited Diseases* (12) is an up-to-date version of this encyclopedic treatment of inborn errors on metabolism. Also see reviews of Kirkman (14d) and Raivio & Seegmiller (195).

Table 2 Progress in Identification of Mendelian Traits and in the Nosology of Mendelian Disease

	Verschuer	McKusick's *Mendelian Inheritance in Man*			
	1958	(1966 ed.)	(1968 ed.)	(1971 ed.)	(Jan. 1973)
Autosomal Dominant	285	269 (+568)	344 (+449)	415 (+528)	498 (+588)
Autosomal Recessive	89	237 (+294)	280 (+349)	365 (+418)	419 (+448)
X-linked	38	68 (+51)	68 (+55)	86 (+64)	92 (+65)
Totals	412	574 (+913)	692 (+853)	866 (+1010)	1009 (+1101)
		487	1545	1876	2110

Insofar as is known, the numbers refer to separate loci. The numbers in brackets refer to loci "in limbo"; i.e., monogenic inheritance of a particular trait has, with some reason, been suggested but proof is not complete.

The increase in number of known recessives in man at a faster pace than known dominants (Table 2) is attributable to the burgeoning of clinical biochemical genetics; almost without exception inborn errors of metabolism in the Garrodian sense, i.e. disorders with deficiency of an enzyme as the "cause", are recessive, either autosomal or X-linked. One of the few exceptions to this statement is acute intermittent porphyria (AIP) (17600), an autosomal dominant.[2] The enzyme deficient in this condition (226), uroporphyrinogen I synthetase (otherwise known as porphobilinogen (PBG) deaminase), is apparently critically rate-limiting so that partial deficiency suffices to result in clinical

[2] Here as elsewhere the five digit number refers to the entry in McKusick's *Mendelian Inheritance in Man* (3rd ed., 1971). In those instances where the last digit is another than 0, the number has, as a rule, been assigned since the 1971 edition.

disease when the pathway is put under stress. It is stressing of the pathway that may account for the intermittency and for the precipitating effects of barbiturates and other factors. Another example of an autosomal dominant enzymopathy may be hereditary angioedema (10610) in which the deficiency lies in the inhibitor of the first component of complement (Cl esterase inhibitor). Why the phenotype (which like AIP is acute and intermittent) should be expressed in the heterozygote is not clear. Kirkman (5) discusses the biochemical basis of dominance.

Progress in the identification of enzymatic functions deficient in specific inborn errors of metabolism is indicated by the fact that the table published in the 1970 review (158) contained 92 entries, whereas an updated table contains 127. In over one-fourth of all proved recessives (autosomal and X-linked) the enzyme defect has now been characterized. For many of the recessive entities represented by the numbers in Table 2 the disease was first established (or first established as a recessive) when the enzyme deficiency was found. (In some instances the proband was an isolated case but the presence of parental consanguinity and of an intermediate level of relevant enzyme activity in both parents served to establish the mode of inheritance—an example of the surrogate approaches to formal genetics to which the human geneticist must resort.) Although the subtitle of *Mendelian Inheritance in Man* (9) emphasizes that these are catalogs of *phenotypes*, this caveat being necessary in order to avoid confusing the terms dominant and recessive, the increase in the number of entries in the last few years has been mainly in the differentiation of entities *within* a known phenotype, rather than discovery of new phenotypes.

It is remarkable how rapidly after initial clinical description, some of the inborn errors of metabolism have been characterized enzymatically. A recent example is lactosyl ceramidosis in which the entire process took Dawson & Stein (71) less than a year. This facility results from the systematic approach that is successful in many of these disorders. In the case of lactosyl ceramidosis much was already known about the pathway involved, including diseases with defects at other sites (Fig. 3). The defect in ceramidase in Farber lipogranulomatosis has been discovered recently (227).

A great asset to the study of inborn errors of metabolism has been the culture of fibroblasts. It was perhaps not anticipated that such a large proportion of the enzymes deficient in inborn errors of metabolism would be present in normal fibroblasts cultured from skin biopsies—even though no significant level of enzyme activity was found on direct assay of skin. Exceptions include phenylalanine hydroxylase, histidase and hepatic glucose-6-phosphatase—so that phenylketonuria, histidinemia, and type I glycogen storage disease cannot be diagnosed or otherwise investigated in fibroblast culture. (Raivio & Seegmiller (195) gave a table of identified enzymopathies and indicated, where known, whether the defect is expressed in fibroblasts and whether heterozygotes are detectable.) The development of the prenatal diagnosis of inborn errors of metabolism through study of the fibroblasts grown from amniotic fluid depends on the wide enzymatic repertoire of cultured fibroblasts. Some of the enzymes

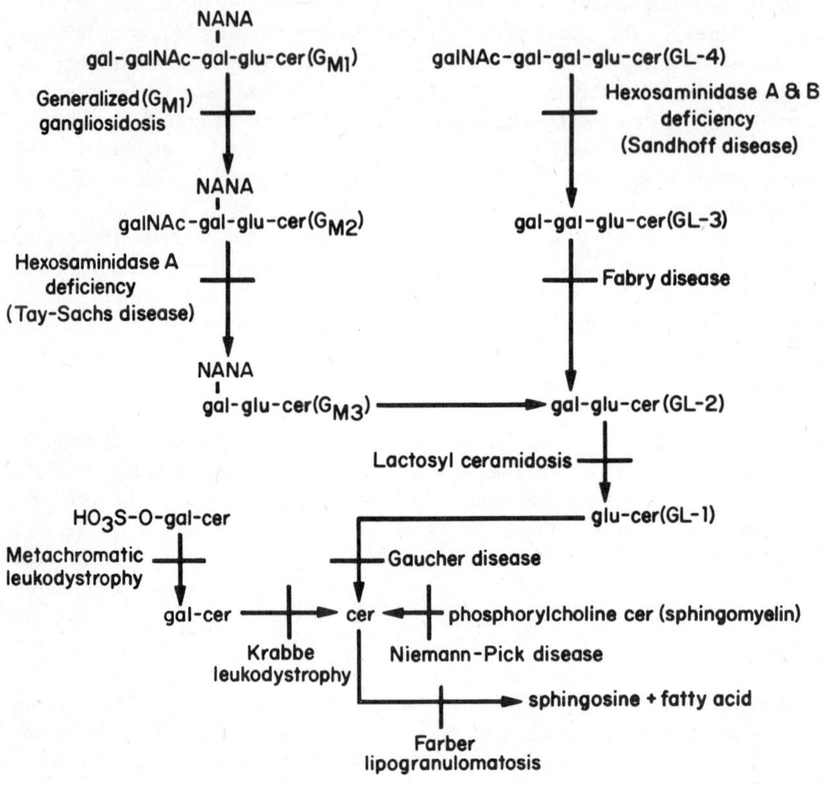

Figure 3 Site of enzymatic defect in 10 lipidoses, all lysosomal diseases. Abbreviations: cer = ceramide; glu = glucose; gal = galactose; galNAc = N-acetyl-galactosamine; G_{Ml} = monosialic ganglioside; GL-2 = glycolipid with two member oligosacchaccharide chain; NANA = N-acetylneuraminic acid. (After Dawson & Stein (71) with additions.)

which are normally not expressed in fibroblasts may be made evident in PHA-stimulated cultured lymphocytes (101); see p.459. Although this approach does not help prenatal diagnosis, it suggests that manipulation of the cultured fibroblasts may elicit expression of enzyme in normal cells. Any such method is unlikely to be effective, however, in disorders such as albinism and McArdle disease[3], in which the enzyme defective is present exclusively in specialized cells, melanocyte and muscle, respectively.

[3] It is our idiosyncrasy to prefer use of the eponym in the nonpossessive form. The eponym is merely a "handle", a convenient label. Frequently, perhaps more often than not, the person whose name is used did not first describe the condition or did not describe it in full detail, and sometimes did not describe it at all.

In most inborn errors of metabolism residual enzyme activity has been found (195), a fortunate circumstance since it opens the possibility that treatment by methods that increase the functional level of the enzyme by increasing its synthesis, augmenting its efficiency or retarding its degradation may be possible (see p.462). (When residual activity is found, another enzyme may be serving the same or a similar function, comparable to the Hb A – Hb A_2 model. Hexosaminidase A and B are examples; only A is absent in Tay-Sachs disease. Because of an increase in hexosaminidase B in that disorder, assays of total hexosaminidase activity failed to show deficiency.)

At least two mutant enzymes with increased activity are known in man: G6PD Hektoen and pseudocholinesterase Cynthiana (255). The heightened activity of the first is due to the presence of an increased amount of enzyme protein; the specific activity of the protein is normal and the rate of decay is normal. This might suggest a regulatory mutation. However, Yoshida (253) found an amino acid substitution, tyrosine for histidine. This is, then, an illustration of the structure-rate hypothesis, or principle, that the structure of a protein may determine the rate at which it is synthesized. Thus, as in acute intermittent porphyria, putative regulatory mutation is disproved when the truth is known. (As far as revealing genetic principles is concerned, G6PD is for enzymopathies what the hemoglobin variants have been for proteins in general.) The nature of the increased activity in pseudocholinesterase Cynthiana (255) is unknown.

Shortened lifespan because of an increased rate of decay, rather than defective catalytic function per se, is the basis of enzyme deficiency in some conditions. The red cell enzymopathies that lead to nonspherocytic hemolytic anemia (including some G6PD variants) are notable examples (236). Normally enzymes of the anucleate red cell must survive the 120-day lifespan of the red cell.

Some variant enzymes owe their defective function to reduced affinity for substrate. These are so-called K_m mutants, K_m being the concentration of substrate, which results in half maximal velocity of the enzyme reaction. K_m mutants can be missed if, as is often the practice, the assay system makes use of excessive amounts of substrate. Scriver & Cameron (211) described a patient with clinically typical hypophosphatasia in whom normal plasma alkaline phosphatase was found by the usual tests, which use high substrate concentrations. At low (and physiologic) substrate concentrations the patient's plasma hydrolyzed phosphoethanolamine more slowly than did normal plasma. Presumably this is a form of hypophosphatasia allelic to the typical form.

Low K_m mutants are undoubtedly frequent among inborn errors of metabolism. As Yoshida (254) pointed out, the severity of deficiency of red cell G6PD activity as assayed in vitro correlates poorly with clinical severity.

The lysosomal diseases have been particularly actively studied. The category includes the mucopolysaccharidoses (abbreviated MPS; of which the Hurler syndrome, MPS I, is the prototype), Tay-Sachs disease, and more than a dozen others (including those shown in Fig. 3). With the notable exceptions cited below, each has a mutation-determined deficiency of a single specific lysosomal acid hydrolase such that a degradative function of the lysosome is defective. A storage disease results.

Mucopolysaccharidoses are the result of defective degradation of glycosaminoglycans because of functional deficiency of the lysosomal enzyme, which cleaves the sidechains at one or another point along their length. Study of the mucopolysaccharidoses and the related mucolipidoses have illustrated genetic heterogeneity, the nosology of genetic disease, allelism, nonallelism, and genetic compounds (159), the use of fibroblasts in the elucidation of genetic disease, contribution of rare inborn errors to the understanding of the normal (109, 177), prenatal diagnosis, and even therapy by enzyme replacement. All these aspects have been recently reviewed comprehensively elsewhere (10).

In each of five different mucopolysaccharidoses, Neufeld and her colleagues (176) demonstrated deficiency of a specific protein ("corrective factor") produced by normal fibroblasts and present in normal urine and in medium that normal fibroblasts have grown. This fact was first demonstrated by mixing fibroblasts of different genotype or by "treating" diseased fibroblasts with medium or protein concentrates from urine. The enzymatic nature of the corrective factor absent in four of the five mucopolysaccharidoses is now known, the most recently identified being sulfo-iduronate sulfatase as the enzyme deficient in the Hunter syndrome (154, 176).

Some surprises were turned up by these studies. Two disorders that on clinical grounds had been considered quite distinct were found to have deficiency of the same enzyme, *α-L*-iduronidase (20, 248). These are the Hurler syndrome (MPS I, a severe disease with progressive intellectual and physical deterioration and death usually before age 10 years) and the Scheie syndrome (formerly MPS V, a relatively mild disorder compatible with an effective professional career and not known to reduce life-span). These disorders are probably the result of homozygosity of alleles and thus comparable to SS disease (sickle cell anemia) and CC disease, allelic severe and mild diseases, respectively, among the hemoglobinopathies. If this interpretation is correct, then there should be mucopolysaccharidosis cases that represent the genetic compound analogous to SC disease among the hemoglobinopathies. Presumed Hurler-Scheie cases have indeed been identified (159). The phenotype is intermediate, parental consanguinity is absent, and deficiency of *α-L*-iduronidase is found. All three forms of iduronidase deficiency —Hurler syndrome, Scheie syndrome, and the Hurler-Scheie compound— excrete an excess of both dermatan sulfate and heparan sulfate in the urine. Iduronide is a constituent of both mucopolysaccharides.

Another surprise: Whereas the Hurler and Scheie syndromes, previously thought to be separate entities proved to have a defect in the same enzyme, the opposite situation developed with Sanfilippo syndrome (MPS III). Previously MPS III was considered to be a single entity characterized by severe progressive mental retardation with only mild somatic features and heparan sulfaturia. It turned out that there are two seemingly phenotypically identical forms of the disease, one with deficiency of heparan sulfate sulfatase (137) and one with deficiency of N-acetyl-alpha-*D*-glucuronidase (179).

Spranger & Wiedemann (223) gave the designation mucolipidosis to a group of disorders that appear to combine features of the mucopolysaccharidoses and sphingolipidoses, i.e., both mucopolysaccharides and glycolipids are thought to accumulate, although there is some disagreement on this point (145a). At least three mucolipidoses are recognized (223). The best studied of these is I-cell disease (so called for the striking cytoplasmic inclusions), also known as mucolipidosis II in the Spranger-Wiedemann classification (223).

Wiesmann et al (249) pointed out that concentrations of many lysosomal enzymes (but not acid phosphatase and α-glucosidase) are high in the medium in which I-cell disease fibroblasts have grown and not in the fibroblasts themselves. Furthermore, the levels of these enzymes are elevated in the blood and urine of I-cell patients. They (249) suggested that the defect in I-cell disease is one that results in leaky lysosomes. The lack of multiple lysosomal enzymes results in storage diseases with unusual heterogeneity of material stored.

More recently Hickman & Neufeld (109) have proposed that the loss of enzymes from cells is not due to a defect in the lysosomal membrane that renders it leaky. Rather they suggest that normally many lysosomal enzymes are excreted from the cell where they are synthesized and then re-enter cells where they serve their degradative role in the acid milieu of the lysosome; and that for re-entry the enzymes must have a specific chemical constitution, perhaps particular carbohydrate side-chains essential to recognition by and transport into the cell. They conceive that the defect in mucolipidosis II is deficiency of the enzyme which modifies a cluster of enzymes in a way that re-entry is made possible. They provide experimental support for their hypothesis by showing that the iduronidase demonstrated in the tissue culture medium of ML II fibroblasts by use of the artificial test substrate phenyliduronide does not correct the metabolic defect of Hurler fibroblasts, whereas the iduronidase in the medium of normal fibroblasts is effective.

Whereas diseases caused by enzyme deficiency are usually recessive (see p.445), dominant disorders are more likely to have a change in a structural or other nonenzymic protein (e.g., the difference in the nature of the defect in the dominant and recessive methemoglobinemias). The converse—that change in a structural or other nonenzymic protein will almost certainly be dominant—cannot be said. An abnormal structure of collagen has been found in two forms of the Ehlers-Danlos syndrome, both of which are inherited as recessives and each of which in fact has deficiency of an enzyme concerned in the posttranslational modification of the collagen polypeptide chains. The Ehlers-Danlos (E-D) syndrome is characterized by joint hypermobility and by excessive stretchability, fragility, and bruisability of skin. Seven varieties of disease can be distinguished (10) on the basis of the severity of individual ones of these features, the mode of inheritance, and (recently in the case of two of the seven) the biochemical defect. One form, E-D VI, has deficiency of protocollagen lysyl hydroxylase, the enzyme that catalyses the hydroxylation of lysyl residues that have been incorporated into the protocollagen polypeptide chain (136, 192). As a result the collagen of these patients is deficient in hydroxylysine and is defective in cross-linkage because hydroxylysine is involved in certain of the crosslinks. This might be called the ocular form of E-D because, in addition to joint and skin changes like those in some of the other forms of E-D, the patients have such striking ocular fragility that rupture of the globe and/or retinal detachment may occur with relatively mild trauma. Furthermore, scoliosis is especially severe.

The other enzymatically defined form, E-D VII, has a defect (148) in procollagen peptidase, the enzyme that cleaves the registration peptide off the end of procollagen. The primitive collagen polypeptide chain synthesized on the ribosome is about 20% longer than that of mature collagen (53). The extra piece seems to function in bringing polypeptide chains into register for formation of the triple helix—hence its designation, coordination or registration peptide. It has cysteine residues to allow crosslinking whereas mature collagen lacks –S–S– linkages and achieves crosslinking through other mechanisms. Persistence of some procollagen, easily identified by its sulfur content and larger molecular size, is the chemical finding in patients with E-D VII. Profound joint hypermobility is the leading clinical feature. The patients are dwarfed. The same enzymatic defect was earlier described in cattle with a disorder characterized mainly by extreme fragility of the skin (143). The mutation in cattle may lead to more complete deficiency of the enzyme than is the case in the human disease.

The Ehlers-Danlos syndrome might be called a primary disorder of connective tissue, whereas conditions such as homocystinuria and alkaptonuria are secondary disorders of connective tissue because the primary defect is elsewhere. Danks et al (69, 70) showed that the Menkes syndrome ("kinky hair syndrome"), an X-linked recessive disorder, has as its basic defect impairment of intestinal absorption of copper. Copper deficiency appears to account for the striking abnormalities of collagen and elastin in Menkes syndrome. Lysyl oxidase, which catalyzes oxidative deamination of lysine to form allysine as the first step in crosslinking of both collagen and elastin, is a copper dependent enzyme. (Experimental copper deficiency results in connective tissue manifestations including rupture of the aorta). Other widespread abnormalities in the Menkes syndrome (as in sheep that graze in copper-poor areas) are explained by impairment of other Cu-dependent enzymes.

Matching the rapid advances in identification of deficiencies of enzyme function in inborn errors of metabolism is the flood of inferential genetic information provided by studies of the amino acid sequence of proteins, which is one of the main approaches (cell hybridization, see below, is another) that substitutes for controlled matings in human genetics. The principles on which genetic conclusions rest are (*a*) the one-gene-one-polypeptide law and (*b*) the principle of colinearity of the coding bases in DNA and RNA and the amino acids of proteins determined by them. *The Atlas of Protein Sequence and Structure* (2) is rich in information from which genetic inferences can be drawn. In addition to assembling most of the known data the book contains useful reviews on tracing biochemical evolution, gene duplication in evolution, the evolution of specific systems such as the immunoglobulins, varieties of mutations revealed by the human hemoglobins, etc. Ohno's *Evolution by Gene Duplication* (11) is a thought-provoking and plausible analysis.

The hemoglobins have probably been the richest single source of information on the genetics of man. At least four areas of interest are particular developments in the last three years.

The several hemoglobins Lepore are the consequence of microchromosomal aberrations arising through unequal crossing-over as diagrammed in Fig. 4. The resultant nonalpha chain of hemoglobin has features of the delta chain at the NH_2-end and of the beta chain at the COOH-end. The different hemoglobins Lepore differ at the apparent site of anomalous crossing-over and therefore in the proportions of the δ and β chains represented. Recently two examples of "anti-Lepore"—the complementary situation—have been found. These are Hb P_{Congo} (146) and Hb Miyada (180). As diagrammed in Fig. 4, even homozygotes have some normal delta and beta chains, but have a unique nonalpha chain that has the amino acid sequence of the beta chain at the NH_2-end and of the delta chain at the COOH-end.

Three variant hemoglobins with elongated chains have been found. The α chain of Hb Constant Spring (CS) has 31 additional residues on the COOH-end (56); the β chain of Hb Tak has 10 additional residues on the COOH-end. Hunt & Dayhoff (121) found that the additional carboxy-terminal segment of the Hb CS α chain had 9 residues in common with the 68–98 AA segment of the normal alpha chain, and suggested that this may have been its origin. (They proposed that the additional segment of Hb Tak β chain may have originated from the normal β chain through a similar mechanism, partial gene duplication.) In arriving at the conclusion on Hb CS, they scanned many proteins of known sequence in 31 residue lengths. About 32,500 such segments were present in the data surveyed, of which 111 were in the α Hb chain. Thus, there was one chance in 293 that the extra segment would be found to match closest with a segment in the α chain. Two other segments were found to have 9 identities: (*a*) the 68–98 segment of mouse α Hb chains and (*b*) the 161–191 segment of Myxobacter trypsin-like enzyme. The derivation of the extra piece in Hb CS must be very ancient to allow for its *difference* from the normal α and β chains. These findings

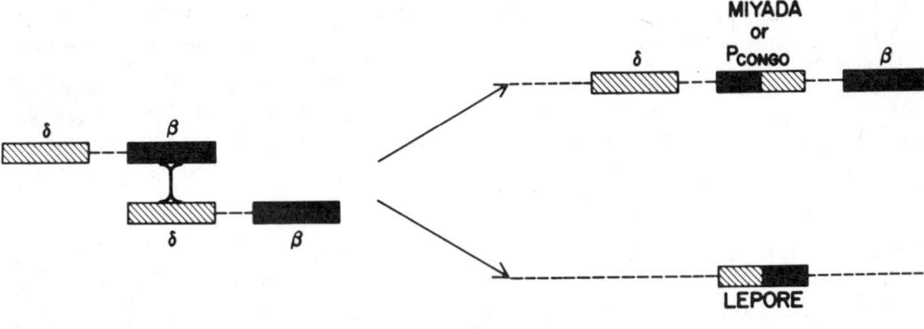

Figure 4 Schematic representation of presumed origin of genes for Lepore and anti-Lepore hemoglobins: nonhomologous pairing and unequal crossing-over.

and conclusions are consistent with the prevailing belief that the normal messenger RNAs for the hemoglobin chains have a length greater than that needed to code for the actual protein. Thus, Hb CS and Hb Tak may be the consequence of terminator mutations and the reading out of occult anachronistic sequences. Hb Wayne (215), a third hemoglobin with an elongated chain, is of particular interest because it appears to represent the first frame-shift mutation identified in man. The last three amino acids in the alpha globin chain (lys-tyr-arg) have been replaced by asn-thr-val-lys-leu-glu-pro-arg, so that the chain is 146 residues long rather than the normal 141. Reference to the code and to the sequence of Hb Constant Spring indicated that deletion of a single nucleotide in the codon for residue 139 of the alpha chain would result in the sequence of Hb Wayne through frame-shift.

Although the beta hemoglobin locus is single in man the evidence is now conclusive [see review of Hollán et al (114)] that the gamma hemoglobin locus is multiple probably in all men (114), and the alpha hemoglobin locus is multiple in some and not in others. The occurrence of only mutant hemoglobin in homozygotes for Hb J Tongariki (17) indicates that in Melanesians the alpha locus is single. On the other hand, in a remarkable Hungarian family in which some persons had two alpha chain mutants, Hb G Buda and Hb J Pest, the presence of normal hemoglobin A indicates that two alpha loci must be present. The proportion of mutant hemoglobin in persons heterozygous for an alpha mutant also supports the presence of two alpha loci.

Hemoglobin Olympia is unique among hemoglobins in that the precise code change is uniquely inferrable. This hemoglobin has substitution of methionine for valine at beta 20 (225). From reference to the code dictionary it can be seen that this means that GUG is present in the wild-type cistron and AUG in the mutant gene. Hemoglobin Olympia is an "electrophoretically silent" mutation. Of the 2,200 possible amino acid substitutions in hemoglobin A, only about one-third would produce change in electrical charge.

IMMUNOGENETICS

Understanding of histocompatibility and specifically the HL-A complex has advanced rapidly in the last three years. Reviews were provided by Kissmeyer-Nielsen et al (132) and Bodmer (31). The immune response locus (or loci), which is part of the HL-A complex, was reviewed by Benacerraf & McDevitt (21) and work bearing on Ir genes in man was published by Levine et al (147) and by Marsh et al (153a).

LINKAGE AND ASSIGNMENT

A review of human chromosome mapping was published in this series (196). Edwards (82) discussed estimating recombination fraction for X-chromosome loci under various conditions of ascertainment and provided lod tables. Substantial progress has been made in identifying human autosomal linkage groups and making new chromosomal assignments. Table 3 lists first the autosomal assignments, then the known autosomal linkages not yet assigned to a particular autosome, third the possible autosomal linkages and assignments, and finally the X-linkage group. The numbers in parentheses after the traits refer to those given in McKusick's *Mendelian Inheritance in Man* (3rd ed., 1971) with additions.

The rapid expansion in our knowledge of the chromosome No. 1 map indicates the mutual potentiation of family study and cell hybridization. The exact ordering must be considered tentative, however.

The assignment of genes to specific autosomes by cell hybridization is discussed fully by Ruddle (206). An interesting finding by this method is that the adenosine deaminase locus may be on chromosome No. 20. Since by family data ADA appears to be linked to HL-A (83) and to P, and HL-A is linked to PGM_3 (141, 142), then this syntenic group may be on chromosome No. 20. Weitkamp (242) suggested that the order may be P - ADA - HL-A - PGM_3. Edwards (83) had suggested the order HL-A - PGM_3 - ADA - P. The HL-A "locus" is itself a cluster of closely linked loci: The "LA" and "Four" (or 1 and 2) loci and probably other loci concerned with immune response (Ir) and reactivity in mixed lymphocyte culture (MLC). Lamm et al (141) suggested that PGM_3 is likely to be on the "Four" side of the HL-A region.

Gene-assignment by quantitative assay of an enzyme or other protein, either elevated level in a trisomy syndrome or reduced level in a deletion syndrome, is unreliable [see review of former by Pantelakis (184)]. Into this category falls the suggested assignment of phosphohexokinase to chromosome No. 21 (184) on the basis of findings in trisomy 21, and assignment of IgA to the long arm of chromosome 18 on the basis of findings in the 18q- syndrome (246). In the same category is the suggestion that a retinoblastoma locus is on the long arm of chromosome 13 (98, 106, 183, 230). Wilson et al (251) found that deletion of part of the long arm of chromosome 13 results in a variable syndrome but distinctive features are retinoblastoma and hypoplastic or absent thumbs, each present in about a fourth of the more than 40 reported cases. In man, deletion mapping has been disappointing, even treacherous. No autosomal assignment has been fully established by this method, although red cell triosephosphate isomerase was thought to be on the short arm of chromosome No. 5 (41, 43, 221) and earlier a cystic fibrosis locus was thought to be on the short arm of chromosome 5 and a pycnodysostosis locus on the short arm of a G group chromosome.

Theoretically yet another method of gene assignment is that of *in situ* hybridization. By this method Price et al (194) concluded that the β-δ-γ hemoglobin linkage group may be on a group B chromosome and the α hemoglobin locus on chromosome No. 2. Bishop & Jones (27) and Prensky & Holmquist (193) criticized the conclusion, doubting that the sensitivity of the method is adequate to demonstrate localization.

Renwick (197) discussed the analytic methodology for using chromosome variants in family studies aimed at mapping the autosomes. In a family segregating for a "fragile site" on the long arm of chromosome No. 16 Magenis et al (152) could confirm the assignment of the α-haptoglobin locus to that

TABLE 3 Linkages, including specific autosomal assignments where known (7 May, 1973)
(Assembled with assistance of F.H. Ruddle, Ruth Sanger, L.R. Weitkamp and others)

A. *Autosomal linkages*

1. *Chromosomal assignment*

a. Chromosome No. 1 (see refs. 62, 63, 110, 123, 128, 163, 164, 205, 243, 247)

```
             Cae  ?   AmS                               PeC*
cen:1qh+::Fy::AOD:AmP: E2 ?:: PGM**1:::: El1:Rh:::::PGD**
   0     10   10    5    15   15         24   3   24
          ____________  ______________  ____________
               23             34              27
```
(approx. map interval in centimorgans)

Key

cen = centromere
1qh+ = enlarged heterochromatic segment of long arm of 1 (The "uncoiler" region (h+) is by definition on the long arm but whether all or any of the rest of the loci are there or on the short arm is unknown)
Cae = zonular pulverulent cataract (11680)†
Fy = Duffy blood group (11070)
AOD = auriculo-osteodysplasia (10900)
AmS = salivary amylase (10470)
AmP = pancreatic amylase (10471)
(Last two symbolized Amy_1 and Amy_2 by Merritt et al, *Am. J. Hum. Genet.*, 1973.

E_2 = pseudocholinesterase$_2$ (17750)
PGM_1 = phosphoglucomutase$_1$ (17190)
El_1 = elliptocytosis$_1$ (13050)
Rh = Rhesus blood group (11170)
PeC = peptidase C (17000)
PGD = 6-phosphogluconate dehydrogenase (17220)

```
        26%     13%   ? 15%   ?? 35%
    .        .     .       .        .
_____________________________________________
      PGM1        Rh   PGD   PepC   Fy
```

b. Chromosome No. 4 or 5
Adenine B auxotroph, human complement for hamster* (10255)
Esterase regulator* (13335)

c. Chromosome No. 6
Malic enzyme (NADP-dependent form of malate dehydrogenase), soluble* (15425)
Indophenoloxidase B* (14746)

d. Chromosome No. 7
Mannose phosphate isomerase* (15455)
Pyruvate kinase-3, or leukocytic form* (17905)

*Established by cell hybridization.

**Both family and hybrid cells data. Others (with no asterisk) established by family studies.

† Numbers in parentheses refer to McKusick's *Mendelian Inheritance in Man*, where references and details can be found. (Numbers not ending in 0 have been assigned since the 1971 edition.)

e. Chromosome No. 10
Glutamic oxaloacetic transaminase, soluble* (13825)

f. Chromosome No. 11
Lactic dehydrogenase A* (15000) See ref. 33
Esterase A4* (13340)
Puck surface killer antigen - KA, or AL* (14875)
Glutamate-pyruvate transaminase C (13827)

g. Chromosome No. 12
Lactic dehydrogenase B* (15010)
Peptidase B* (16990)
Human complement of hamster glycine auxotroph - Gly A^+ (?serine hydroxymethylase)* (13845)
Glutamate-pyruvate transaminase B (13826)

h. Chromosome No. 14
Nucleoside phosphorylase* (16405)

i. Chromosome No. 16
α-haptoglobin (14010) (long arm)
Adenine phosphribosyltransferase* (10260)

j. Chromosome No. 17
Thymidine kinase* (18830) (long arm)

k. Chromosome No. 19
Glucosephosphate isomerase (phosphohexose isomerase)* (17240)

l. Chromosome No. 20
?Adenosine deaminase* (10270) See 3 loci lined to ADA in syntenic list

m. Chromosome No. 21
Indophenoloxidase A (IPO-A)† (14745)
Antiviral protein AVP† (10745)

2. *Syntenic loci for which chromosomal assignment not achieved*
(neuterized‡ intervals in cM)

a. Lutheran (Lu) blood group locus (11120), secretor (Se) locus (18210) and myotonic dystrophy (Dm) locus (16090) - intervals 13 and 4 cM (see ref. 199).

b. ABO blood group locus (11030), nail-patella (Np) locus (16120), adenylate kinase (AK) locus (10300) - intervals about 13 and 0 cM, respectively.

c. Beta (14190) and delta (14210) hemoglobin loci; interval about 0 cM. (The two gamma loci, γ136A and γ136G, may be linked to beta and delta. See section 3h, below. The homologous loci are linked in the mouse.)

d. Am_2 immunoglobulin locus (14700), Gm immunoglobulin region = γG_1, γG_2, γG_3 (14710), α_1-antitrypsin (Pi) locus (10740) - closely clustered immunoglobulin loci in order $\gamma G_2 : \gamma G_3 : \gamma G_1 : \gamma A_2$: Pi - all intervals about 0, except last, 28 cM. γG_3 and γG_1 known to be closely linked because of Lepore - like myeloma protein. Relative positions of γG_2 and γG_4 are unknown. See ref. 96, 237.

e. Transferrin (Tf) locus (19000) and pseudocholinesterase$_1$ (E_1) locus (17740) - interval about 15 cM.

f. Albumin (Alb) locus (10360) and group-specific component (Gc) locus (13920) - interval about 3 cM.

‡ Average of male and female values.
*Established by cell hybridization.

g. Pelger-Huet locus (16940) and unusual muscular dystrophy locus (15900).
h. MNSs (MNS) blood group locus (11130) and sclerotylosis (Tys) locus (18160) - intervals about 5 cM.
i. HL-A region = HL − A_{LA} and HL-A4 (14280), phosphoglucomutase$_3$ (PGM_3) locus (17210), adenosine deaminase locus (10270), P blood group locus** (11140). LA and 4 are also called 1 and 2. Mixed lymphocyte culture locus may be a separate linked locus of the HL-A region. This linkage group may be on chromosome No. 20. An immune response locus (Ir) may also be in the HL-A complex. See refs. 61, 83, 141, 142, 158.
j. Isocitrate dehydrogenase (14770) and malate dehydrogenase (NAD-dependent form), soluble* (15425) See ref. 219.

3. *Autosomal linkages not yet confirmed*

a. Chromosome 2 — Acid phosphatase (17150)
b. Chromosome 2 or 4 — MNSs (11130)
c. Lewis locus (11110) and acid phosphatase (17150)
d. Chromosome 5p — Triosephosphate isomerase (19045)
e. Chromosome 6 — Gm (14710) (See ref. 35)
f. Chromosome 6p — Hageman factor (23400)
g. By a method of in situ hybridization, Price, Conover & Hirschhorn (194) concluded that the beta-gamma-delta hemoglobin linkage group may be on a group B chromosome and the alpha hemoglobin locus on chromosome No. 2.
h. Gamma hemoglobin locus (14200) and beta-delta hemoglobin region. From the structure of Hb Kenya (an γ - β hybrid), the order may be Gγ-Aγ-δ-β. See ref. 117.
i. β hemoglobin locus and MNSs locus (inconsistent with sections 3b and 3g, above)
j. Lactate dehydrogenases B and C (15015)
k. Chromosome 13q — Retinoblastoma (18020)
l. Chromosome 15 — ABO (11030)
m. ABO blood group locus (11030) and xeroderma pigmentosum locus (27870)
n. Chromosome 18 — IgA (14685)
o. GPI (17240) and ABO (inconsistent with 1k and 3l)
p. Chromosome 21 or 22 — Pycnodysostosis (26580)
q. Chromosome 21 — Phosphohexokinase (17230)
r. Chromosome 21 — Ag lipoprotein (15200)
s. α-haptoglobin and catalase (11550)
t. Colonic polyposis (17510) and Duffy (11070)
u. Dombrock blood group (11060) and MNSs

B. *X-borne linkages*

1. Colorblindness loci (30380, 30390) and G6PD locus (30590)
2. Colorblindness loci and hemophilia A (30690)
3. G6PD locus and hemophilia A locus
4. Xg blood group locus (31470) and locus for X-linked ichthyosis (30810)
5. Xg blood group locus and ocular albinism locus (30050)
6. Xg blood group locus and angiokeratoma locus (30150)
7. Xm serum protein locus (31490) and the locus for the Hunter syndrome (30990)

**Both family and hybrid cell data. Others (with no asterisk) established by family studies.

8. Xm and colorblindness loci
9. Deutan and protan colorblindness loci
10. Xg blood group locus and locus for retinoschisis (31270)
11. Xg and mental retardation with or without hydrocephalus (30950) (awaiting confirmation)
12. Colorblindness loci and muscular dystrophy with contractures (31030)

Tentative Map of Xq (long arm of X chromosome)

```
              ?        ?           ?                        ?
(centromere) O-----PGK-----HGPRT-----heA-G6PD-mdc-cbD-cbP--
```

Tentative Map of ?Xp (short arm of X chromosome)

```
  rs--oa--mr--Xg--ich---Fa
-17 -17 -11       11    24  (numbers = interval from Xg in cM)
```

Polarity arbitrarily represented; for example, all loci may be to the right of Xg assuming that Xg is near the distal end of Xp and the centromere on the right end of this diagram. Xg is known not to be on the distal end of Xq.

Key

rs = retinoschisis (31270)
oa = ocular albinism (30050)
Xg = Xg blood group (31470)
mr = mental retardation (30950)
ich = ichthyosis (30810)
Fa = Fabry disease, angiokeratoma, or α-galactosidase deficiency (30150)*
HGPRT = hypoxanthine guanine phosphoribosyl transferase (30800)*
heA = hemophilia A (30670)
G6PD = glucose-6-phosphate dehydrogenase (30590)*
cbD = deutan colorblindness (30380)
cbP = protan colorblindness (30390)
mdc = muscular dystrophy with contractures (31030)
PGK = phosphoglycerate kinase (31180)*

Other pairs of X-linked loci proved to be not closely linked include:

Colorblindness loci and retinitis pigmentosa (31260)
G6PD and thyroxine-binding globulin (31420)
hemophilia A (30670) and hypophosphatemia (30780)
HGPRT (30800) and G6PD (30590)
ichthyosis (30810) and both colorblindness loci
ichthyosis and G6PD
Xg and Addison disease with cerebral sclerosis (30010)
Xg and agammaglobulinemia (30030)
Xg and Becker muscular dystrophy (31010)
Xg and choroideremia (30310)
Xg and colorblindness loci
Xg and Duchenne muscular dystrophy (31020)
Xg and ectodermal dysplasia, anhidrotic (30510)
Xg and G6PD
Xg and hemophilia A
Xg and hemophilia B (30690)
Xg and HGPRT
Xg and keratosis follicularis (30880)
Xg and mucopolysaccharidosis II (30990)
Xg and retinitis pigmentosa (31260)
Xg and testicular feminization (31370)
Xg and thrombocytopenia (31390)
Xg and thyroxine-binding globulin (31420)
Xg and Xm (31490)

(Data on the above linkages with *Xg* and others are tabulated by Race and Sanger, *Blood Groups in Man*, 5th ed.)

---

*Studied by cell hybridization as well as by the kindred method.

chromosome. Renwick (197) presented an analysis of the data of Weitkamp et al (244) suggesting assignment of the red cell acid phosphatase locus to chromosome No. 2. In a family in which three cytogenetic markers were segregating, Jacobs et al (123) could not demonstrate linkage between the Duffy locus and the break point in the long arm of chromosome No. 1 involved in formation of a 1q+?; Cq-translocation. This suggests that the break point was located toward the distal end of the long arm, the Duffy locus being near the centromere. Data from four large kindreds were published by Wikramanayake et al (250); no clear assignment was achieved.

Bender & Burkhardt (23) and Hamerton (ref. 7, vol. 2, p. 377) assembled in tabular form information on exclusions of gene assignment achieved through the study of chromosome aberrations. When a person with a deficiency is heterozygous at a particular locus, the locus cannot be on the chromosome segment that is deficient. For example, it has been shown by this approach that the HL-A "locus" cannot be on the segment of the short arm of chromosome 5 deleted in the cri-du-chat syndrome (24). Unfortunately, this exclusion catalog is not being updated (22).

Fialkow et al (93) presented support for synteny of 6-PGD and Rh; a patient with chronic myeloid leukemia was found to be hemizygous at both loci but not at several others tested.

Those loci that are assigned to specific chromosomes by cell hybridization should be tested in linkage studies by the standard pedigree method. Of course, allelic variability is a necessary condition for kindred studies. Edwards et al (84) demonstrated inherited electrophoretic variants of nucleoside phosphorylase, and Ricciutti & Ruddle (205) showed that the locus is on chromosome No. 14. Ruddle's group (205) also assigned the adenine phosphoribosyltransferase locus to chromosome No. 16. Mapping of APRT against haptoglobin which is known to be on the long arm of 16 (152, 203) should be done but will require allelic variation in APRT. Kelley et al (130) had reported a variation in heat stability of APRT, the ratio of frequencies of the heat-labile and heat-stable alleles being 85:15, a value highly favorable for linkage study. However, Mowbray et al (169) found a continuous and unimodal distribution for residual red cell APRT activity after heating. They studied fresh blood and suggested that the results of Kelley et al (130) may have been altered by their use of old samples subjected to variable amounts of freezing and thawing. Mowbray et al (169) did find rare examples of electrophoretic variation in APRT.

Ohno's principle of evolutionary conservatism of the X chromosome asserts that there is a homology of the X chromosome among mammals. The examples of X chromosome homology are sufficiently numerous that the finding of X-linked testicular feminization in mice (151) and demonstration of apparent physiologic homology to the disease in man (103) makes X-linked recessive rather than male-limited autosomal dominant inheritance a virtual certainty for man. X-linked dominant hypophosphatemic rickets is one of the most recently discovered (85) examples of a mouse disorder homologous to a human disease; phosphoglycerate kinase is X-linked in man (50), the Chinese hamster (247) and the kangaroo (64). Although the homology of the two H-2 loci with the two HL-A loci and the linkage homology of closely linked hemoglobin and immunoglobulin loci can be pointed to, it seems improbable that autosomal linkage homology will be found between mammals so far apart as mice and men, although some homologies may be found among higher primates. (The finer delineation of the chromosomes by the new "banding" techniques suggests linkage homology of higher primates (73) but not of mouse and man.) Thus, whereas glucosephosphate isomerase (GPI), which appears to be on chromosome No. 19 in man (160), is loosely linked to the beta homoglobin locus in mice, this is no reason to suppose that the beta hemoglobin locus is carried by chromosome No. 19 of man. [Ritter et al (201) had two families that gave weak support to linkage of the ABO and GPI loci.]

From study of an X-autosome translocation, Grzeschik et al (108) concluded that the G6PD and HGPRT loci are on the short arm and PGK on the long arm of the X chromosome. On the other hand, Ricciuti & Ruddle (194a), from studies of the same anomalous chromosome KOP t (Xq−; 14q+) in cell hybrids concluded that all three are on the long arm. From a comparison of the findings in KOP with those in two other X- autosome translocations, they (205) concluded further that the order on the long arm is as follows: centromere –PGK –HGPRT –G6PD.

Ferguson-Smith (91) thinks the Xg locus may be at the distal end of the short arm of the X chromosome. In an Xg(a−) XX child with true hermaphroditism the father was Xg(a+). Ferguson-Smith suggested that male-determining factors on the short arm of the Y chromosome may have replaced a terminal segment of the X short arm. From a case of isochromosome of the short arm of the X chromosome de la Chapelle et al (75) also concluded that the Xg locus may be on the short arm and may be incompletely lyonized. In the isochromosome case the Xg locus was present in triplicate if it is on the short arm. Although the abnormal X chromosome was inactivated in all cases, the patient was weakly Xg(a+).

The polarity of the loci linked to Xg is unknown. Certain families with an anomalous distribution of Xg types have led to the suspicion that the Xg locus is on the end of the short arm. Pearson (190a) has evidence that it is not on the distal third of the long arm. A mother with a balanced t(3p+;Xq−) was Xg(a+), the father was Xg(a−) and the unbalanced Xq− daughter was Xg(a+).

CLINICAL GENETICS

Diagnosis

Heterozygote detection in homocystinuria was studied by Goldstein et al (102) using PHA-stimulated lymphocytes. Even though resting lymphocytes do not normally show the relevant enzyme, cystathionine synthetase, stimulated lymphocytes do; thus an enzymatic diagnosis of homozygosity is possible and the heterozygote can be demonstrated. Nadler & Egan (175) used the same method for detecting the heterozygote for lysosomal acid phosphatase deficiency, and the first use seems to have been in glycogen storage disease II by Hirschhorn et al (113).

Hill & Puck (111) devised a new method for detection of galactosemia in fibroblasts. They reasoned that if galactosemic fibroblasts have one main route for galactose metabolism to trichloroacetic acid—insoluble material while normal fibroblasts have two, inhibition of the one remaining pathway in galactosemic cells would facilitate the detection of the defect. *D*-galactono-γ-lactone was used for inhibition of the alternative pathway. Radioautography was chosen as the method of cell study because its sensitivity requires only a small number of cells—an advantage especially in prenatal diagnosis. Galactosemia and nongalactosemia could be distinguished with as few as 100–1000 cells. The principle underlying this method—inhibition of alternative pathways—should be useful in the detection of other inborn errors of metabolism and should be useful especially in prenatal diagnosis.

Screening of neonates for inborn errors of metabolism, especially phenylketonuria, has now had massive application (15d, 150). Screening for conditions that represent risk factors in (or precursors of) degenerative disorders of later life is in an exploratory stage. Two risk factors potentially identifiable by biochemical means in the newborn are hyperlipidemia (139) and α_1-antitrypsin deficiency (13d). These predispose to atherosclerosis and pulmonary emphysema, respectively, and are frequent abnormalities. In Northern Europeans α_1-antitrypsin deficiency in homozygous state has a frequency of about 1 in 4000 and heterozygotes, who may also be vulnerable to pulmonary disease, have a frequency of 3% or more. In children, Sharp & colleagues (217) described cirrhosis associated with α_1-antitrypsin deficiency. Material that appears to be α_1-antitrypsin (an α_1-globulin glycoprotein), accumulates in the rough endoplasmic reticulum of the hepatocytes, especially in the periportal areas of lobules. The electrophoretic variant of α_1-antitrypsin which is associated with cirrhosis (and with emphysema) is that determined by the so-called Pi^{z} allele, Pi^{m} being the main wildtype allele. (Pi, the designation for the system, means protease inhibitor.) Accumulation of α_1-antitrypsin in the liver may be the result of impaired transport of the structurally altered glycoprotein determined by the Pi^{z} gene. Puzzlement over the disparate phenotypes—cirrhosis in childhood, emphysema in adults—produced by the same genotype is lessened by the description of children (100) and adults (58) with both phenotypes. There is probably an important interaction of genotype and environment in determining the disorder, be it pulmonary or hepatic disease. When the environmental factors are better understood, preventive measures may be possible and make screening for detection of susceptible persons highly worthwhile. The full pathogenetic significance of the several Pi variants is not yet known.

Prognosis (*Genetic Counseling*)

The impact of genetic counseling (16) has been examined from two points of view (*a*) The psychosocial: How much and what of genetic counseling does the consultand or counselee[4] understand? What course of action does he pursue? (*b*) The statistical and populational: assuming 100% effectiveness of genetic counseling, e.g. that based on prenatal diagnosis, how would gene frequencies be influenced?

Prenatal diagnosis by amniocentesis showed a rapid advance as a result of intense interest born out of changed legal, social, and personal views on abortion (4, 92, 166, 167). The use of the linkage principle in prenatal diagnosis and genetic counseling was discussed from a theoretical point of view (156, 198) and was used in a case of myotonic dystrophy by Schrott et al (210).

Because of the relatively high frequency of sickle cell anemia (168) and the impracticality of other methods of prevention, prenatal diagnosis has been attempted. A problem is the fact that little beta hemoglobin chain (the site of the defect) is synthesized in utero. There are now, however, methods sufficiently sensitive to detect sickle hemoglobin in the midtrimester of pregnancy (116, 129), provided that a small sample of fetal blood can be obtained. The last is no simple task. Kan et al (129), who used the incorporation of radioactive leucine as the means of identifying Hb S, thought that mixed placental blood could be used.

Treatment

As indicated by a report on a combined study (54) in several centers in England and Baltimore, the efficacy of anti-D gammaglobulin is well established in the prevention of Rh-hemolytic disease.

Di Ferrante et al (77) first attempted enzyme replacement in the mucopolysaccharidoses by intravenous infusion of plasma or plasma derivatives. The rationale is enzyme replacement and was based on the studies of Neufeld and colleagues (p.448). Relatively extensive experience has accumulated in a short time. Improvement appears to be genuine (155) [although some (74, 88) question this], but improvement from infusions each 4 or 6 weeks is not maintained after about one year of treatment.

Knudson et al (134) used leukocyte infusions in the treatment of mucopolysaccharidoses. They would seem to have important immunologic drawbacks.

Kidney transplantation is a possibility (174) for enzyme replacement in inborn errors of metabolism and has been proposed in the mucopolysaccharidoses but at this writing has not to our knowledge been tried. It has been used in Fabry disease (55, 76) where the renal failure produced by the disease process was itself adequate justification for transplant. Benefit due to enzyme replacement by the grafted tissue probably occurs in Fabry disease (76, 191) although the views on this differ [see Clarke et al (55) and reply by Krivit et al (138)]. The renal failure of cystinosis (153) and of familial Mediterranean fever (FMF) (57) have also been treated by renal transplant. Although the nature of the enzyme defect is

---

[4] *Counselee* is a readily understood designation. *Consultand* (used by many as synonymous with *counselee*) is a term introduced by Murphy (171, 172) to refer to the person whose genotype is of primary relevance to the genetic counseling problem at hand. It may or may not be the same person as the counselee.

unknown in these conditions, systemic benefit may have occurred. At least the tubule cells of the transplanted kidney apparently do not reaccumulate cystine although deposits occur in mesangial cells. Polycystic renal disease is another hereditary disorder which has been treated by renal transplantation (144) with good results.

Transplantation has also been used to replace the defective immune mechanism in selected genetic disorders. For example: chronic mucocutaneous candidiasis exhibits a defect in cellular immunity (245). Valdimarsson et al (235) achieved good treated results in an affected 12 year old boy by infusion of leukocytes from a healthy HL-A identical brother.

The possibility of modifying proteins to improve their function or survival has been realized in the case of sickle cell anemia (224). Urea in high concentrations inhibits sickling in vitro by disrupting hydrophobic interactions of hemoglobin molecules (214). Urea had been infused into sickle cell anemia patients in the treatment of crises (181). It is doubtful that the amount given was adequate to accomplish much. Since in solution urea is in equilibrium with cyanate, Cerami & Manning (49) used cyanate, which, they found, reacts irreversibly and specifically with the valine at position 6 in the beta polypeptide chain. In vitro they found that amino terminal carbamylation of hemoglobin has definite antisickling effects. Furthermore, partial carbamylation of hemoglobin in cells obtained from a patient with sickle cell anemia resulted in increased red cell survival when reinfused (99). Clinical trials are underway. This is an important model of what it may be possible to accomplish in the treatment of enzymopathies (see p.447) by modification of enzyme in CRM+ disorders.

Shope papilloma virus induces arginase in chronic carriers, e.g., workers with the virus. Rogers (204) could show that the arginase "induced" in the rabbit is virus arginase not rabbit arginase; thus, DNA coding for the structure of arginase is presumably introduced. On this rationale, virus was administered to patients with argininemia due to deficiency of arginase (231). It is doubtful that any benefit was produced.

Pharmacogenetics—the genetics of the metabolism of and response to drugs—is a further aspect of genetics and therapy. The field was reviewed by La Du (140).

Prevention

In the hands of Kaback and his colleagues (127), Tay-Sachs disease has proved to be a genetic disorder susceptible to control, albeit not complete eradication. The characteristics of the disease which make it favorable for this approach are (*a*) occurrence predominantly in a defined ethnic group, Ashkenazic Jews, (*b*) availability of tests for identifying heterozygotes, and (*c*) capability of diagnosis of the homozygous state in the mid-trimester fetus. Practical feasibility has been demonstrated by a study in the Baltimore-Washington area where the disease may have a frequency of 1 in 3600 to 5000 among Ashkenazic Jews.

Summary

The following quantitative measures of progress in human genetics are listed:

(*a*) The enzymatic defect is now known in about 125 inborn errors of metabolism, about a fourth of all proved recessives (each locus being wittingly counted only once).

(*b*) All the chromosomes are uniquely identifiable by morphological means.

(*c*) At least one locus has been assigned confidently to 14 of the 22 autosomes and in three instances (Nos. 1, 16, and 17) to a specific part (the long arm) of an autosome.

(*d*) Relatively extensive mapping of the X chromosome and of chromosome No. 1 has been achieved. Some loci (PGK, HGPRT, G6PD, in that order, beginning at the centromere) are assigned with fair confidence to the long arm of the X chromosome.

Note added in proof: Several additions to, and some revisions of, this linkage list were made in the *International Workshop on Human Gene Mapping*, New Haven, Conn., June 1973 (report edited by F. H. Ruddle et al and published by National Foundation March of Dimes).

Literature Cited: Monographs

1. Bergsma, D., Borgaonkar, D.S., Shah, S.A. (eds.) 1972. *Advances in Human Genetics and their Impact on Society*. Birth Defects: Orig. Art. Series 8, No. 4
2. Dayhoff, M.O. 1972. *Atlas of Protein Sequence and Structure 1972* Vol. 5. Washington, D.C.: National Biomedical Research Foundation
3. de Grouchy, J., Ebling, F.J.G., Henderson, I.W. (eds.) 1972. *Human Genetics*. Proc. IVth Int. Cong. Human Genet., Paris, Sept., 1971. Amsterdam: Excerpta Med.
4. Dorfman, A. (ed.) 1972. *Antenatal Diagnosis*. Chicago: Univ. Chicago Press
5. Fraser, F.C., McKusick, V.A. (eds.) 1970. *Congenital Malformations*. Proc. 3rd Int. Conf., The Hague, 1969 New York: Excerpta Med.
6. Giannelli, F. 1970. *Human Chromosome DNA Synthesis*. Vol. 5, Monographs in Human Genetics. Basel: S. Karger
7. Hamerton, J.L. 1971. *Human Cytogenetics*. Vol. 1. General Cytogenetics. Vol. 2. Clinical Cytogenetics. New York: Academic
8. Harris, H., Hirschhorn, K. (eds) 1972. *Advances in Human Genetics*. Vol. 3 (*a*) Nadler, H.L.: Prenatal detection of genetic disorders, p.1. (*b*) O'Brien, J.S.: Ganglioside storage disease, p.39. (*c*) Bloom, A.D.. Induced chromosomal aberrations in man, p.99. (*d*) Ruddle, F.H.: Linkage analysis using somatic cell hybrids, p.173. (*e*) Comings, D.E.: The structure and function of chromatin, p.237.
9. McKusick, V.A. 1971. *Mendelian Inheritance in Man. Catalogs of Autosomal Dominant, Autosomal Recessive and X-linked Phenotypes*. 3rd ed. Baltimore: Johns Hopkins University
10. McKusick, V.A. 1972. *Heritable Disorders of Connective Tissue*. 4th ed. St. Louis: C.V. Mosby
11. Ohno, S. 1970. *Evolution by Gene Duplication*. New York: Springer-Verlag
12. Stanbury, J.B., Wyngaarden, J.B., Fredrickson, D.S. 1972. *The Metabolic Basis of Inherited Diseases*. 3rd ed. New York: McGraw-Hill
13. Steinberg, A.G., Bearn, A.G. (eds.) 1970. *Progress in Medical Genetics*.

New York, London: Grune & Stratton Vol. VII. (*a*) Migeon, B.R., Childs, B.: Hybridization of mammalian somatic cells, p. 1. (*b*) Hsia, D.Y.-Y.: Phenylketonuria and its variants, p. 29. (*c*)Ruddy, S., Austen, K.F.: Inherited abnormalities of the complement system in man, p. 69. (*d*) Fagerhol, M.: The Pi system—inherited variants of serum α_1-antitrypsin, p. 96. (*e*) Milne, M.D.: Genetic aspects of renal diseases, p. 112. (*f*) Sutton, H.E.: The haptoglobins, p. 163. (*g*) Parker, W.C.: Some legal aspects of genetic counseling, p. 217

14. Steinberg, A.G., Bearn, A.G. (eds.) 1972. *Progress in Medical Genetics*. Grune & Stratton: New York and London, Vol. VIII (*a*) Fenner, F.: Genetic aspects of viral diseases of animals, pp. 1–60. (*b*) German, J.: Genes which increase chromosomal instability in somatic cells and predispose to cancer, pp. 61–102. (*c*) Morton, N.E.: The future of human population genetics, pp. 103–124. (*d*) Kirkman, H.N.: Enzyme defects, pp. 125–168. (*e*) Clarke, C.A.: Prevention of Rh isoimmunization, pp. 169–224. (*f*) Brady, R.O., Kolodny, E.H.: Disorders of ganglioside metabolism, pp. 225–242. (*g*) Scott, C.I., Jr.: The genetics of short stature, pp. 243–300.
15. Steinberg, A.G., Bearn, A.G. (eds.) 1973. *Progress in Medical Genetics*. Grune & Stratton: New York and London, Vol. IX (*a*) Miller, O.J., Miller, D.A., Warburton, D.: Application of new staining techniques to the study of chromosomes, p. 1. (*b*) Fowler, R.E., Edwards, R.G.: The genetics of early human development, p. 49. (*c*) Knudson, A.G., Jr., Strong, L.C., Anderson, D.E.: Heredity and cancer in man, p. 113. (*d*) Clow, C.L., Fraser, F.C., Laberge, C., Scriver, C.R.: On the application of knowledge to the patient with genetic disease, p. 159. (*e*) Federman, D.D.: Genetic control of sexual difference, p. 215. (*f*) Lees, R.S., Wilson, D.E., Schonfeld, G., Fleet, S.: The familial dyslipoproteinemias, p. 237. (*g*) Vesell, E.S.: Advances in pharmacogenetics, p. 291.
16. Stevenson, A.C., Davison, B.C.C.: 1970. *Genetic Counselling*. Philadelphia, Lippincott

Literature Cited:Articles

17. Abramson, R.K., Rucknagel, D.L., Shreffler, D.C. 1970. Two Melanesian homozygotes for Hb J Tongariki. *Science* 169:194–96
18. Allderdice, P.W., Miller, O.J., Miller, D.A., Breg, W.R., Gendel, E., Zelson, C. 1971. Familial translocation involving chromosomes 6, 14 and 20, identified by quinacrine fluorescence. *Humangenetik* 13:205–09
19. Arrighi, F.E., Hsu, T.C. 1971. Localization of heterochromatin in human chromosomes. *Cytogenetics* 10:81–86
20. Bach, G., Friedman, R., Weissmann, B., Neufeld, E. F. 1972. The defect in the Hurler and Scheie syndromes: deficiency of *α-L*-iduronidase. *Proc. Nat. Acad. Sci.* 69:2048–51
21. Benacerraf, B., McDevitt, H.O. 1972. Histocompatibility-linked immune response genes. *Science* 175:273–79
22. Bender, K. (Freiburg, West Germany): Personal communication, 1973
23. Bender, K., Burchkardt, K. 1970. On the localization of genes on certain autosomes of man through chromosome aberrations III. Exclusion of the possibility of gene assignment. *Humangenetik* 9:75–85
24. Bender, K., Schindera, F., Kissmeyer-Nielsen, F. 1970. Localization exclusion of the HL-A genes from the short arm of human chromosome 5. *Humangenetik* 11:78–80
25. Berger, R. 1971. Une nouvelle technique d'analyse du caryotype. *C. R. Acad. Sci.* 273 (Series D): 2620–22
26. Bhasin, M.K., Foerster, W. 1972. A simple banding technique for identification of human metaphase chromosomes. *Humangenetik* 14:247–50
27. Bishop, J.O., Jones, K.W. 1972. Chromosomal localization of human haemoglobin structural genes. *Nature* 240:149–50
28. Bloom, A.D., Choi, K.W., Lamb, B.J. 1971. Immunoglobulin production by human lymphocytoid lines and clones: absence of genic exclusion. *Science* 172:382–83
29. Bobrow, M., Collacott, H.E.A.C., Madan, K. 1972. Chromosome banding with acridine orange. (Letter) *Lancet* 2:1311
30. Bobrow, M., Madan, K., Pearson, P.L. 1972. Staining of some specific regions of human chromosomes, particularly the secondary constriction of no. 9. *Nature N.B.* 238:122–24

31. Bodmer, W.F. 1972. Evolutionary significance of the HL-A system. *Nature* 237:39–45
32. Bomsel-Helmreich, O. 1970. Fate of heteroploid embryos. *Advan. Biosci.* 6:381–403
33. Boone, C., Chen, T., Ruddle, F.H. 1972. Assignment of three human genes to chromosomes (LDH-A to 11, TK to 17 and IDH to 20) and evidence for translocation between human and mouse chromosomes in somatic cell hybrids. *Proc. Nat. Acad. Sci.* 68: 510–14
34. Borgaonkar, D.S. (Baltimore) 1973. Personal communication
35. Borgaonkar, D.S., Bias, W.B., Chase, G.A., Sadasivan, G., Herr, H.M., Golomb, H.M., Bahr, G.F., Kunkel, L.M. 1973. Identification of a C6/G21 translocation chromosome by the Q-M and Giemsa banding techniques in a patient with Down's syndrome, with possible assignment of Gm locus. *Clin. Genet.* 4:53–57
36. Borgaonkar, D.S., Hollander, D.H. 1971. Quinacrine fluorescence of the human Y chromosome. *Nature* 230:52
37. Borgaonkar, D.S., McKusick, V.A., Herr, H.M., de los Cobos, L., Yoder, O.C. 1969. Constancy of the length of human Y chromosome. *Ann. Genet.* 12:262–64
38. Breg, W.R. 1972. Quinacrine fluorescence method for identifying metaphase chromosomes, with special reference to photomicrography. *Stain Techn.* 47:87–93
39. Breg, W. R., Miller, D.A., Allderdice, P.W., Miller, O.J. 1972. Identification of translocation chromosomes by quinacrine fluorescence. *Am. J. Dis. Child.* 123:561–64
40. Britten, R.J., Kohne, D.E. 1968. Repeated sequences in DNA. *Science* 161:529–40
41. Brock, D.J.H., Singer, J.D. 1970. Red-cell triosephosphate isomerase and chromosome 5. (Letter) *Lancet* 2: 45–46
42. Brown, S.W., Chandra, H.S. 1973. Inactivation system of the mammalian X chromosome. *Proc. Nat. Acad. Sci.* 70:195–99
43. Buckton, K.E., Jacobs, P.A., Rae, L.A., Newton, M.S., Sanger, R. 1971. An inherited X-autosome translocation in man. *Ann. Human Genet.* 35: 171–78
44. Bühler, E.M., Muller, H., Stalder, G.R. 1971. A strongly fluorescing abnormal chromosome in a malformed child. *Humangenetik* 12:64–66
45. Caspersson, T., Gahrton, G., Lindsten, J., Zech, L. 1970. Identification of the Philadelphia chromosome as a number 22 by quinacrine mustard fluorescence analysis. *Exp. Cell Res.* 63:238–40
46. Caspersson, T., Hultén, M., Lindsten, J., Therkelsen, A.J., Zech, L. 1971. Identification of different Robertsonian translocations in man by quinacrine mustard fluorescence analysis. *Hereditas* 67:213–20
47. Caspersson, T., Lindsten, J., Zech, L., Buckton, K.E., Price, W.H. 1972. Four patients with trisomy 8 identified by the fluorescence and Giemsa banding techniques. *J. Med. Genet.* 9:1–7
48. Caspersson, T., Lomakka, G., Zech, L. 1971. The 24 fluorescence patterns of the human metaphase chromosomes: Distinguishing characters and variability. *Hereditas* 67:89–102
49. Cerami, A., Manning, J.M. 1971. Potassium cyanate as an inhibitor of the sickling of erythrocytes in vitro. *Proc. Nat. Acad. Sci.* 68:1180–83
50. Chen, S.H., Malcolm, L.A., Yoshida, A., Giblett, E.R. 1971. Phosphoglycerate kinase: An X-linked polymorphism in man. *Am. J. Human Genet.* 23:87–91
51. Chen, T.R., Ruddle, F.H. 1971. Karyotype analysis utilizing differentially stained constitutive heterochromatin of human and murine chromosomes. *Chromosoma* 34:51–72
52. Choi, K.W., Bloom, A.D. 1970. Biochemically marked lymphocytoid lines: establishment of Lesch-Nyhan cells. *Science* 170:89–90
53. Church, R.L., Pfeiffer, S.E., Tanzer, M.L. 1971. Collagen biosynthesis: synthesis and secretion of a high molecular weight collegen precursor (procollagen). *Proc. Nat. Acad. Sci.* 68:3241
54. Clarke, C.A., et al 1971. Prevention of Rh-haemolytic disease: final results of the "high-risk" clinical trial. (A combined study from centers in England and Baltimore.) *Brit. Med. J.* 2:607–09
55. Clarke, J.T.R., Guttmann, R.D., Wolfe, L.S., Beaudoin, J.G., Morehouse, D.D. 1972. Renal allotransplantation in Fabry's disease. *New Eng. J. Med.* 287:1215–18
56. Clegg, J.B., Weatherall, D.J., Milner, P.F. 1971. Haemoglobin Constant

Spring—a chain termination mutant? *Nature* 234:337–40
57. Cohen, A.S., Bricetti, A.B., Harrington, J.T., Mannick, J.A. 1971. Renal transplantation in two cases of amyloidosis. *Lancet* 2:213–16
58. Cohen, K.L., Rubin, P.E., Echevarria, R.A., Sharp, H.L., Teague, P.O. 1973. Alpha-1 antitrypsin deficiency, emphysema, and cirrhosis in an adult. *Ann. Int. Med.* 78:227–32
59. Cohen, M.M. 1971. The chromosomal constitution of 165 human translocations involving D group chromosomes identified by autoradiography. *Ann. Génét.* 14:87–96
60. Comings, D.E. 1972. Evidence for ancient tetraploidy and conservation of linkage groups in mammalian chromosomes. *Nature* 238:455–67
61. Cook, P.J.L., Hopkinson, D.A., Robson, E.B. 1970. The linkage relationships of adenosine deaminase. *Ann. Hum. Genet.* 34:187–88
62. Cook, P.J.L., Noades, J., Hopkinson, D.A., Robson, E.B., Cleghorn, T.E. 1972. Demonstration of a sex difference in recombination fraction in the loose linkage Rh and PGM (1). *Ann. Human Genet.* 35:239–54
63. Cooke, P.J.L., Povey, S., Robson, E.B. 1972. Linkage studies on peptidases A, B, C and D in man. *Ann. Human Genet.* 36: 89–98
64. Cooper, D.W., Vandeberg, J.L., Sharman, G.B., Poole, W.E. 1971. Phosphoglycerate kinase polymorphism in kangaroos provides further evidence for paternal inactivation. *Nature N.B.* 230:155–57
65. Craig-Holmes, A.P., Shaw, M.W. 1971. Polymorphism of human constitutive heterochromatin. *Science* 174:702–04
66. Crandall, B.F., Weber, F., Muller, H.M., et al 1972. Identification of 21 and 22r chromosomes by quinacrine fluorescence. *Clin. Genet.* 3:264–71
67. Crick, F. 1971. General model for the chromosomes of higher organisms. *Nature* 234:25–27
68. Crossen, P.E. 1972. Giemsa banding patterns of human chromosomes. *Clin. Genet.* 3: 169–79
69. Danks, D.M., Cartwright, E., Stevens, B.J., Townley, R.R.W. 1973. Menkes' kinky hair disease: further definition of the defect in copper transport. *Science* 179:1140–41
70. Danks, D.M., Stevens, B.J., Campbell, P.E., Gillespie, J.M., Walker-Smith, J., Bloomfield, J., Turner, B. 1972. Menkes kinky-hair syndrome. *Lancet* 1:1100–02
71. Dawson, G., Stein, A.O. 1972. Lactosyl ceramindosis: catabolic enzyme defect of glycosphingolipid metabolism. *Science* 170:556–58
72. de Grouchy, J. 1970. 21 p- maternel en double exemplaire chez un trisomique 21. *Ann. Génét.* 13:52–55
73. de Grouchy, J., Turleau, C., Roubin, M., Klein, M. 1972. Evolution caryotypiques de l'homme et du chipanze. Etude comparative des topographies de bandes àpres dénaturation ménagée. *Ann. Génét.* 15:79–84
74. Dekaban, A.S., Holden, K.R., Constantopoulos, G. 1972. Effects of fresh plasma or whole blood transfusions on patients with various types of mucopolysaccharidosis. *Pediatrics* 50: 688–701
75. de la Chapelle, A., Schroder, J., Pernu, M. 1972. Isochromosome for the short arm of X: a human 46, XXpi syndrome. *Ann. Human Genet.* 36:79–88
76. Desnick, R.J., Simmons, R.L., Allen, K.Y., Woods, J.E., Anderson, C.F., Najarian, J.S., Krivit, W. 1972. Correction of enzymatic deficiencies by renal transplantation: Fabry's disease. *Surgery* 72:203–11
77. Di Ferrante, N., Nichols, B.L., Donnelly, P.V., Neri, G., Hrgovcic, R., Berglund, R.K. 1971. Induced degradation of glycosaminoglycans in Hurler's and Hunter's syndromes by plasma infusion. *Proc. Nat. Acad. Sci.* 68: 303–07
78. Donahue, R.P., Bias, W.B., Renwick, J.H., McKusick, V.A. 1968. Probable assignment of the Duffy blood group locus to chromosome 1 in man. *Proc. Nat. Acad. Sci.* 61:949–55
79. Drets, M.E., Shaw, M.W. 1971. Specific banding patterns of human chromosomes. *Proc. Nat. Acad. Sci.* 68: 2073–77
80. Dutrillaux, B., de Grouchy, J., Finaz, C., Lejeune, J. 1971. Evidence de la structure fine des chromosomes humains par digestion enzymatique (pronase en particulier). *C.R. Acad. Sci.* 273:587–88
81. Dutrillaux, B., Lejeune, J. 1971. Sur une nouvelle technique d'analyse du caryotype humain. *C. R. Acad. Sci.*, Paris 272:2638–40
82. Edwards, J.H. 1971. The analysis of X-linkage. *Ann. Human Genet.* 34:229–50

83. Edwards, J.H., Allen, F.H., Glenn, K.P., Lamm, L.U., Robson, E.B. 1972. Personal communication. To be published in *Histocompatibility Testing* 1972
84. Edwards, Y.H., Hopkinson, D.A., Harris, H. 1971. Inherited variants of human nucleoside phosphorylase. *Ann. Human Genet.* 34:395–408
85. Eicher, E. (Bar Harbor) 1972. Personal communication
86. Eicher, E.M. 1970. X-autosome translocations in the mouse: total inactivation versus partial inactivation of the X chromosome. *Advan. Genet.* 15: 176–261
87. Emerit, I., Noel, B., Thiriet, M., Loubon, M., Quack, B. 1971. Short arm deletion of chromosome 14. *Humangenetik* 15:33–38
88. Erickson, R.P., Sandman, R., Robertson, W.R.B., Epstein, C.J. 1972. Inefficacy of fresh frozen plasma therapy of mucopolysaccharidosis II. *Pediatrics* 50: 693–701
89. Evans, H.J., Buckton, K.E., Sumner, A.T. 1971. Cytological mapping of human chromosomes: Results obtained with quinacrine fluorescence and the acetic saline-Giemsa techniques. *Chromosoma* 35:301–25
90. Faed, M.J.W., Marrian, V.J., Robertson, J., Robson, E.B., Cook, P.J.L. 1972. Inherited pericentric inversion of chromosome 5: a family with history of neonatal death and a case of the 'cri du chat' syndrome. *Cytogenetics* 11: 400–11
91. Ferguson-Smith, M.A. 1966. X-Y chromosomal interchange in the aetiology of true hermaphroditism and of XX Klinefelter's syndrome. *Lancet* 2: 475–76
92. Ferguson-Smith, M.E., Ferguson-Smith, M.A., Nevin, N.C., Stone, M. 1971. Chromosome analysis before birth and its value in genetic counseling. *Brit. Med. J.* 4:69–74
93. Fialkow, P.J., Lisker, R., Giblett, E.R. Zavala, C., Cobo, A., Detter, J.C. 1972. Genetic markers in chronic myelocytic leukemia: evidence opposing autosomal inactivation and favoring 6-PGD - Rh linkage. *Ann. Human Genet.* 35: 321–26
94. Finaz, C., de Grouchy, J. 1972. Identification of individual chromosomes in the human karyotype by their banding pattern after proteolytic digestion. *Humangenetik* 15:249–52
95. Francke, U. 1972. Quinacrine mustard fluorescence of human chromosomes: characterization of unusual translocations. *Am. J. Human Genet.* 24:189–213
96. Gedde-Dahl, T., Jr., Fagerhol, M.K., Cook, P.J.L., Noades, J. 1972. Autosomal linkage between the Gm and Pi loci in man. *Ann. Human Genet.* 35: 393–400
97. Genest, P., Lejeune, J. 1972. Recherche sur l'origine d'un petit chromosome Y multicentenaire. *Ann. Génét.* 15: 51–53
98. Gey, W. 1970. Dq-, multiple malformations and retinoblastoma. *Humangenetik* 10:362–65
99. Gillette, P.N., Manning, J.M., Cerami, A. 1971. Increased survival of sickle-cell erythrocytes after treatment *in vitro* with sodium cyanate. *Proc. Nat. Acad. Sci.* 68:2791–93
100. Glasgow, J.F.T., Lynch, M.J., Hercz, A., Levison, H., Sass-Kortsak, A. 1973. Alpha$_1$-antitrypsin deficiency in association with both cirrhosis and chronic obstructive lung disease in two sibs. *Am. J. Med.* 54:181–94
101. Goldstein, J.L., Campbell, B.K., Gartler, S.M. 1972. Cystathionine synthase activity in human lymphocytes: Induction by phytohemagglutinin. *J. Clin. Invest.* 51:1034–37
102. Goldstein, J.L., Campbell, B.K., Gartler, S.M. 1973. Homocystinuria: heterozygote detection using phytohemagglutinin-stimulated lymphocytes. *J. Clin. Invest.* 52:218–21
103. Goldstein, J.L., Wilson, J.D. 1972. Studies on the pathogenesis of the pseudohermaphroditism in the mouse with testicular feminization. *J. Clin. Invest.* 51:1647–58
104. Goldstein, S. 1971. The pathogenesis of diabetes mellitus and its relationship to biological aging. *Humangenetik* 12: 83–100
105. Goodman, M., Barnabas, J., Matsuda, G., Moore, G.W. 1971. Molecular evolution in the descent of man. *Nature* 233:604–13
106. Grace, E., Drennan, J., Colver, D., Gordon, R.R. 1971. The 13q-deletion syndrome. *J. Med. Genet.* 8:351–57
107. Grace, E., Sutherland, G.R., Bain, A.D. 1972. Familial insertional translocation. (Letter) *Lancet* 2:23
108. Grzeschik, K.H., Allderdice, P.W., Grzeschik, A., Opitz, J.M., Miller, O.J., Siniscalco, M. 1972. Cytological mapping of human X-linked genes by use of somatic cell hybrids involving an X-autosome translocation. *Proc. Nat. Acad. Sci.* 69:69–73

108a. Hecht, F., Kimberling, W.J. 1971. Patterns of D chromosome involvement in human DqDq and DqGq Robertsonian rearrangements. *Am. J. Hum. Genet.* 23:361–67

109. Hickman, S., Neufeld, E.F. 1972. A hypothesis for I-cell disease: defective hydrolases do not enter lysosomes. *Biochem. Biophys. Res. Commun.* 49: 992–99

110. Hill, C.J., Rowe, S.I., Lovrien, E.W. 1972. Probable genetic linkage between human serum amylase (Amy-2) and Duffy blood group. *Nature* 235: 162–63

111. Hill, H.Z., Puck, T.T. 1973. Detection of inborn errors of metabolism: galactosemia. *Science* 179:1136–39

112. Hilwig, I., Gropp, A. 1972. Staining of constitutive heterochromatin in mammalian chromosomes with a new fluorochrome. *Exp. Cell Res.* 75: 122–26

113. Hirschhorn, K., Nadler, H.L., Waithe, W.I., Brown, B.I., Hirschhorn, R. 1969. Pompe's disease: detection of heterozygotes by lymphocyte stimulation. *Science* 166:1632–33

114. Hollán, S.R., Jones, R.T., Koler, R.D. 1972. Duplication of haemoglobin genes. *Biochemie* 54:639–48

115. Hollander, D.H., Borgaonkar, D.S. 1971. The quinacrine fluorescence method of Y-chromosome identification. *Acta Cytol.* 15:452–54

116. Hollenberg, M.D., Kaback, M.M., Kazazian, H.H., Jr. 1971. Adult hemoglobin synthesis by reticulocytes from the human fetus at midtrimester. *Science* 174:698–702

117. Huisman, T.H.J., Wrightstone, R.N., Wilson, J.B., Schroeder, W.A., Kendall, A.G. 1972. Hemoglobin Kenya, the product of fusion of γ and β polypeptide chains. *Arch. Biochem. Biophys.* 153:850–53

118. Hultén, M., Pearson, P.L. 1971. Fluorescent evidence for spermatocytes with two Y chromosomes in an XYY male. *Ann. Human Genet.* 34:273–76

119. Hungerford, D.A. 1971. Chromosome structure and function in man. I. Pachytene mapping in the male, improved methods and general discussion of initial results. *Cytogenetics* 10:23–32

120. Hungerford, D.A., Ashton, F.T., Balaban, G.B., LaBadie, G.U., Messatzzia, L.R., Haller, G., Miller, A.E. 1972. The C-group pachytene bivalent with a locus characteristic for parachromosomally situated particulate bodies (parameres): a provisional map in human males. *Proc. Nat. Acad. Sci.* 69: 2165–68

121. Hunt, L.T., Dayhoff, M.O. 1972. The origin of the genetic material in the abnormally long human hemoglobin A and B chains. *Biochem. Biophys. Res. Comm.* 47:699–704

122. Jacobs, P.A. (Honolulu) 1973. Personal communication

123. Jacobs, P.A., Brunton, M., Frackiewicz, A., Newton, M., Cook, P.J.L., Robson, E.B. 1970. Studies on a family with three cytogenetic markers. *Ann. Human Genet.* 33:325–36

124. Jacobs, P.A., Ross, A. 1966. Structural abnormalities of the Y chromosome in man. *Nature* 210:352–54

125. Jones, K.W., Corneo, G. 1971. Location of satellite and homogeneous DNA sequences on human chromosomes, *Nature N.B.* 233:268–71

126. Juberg, R.C., Jones, B. 1970. The Christchurch chromosome (Gp-). Mongolism, erythroleukemia and an inherited Gp- chromosome (Christchurch). *New Eng. J. Med.* 282:292–97

127. Kaback, M.M., O'Brien, J.S. 1973. Tay-Sachs: prototype for prevention of genetic disease. *Hospital Practice* 8(3): 107–16

128. Kamaryt, J., Adamek, R., Vrba, M. 1971. Possible linkage between uncoiled chromosome un 1 and amylase polymorphism (Amy-2) loci. *Humangenetik* 11:213–20

129. Kan, Y.W., Dozy, A.M., Alter, B.O., Frigoletto, F.D., Nathan, D.G. 1972. Detection of the sickle gene in the human fetus. *New Eng. J. Med.* 287: 1–4

130. Kelley, W.N., Levy, R.I., Rosenbloom, F.M., Henderson, J.F., Seegmiller, J.E. 1968. Adenine phosphoribosyltransferase deficiency: A previously undescribed genetic defect in man. *J. Clin. Invest.* 47:2281–89

131. Kelly, T.E., Chase, G.A., McKusick, V.A. 1972. High frequency of Tay-Sachs gene in a gentile community: model of Ashkenazic distribution. (Abstract) *Am. J. Human Genet.* 24:37a

132. Kissmeyer-Nielsen, F., Jorgensen, F., Lamm, L.U. 1972. The HL-A system in clinical medicine. *Johns Hopkins Med. J.* 131:385–400

133. Klinger, H.P., Moser, G.C. 1972. Improved chromatin-fluorescence technique. (Letter) *Lancet* 2:1366

134. Knudson, A.G., Jr., DiFerrante, N., Curtis, J.E. 1971. The effect of leukocyte transfusion in a child with mucopolysaccharidosis II. *Proc. Nat. Acad. Sci.* 68:820–23

135. Kohne, D.E. 1970. Evolution of higher-organism DNA. *Quart. Rev. Biophys.* 3:327–75

136. Krane, S.M., Pinnell, S.R., Erbe, R.W. 1972. Lysyl-protocollagen hydroxylase deficiency in fibroblasts from siblings with hydroxylysine-deficient collagen. *Proc. Nat. Acad. Sci.* 69:2899–2903

137. Kresse, H., Neufeld, E.F. 1972. The Sanfilippo A corrective factor. Purification and mode of action. *J. Biol. Chem.* 247:2164–70

138. Krivit, W., Desnick, R.J., Bernlohr, R.W., Wold, F., Najarian, J.S., Simmons, R.L. 1972. Enzyme transplantation in Fabry's disease (Editorial) *New Eng. J. Med.* 287:1248–49

139. Kwiterovich, P.O., Levy, R.I., Fredrickson, D.S. 1973. Neonatal diagnosis of familial type - II hyperlipoproteinaemia. *Lancet* 1:118–21

140. La Du, B.N. 1972. Pharmacogenetics: defective enzymes in relation to reactions to drugs. *Ann. Rev. Med.* 23: 453–68

141. Lamm, L.U., Kissmeyer-Nielsen, F., Svejgaard, A., Petersen, G.B., Thorsby, E., Mayr, W., Hodman, C. 1972. On the orientation of the HL-A region and the PGM_3 locus in the chromosome. *Tissue Antigens* 2:205–14

142. Lamm, L.U., Svejgaard, A., Kissmeyer-Nielsen, F. 1971. PGM_3:HL-A is another linkage in man. *Nature N.B.* 231:109–10

143. Lapière, C.M., Lenaers, A., Kohn, L.D. 1971. Procollagen peptidase: an enzyme excising the coordination peptides of procollagen. *Proc. Nat. Acad. Sci.* 68:3054–58

144. Lazarus, J.M., Bailey, G.L., Hampers, C.L., Merrill, J.P. 1971. Dialysis and transplantation in polycystic renal disease. *J.A.M.A.* 217:1821–24

145. Lee, C.L.Y., Welch, J.P., Winsor, E.J.T. 1972. Banding patterns in human chromosomes: production by proteolytic enzymes. *J. Hered.* 63: 296–97

145a. Leroy, J.G., Ho, M.W., MacBrinn, M.C., Zielke, K., Jacob, J., O'Brien, J.S. 1972. I-Cell disease: biochemical studies. *Pediat. Res.* 6:752–57

146. Lehmann, H., Charlesworth, D. 1970. Observatioms on hemoglobin P (Congo type). *Biochem. J.* 118:12–13P

147. Levine, B.B., Stember, R.H., Fotino, M. 1972. Ragweed hay fever: genetic control and linkage to HL-A haplotypes. *Science* 178:1201–03

148. Lichtenstein, J.R., Martin, G.R., Kohn, L.D., Byers, P.H., McKusick, V.A. 1973. A defect in the conversion of procollagen to collagen in one form of the Ehlers-Danlos syndrome. *Science.* In press

149. Licznerski, G., Lindsten, J. 1972. Trisomy 21 in man due to maternal nondisjunction during the first meiotic division. *Hereditas* 70:153–54

150. Littlefield, J.W. 1972. Genetic screening. (Editorial) *New Eng. J. Med.* 286: 1155–56

151. Lyon, M.F., Hawkes, S.G. 1970. X-linked gene for testicular feminization in the mouse. *Nature* 225:1217–19

152. Magenis, R.E., Hecht, F., Lovrien, E.W. 1970. Heritable fragile site on chromosome 16: probable localization of haptoglobin locus in man. *Science* 170:85–87

153. Mahoney, C.P., Striker, G.E., Hickman, R.O., Manning, G.B., Marchioro, T.L. 1970. Renal transplantation for childhood cystinosis. *New Eng. J. Med.* 283:397–402

153a. Marsh, D.G., Bias, W.B., Hsu, S.H., Goodfriend, L. 1973. Association of the HL-A7 cross-reacting group with a specific reaginic antibody response in allergic man. *Science* 179:691–93

154. Matalon, R., Dorfman, A. (Chicago) 1973: Personal communication

155. Maumenee, I.H. (Baltimore) 1972: Personal communication

156. Mayo, O. 1970. The use of linkage in genetic counselling. *Human Hered.* 20: 473–85

157. McClure, H.M. 1972. Trisomy in a chimpanzee. *Am. J. Pathol.* 67:413–16

158. McKusick, V.A. 1970. Human genetics. *Ann. Rev. Genet.* 4:1–46

159. McKusick, V.A., Howell, R.R., Hussels, I.E., Neufeld, E.F., Stevenson, R. 1972. Allelism, non-allelism and genetic compounds among the mucopolysaccharidoses: hypotheses. *Lancet* 1: 933–36

160. McMorris, F.A., Chen, T.R., Ricciuti, F., Tischfield, J., Creagan, R., Ruddle, F. 1973. Chromosome assignments in man of the genes for two hexosephosphate isomerases. *Science* 179:1129–31

161. Meisner, L.F., Inhorn, S.L. 1972. Normal male development with Y chromosome long arm deletion (Yq−). *J. Med. Genet.* 9:373–77

162. Merril, C.R., Geier, M.R., Petricciani, J.C. 1971. Bacterial virus gene expression in human cells. *Nature* 233: 398–400
163. Merritt, A.D., Lovrien, E.W., Rivas, M.L., Conneally, P.M. Human amylase loci: genetic linkage with the Duffy blood group locus and assignment to linkage group I. *Am. J. Human Genet.* In press
164. Merritt, A.D., Rivas, M.L., Ward, J.C. 1972. Evidence for close linkage of human amylase loci. *Nature N.B.* 239: 243–44
165. Miller, O.J., Miller, D.A., Warburton, D. 1973. The application of new staining techniques to the study of human chromosomes. Steinberg, A.G., Bearn, A.G., eds.: *Progress in Medical Genetics*, Vol. IX. In press
166. Milunsky, A., Littlefield, J.W. 1972. The prenatal diagnosis of inborn errors of metabolism. *Ann. Rev. Med.* 23: 57–76
167. Milunsky, A., Littlefield, J.W., Kanfer, J.N., Kolodny, E.H., Shih, V.E., Atkins, L. 1970. Prenatal genetic diagnosis. *New Eng. J. Med.* 283:1370–81, 1441–46, 1498–1503
168. Motulsky, A.G. 1973. Frequency of sickling disorders in U.S. Blacks. *New Eng. J. Med.* 288:31–33
169. Mowbray, S., Watson, B., Harris, H. 1972. A search for electrophoretic variants of human adenine phosphoribosyl transferase. *Ann. Human Genet.* 36: 153–62
170. Murdoch, J.L., Walker, B.A., Hall, J.G., Abbey, H., Smith, K.K., McKusick, V.A. 1970. Achondroplasia—a genetic and statistical survey. *Ann. Human Genet.* 33:227–44
171. Murphy, E.A. 1970. The ENSU scoring system in genetic counselling. *Ann. Human Genet.* 34:73–78
172. Murphy, E.A. (Baltimore) 1972. Personal communication
173. Mutton, D.E., Daker, M.G. 1973. Pericentric inversion of chromosome 9. *Nature N.B.* 241:80
174. Nadler, H.L. 1972. Allotransplantation for the treatment of inborn errors of metabolism. (Editorial) *Ann. Intern. Med.* 77:314–16
175. Nadler, H.L., Egan, T.J. 1970. Deficiency of lysosomal acid phosphatase. A new familial metabolic disorder. *New Eng. J. Med.* 282:302–07
176. Neufeld, E.F. (Bethesda) 1972. Personal communication
177. Neufeld, E.F., Fratantoni, J.C. 1970. Inborn errors of mucopolysaccharide metabolism. *Science* 169:141–46
178. Niebuhr, E. 1972. Dicentric and monocentric Robertsonian translocations in man. *Humangenetik* 16:217–26
179. O'Brien, J.S. 1972. Sanfilippo syndrome: profound deficiency of alpha-acetylglucosaminidase activity in organs and skin fibroblasts from type B patients. *Proc. Nat. Acad. Sci.* 69: 1720–22
180. Ohta, Y., Yamaoka, K., Sumida, I., Yanase, T. 1971. Haemoglobin Miyada, a beta-delta fusion peptide (anti-Lepore) type discovered in a Japanese family. *Nature* 234:218–20
181. Opio, E., Barnes, P.M. 1972. Intravenous urea in treatment of bone-pain crises of sickle-cell disease. *Lancet* 2: 160–61
182. O'Riordan, M.L., Robinson, J.A., Buckton, K.E., Evans, H.J. 1971. Distinguishing between the chromosomes involved in Down's syndrome (trisomy 21) and chronic myeloid leukaemia Ph^1 by fluorescence. *Nature* 230: 167–68
183. Orye, E., Delbeke, M.J., Vandenabeele, B. 1971. Retinoblastoma and D-chromosome deletions. (letter) *Lancet* 2:1376
184. Pantelakis, S.N., Karaklis, A.G., Alexiou, D., Vardas, E., Valaes, T. 1970. Red cell enzymes in trisomy 21. *Am. J. Human Genet.* 22:184–93
185. Pardue, M.L., Gall, J.G. 1970. Chromosome localization of mouse satellite DNA. *Science* 168:1356–58
186. Paris Conference 1971: *Standardization in human cytogenetics*. Birth Defects: Orig. Art. Series, 8 (No. 7): 1–46, 1972. The National Foundation, New York. Also see *Cytogenetics* 11:313–62, 1972
187. Patil, S.R., Merrick, S., Lubs, H.A. 1971. Identification of each human chromosome with a modified Giemsa stain. *Science* 173:821–22
188. Paul, J. 1972. General theory of chromosome structure and gene activation in eukaryotes. *Nature* 238:444–46
189. Pearson, P. 1972. The use of new staining techniques for human chromosome identification. *J. Med. Genet.* 9:264–75
190. Pearson, P.L., Bobrow, M. 1970. Definitive evidence for the short arm of the Y chromosome associating with the X chromosome during meiosis in the human male. *Nature* 226:959–61
190a. Pearson, P.L. (Leiden). 1973. Personal communication

191. Philippart, M., Franklin, S.S., Gordon, A. 1972. Reversal of an inborn sphingolipidosis (Fabry's disease) by kidney transplantation. *Ann. Int. Med.* 77: 195–200
192. Pinnell, S.R., Krane, S.M., Kenzora, J., Glimcher, M.J. 1972. A new heritable disorder of connective tissue with hydroxylysine-deficient collagen. *New Eng. J. Med.* 286:1013–20
193. Prensky, W., Holmquist, G. 1973. Chromosomal localization of human haemoglobin structural genes: techniques queried. *Nature* 241:44–45
194. Price, P.M., Conover, J.H., Hirschhorn, K. 1972. Chromosomal localization of human haemoglobin structural genes. *Nature* 237:340–42
194a. Ricciuti, F.C., Ruddle, F.H. 1973. Assignment of three genes loci PGK, HGPRT, G6PD) to the long arm of the human X-chromosome by somatic cell genetics. *Genetics*. In press
195. Raivio, K.O., Seegmiller, J.E. 1972. Genetic diseases of metabolism. *Ann. Rev. Biochem.* 41:543–76
196. Renwick, J.H. 1971. The mapping of human chromosomes. *Ann. Rev. Genet.* 5:81–120
197. Renwick, J.H. 1971. Assignment and map-positioning of human loci using chromosomal variation. *Ann. Hum. Genet.* 35:79–97
198. Renwick, J.H., Bolling, D.R. 1971. An analysis procedure illustrated on a triple linkage of use for prenatal diagnosis of myotonic dystrophy. *J. Med. Genet.* 8:399–406
199. Renwick, J.H., Bundey, S.E., Ferguson-Smith, M.A., Izatt, M.M. 1971. Confirmation of linkage of the loci for myotonic dystrophy and ABH secretion. *J. Med. Genet.* 8:407–16
200. Ris, H., Kubai, D.F. 1970. Chromosome structure. *Ann. Rev. Genet.* 4: 263–94
201. Ritter, H., Tariverdian, G., Arnold, H., Blume, K.G., Schroter, W., Zimmerschitt, E., Brittinger, G., Konig, E., Wendt, G.G. 1971. Evidence for linkage between the loci for the ABO-system and the locus for phosphoglucoseisomerase (PGI). *Humangenetik* 11:349–50
202. Robinson, J.A. 1973. Origin of extra chromosome in trisomy 21. *Lancet* 1: 131–33
202a. Robson, E.B., Cook, P.J.L., Corne, G., Hopkinson, D.A., Noades, J., Cleghorn, T.E. 1973. Linkage data on *Rh*, PGM_1, *PGD*, *peptidase C*, and *Fy* from family studies. *Ann. Human Genet.* 36: 393–99
203. Robson, E.B., Polani, P.E., Dart, S.J., Jacobs, P.A., Renwick, J.H. 1969. Probable assignment of the alpha locus of haptoglobin to chromosome 16 in man. *Nature* 223:1163–65
204. Rogers, S. 1971. Gene therapy: a potentially invaluable aid to medicine and mankind. *Res. Commun. Chem. Pathol. Pharmacol.* 2:587–600
204a. Rowley, J.D. 1973. A new consistent chromosomal abnormality in chronic myelogenous leukemia identified by quinacrine fluorescence and Giemsa staining *Nature* 243:290–93
205. Ruddle, F.H. (New Haven) 1973. Personal communication
206. Ruddle, F.H. 1973. Linkage analysis in man by somatic cell genetics. *Nature* 242:165–69
207. Ruddle, F., Ricciuti, F., McMorris, F.A., Tischfield, J., Creagan, R., Darlington, G., Chen, T. 1972. Somatic cell genetic assignment of peptidase C and the Rh linkage group to chromosome A-1 in man. *Science* 176:1429–31
208. Saunders, G.F., Hsu, T.C., Getz, M.J., Simes, E.L., Arrighi, F.E. 1972. Locations of a human satellite DNA in human chromosomes. *Nature N.B.* 236: 244–46
209. Scheres, J.M.J.C. 1972. Identification of two Robertsonian translocations with a Giemsa banding technique. *Humangenetik* 15:253–56
210. Schrott, H.G., Karp, L., Omenn, G.S. 1973. Prenatal prediction in myotonic dystrophy: guidelines for genetic counseling. *Clin. Genet.* 4:38–45
211. Scriver, C.R., Cameron, D. 1969. Pseudohypophosphatasia. *New Eng. J. Med.* 281:604–06
212. Seabright, M. 1971. A rapid banding technique for human chromosomes. *Lancet* 2:971–72
213. Seabright, M. 1972. The use of proteolytic enzymes for the mapping of structural rearrangements in the chromosomes of man. *Chromosoma* 36: 204–10
214. Segel, G.B., Feig, S.A., Mentzer, W.C., McCaffrey, R.P., Wells, R., Bunn, H.F., Shohet, S.B., Nathan, D.G. 1972. Effects of urea and cyanate on sickling *in vitro*. *New Eng. J. Med.* 287:59–64
215. Seid-Akhavan, M., Winter, W.P., Abramson, R.K., Rucknagel, D.L. 1972. Hemoglobin Wayne: a frameshift variant occurring in two distinct forms. (Abstract). Meeting of *Am. Soc. Hemat.*, Hollywood, Fla., Dec. 3–6

216. Shapiro, N.I., Khalizev, A.E., Luss, E.V., Marshak, M.I., Petrova, O.N., Varshaver, N.B. 1972. Mutagenesis in cultured mammalian cells. I. Spontaneous gene mutations in human and Chinese hamster cells. *Mutat. Res.* 15: 203–14
217. Sharp, H.L., Bridges, R.A., Krivit, W., Freier, E.F. 1969. Cirrhosis associated with alpha-1-antitrypsin deficiency: a previously unrecognized inherited disorder. *J. Lab. Clin. Med.* 73:934–39
218. Shaw, M.W. (Houston) 1972. Personal communication
219. Shows, T.B. 1972. Genetics of human-mouse somatic cell hybrids: linkage of human genes for isocitrate dehydrogenase and malate dehydrogenase. *Biochem. Genet.* 7:193–204
220. Sly, W.S., Quinton, B.A., McAlister, W.H., Rimoin, D.L. 1973. Beta glucuronidase deficiency: report of clinical, radiologic, and biochemical features of a new mucopolysaccharidosis. *J. Pediat.* 82:249–57
221. Sparkes, R.S. 1970. Red-cell triosephosphate isomerase and chromosome 5. (Letter) *Lancet* 2:570
222. Sperling, K., Wiesner, R. 1972. A rapid banding technique for routine use in human and comparative cytogenetics. *Humangenetik* 15:349–54
223. Spranger, J.W., Wiedemann, H.R. 1970. The genetic mucolipidoses. Diagnosis and differential diagnosis. *Humangenetik* 9:113–39
224. Stamatoyannopoulos, G. 1972. The molecular basis of hemoglobin disease, *Ann. Rev. Genet.* 6:47–70
225. Stamatoyannopoulos, G., Nute, P.E., Adamson, J.W., Bellingham, A.J., Funk, D. 1973. Hemoglobin Olympia (β20 valine$\rightarrow$methionine): an electrophoretically silent variant associated with high oxygen affinity and erythrocytosis. *J. Clin. Invest.* 52:342–49
226. Strand, L. J., Meyer, U.A., Felsher, B.F., Redeker, A.G., Marver, H.S. 1972. Decreased red cell uroporphyrinogen I synthetase activity in intermittent porphyria. *J. Clin. Invest.* 51: 2530–36
227. Sugita, M., Dulaney, J.T., Moses, H.W. 1972. Ceramidase deficiency in Farber's disease (lipogranulomatosis). *Science* 178:1100–02
228. Sumner, A.T. 1972. A simple technique for demonstrating centromeric heterochromatin. *Exp. Cell Res.* 75:304–06
229. Sumner, A.T., Evans, H.J., Buckland, R.A. 1971. New technique for distinguishing between human chromosomes. *Nature N.B.* 232:31–32
230. Taylor, A.I. 1970. Dq-, Dr and retinoblastoma. *Humangenetik* 10:209–17
231. Terheggen, H.G., Lavinha, F., Colombo, J.P., van Sande, M., Lowenthal, A. 1972. Familial hyperargininemia. *J. Génét. Hum.* 20:69–84
232. Thomas, C.A., Jr. 1971. The genetic organization of chromosomes. *Ann. Rev. Genet.* 5:237–56
233. Turleau, C., de Grouchy, J., Klein, M. 1972. Phylogénie chromosomique de l'homme et des primates hominiens. Essai de reconstitution du caryotype de l'ancêtre commun. *Ann. Génét.* 15: 225–40
234. Turleau, C., de Grouchy, J. 1972. Caryotypes de l'homme et du chimpanzé. Comparaison de la topographie des bandes. Mécanismes évolutifs possibles. *C.R. Acad. Sci. Paris* 274:2355–57
235. Valdimarsson, H., Moss, P.D., Holt, P.J.L., Hobbs, J.R. 1972. Treatment of chronic mucocutaneous candidiasis with leukocytes from HL-A compatible siblings. *Lancet* 1:469–72
236. Valentine, W.N. 1972. Red cell enzyme deficiencies as a cause of hemolytic disorders. *Ann. Rev. Med.* 23:93–100
237. Van Loghem, E. 1971. Formal genetics of the immunoglobulin systems. *Ann. N.Y. Acad. Sci.* 190:134–49
238. Wang, H.C., Fedoroff, S. 1972. Banding in human chromosomes treated with trypsin. *Nature N.B.* 235:52–54
239. Warren, R.J., Rimoin, D.L., Summitt, R.L. 1973. Identification by fluorescent microscopy of the abnormal chromosomes associated with the G-deletion syndromes. *Am. J. Human Genet.* 25: 77–81
240. Washburn, S.L., McCown, E.R. 1972. Evolution of human behavior. *Social Biol.* 19:163–70
241. Weisblum, B., DeHaseth, P. 1972. Quinacrine - a chromosome stain specific for deoxyadenylate-deoxythymidylate-rich regions in DNA. *Proc. Nat. Acad. Sci.* 69:629–32
242. Weitkamp, L.R. 1971. Further data on the genetic linkage relations of the adenosine deaminase locus. *Human Hered.* 21:351–56
243. Weitkamp. L.R., Guttormsen, S.A., Greendyke, R.M. 1971. Genetic linkage between a locus for 6-PGD and the Rh locus: Evaluation of possible

heterogeneity in the recombination fraction between sexes and among families. *Am. J. Human Genet.* 23:462–70

244. Weitkamp, L.R., Janzen, M.K., Guttormsen, S.A., Gershowitz, H. 1969. Inherited pericentric inversion of chromosome number two: a linkage study. *Ann. Human Genet.* 33:53–59

245. Wells, R.S., Higgs, J.M., MacDonald, A., Valdimarsson, H., Holt, P.J.L. 1972. Familial chronic muco-cutaneous candidiasis. *J. Med. Genet.* 9: 302–10

246. Wertelecki, W., Gerald, P.S. 1971. Clinical and chromosomal studies of the 18q- syndrome. *J. Pediat.* 78:44–52

247. Westerveld, A., Meera Khan, P. 1972. Evidence for linkage between human loci for 6-phosphogluconate dehydrogenase and phosphoglucomutase (1) in man-chinese hamster somatic cell hybrids. *Nature* 236:30–32

248. Wiesmann, U., Neufeld, E.F. 1970. Scheie and Hurler syndromes: apparent identity of the biochemical defect. *Science* 169:72–74

249. Wiesmann, U.N., Lightbody, J., Vasselka, F., Herschkowitz, N.N. 1971. Multiple lysosomal enzyme deficiency due to enzyme leakage? (letter). *New Eng. J. Med.* 284:109–10

250. Wikramanayake, E., Renwick, J.H., Ferguson-Smith, M.A. 1971. Chromosomal heteromorphisms in the assignment of loci to particular autosomes: a study of four pedigrees. *Ann. Génét.* 14: 245–56

251. Wilson, M.G., Melnyk, J., Towner, J.W.J. 1969. Retinoblastoma and deletion D (14) syndrome. *J. Med. Genet.* 6:322–27

252. Wilson, A.C., Sarich, V.M. 1969. A molecular time scale for human evolution. *Proc. Nat. Acad. Sci.* 63:1088–93

253. Yoshida, A. 1970. Amino acid substitution (histidine to tyrosine) in a glucose-6-phosphate dehydrogenase variant (G6PD Hektoen) associated with over-production. *J. Mol. Biol.* 52: 483–89

254. Yoshida, A. 1973. Hemolytic anemia and G6PD deficiency. *Science* 179: 532–37

255. Yoshida, A., Motulsky, A.G. 1969. A pseudocholinesterase variant (E Cynthiana) associated with elevated plasma enzyme activity. *Am. J. Human Genet.* 21:486–98

256. Zech, L. 1969. Investigations of metaphase chromosomes with DNA binding fluorochromes. (abstract) *Exp. Cell Res.* 58:463

257. Zur Hausen, H., Diehl, V., Wolf, H., Schulte-Holthausen, H., Schneider, U. 1972. Occurrence of Epstein-Barr virus genomes in human lymphoblastoid cell lines. *Nature N.B.* 237: 189–90

REPRINTS

The conspicuous number aligned in the margin with the title of each article in this volume is a key for use in ordering reprints.

Available reprints are priced at the uniform rate of $1 each postpaid. Payment must accompany orders less than $10. A discount of 20% will be given on orders of 20 or more. For orders of 200 or more, any Annual Reviews article will be specially printed.

The sale of reprints of articles published in the Reviews has been expanded in the belief that reprints as individual copies, as sets covering stated topics, and in quantity for classroom use will have a special appeal to students and teachers.

AUTHOR INDEX

E

F

G

I

J

T

SUBJECT INDEX

CUMULATIVE INDEXES

CONTRIBUTING AUTHORS VOLUMES 3-7

CHAPTER TITLES VOLUMES 3-7